THE ROUTLEDGE RESEARCH COMPANION TO NINETEENTH-CENTURY BRITISH LITERATURE AND SCIENCE

Tracing the continuities and trends in the complex relationship between literature and science in the long nineteenth century, this companion provides scholars with a comprehensive, authoritative and up-to-date foundation for research in this field. In intellectual, material and social terms, the transformation undergone by Western culture over the period was unprecedented. Many of these changes were grounded in the growth of science. Yet science was not a cultural monolith then any more than it is now, and its development was shaped by competing world views. To cover the full range of literary engagements with science in the nineteenth century, this companion consists of twenty-seven chapters by experts in the field, which explore crucial social and intellectual contexts for the interactions between literature and science, how science affected different genres of writing, and the importance of individual scientific disciplines and concepts within literary culture. Each chapter has its own extensive bibliography. The volume as a whole is rounded out with a synoptic introduction by the editors and an afterword by the eminent historian of nineteenth-century science Bernard Lightman.

John Holmes is Professor of Victorian Literature and Culture at the University of Birmingham, UK.

Sharon Ruston is Professor of Romanticism at Lancaster University, UK.

THE ROUTLEDGE RESEARCH COMPANION TO NINETEENTH-CENTURY BRITISH LITERATURE AND SCIENCE

Edited by
John Holmes and Sharon Ruston

Routledge
Taylor & Francis Group

LONDON AND NEW YORK

First published 2017
by Routledge
2 Park Square, Milton Park, Abingdon, Oxon OX14 4RN

and by Routledge
711 Third Avenue, New York, NY 10017

Routledge is an imprint of the Taylor & Francis Group, an informa business

© 2017 John Holmes and Sharon Ruston

British Library Cataloguing in Publication Data
A catalogue record for this book is available from the British Library

Library of Congress Cataloging in Publication Data
A catalog record for this book has been requested

ISBN: 978-1-4724-2987-2 (hbk)
ISBN: 978-1-315-61333-8 (ebk)

Typeset in Bembo
by Florence Production Ltd, Stoodleigh, Devon, UK

MIX
Paper from
responsible sources
FSC
www.fsc.org   FSC® C013056

Printed and bound in Great Britain by
TJ International Ltd, Padstow, Cornwall

CONTENTS

Contents

Contents

FIGURES

CONTRIBUTORS

David Amigoni is Professor of Victorian Literature at Keele University. He is the author of *Colonies, Cults and Evolution: Literature, Science and Culture in Nineteenth-Century Writing* (2007), editor of *Life Writing and Victorian Culture* (2006) and co-editor of *Charles Darwin's 'Origin of Species': New Interdisciplinary Essays* (1995).

Suzy Anger is Associate Professor of English at the University of British Columbia. She is the author of *Victorian Interpretation* (2005), editor of *Knowing the Past: Victorian Literature and Culture* (2001) and co-editor of *Victorian Science as Cultural Authority* (vol. 2 of *Victorian Science and Literature*, 2011). She was President of the Northeast Victorian Studies Association from 2010 to 2014.

Debbie Bark is a Teaching Fellow in English Literature at the University of Reading. She is the editor of *The Collected Works of Ann Hawkshaw* (2014).

Michelle Boswell is a member of the Faculty in Humanities at Thomas Jefferson High School for Science and Technology in Virginia. She was formerly a Lecturer in Professional Writing at the University of Maryland. She completed her doctorate at the University of Maryland on 'Beautiful Science: Victorian Women's Scientific Poetry and Prose' in 2014.

Adelene Buckland is Senior Lecturer in English Literature at King's College London. She is the author of *Novel Science: Fiction and the Invention of Nineteenth-Century Geology* (2013).

Gowan Dawson is Professor of Victorian Literature and Culture at the University of Leicester. He is the author of *Darwin, Literature and Victorian Respectability* (2007), *Show Me the Bone: Reconstructing Prehistoric Monsters in Nineteenth-Century Britain and America* (2016) and co-author of *Science in the Nineteenth-Century Periodical: Reading the Magazine of Nature* (2004). With Bernard Lightman, he is general editor of *Victorian Science and Literature* (8 vols, 2011–12) and co-editor of *Victorian Scientific Naturalism: Community, Identity, Continuity* (2014).

Christine Ferguson is Professor of English Literature at the University of Stirling. She is the author of *Language, Science, and Popular Fiction in the Victorian Fin-de-Siècle: The Brutal Tongue* (2006) and *Determined Spirits: Eugenics, Heredity, and Racial Regeneration in Anglo-American Spiritualist Writing, 1848–1930* (2012) and the co-editor of *Spiritualism, 1840–1930* (2014).

Peter Garratt is Senior Lecturer in Victorian Literature at Durham University. He is the author of *Victorian Empiricism: Self, Knowledge and Reality in Ruskin, Bain, Lewes, Spencer and George Eliot* (2010) and editor of *The Cognitive Humanities: Embodied Mind in Literature and Culture* (2016).

Barri J. Gold is Professor and Head of English at Muhlenberg College. She is the author of *ThermoPoetics: Energy in Victorian Literature and Science* (2010).

Pamela Gossin is Professor of History of Science and Interdisciplinary Literature and Science studies at the University of Texas at Dallas. She is the author of *Thomas Hardy's Novel Universe: Astronomy, Cosmology and Gender in the Post-Darwinian World* (2007) and editor of the *Encyclopedia of Literature and Science* (2002).

John Holmes is Professor of Victorian Literature and Culture at the University of Birmingham. He is the author of *Dante Gabriel Rossetti and the Late Victorian Sonnet Sequence* (2005) and *Darwin's Bards: British and American Poetry in the Age of Evolution* (2009), and editor of *Science in Modern Poetry: New Directions* (2012). He was Chair of the British Society for Literature and Science from 2012 to 2015.

Alice Jenkins is Professor of Victorian Literature and Culture at the University of Glasgow and was the founding Chair of the British Society for Literature and Science. She is the author of *Space and the 'March of Mind': Literature and the Physical Sciences in Britain, 1815–1850* (2007) and editor of *Michael Faraday's Mental Exercises: An Artisan Essay-circle in Regency London* (2008).

Meegan Kennedy is Associate Professor in English and History and Philosophy of Science at Florida State University. She is the author of *Revising the Clinic: Vision and Representation in Victorian Medical Narrative and the Novel* (2010).

Bernard Lightman is Professor of Humanities at York University, Toronto. He edited the journal *Isis* from 2004 to 2014. He is the author of *The Origins of Agnosticism: Victorian Unbelief and the Limits of Knowledge* (1987), *Victorian Popularizers of Science: Designing Nature for New Audiences* (2007) and *Evolutionary Naturalism in Victorian Britain: The 'Darwinians' and Their Critics* (2009). With Gowan Dawson, he is general editor of *Victorian Science and Literature* (8 vols, 2011–12) and co-editor of *Victorian Scientific Naturalism: Community, Identity, Continuity* (2014). Other co-edited volumes include *Figuring it Out: Science, Gender and Visual Culture* (2006) and *Science in the Marketplace: Nineteenth-Century Sites and Experiences* (2007).

Andrew Mangham is Associate Professor in English Literature at the University of Reading. He is the author of *Violent Women and Sensation Fiction: Crime, Medicine and Victorian Popular Fiction* (2007) and *Dickens's Forensic Realism: Truth, Bodies, Evidence* (2016), editor of *The Cambridge Companion to Sensation Fiction* (2013), and co-editor of *The Female Body in Medicine and Literature* (2011).

Iwan Rhys Morus is Professor of the History of Science at Aberystwyth University. He is the author of *Frankenstein's Children: Electricity, Exhibition, and Experiment in Early-Nineteenth-Century London* (1998), *When Physics Became King* (2005) and *Shocking Bodies: Life, Death and Electricity in Victorian England* (2011).

Ralph O'Connor is Professor in the Literature and Culture of Britain, Ireland and Iceland at the University of Aberdeen. He is the author of *The Earth on Show: Fossils and the Poetics of*

Popular Science, 1802–1856 (2007), editor of *Science as Romance* (vol. 7 of *Victorian Science and Literature*, 2012), and co-editor of *Uncommon Contexts: Encounters between Science and Literature, 1800–1914* (2013).

Stella Pratt-Smith is the author of *Transformations of Electricity in Nineteenth-Century Literature and Science* (2016).

Sadiah Qureshi is Senior Lecturer in Modern History at the University of Birmingham. She is the author of *Peoples on Parade: Exhibitions, Empire and Anthropology in Nineteenth-Century Britain* (2011).

Julia Reid is a Lecturer in Victorian Literature at the University of Leeds. She is the author of *Robert Louis Stevenson, Science, and the Fin de Siècle* (2006) and the editor of Stevenson's *The Amateur Emigrant* (2016).

Adam Roberts is Professor of Nineteenth-century Literature at Royal Holloway, University of London. He is the author of *Science Fiction (New Critical Idiom)* (2000), *The History of Science Fiction* (2007) and several science fiction novels.

Sharon Ruston is Professor of Romanticism at Lancaster University. She is the author of *Shelley and Vitality* (2005) and *Creating Romanticism: Case Studies in Literature, Science and Medicine in the 1790s* (2013), editor of a special issue of *Essays and Studies* on 'Literature and Science' (2008) and co-editor of *The Collected Letters of Sir Humphry Davy* (forthcoming with Oxford University Press).

Anna Katharina Schaffner is a Reader in Comparative Literature and Medical Humanities at the University of Kent. She is the author of *Modernism and Perversion: Sexual Deviance in Sexology and Literature, 1850–1930* (2012) and *Exhaustion: A History* (2016) and co-editor of *Modernist Eroticisms: European Literature after Sexology* (2012).

Charlotte Sleigh is Professor of Science Humanities at the University of Kent. She is the author of *Six Legs Better: A Cultural History of Myrmecology* (2007) and *Literature and Science* (2010) and currently edits the *British Journal of the History of Science*.

Jonathan Smith is William E. Stirton Professor of English at the University of Michigan–Dearborn. He is the author of *Fact and Feeling: Baconian Science and the Nineteenth-Century Literary Imagination* (1994) and *Charles Darwin and Victorian Visual Culture* (2006) and the co-editor of *Negotiating Boundaries* (vol. 1 of *Victorian Science and Literature*, 2011).

Gregory Tate is a Lecturer in Victorian Literature at the University of St Andrews. He is the author of *The Poet's Mind: The Psychology of Victorian Poetry 1830–1870* (2012).

Paul White is a senior editor at the Darwin Correspondence Project. He is the author of *Thomas Huxley: Making the "Man of Science"* (2002).

Martin Willis is Professor of English Literature at Cardiff University, Honorary Senior Lecturer at the Cardiff School of Medicine and Chair of the British Society for Literature and Science. He is the author of *Mesmerists, Monsters and Machines: Science Fiction and the Cultures of Science in the Nineteenth Century* (2006), *Vision, Science and Literature, 1870–1920: Ocular Horizons* (2011) and *Literature and Science: A Readers' Guide to Essential Criticism* (2015), and is the founding editor of the *Journal of Literature and Science*.

ACKNOWLEDGEMENTS

The number of people whose advice helps to shape a book of this scale and scope is huge. We are sadly not able to thank everyone who has read, heard or commented on individual chapters, or everyone who has made suggestions as to the shape of the book as a whole, though we are grateful to them all. We would like to express our thanks to the editorial and production teams at Routledge, and to the copy-editors at Florence Production, who have helped to bring this book into existence. Moreover, there are two people aside from the contributors themselves whose input has been fundamental to producing the book as it stands. The first is Ann Donahue, who as commissioning editor at Ashgate proposed the book to us in the first place. Ann's commitment to the project and the support she has given us have been exemplary. We are delighted as well as grateful that she gave us the opportunity to edit this book, which we hope and believe will be of real value to the field. The second person we must thank is Andrew Lacey, who copy-edited the whole book and prepared the index with consummate skill and precision ahead of submission. In the process, he spared us from having to attempt to do this ourselves, as well as doing it far better than we would have done.

INTRODUCTION

Literature and science in the nineteenth century

John Holmes

University of Birmingham

Sharon Ruston

Lancaster University

In 1898, the naturalist Alfred Russel Wallace took stock of scientific progress over the course of the nineteenth century. By calibrating his qualitative judgements quantitatively, he sought to produce a list of the most significant scientific and technological discoveries of his time. His conclusions amply bore out the title of his book *The Wonderful Century*. According to Wallace's estimates, the nineteenth century had seen two dozen major scientific discoveries and technological inventions, against no more than fifteen in all preceding centuries put together. In the physical sciences, the principle of conservation of energy had been established as the first law of thermodynamics. The molecular theory of gases had been put forward and confirmed. The speed of light had been measured, as had the atomic weights of the different elements. In geology, uniformitarian principles had explained the Earth's physical form and its history. The Ice Age and the antiquity of humankind had been established as historical or rather prehistoric facts. The theory of organic evolution had transformed biology, a process in which Wallace himself had had a hand, though with characteristic modesty and generosity he attributed this seismic shift in scientific and popular opinion principally to Charles Darwin. Cell theory, embryology, the germ theory of disease and the discovery of the part played by white blood cells in fighting infection had between them revolutionized microbiology and medicine. In practical terms, antiseptics and anaesthetics had revolutionized surgery. New technologies had allowed people to travel and trade at speeds and in numbers barely imaginable before, to send messages and even their own voices across the world in an instant, to light their streets and houses cleanly and efficiently, to record visual impressions as photographs and sounds on phonographs, to see inside objects and organisms with Röntgen rays (now called X-rays) and even to decipher the chemical composition of the stars. The reach and sophistication of theoretical science had grown so vast that by Wallace's estimate even the scientific revolution of the seventeenth century was slight by comparison, while the technological advances of the century had created a modern civilization whose material world was utterly unlike that of a hundred years before. (For the tabulation of these discoveries, see Wallace 154–5; for his fuller account of them, see 1–156.)

1

Wallace's ranking of the importance of given discoveries, like his contemporary Francis Galton's attempts to measure beauty and genius, are open to question and even mockery. It is not clear how it is possible to calibrate the relative significance of spectrum analysis and Harvey's discovery of the circulation of the blood, nor is it immediately apparent why the friction match, invented in the 1820s, should count as a major invention on its own, while the earlier barometer and thermometer only amount to one major invention between them. But Wallace's list remains a compelling witness not only to the scale of scientific achievement of the nineteenth century but also to its cultural centrality. Through new technologies, science transformed the lives of millions of people across the globe, nowhere more so than in Britain, where the Industrial Revolution drove social change at an unprecedented pace. Over the same period, science established itself as the hallmark of modern civilization, with a cultural authority to rival the Church, the Classics and the arts.

In this book we aim, as Wallace did, to take stock of the full significance of science for nineteenth-century Britain, as it is manifest in the literature of this period. Research on literature and science in the nineteenth century has long been a distinguished strand within literary scholarship. In the 1870s, Edward Dowden pioneered academic work on what he called 'The Scientific Movement in Literature' while teaching at Trinity College, Dublin. In the 1930s, the American critics Carl Grabo, Lionel Stevenson and Joseph Warren Beach probed the relationship between science and nineteenth-century poetry. In the 1980s, Gillian Beer, George Levine and Sally Shuttleworth teased out new and subtle understandings of the relationship between evolutionary theory, psychology and literary structures in Victorian fiction and scientific writing. Over the last twenty years, this line of investigation has grown into a major field of enquiry in its own right. The pioneering Society for Literature and Science (SLS) – now the Society for Literature, Science and the Arts (SLSA) – which was founded in America in the 1980s, now has branches covering Europe (since 2000) and the rest of the world (since 2015). The more overtly historicist British Society for Literature and Science (BSLS) was launched in 2006, while the International Union for the History and Philosophy of Science and Technology established its own Commission on Science and Literature (CoSciLit) in 2014. Between them, these learned societies have a global reach and, while they cover all places and periods, the study of interactions between science and literature in Britain in the nineteenth century holds a prominent place within their research. There are journals that deal specifically with these topics too, such as the British-based *Journal of Literature and Science* (founded in 2007) and the American *Configurations* (founded in 1993). At the same time, science and the history of science are increasingly unavoidable points of reference for scholars working on nineteenth-century literature within societies and communities that do not define themselves in these overtly interdisciplinary terms. These societies too have thrived and proliferated over recent decades, including the Australasian Victorian Studies Association (AVSA, founded 1973), the British Association for Romantic Studies (BARS, 1989), the North American Society for the Study of Romanticism (NASSR, 1991), the British Association for Victorian Studies (BAVS, 2000) and the North American Victorian Studies Association (NAVSA, 2002). Finally, historians of science, represented by the History of Science Society (HSS, 1924) and the British Society for the History of Science (BSHS, 1947), have increasingly woken up to the possibilities of using literary sources and literary methods in their analysis of the history of nineteenth-century science itself. This book represents each of these different scholarly communities, with chapters written by historians as well as literary critics, by scholars from Britain, contintental Europe and North America, by romanticists and Victorianists. It aims to serve them all by providing a detailed account of the field, tracing research done and projecting work still to do, and offering both firm knowledge and productive speculation about all aspects of the relationship between science and literature in Britain across the nineteenth century.

One crucial reason for the massive expansion in research into this relationship has been the breaking apart of received canons of both 'literature' and 'science'. Until recently, most research on literature and science in the nineteenth century paid heed only to a small number of canonical writers, the Shelleys, Alfred Tennyson, George Eliot and Thomas Hardy being chief among them. Literature itself was taken to cover only creative writing, principally novels and poetry, or occasionally criticism, but not the wide range of non-fiction genres included within that term in the nineteenth century itself. Lately, literary scholars have expanded their remit, taking in scientific writing itself, whether for professional, elite, popular or juvenile audiences, as well as poetry, fiction and theatre, by otherwise neglected writers, many of them women. (For generous selections from the full spectrum of literature, in this broad sense, that engaged with science in the nineteenth century, see Hawley, ed., and Dawson and Lightman, eds.) If the literary canon has been opened up, so too has the canon of accepted science. Although Darwin is still acknowledged and studied as a major presence within Victorian culture and a deft author in his own right, he no longer holds the near-monopoly he once enjoyed, neither within the history of evolution nor within studies of Victorian literature and science. Scholarly work on the more familiar and accessible nineteenth-century sciences, such as evolution, geology and psychology, has continued apace, but so has work on thermodynamics, chemistry and mathematical logic. As the Whig history of science, imagined as the progress towards ever more correct ideas, has been repudiated, so have marginal and even occult sciences such as phrenology and mesmerism crept back into view. On the other hand, as our own scientific preoccupations change, our attention is directed to aspects of nineteenth-century science that may previously have been marginalized, such as ecology or the science of sexual identity. That Wallace closes *The Wonderful Century* not with a paean to his age and its achievements but with a warning of the environmental degradation brought about by capitalism (369–81) gives his book a fresh piquancy. His long polemic 'Vaccination a Delusion', reprinted in his book (Wallace 213–315), has its own latter-day echoes.

Altogether, the picture of the interactions between science and literature in the nineteenth century is immeasurably richer than it was only fifteen or twenty years ago. It is this picture that this book sets out to reveal, for the benefit of researchers entering the field at any point and from any angle. In order to achieve this, we have taken transects through this field from different sides. In the first, principally historical, section, we survey the field as a whole, tracing key cultural trends within the period that reverberate within current research: the globalization of science through the mechanisms of imperialism, the emergence of distinctive scientific cultures and institutions, the relationship between science and religion, and the role of women in writing about science in the nineteenth century. For the second section, we have taken 10 distinct literary genres to explore how each of them engages with and bears on science in the period. In this we take a lead from the new, inclusive scholarship without abandoning a commitment to the distinctive value of literature in the narrower sense. Chapters on novels, science fiction, poetry and literary criticism stand alongside chapters on scientific biography, several genres of science writing, including professional scientific prose, science popularization, writing for the periodical press and scientific literature for children, and the theatrical staging of science. For the third and fourth sections, we approach the relationship between literature and science from the sciences themselves, taking the most significant disciplines and developments within first the mathematical and physical sciences and then the biological and human sciences, and tracing their ramifications through the literature of the period.

Inevitably, a topic as vast as science in the nineteenth century, and as science within nineteenth-century literature, cannot be set down in its totality. Science, while invariably shaped by different national contexts and cultural moments, can also cross national and historical boundaries. To understand the place and forms of science in Britain in the nineteenth century it is important to recognize that much of the science itself originated elsewhere, and that it forms part of a continuum that cannot be neatly carved at 1800 and again a hundred years further on. The scientific culture of nineteenth-century Britain was built on specimens and data brought home from across a vast empire and through still-famous acts of exploration: James Cook's voyages to the Pacific in the 1770s, sponsored by the Royal Society and initially accompanied by Joseph Banks; Mungo Park's expeditions to the Niger River in the 1790s and early 1800s, with Banks's patronage; Darwin's voyage round South America and across to the Galapagos as the naturalist on board the *Beagle* in the 1830s; Sir John Franklin's voyages into the Arctic, and those sent out to look for him, his men and his ships after all traces of them were lost in 1845, as they searched for the Northwest Passage; Joseph Hooker's botanical expedition to the Himalayas in the late 1840s; Burton's and Speke's journeys into the hinterland of east Africa in the 1850s; and so on. Indeed, British science and imperial expansion were tightly knotted together, as these examples imply, whether through the role of the navy in exploration or through the role of the explorers themselves in charting out territory that would be claimed for Britain's sphere of interest.

Huge as it was, the British Empire was not hermetically sealed, and British science was in dialogue and even in direct competition with science from Continental Europe and the United States. The discovery of oxygen in the 1770s, like that of the planet Neptune in the 1840s, was played out through rivalries between British and Continental – principally French – scientists. British resistance to evolutionism in the early nineteenth century was motivated in part by suspicion of French republicanism during a time when Britain and France were at war. After the defeat of Napoleon in 1815, the international reputation of Georges Cuvier and Étienne Geoffroy Saint-Hilaire drew many aspiring British biologists, both real and fictional, to the Jardin des Plantes in Paris. Robert Grant and Richard Owen defined their own positions on comparative anatomy and its implications in terms drawn from Parisian science, with Grant following Geoffroy and Owen aspiring to be the English Cuvier. The naturalists Sarah and Edward Bowdich apprenticed themselves to Cuvier before setting off to explore the Gambia (Orr). Eliot's Tertius Lydgate tellingly conducts his studies 'in London, Edinburgh, and Paris' (Eliot 174), rather than Oxford or Cambridge, before arriving in Middlemarch to practise medicine. At much the same time the doctor and poet Thomas Gordon Hake completed his medical and scientific training in Paris, although it would be several decades before this education resurfaced in his verse (see Holmes). If France led the way in evolution and medical research, other European countries were the fountainheads of other sciences. In Italy, for example, Volta and Galvani made major advances in the study of electricity, which would underpin its use in both technology and medical theory from the late eighteenth century onwards. Non-Euclidean geometry, cell theory and sexology were pioneered in Germany and the Austrian Empire. America too played a shaping role in scientific and technological developments, from Benjamin Franklin's famous experiments with electricity to the commercial extraction of petroleum, from Louis Agassiz's development at Harvard of his theory of the Ice Age to Asa Gray's championing of a theistic version of Darwinian evolution, also at Harvard and in the face of stiff resistance from Agassiz himself.

As with science, so did European and American literature likewise help to shape British literature. Nineteenth-century poets from Erasmus Darwin, Percy Shelley and John Keats to George Meredith, Algernon Swinburne and John Addington Symonds typically drew on

Greek, German or even Egyptian myths in articulating their responses to science. The scientific thinking of the chemist John Tyndall, like that of Coleridge and Carlyle, cannot be dissevered from his intellectual engagement with German transcendental idealism. When the journal *Nature* was launched in 1869, it opened not with a scientific paper but with a series of aphorisms from Goethe, translated by T. H. Huxley. Naturalism may have been a marginal movement within British fiction, but the story of the biologization of the novel cannot be told without reference to Zola. For late Victorian radicals such as Swinburne, the polemics of Giuseppe Mazzini and the poetry of Walt Whitman shaped not only their politics but their attitude to science as well.

It is beyond the scope of this book to explore literature and science across nineteenth-century Europe, or to trace in detail the global networks that helped to fashion science and literature in Britain in the nineteenth century. Instead, we have opted to let the authors of individual chapters bring in these contexts as and where they bear on their specific topics. One continuity that we have chosen to stress, however, is that of scientific and literary developments throughout the nineteenth century itself. Much scholarly work on nineteenth-century literature and science focuses on either the Romantic period or the Victorians. This creates the impression of a century that can be carved at the joint somewhere around 1830. Wallace recognized no such break in his *The Wonderful Century*, celebrating the early nineteenth-century chemistry of William Hyde Wollaston and Humphry Davy alongside the Victorian biology of Darwin. His book illustrates continuity in other ways too. In his record of the failures of his century, Wallace includes the neglect of phrenology and mesmerism, early nineteenth-century sciences that by the end of the century were obsolete for many but remained tenaciously current for others, including Wallace himself (159–212). Other modes of scientific thinking morphed as the century wore on. Natural theology was encapsulated by William Paley at the beginning of the century and carved in granite in the eight volumes of the Bridgewater Treatises in the 1830s. Secularized by Darwin in his adaptationist explanation of evolution, it was refashioned by Wallace himself after his conversion to spiritualism. Where Wallace insisted that some form of intervention by higher spirits was required to account for human evolution, other, more conventional thinkers such as Thomas Rymer Jones, Frank Buckland and Frederick Le Gros Clark carried the older, Paleyan tradition forward in spite of the case for evolution.

For all that the chain of thought remains unbroken across the century in science and in literature, there is nonetheless a change in the culture and conceptualization of science, which first becomes fully apparent in the early 1830s. The foundation of the British Association for the Advancement of Science (BAAS) in 1831 signalled the beginning of collective action by scientists. At much the same moment, the Royal Society began reforming both itself and its leading journal, the *Philosophical Transactions*, to reassert its scientific credentials. The BAAS was inclusive and public, the Royal Society a select club that became the preserve of a specifically scientific elite, yet both played important roles in the emergence of the professional scientist as a recognizable and authoritative figure within society. Both stimulated scientific research and dissemination too, with dissemination in particular increasingly widespread in the age of the new steam presses, which revolutionized book and periodical production (Secord 14–21). Other '*sites* and *experiences*' (Fyfe and Lightman 4) of science likewise changed rapidly in the early nineteenth century. At the end of the previous century, Edinburgh offered the best education in medicine, but early in the new century a number of new institutions began to promote medical and scientific knowledge. The number of museums increased, as did lectures for working men at Mechanics' Institutes, and public galleries of science in the regions as well as in London. Learned and polite society, including women, could go to lectures at newly established sites such as the Royal Institution.

From the 1830s onwards, there was ever more science being done in more places by more people and before a wider reading public. The formal opportunities for education in the sciences expanded too. The establishment of London University (later University College, London) in 1826 and the London Royal Polytechnic Institution in 1838 opened up new possibilities for secular education in the sciences. Cambridge and especially Oxford took time to catch up, but in the 1850s Oxford finally committed itself to building a new museum of natural history, which became a centre, along with the adjacent laboratories, of science teaching at the university. By the end of the 1870s, women too had the opportunity to study science at Oxbridge, UCL and elsewhere. The poet and philosopher Constance Naden was an outstanding product of this educational revolution, studying German, French and Botany at the Birmingham and Midland Institute before going on to gain first-class certificates in organic chemistry, physiology, geology and physics at Mason Science College, later incorporated into the University of Birmingham (Hughes 17, Naden xii).

If the means of the production and distribution of science began to change around the start of the Victorian age, so too did conceptions of science itself. The nineteenth century witnessed the transition from the 'man of science' to the 'scientist'. In his 1802 preface to *Lyrical Ballads*, William Wordsworth set up an opposition between 'Poetry and Matter of Fact, or Science' (1:135). Even within Wordsworth's poetry this is a false distinction. As critics have discussed, Wordsworth expressly adopts a scientific method in his *Lyrical Ballads*: these so-called 'experiments' (1:119) are intended to test out certain hypotheses, and they proceed by means of accurate observation and expression to trace the 'the primary laws of our nature' (1:124; see Wyatt; Smith; Ruston). Indeed, Coleridge identified Wordsworth's own tendency to 'matter-of-factness' (1993 2:216) in the *Biographia Literaria*. But it was not only poetry that could model itself on science; science too embraced the concerns of poetry. The early nineteenth-century 'man of science' had an extensive purview. Knowledge of the physical world was not cut off from moral, metaphysical and religious ideas. In *A Discourse, Introductory to a Course of Lectures on Chemistry*, delivered to the newly established Royal Institution on 21 January 1802, Humphry Davy argued that the 'man of true genius' would not put limits on his knowledge or the methods he would use to acquire it, urging him instead to:

> [M]ake use of all the instruments of investigation which are necessary for his purposes; and in the search of discovery, he will rather pursue the plans of his own mind than be limited by the artificial divisions of language.
>
> *(2:315)*

Davy referred here to the new scientific disciplines that were beginning to form and become distinct from each other. By contrast, he proposed that the best way forward is inclusive and idiosyncratic. It is also clearly imaginative. Davy's personal world view, grounded on mutual dependency and deep connection, is apparent here. Expressing these sentiments at the very start of the nineteenth century, he was in tune with Wordsworth and Coleridge, perhaps unsurprisingly, as they had been members of the same artistic and scientific circle.

The kind of science that Wordsworth objected to was equally complained of by Davy. His lecture continues:

> Following extensive views, [the 'man of true genius'] will combine together mechanical, chemical, and physiological knowledge, whenever this combination may be essential; in consequence his facts will be connected together by simple and obvious analogies,

and in studying one class of phaenomena more particularly, he will not reject its relations to other classes.

(2:315)

The 'man of true genius' will extend beyond his immediate scientific field to include others whenever this is necessary; he will not impose disciplinary limits on himself. This method will encourage us to see the analogies present in the world. In his poem 'Memorials of a Tour in Italy, 1837', Wordsworth laments:

[A] chilled age, most pitiably shut out
From that which *is* and actuates, by forms,
Abstractions, and by lifeless fact to fact
Minutely linked with diligence uninspired,
Unrectified, unguided, unsustained,
By godlike insight.

(I, ll. 325–30)

Davy credited the man of science with an 'active and powerful' (2:321) mind rather than a passive one. Coleridge's essays on the 'Principles of Method', published in the revised and extended edition of *The Friend* in 1818, reinforced this same ideal. Coleridge despaired of botany, which he called 'Little more than an enormous nomenclature; a huge catalogue' (1969–2002 469). He criticized Erasmus Darwin in particular for looking for 'antithesis' where he should have looked for 'analogy', for 'contrast' rather than 'resemblance' and for failing to acknowledge 'the harmony between the vegetable and animal world' (Coleridge 1969–2002 470). To avoid this kind of stultified science, Coleridge advocated that the 'science of Method' be taught:

METHOD, therefore, becomes natural to the mind which has been accustomed to contemplate not *things* only, or for their own sake alone, but likewise and chiefly the *relations* of things, either their relations to each other, or to the observer, or to the state and apprehension of the hearers.

(1969–2002 451; see also Smith 64)

Davy, Wordsworth and Coleridge all reject a scientific practice that encourages a passive mind, where what is needed is an active mind to explore the connections, resemblances, and harmonies between all parts of the living world, and between this world and the mind itself.

When William Whewell coined the word 'scientist' at the BAAS meeting in 1833, he was thinking along similar lines to Davy and the Romantic poets. Whewell praised Mary Somerville's book *On the Connexion of the Physical Sciences* for evincing these same principles, revealing continuities with Romantic thought concerning what Coleridge called in 'The Eolian Harp' 'the one Life within us and abroad' (l. 26). He could not help observing nonetheless that '[t]he tendency of the sciences has long been an increased proclivity to separation and dismemberment' ([Whewell] 58–9). Whewell lamented the increasing specialization, which meant that a chemist might only be an electrochemist and that poets, such as Goethe, would rarely be successful in their incursions into 'the fields of experimental science' (59). By his own account, when he proposed the term 'scientist' it was 'by analogy with *artist*' ([Whewell] 59), in an attempt to underscore and reaffirm the resemblance between the two. Yet, despite this, the term came to stand instead for an opposition, with the result that, as Richard Yeo notes, many nineteenth-century scientists themselves, such as Michael Faraday and Huxley, eschewed the term because, like Davy, they 'preferred to think of their work as part of broader philosophical, theological, and moral concerns' (6). As a result, it was not until the end of the century that 'scientist' attained the wider currency that it holds today.

The eventual replacement of the 'man of science' with the 'scientist' by the end of the nineteenth century is more than simply a matter of nomenclature. But it would be a travesty to reduce it either to a straightforward transition from a Romantic and holistic conception of science to a technocratic or scientistic position like that described many decades later by C. P. Snow as one of two sundered cultures. Throughout the century scientists claimed a right to speak on moral, cultural and religious matters. They insisted too on the centrality of imagination to their own methods. In a well-known paper first given at the BAAS in Liverpool in 1870, Tyndall reasserted the vital role of the imagination in science. He prefaced the published text of his essay with a quotation from a presidential address to the Royal Society given by the surgeon Sir Benjamin Brodie eleven years earlier in which Brodie – a contemporary of Davy – called the imagination at once 'the source of poetic genius' and 'the instrument of discovery in science' (quoted in Tyndall 2:101–2). The physicist Oliver Lodge went further. Even as late as 1910, he was arguing that scientists needed to take account of the 'intuitions of genius' (Lodge 197), including those of poets such as Tennyson, in formulating their hypotheses about 'the relation between man and the rest of the Universe'.

On the other hand, Wordsworth's opposition to 'Poetry and Matter of Fact, or Science' and Davy's resistance to specialization suggest that the seeds of Snow's 'Two Cultures' – as an idea, if not a reality – were sown very early in the century. This discourse of opposition would recur throughout the century too. In an essay on Pope published in the *North British Review* in 1848, Thomas De Quincey distinguished between the literature of power and the literature of knowledge, contrasting the persistence of literary works from the *Iliad* onwards with the ephemeral nature of even the greatest works of science, which were always open to be superseded by new theories and discoveries. In the early 1880s, Huxley and Matthew Arnold made contrasting cases for education to be grounded in the sciences and the humanities, Huxley giving his talk 'Science and Culture' at the opening of Mason College in Birmingham in 1880 and Arnold replying two years later from Cambridge in his Rede lecture on 'Literature and Science'. What the debate between Huxley and Arnold reveals is not so much the separation of science from the rest of culture as a growing confidence on the part of scientists that they can and should take a leading role in culture, if necessary at the expense of the humanities and – famously, in Huxley's case – of the church.

This positioning of science as an authoritative and distinctive cultural discourse was achieved on the back of the innovations of the 1830s: the foundation of the BAAS, the reform of the Royal Society and the conceptualization of the 'scientist', even if that label took a long time to be fixed. It was bolstered by the introduction of the positivist philosophy theorized in the same decade in France by Auguste Comte. Comte argued that positive knowledge provided by sensory experience was the only robust form of knowledge. In so doing, he privileged science over theology and metaphysics, plotting as he did so a historical trajectory in which metaphysics replaced theology to be replaced in turn by science. In the 1840s and 1850s, secular British thinkers and writers including John Stuart Mill, Harriet Martineau, George Eliot, George Henry Lewes and Herbert Spencer were drawn to Comte's confident prophecy of an age of positive knowledge untrammelled by religious or metaphysical prejudice. Like Comte himself, they took this as a licence to extend the reach of science, not only challenging the cultural authority of theology but also claiming new territory previously held within the purview of humanist disciplines like literature and philosophy. Along with Alexander Bain, Spencer and Lewes sought to develop a new scientific approach to psychology. Following Comte's lead, Spencer pioneered sociology on scientific principles too, and sought to unify all positive knowledge into one synthetic philosophy.

As a philosophy, Comtean positivism never dominated the mainstream of science in Britain. Indeed many British scientists actively denied or repudiated Comte's influence, particularly after the elaboration and reification of positivism into its own religion of humanity, dismissed by Huxley as an 'incongruous mixture of bad science with eviscerated papistry' (255). Yet Comte's assertion of the primacy of positive, scientific knowledge was instrumental in reshaping the role of science in Victorian culture. It underpinned the campaigns of Huxley and Tyndall for naturalism and against natural theology in science and of Eliot and Lewes for scientific realism in fiction. More generally, it drove forward the move to give science a central place within education and culture at large. When Henry Acland, reader in anatomy at the University of Oxford, warned his fellow academics in 1848 that they had better 'add some study of the fundamental arrangements of the Natural world' (39) to the education offered by the university, as otherwise 'if we do not *add* it, we may live to see . . . such knowledge *substituted* for our present system', the spectre haunting Oxford was positivism. Five years later, Oxford's Convocation agreed to build the Oxford University Museum of Natural History, with Acland taking a leading role in its design alongside the architect Benjamin Woodward, the stone carvers James and John O'Shea, the geologist John Phillips, the chemist Charles Daubenay, the critic John Ruskin and several members of the Pre-Raphaelite Brotherhood and their circle. The building itself is an explicitly Christian temple to God's creation and a uniquely rich example of a collaboration between scientists and artists, yet it was built, at least in part, in response to the rise of secular science under the star of positivism. When it opened in 1860, in the wake of Darwin's and Wallace's discovery of the principle of natural selection, the museum was a strong centre of resistance, under the leadership of Phillips, its first keeper, to the secularization of science. By the end of the century, it had become a centre for research and teaching in the new, secular evolutionary biology, with the anthropologist E. B. Tylor as keeper and the staunch Darwinians E. Ray Lankester and E. B. Poulton heading up the professoriate, and the scientific naturalism, if not the metaphysics, of positivism clearly in the ascendancy.

To return to *The Wonderful Century*, although Wallace takes in the whole of the nineteenth century in his book, a disproportionate number of the discoveries and the scientists he celebrates date from the Victorian period, and it is hard to imagine an equivalent book being written in the same way a hundred years before. In Wallace's book, science does not stand as one cultural activity, one mode of thought, one source of wonder among many. Instead it stands alone, as the defining glory – and at times the failure – of the age. Wallace was an eccentric, an outsider who felt no need to trim his scientific cloth to prevailing fashions, and who never held an academic post. Yet in many ways he epitomizes the contradictory place of science and its relation to literature in nineteenth-century culture. Like Davy, he resisted specialization, roaming widely across sciences, spiritualism and socialism in his writing, and unifying them into an idiosyncratic world-picture. On the other hand, his celebration of science in his time is positivistic, even scientistic, and he is as scrupulous in his adherence to scientific standards, as he sees them, in his spiritualist and political writings as he is in his natural history. And, though he is remembered as a scientist, he made his living principally as an author. His travel books and volumes of essays – like Davy's lectures and Darwin's 'one long argument' (86) *On the Origin of Species* – are themselves literature as well as science. *The Wonderful Century* itself opens each chapter with one or more verse epigraphs, most of them by nineteenth-century poets, including Mathilde Blind, whose poem *The Ascent of Man* was reprinted the year after with a prefatory note by Wallace himself. Science became more and more productive and assertive over the course of the nineteenth century, but even at the close it remained in a close symbiotic relationship with literature. Nineteenth-century science defined itself both in alliance with and in opposition to literature; it shaped and was shaped by imaginative writing in all its forms; and it existed then and survives today as written literature in its own right.

The chapters in this collection deal with both popular and canonical writers, and with many different forms of writing. Many of them examine – and contest – perceived oppositions associated with literature and science, in particular subjectivity versus objectivity, truth versus fiction, fact versus the imagination, the real versus the ideal. While contributing authors have been given the freedom to pursue their topics as they see fit, in general, their chapters offer the reader a sense of the key primary material that deals with the topic and some idea of the historical context in which the topic operated, and point towards the body of scholarship and criticism that has commented upon it. Some of the chapters explicitly suggest possible ways to pursue further study in the particular area in question, offering leads for future research and considering what remains to be done. Where the bibliographies do not cover the critical field, lists of further reading are also given.

The first four chapters, grouped together in the first section, set out key historical and cultural contexts for the study of literature and science of the nineteenth century. Sadiah Qureshi's chapter on 'Science, Empire and Globalization in the Nineteenth Century' expands on the theme of the global reach and networks of science in an age of empire. Qureshi considers how the expanded range of science, and the ideological frame of imperialism, determined how the nineteenth century sought to order its conceptions of space, life and knowledge. Turning back to the heart of empire, Martin Willis charts the rise and proliferation of institutions dedicated to producing and disseminating science in nineteenth-century Britain. In 'Scientific Cultures and Institutions', Willis argues for the centrality of poets and novelists, performance and writing, within institutions such as the Royal Institution, the British Association for the Advancement of Science and the X-Club of the Royal Society. The cultural hegemony wished for by the Royal Society was no longer possible after the proliferation of knowledge-producing bodies such as the Royal Polytechnic, which had not only laboratory space but also a stage on which plays could be performed.

Paul White's chapter on 'Science and Religion' acknowledges that science became an ideological weapon in the discussion of the role of religion in a number of sciences of the period, including, notably, geology, biology, physics and anthropology. White confirms Willis's thesis that the control hitherto exercised by such institutions as the Royal Society and the Royal College of Physicians was threatened during the century. There were attempts to integrate religious belief into science, whether in phrenology, thermodynamics or geology, while the continued persistence of natural theology testified to the desire to make this work. As White points out, the relation between religion and science is still very much a live topic today and one that continues to be played out in literary texts. In 'Women and Science', Michelle Boswell considers a topic touched on by most of the authors in this volume but looked at here in relief across the genres and across the century. Boswell's chapter notes that women participated in a diverse range of genres and in many different scientific disciplines. They practised, popularized and published science. In a nod to Willis's chapter, Boswell observes that there were certain specific institutions that encouraged women to publish science writing. She also recognizes that some of the published writings by women were situated within a conventional religious framework. As for many of the contributors to this volume, genre is for Boswell, and for the women who use it, an 'interpretative tool' rather than a fixed taxonomy.

Charlotte Sleigh's chapter, 'The Novel as Observation and Experiment', is a useful opening chapter for the second section of this collection, on how different genres of writing in the nineteenth century engaged with science. Sleigh sets out a number of important definitions.

Indeed, as with a number of other chapters, it is apparent that, in their struggle for identity in the nineteenth century, literature and science often define themselves against one another. This is a point that comes out in both Adam Roberts's and Gregory Tate's chapters too, on science fiction and poetry, respectively. On the other hand, Sleigh argues that the novel becomes a space where science's claims can be 'rehearsed and trialled', a claim Tate makes for poetry too. Sleigh finds comparison between the nineteenth-century novel and science in the way that they both ask '[w]hat things are real, and how can we know about them?' Both novels and science offer ways of seeing the world, and Sleigh points out that there is a 'political imperative' in this, because interpretation offers control. Perhaps surprisingly, Sleigh argues that the work of H. G. Wells could be 'Britain's best example of naturalist fiction', thus concurring with the general tenor of Roberts's chapter 'From Gothic to Science Fiction', which challenges some of the traditional views of these genres. Rather than agree with the idea that the Gothic is opposite to reason, which means that it would be *de facto* opposite to science, Roberts finds that, in this instance too, literature and science are 'mutually reinforcing'. Roberts, like Sleigh, is concerned with ideas of perspective, something that comes up repeatedly in the volume as a whole, concentrating in this chapter on such ideas as scale and vastness. This is only one area where the Gothic and science intersect; for example, new perceptions of the universe encourage a sense of human insignificance that had long been a preoccupation for Gothic and science fiction writers.

Tate's chapter, 'Poetry and Science', makes a special case for poetry as the form most often held opposite to science. Using examples from both canonical and less well known poets, Tate sets out to interrogate notions of poetic subjectivity against scientific objectivity. In so doing, he compares Wordsworth to Hopkins. Tate argues that a focus on poetry means that the traditional division between 'Romantic' and 'Victorian' is complicated and even diminished, noting that a careful observation and description of nature was among the objects shared by both. He considers the role that quoting poetry offers to scientific writing, finding that, in this regard, poetry is the established form contributing an authority that science does not yet have. Tate also looks at poetry written by natural philosophers, focusing on Davy. As for Sleigh on the novel, poetry is found to be a space – unlike science – in which different perspectives can be compared and even reconciled.

Peter Garratt and David Amigoni consider in their respective chapters the cultural authority that is offered by specific genres. They both also explicitly examine neglected or forgotten texts. Garratt's chapter on 'Scientific Literary Criticism' shows how, at the same time as science was professionalizing, the role of the literary critic was also becoming a professional one. One means by which to achieve this was to follow a 'scientific model of criticism'. The body of criticism that Garratt discusses here emerged in the nineteenth century written by people who crossed the worlds of both science and literature. Amigoni's chapter, 'Writing the Scientist: Biography and Autobiography', claims that in these two genres literature and science come together most clearly. These are scientific lives narrated in literary form, with private archive materials (letters, notebooks and other manuscript sources) used to bolster these texts' claims to cultural authority in the nineteenth century. Amigoni is particularly interested in the interplay of texts written in dialogue with or in response to others, and Jonathan Smith's chapter, 'Writing Science: Scientific Prose' likewise discovers that scientific prose is often written in dialogue with other forms of writing. Specifically, Smith argues that scientific prose employs the language and techniques most associated with literary writing. It is important to acknowledge, he writes, that such devices are not mere ornaments but 'constitutive of knowledge-making'. Science writing can be just as poetic and imaginative as literary fiction and, equally, literary devices are 'part of the knowledge-making process' of scientific prose.

The last four chapters in this section all concern texts that have not been taken sufficiently seriously by studies of literature and science in the past. Ralph O'Connor's 'Science for the General Reader' finds that so-called 'popular' science writing has been acknowledged as important but not analysed as a literary text in its own right. O'Connor's argument dovetails with Smith's, not simply for this reason but also because it again raises questions of genre. In the texts that O'Connor examines in his chapter he often finds generic hybridity and sometimes questions whether we know what the genre is at all. Like Smith, O'Connor argues that the texts he examines are productive of knowledge itself rather than being a by-product of knowledge proper. Gowan Dawson, like O'Connor, raises questions of readership and audience in his chapter, 'Science in the Periodical Press'. He too finds that the hybrid 'magazine text' found in nineteenth-century periodicals 'blurs generic distinction' and reveals these very distinctions to have been anachronistically imposed. Nineteenth-century periodicals have long been acknowledged to be truly interdisciplinary forms, bringing science and literature into 'proximity' with each other, but Dawson presents some convincing examples to show how resituating a text in its original periodical context changes its meaning.

Debbie Bark's chapter, 'Science for Children', looks at another neglected and popular form. Indeed, as she argues, the popularity of these texts for children reveals science's 'cultural currency' during the nineteenth century. This chapter demonstrates science as an 'everyday activity' rather than authorized by an elite institution. Like O'Connor and Dawson, Bark is alive to the market forces of the book-buying public. She finds children's literature to be varied in its representation of nature, characterizing it as a 'vehicle for scientific enquiry and debate'. She shows that women could exercise authority as authors of this literature, and that such texts can equally provide a space for subversion and a challenge to authority. Iwan Rhys Morus's chapter, 'Staging Science', neatly concludes this section and fits with the contributions of Smith, O'Connor and Dawson in particular. For Morus, science is a 'set of practices rather than a body of knowledge' and the performance of science, such as seeing Pepper's Ghost at the Royal Polytechnic, is one of these practices. Morus is insistent that such performances must be recognized as themselves 'knowledge making' and discusses the ways in which the performer demonstrates 'embodied practice in knowledge making'. Such spectacles were hugely popular and highly organized. They show in relief the kinds of strategies used by all to discriminate between illusion and reality, fact and fiction. In this instance, the stage is seen as a 'cultural space' in which the 'highly contested nature of scientific authority' is played out.

The third and fourth sections of the book are divided into scientific disciplines as they emerged in the nineteenth century. The third section concentrates on the mathematical and physical sciences. The first two chapters of this section deal with disciplines with far older identities. In her chapter on 'Mathematics', Alice Jenkins considers a discipline that has been neglected in literature and science studies (and particularly by those of a historicist nature) despite the fact that it had a huge and demonstrable impact on Victorian culture. Education in mathematics was such that it might be regarded as illustrative of a 'shared experience' or 'common knowledge' for men of a certain class, much as Dawson suggests might be the case for periodical circulation. Jenkins discusses ways in which mathematics might and might not fit easily into methodological models of literature and science studies, neatly picking up ideas voiced in previous chapters. She briefly considers the ways that journals, networks, institutions and popular writings have affected mathematics. Similarly to Tate, she focuses on people who were both mathematicians and poets, examining shared features such as the pleasure of concision in expression and of counting. Conversely, Pamela Gossin's chapter on 'Astronomy' considers a science that has long been a popular subject in literature and science studies. Gossin reflects upon many forms of the 'literature of astronomy' in the first post-Newtonian century, including poetry, novels, popular and technical texts, cosmological treatises, histories of astronomy, astronomical biographies and literary criticism. While most members of the 'generally educated public' gained only 'a passing, elementary textbook-level understanding of astronomical phenomena and their cosmological consequences', Gossin shows that many nineteenth-century literary writers – such as Tennyson, Hardy and Eliot – developed a sophisticated and sensitive knowledge of the collective achievements of the Herschels and other contemporary astronomical discoveries through accounts in the popular press and other public 'marketplaces'.

Davy looms large in this volume as a whole, featuring in particular in Adelene Buckland's chapter on 'Geology' and Sharon Ruston's on 'Chemistry'. Buckland, like Gossin, considers a science that cannot straightforwardly see or comprehend its subject: deep time. New methods were needed to accommodate this. Geology was a fashionable though controversial science in the nineteenth century, and Buckland's chapter returns to ideas of perception as geology struggles with the idea that human perception of the natural world might be inadequate or misleading. For Ruston, the science of chemistry, which prided itself on its modernity, remained closely linked to its precursor alchemy throughout the nineteenth century. Both Ruston and Buckland quote Coleridge on the primary imagination and Davy on the creativity of the chemist in their discussions of how the scientific imagination was used to conceptualize the act of creation. In her chapter, 'Thermodynamics', Barri J. Gold notes that this new science, like chemistry, acknowledged that matter can neither be created nor destroyed, but instead changes into different forms. Thermodynamics, like mathematics, has been rather neglected by literature and science scholars, and Gold reveals that there is much work still to be done. There are some fascinating peculiarities about this field of work, which mean that it forces us to 'rethink' the 'rules' of literature and science. For example, Gold asks us to think about how we might discuss the ideas of thermodynamics given that for much of the century it was barely known and had not yet been named.

There is some useful overlap between the final two chapters in this section, Stella Pratt-Smith's on 'Electricity' and Meegan Kennedy's on 'Technology'. Both consider sources of energy, including electricity, steam and gas. As with some earlier chapters, Pratt-Smith considers a force that can neither be seen nor touched and which can only be 'obliquely observed' by humans. The 'unusually broad range of concerns' with which electricity is engaged encourages the idea that it possesses 'elusive, mysterious, or figurative qualities'. This rhetoric that accompanies electricity is not merely embellishment but forms an integral part of our understanding of it. Kennedy's wide-ranging chapter demonstrates the sheer scale of technological advancement in the nineteenth century and its deep engagement with the century's literature. Her chapter is also witness to just how fruitful technology has been to critical studies. Recalling Gossin's and Buckland's chapters, Kennedy also focuses briefly on questions of perception and vision; in this instance comparing the 'changed scale of vision' now possible with improved microscopes, telescopes and the new Röntgen or X-rays.

Section four concerns the biological and human sciences. There are interesting connections to be drawn with chapters in the previous section, revealing some common ways of understanding the world throughout the century. In 'Natural History, Evolution and Ecology', John Holmes perceives a shift in world view from fixity (with natural history) to flux (with evolution), a final recognition – as also noted by Gold and Ruston – that the world is perpetually in transformation. The study of literature is 'integral to our understanding of natural history', which was seen to be morally improving. What is at stake in this debate is 'the nature of nature itself', nothing less than our view of the natural world and our place within it in relation to other living things. The sheer volume of criticism on evolution testifies to its importance as a subject. The sweep of evolutionary history was, similar to earth and planetary history, immense, and to be imagined rather than witnessed. Julia Reid also considers the relation of the present to the past in her chapter on 'Archaeology and Anthropology'. There are links to be made with Buckland's topic; for example, just as there were amateur geologists, so did amateur archaeology flourish in the nineteenth century. Reid argues that because archaeology and anthropology were not fully professionalized, even by the end of the century, the boundaries between scientific and literary writing on these subjects were blurred. She notes the generic instability of much archaeological writing and the common endeavour between this, anthropological writing and literary fiction to explore the relationship between the present and the past.

Andrew Mangham's chapter on 'Medical Research' considers the writer as scientist. With clear links back to Sleigh's chapter on the novel, Mangham notes the number of novelists who consider their work as experimental. His chapter is concerned with methodologies primarily and he finds, as Sleigh does, that the novel is a space for self-scrutiny, reflection and the consideration of ethics. He opens by noting that critics have perceived the influence of medicine on novel writing in many – and even opposing – ways. Similarly, Suzy Anger in 'Sciences of the Mind' recognizes that this science, newly emerged at the end of the nineteenth century, takes multiple forms in the literary texts of the period. Eliot is again held up as a model of 'scientific insight', exploring the psychology of her characters. Pseudo-sciences such as phrenology and mesmerism enabled questions about the self, free will and volition, encouraging a medical interest in automatism, the power of suggestion, the mind's unconscious processes and the possibility of dual personality. In Anger's chapter there is a perceived shift towards the scientific explanation too of such practices as spiritualism and telepathy.

Anna Katharina Schaffner's chapter on 'Sexology' perceives a similar shift from Christian to biological and psychological models, arguing that sexual perversions came to be no longer primarily considered as sinful but instead assessed as pathological. Schaffner's chapter demonstrates that the boundaries between scientific and literary discourses on sexuality are much less firmly drawn than we might expect. In fact, scientific studies of sex are found to be 'substantially shaped by the literary imagination'. In particular, many British theorists of sexuality were also literary critics and writers, and literary concepts frequently informed their theoretical models. Christine Ferguson picks up on the idea of so-called 'rejected knowledge', touched on by Anger, to argue that occult practices were everywhere in nineteenth-century life, from local fairs to public lectures and periodical discussions. Their proponents found a way to update and revise ancient occult science, making it appear to cohere with, rather than challenge, the new world view that evolution and geology put forward.

To close, we invited the historian of nineteenth-century science Bernard Lightman to consider how the research distilled in this volume – both the fresh contributions made by the authors themselves and the existing scholarship which they summarize and assess – bears not only on the study of literature and science itself but also on the history of science. Lightman's conclusion is that the 'study of Victorian science has truly become an interdisciplinary enterprise', in which historians of science and literature and science scholars each need to make full and careful use of one another's work. We agree. It is our hope that this *Research Companion* will provide critics and scholars working on science and literature in the nineteenth century with a comprehensive and rigorously historicist grounding in their own field. At the same time, we hope that it offers historians of nineteenth-century science too a rich fund of literary sources and critical scholarship to draw on in their research – research that is, as Lightman suggests, ultimately undisseverable from the study of literature and science.

Bibliography

Primary texts

Acland, Henry. *Remarks on the Extension of Education at the University of Oxford*. Oxford: Parker, 1848.

Coleridge, Samuel Taylor. 'The Friend'. Ed. Barbara E. Rooke. *The Collected Works of Samuel Taylor Coleridge*. Ed. Katherine Coburn. 16 vols. London: Routledge and Kegan Paul, 1969–2002. Vol. 4:1.

Coleridge, Samuel Taylor. *Poems*. Ed. H. J. Jackson. Oxford: Oxford University Press, 1985.

Coleridge, Samuel Taylor. *Biographia Literaria*. Eds J. Engell and W. Jackson Bate. Princeton, NJ: Princeton University Press, 1993.

Darwin, Charles. *Autobiographies*. Eds. Michael Neve and Sharon Messenger. London: Penguin, 2002.

Davy, Humphry. *The Collected Works of Sir Humphry Davy*. Ed. John Davy. 9 vols. London: Smith, Elder, 1839–40.

Dawson, Gowan, and Bernard Lightman, eds. *Victorian Science and Literature*. 8 vols. London: Pickering and Chatto, 2011–12.

Eliot, George. *Middlemarch*. Ed. W. J. Harvey. Harmondsworth: Penguin, 1965.

Hawley, Judith, ed. *Literature and Science, 1660–1834*. 8 vols. London: Pickering and Chatto, 2003–04.

Hughes, William R. *Constance Naden: A Memoir*. London: Bickers, 1890.

Huxley, Thomas H. *Science and Christian Tradition*. London: Macmillan, 1894.

Lodge, Oliver. *Reason and Belief*. London: Methuen, 1910.

Naden, Constance C. W. *Induction and Deduction: A Historical and Critical Sketch of Successive Philosophical Conceptions Respecting the Relations Between Inductive and Deductive Thought and Other Essays*. Ed. R. Lewins. London: Bickers, 1890.

Tyndall, John. *Fragments of Science*. 8th edn. 2 vols. London: Longmans, Green, 1892.

Wallace, Alfred Russel. *The Wonderful Century: Its Successes and Its Failures*. New York, NY: Dodd, Mead, 1898.

Whewell, William. '*On the Connexion of the Physical Sciences*. By Mrs. Somerville'. *Quarterly Review* 51 (1834): 54–68.

Wordsworth, William. *Prose Works*. Eds. W. J. B. Owen and J. W. Smyser. 3 vols. Oxford: Clarendon, 1974.

Secondary texts

Fyfe, Aileen, and Bernard Lightman, eds. *Science in the Marketplace: Nineteenth-Century Sites and Experiences*. Chicago and London: University of Chicago Press, 2007.

Holmes, John. '*The New Day*: Dr Hake and the Poetry of Science'. *Journal of Victorian Culture* 9 (2004): 68–89.

Orr, Mary. 'Pursuing Proper Protocol: Sarah Bowdich's Purview of the Sciences of Exploration'. *Victorian Studies* 49 (2007): 277–85.

Ruston, Sharon. *Creating Romanticism: Case Studies in the Literature, Science and Medicine of the 1790s*. Basingstoke: Palgrave Macmillan, 2013.

Secord, James A. *Visions of Science: Books and Readers at the Dawn of the Victorian Age*. Oxford: Oxford University Press, 2014.

Smith, Jonathan. *Fact and Feeling: Baconian Science and the Nineteenth-Century Literary Imagination*. Madison, WI: University of Wisconsin Press, 1994.

Wyatt, John. *Wordsworth and the Geologists*. Cambridge: Cambridge University Press, 1995.

Yeo, Richard. *Defining Science: William Whewell, Natural Knowledge, and Public Debate in Early Victorian Britain*. Cambridge: Cambridge University Press, 1993.

Further reading

Beach, Joseph Warren. *The Concept of Nature in Nineteenth-Century English Poetry*. New York, NY: Macmillan, 1936.

Beer, Gillian. *Darwin's Plots: Evolutionary Narrative in Darwin, George Eliot and Nineteenth-Century Fiction*. 3rd edn. Cambridge: Cambridge University Press, 2009.

Dowden, Edward. *Studies in Literature 1789–1877*. London: Kegan Paul, 1878.

Grabo, Carl Henry. *A Newton Among Poets: Shelley's Use of Science in 'Prometheus Unbound'*. Chapel Hill: University of North Carolina Press, 1930.

Levine, George. *Darwin and the Novelists: Patterns of Science in Victorian Fiction*. Cambridge, MA: Harvard University Press, 1988.

Shuttleworth, Sally. *George Eliot and Nineteenth-Century Science: The Make-Believe of a Beginning*. Cambridge: Cambridge University Press, 1984.

Stevenson, Lionel. *Darwin Among the Poets*. Chicago, IL: University of Chicago Press, 1932.

I
Contexts

1

SCIENCE, EMPIRE AND GLOBALIZATION IN THE NINETEENTH CENTURY

Sadiah Qureshi

University of Birmingham

The modern sciences are a global enterprise.[1] Scientists work in national and international contexts; whether in their educational and technical training, research teams, peer-reviewed journals, laboratories or mega-projects such as the CERN hadron collider, which is so large that it spans the Franco-Swiss national border, science is the transregional production of natural knowledge. The nineteenth century was a key period in the internationalization and globalization of the sciences. One barometer of this shift might be the international conference or world congress since, after all, despite a vast network of correspondence, 'The Republic of letters never assembled' (Alder 19). While national scientific societies had existed since the 1660s, the first transnational scientific conference took place much later in 1798, when Franz Xaver von Zach called together his colleagues to discuss astronomical and meteorological standards. The second such event also took place in 1798 as the revolutionary French State met to discuss the length of the metre. From these early origins, the number of international conferences rapidly expanded between the late 1860s and 1880s. Modelled on diplomatic assemblies, many early congresses met to discuss adopting standard measurements of experimental factors from time to electrical currents. However, they had a lasting impact by setting patterns of international collaboration and conversation that are fundamental to the sciences.[2] In conjunction, the sciences made claims to globally applicable knowledge. In the nineteenth century, such claims both coincided with and constituted forms of globalization as the sciences offered frameworks for making an increasingly globalized world, whether through new ways of reordering space or through creating human hierarchies. Yet, such ways of knowing were not uncontested.

By exploring scientific attempts to order space, life and knowledge, this chapter argues that the nineteenth-century witnessed a significant period of internationalization and expansion that laid the foundations of modern globalized science. In doing so, it pays particular attention to imperialism. Nineteenth-century sciences depended upon practices that both constituted and benefited imperialism. The sciences made knowledge claims in the service of empire, both implicitly and explicitly. Likewise, the material practices underlying scientific research often depended upon imperial infrastructure, whether in mapping new territories, providing suitable sites for fieldwork, brokering knowledge or collecting specimens. Drawing on examples from the physical and biological sciences throughout, this chapter focuses on the material practices of nineteenth-century science. In doing so, it traces how imperialism provided practitioners with career opportunities, access to resources, networks for collaboration and competition and locations to conduct research. This offers glimpses into how the sciences became global both in terms of their theoretical scope and everyday practice and how they constituted globalization through reshaping the modern world.

Ordering space

The sciences helped reorder space in the nineteenth century through mapping vast swathes of the globe. Whether plotting shipping routes, outlining nations or claiming ownership of colonies, mapping is a political process. As a form of information acquisition and representation, mapping helps make places knowable and can be instrumental for activities such as trade, governance and scientific exploration. As Matthew Edney's wonderful work shows, one of the most ambitious mapping projects to take place in the nineteenth century was the mapping of India. William Lambton surveyed the Kingdom of Mysore in 1799, after the British defeated and killed Tipu Sultan at the Battle of Srirangapatam (Seringapatam) in the same year. Lambton then began surveying Madras in 1802. These early investigations are often credited with initiating the trigonometric survey because Lambton used triangulation to calculate distances. The process works by establishing an accurate base line from which a geometric net of triangles is built up. The net can then be used to make measurements that form the basis for new maps. By 1815 Lambton had triangulated Southern India. He was joined three years later by Captain George Everest. At this stage, the project was named the Great Trigonometrical Survey of India and continued into northern territory. When Lambton died in 1823, Everest became the head of the project until the final measurements were taken in 1841.

Intelligence gathering, whether through mapping or other means, has been argued to have been central to deciding the success or failure of state institutions in India (Bayly, 1997). Although the colonial authorities never achieved total control, projects such as the trigonometric survey did provide useful knowledge of the subcontinent and underpinned political claims about how India ought to be ruled and by whom. For example, colonial authorities claimed that the work 'could not be undertaken by the Indians themselves' but was necessary for the functioning of canals, military roads and trade (Edney, 319). Despite such claims, the work was heavily dependent on the knowledge of Indians. For example, Radhanath Sikdar joined the survey in 1840 and helped measure Peak XV. When it was found to be the highest mountain in the world, it was renamed the more familiar Mount Everest (Edney, 262). Later, Nain Singh Rawat helped survey Nepal, Tibet and Kashmir during the 1860s and 1870s. Unusually, Rawat's contributions were accredited in the period with the award of a medal from the Royal Geographical Society in 1877. The award is a rare example of elite institutions such as the RGS explicitly acknowledging their debts to indigenous knowledge in making scientific research possible. Both Sikdar and Rawat relied on their local networks of informants to provide intelligence and expertise that were incorporated into the larger mapping project (Driver; Raj). It is crucial that we remember the extent to which local and indigenous knowledge formed the bedrock of such ventures especially given their lasting historical significance. After all, the Great Trigonometrical Survey helped create and consolidate the very notion of India as a geographical whole.

Reconnaissance through mapping, fieldwork and exploration was often integral to global political and military activity. For example, between 1798 and 1801 Napoleon began an expedition to Egypt – then within the Ottoman Empire and strategically important for Eurasian trade – in search of a new colony. Although Napoleon failed to establish an Egyptian colony, his methods of reconnaissance were important for the development of globalized science (Gillispie). The expedition included a Commission of Science and Arts comprising 151 engineers, architects and medically trained personnel. Their expertise was crucial to Napoleon's imperial ambitions. The voyage proved important for many careers, especially given that many of the men had only just graduated or were still studying at the École Polytechnique. The anatomist Étienne Geoffroy Saint-Hilaire and entomologist Jules-Cesar L. Savigny both established their reputations and challenged doyens in their field, such as Georges Cuvier, on the basis of their experiences in Egypt. In Cairo, Napoleon established the Institute of Egypt, modelled on the National Institute of France, to oversee the conduct and publication of much of the research from the expedition (see Reid's related discussion on archaeological expeditions in Chapter 23 below). Most famously, the expedition led to the discovery of the Rosetta Stone, which, when deciphered in 1822, allowed ancient Egyptian hieroglyphs to be read for the first time in the modern world and had profound repercussions for archaeological studies in the region. Most importantly, the research from the expedition was published in the expensive and lavishly illustrated *Description de L'Egypte* between 1809 and 1828. Presented as a tour down the Nile, from Philae to Thebes, through the Valley of the Kings, to the Great Pyramids, the work remains one of the best examples of a vast encyclopedic enterprise intended to create and disseminate scientific knowledge to a broader public. Napoleon's campaign exemplifies the use of large-scale, systematic, state-sponsored exploration in the late eighteenth and nineteenth centuries.

Globalized fieldwork was also fundamentally important to the physical sciences. This is exemplified by astronomical fieldwork such as eclipse expeditions (Pang; Ratcliff). Solar eclipses allow astrophysicists exceptional opportunities to observe the sun's corona, analyse the chemical composition of the outer atmosphere and search for planets. Alex Pang's history of eclipse expeditions showcases how astrophysicists spent months planning how best to take advantage of opportunities to see eclipses across the globe. He explores why questions of how best to observe and record data, who would make a reliable witness, where best to see the full eclipse and how to get there all received sustained attention. As Pang argues, the details of such organization are fundamentally important. It is easy to imagine fieldwork as an ad hoc affair but this erases the labour that had to be invested in setting up field sites and observatories, whether mobile or more permanent. Astronomers had long observed total solar eclipses but in the 1860s scientific societies such as the Royal Astronomical Society, the Royal Society of London and the British Association for the Advancement of Science spearheaded efforts to attract state support for a series of eclipse expeditions. Observations of the 1870 solar eclipse attracted considerable state funding. By 1888 a Permanent Eclipse Committee was set up by the Royal Astronomical Society that, in partnership with the Royal Society in 1894, became the Joint Permanent Eclipse Committee. Eclipse expeditions were undertaken in India, North America and the Caribbean and on Pacific Islands. Many of these field sites had a European imperial presence that was vital for fieldwork. Colonial governments often provided much needed help and expertise, as in 1871, when the trigonometric survey of India was ordered to help an expedition by providing two assistants, some expenses and other aid. Colonial states also provided a vital infrastructure, such as railways and the telegraph. In the later nineteenth-century expeditions, the expansion of mass tourism also contributed to making places such as India more accessible.

Telegraphy became one of the most important new technologies that made it possible to communicate across vast distances speedily. Electric telegraphy was developed in 1837 but, after a series of failures, the first cable to successfully span the Atlantic Ocean was laid in 1861. By the 1870s telegraph cables had created a network that spanned the globe to connect together places as far apart as Gibraltar, India, Singapore, Hong Kong, Australia, New Zealand, Africa's east coast, South America's east and west coasts and the West Indies. In the 1880s and 1890s, the Far East and the west coast of Africa were added to the web of cable routes (Hunt). The total length of the world's submarine cables managed by both the government and private enterprise jumped from barely 4,400 kilometres in 1865 to 50,865 kilometres by 1870, 245,329 kilometres by 1890 and over an astonishing 406,307 kilometres by 1903 (Wenzlhuemer, 119). As Bruce Hunt has shown, the history of how this network developed beautifully exemplifies important aspects of science's global reach. Significantly, he shows that telegraphy had an impact on both the *theoretical* concerns and *practice* of physics. In the early to mid-nineteenth century, electromagnetic phenomena were explained by appealing either to field theory or action-at-a-distance. He argues that action-at-a-distance became dominant in France and Germany and field theory in Britain precisely because of telegraphy. For instance, many British cables were underwater and so British physicists were especially interested in the propagation of signals, leakage and interference. In both France and Germany, action-at-a-distance was of more interest to physicists who did not encounter such problems. The move to standardize the measurement of electrical resistance using the electrical ohm has also been attributed to the role of telegraphy in providing a commercial impetus to attaining consensus. By the 1870s, British engineers commonly used the ohm and it subsequently replaced alternatives across Europe. Meanwhile, telegraph cables became known as the 'nerves of empire' because of their fundamental role in administering and maintaining imperial rule. The British state invested heavily in the new technology partly as a means to aid colonial administration, communication and control. The cables were also a material example of political interests since the gutta-percha used to manufacture the outer layers was produced from the sap of trees harvested in colonies such as Singapore and Malaysia.

Thus, while mapping and exploration provided politically relevant geographical expertise, new technologies helped reorder space by embedding new forms of communication and interconnectedness. Yet, the sciences also made knowledge claims that fundamentally transformed how life itself was understood.

Ordering life

Defining who counted as human and creating racial hierarchies to place them in was one of the most important and globalized attempts to order life in nineteenth-century Europe. For instance, throughout the eighteenth and early nineteenth centuries, in line with Christian orthodoxy, humans were typically considered a single species that originated with an act of divine creation and which might be split into a number of varieties or races (Wheeler). Three, each corresponding to the sons of Noah, were most commonly used, but five and seven were also proposed. Human varieties were distinguished by a wide variety of factors including clothing, religion, language, physical characteristics, geographical origins and nature of social, political and economic organization. Explanations for these differences tended to focus on environmental or post-diluvial migration. By the mid-nineteenth century, European scholars interested in these questions tried to redefine the physical, social and cultural criteria used to classify humans in Europe, giving rise to the new discipline of anthropology (again, see Reid, present volume, Chapter 23).

Such debates depended upon globalized practices of collecting and displaying humans, dead or alive. For example, throughout the nineteenth century, foreign peoples were imported into the major cities of Europe and North America to be displayed at exhibitions and world fairs (Qureshi). In cities such as Paris, London, Berlin and Chicago, members of the public could attend shows in which foreigners sang, danced and performed cultural rites as exemplars of their ethnic origin. At first such shows tended to be small-scale and might involve only an individual or a small group managed by an entrepreneur (Iwan Rhys Morus's contribution to this volume explores the importance of 'Staging Science' further). By the end of the century, foreign peoples were being professionally imported dozens at a time and displayed within mock native villages at world fairs such as the 1889 Exposition Universelle in Paris. Anthropologists went to great lengths to take advantage of performers for their research. At the 1886 Colonial and Indian Exhibition in London, the Anthropological Institute held a series of conferences at the fair, headed by their President Francis Galton; experiments were performed on a San man to test his strength, and Robert James Mann, an expert on South African ethnology, curated the ethnological collection from Natal. When the exhibition ended, published papers appeared in scientific journals such as the *Journal of the Anthropological Institute of Great Britain and Ireland*. Exhibitions became important sites for scientific research into race by making performers into experimental subjects and providing training grounds for practitioners. American anthropologists at the 1889 Exposition Universelle in Paris were so impressed by the villages of displayed peoples at the base of the Eiffel Tower that they emulated them for the 1893 World's Columbian Exposition in Chicago and the 1904 fair in St Louis (Rydell).

Scientific knowledge of human difference depended upon globalized, and often imperial, voyages and expeditions. In the later nineteenth century, new practitioners argued for paid professional positions to be established in anthropology as a means of securing the future of the fledgling discipline. In 1903, Alfred Cort Haddon's presidential address to the Anthropological Institute of Great Britain argued that the discipline's future depended upon 'trained observers and fresh investigations in the field' (Haddon). Haddon disavowed anthropology's founding fathers in the hopes of founding a new generation of *trained* professionals conducting extended in situ observations. A decade later, William H. R. Rivers also campaigned for his peers to adopt 'intensive work' (Rivers 7). Haddon and Rivers were especially concerned that precious and irreplaceable anthropological knowledge was vanishing as elders who had known life before colonialism joined their ancestors. Significantly, Rivers explicitly argued that the infrastructure provided by imperialism and missionary work were crucial since the 'most favorable moment' for investigations was

> ten to thirty years after a people has been bought under the influence of official and missionary. Such a time is sufficient to make intensive work possible, but not long enough to have allowed serious impairment of the native culture.
>
> *(Rivers 7)*

For too long, the polemical denunciations of late nineteenth-century anthropologists held sway in histories of anthropology and the early nineteenth century was often disparaged as an era of 'armchair anthropology' (Compare Stocking; Kuklick; Barth *et al.*). Efram Sera-Shriar has shown that earlier practitioners were immensely sensitive to the value of first-hand experience in ensuring the reliability of their information. Even if they did not travel personally, wherever possible they incorporated first-hand evidence whether through using questionnaires sent to colonial officials or preferentially referencing the work of missionaries conversant in local languages (Sera-Shriar). Thus, throughout the history of anthropology, concerns regarding the accuracy of ethnographic information were intimately tied to broader conditions of travel created by imperial presences.

Crucially, fieldwork relied on the *existing* infrastructure provided by both formal and informal imperialism, such as military activity and trade and missionary networks. The second voyage of HMS *Beagle* between 1831 and 1836 provides the most famous example of how closely scientific activity depended upon broader military and political activity. The voyage was intended to conduct surveys of South America, make meteorological observations, map coral reefs and record accurate longitudes. The voyage is better known for being the formative experience of Charles Darwin's life (Browne). He did not embark on the voyage a transmutationist. He had been trained in the older natural theological tradition explored by John Holmes in his chapter on natural history in this volume. While voyaging, Darwin was able to amass specimens and witness geological processes that later underpinned his conviction that the earth's vast antiquity made it possible for new forms of life to emerge through the accumulation of minute changes.

When Darwin finally plucked up the courage to publish his ideas on evolution through natural selection in *On the Origin of Species* (1859), he drew heavily upon his voyage to assert his expertise. The title page referred to his membership of the 'ROYAL, GEOLOGICAL, LINNEAN, ETC. SOCIETIES', his authorship of 'JOURNAL OF RESEARCHES DURING H. M. S. BEAGLE'S VOYAGE ROUND THE WORLD' and his Cantabrigian MA. These details signalled Darwin's gentlemanly standing and membership of London's most elite learned societies. The first lines of the book declared that 'WHEN on board H.M.S. "Beagle," as naturalist, I was much struck with certain facts . . . These facts seemed to me to throw some light on the origin of species—that mystery of mysteries' (Darwin, title page and 1). These details helped Darwin distance himself from the radical and materialist theories of evolution that had been in circulation throughout the early nineteenth century. For example, Jean-Baptiste Lamarck's theories of transmutation through the inheritance of acquired characteristics were published in 1809 and quickly became associated with French radicalism and resistance. In Britain, campaigners adopted French work on evolution to help make the case for political reform across the Channel (Desmond). Later, in 1844, the anonymous publication of the *Vestiges of the Natural History of Creation* proposed that creation operated through the natural law of evolution through development. By 1853, the book had been through ten editions and generated significant speculation regarding the author's identity and the nature of natural development across the whole of Victorian society (Secord). By contrast, the title page to the *Origin* immediately established Darwin's gentlemanly and scientific credentials. Likewise, by explicitly incorporating his time voyaging, Darwin laid claim to in situ expertise that conferred authority before professionalization made a formal career as a scientist possible in the late nineteenth century.

The immediate controversy surrounding Darwin's ideas on evolution, particularly with respect to humans, is well known within the context of Europe. Much of the work on his reception has focused on the ostensible conflict between science and religion by focusing on Christian opposition to materialist explanations of the natural world's origins and accounts of humanity's place in nature. We know that the rise of modern creationism has masked how theological responses to evolutionary ideas were complex and far more accommodating than many people realize (Numbers). Nonetheless, Darwin's global reception has only just begun to be explored. As Marwa Elshakry's path-breaking work has shown, Muslims in the Middle East drew on classical Islamic philosophy and cosmology to respond to evolutionary theories (Elshakry). Significantly, many Muslim theologians embraced evolutionary ideas as a means of demonstrating that Islam was not in conflict with modern science. For example, she shows that, in *Science and Civilization in Christianity and Islam* (1902), Muhammad Abduh, the Grand Mufti of Egypt, explicitly contrasted Islam's potential for incorporating new scientific ideas with the hostility of Christian theology. In such writing, co-opting evolutionary ideas was a means of defending Islam as a rational religion against racialized accusations of superstitious ignorance. Significantly, his work intertwined discussions of evolution and critical exposition of the Qur'an. It is important that such work was being written by a theologian of the Grand Mufti's status as it reminds us that assumptions of inherent conflict between science and religion are not only problematic but are often based on narrow understandings of religious belief and practice.

Whether reordering space or life, the sciences often drew upon multiple sources to make natural knowledge. Nonetheless, the nineteenth century witnessed significant attempts to marginalize alternative epistemologies whether through silence, erasure or appropriation. It is essential that these ways of disregarding knowledge and asserting authority are acknowledged in histories of science and the modern world.

Ordering knowledge

Ordering globalized scientific knowledge has often been characterized as a process of extraction from the periphery and classification in the metropole; however, this underestimates the extent to which local knowledge was crucial to European attempts at classification. For example, the Royal Botanic Gardens at Kew were the most important botanical gardens within the British Empire (Drayton). Directed by William Hooker and then his son Joseph Dalton Hooker, Kew became a global repository of botanical specimens. Some of these were collected personally by the Hookers on expeditions, while others were sent by correspondents. Joseph Hooker hoped to publish a flora of the entire southern oceans and was forced to reply to local collectors such as the New Zealand missionary Reverend William Colenso. They corresponded between 1841 and Colenso's death in 1899 (Endersby). Colenso sent specimens to Kew, where Hooker classified, named and published accounts of them, thereby integrating them into European botanical knowledge.

The relationship between Hooker and Colenso might easily be characterized as an example of specimens being collected in the periphery by locals (whether by colonial officials or colonized subjects) and then sent to a metropolitan centre to be processed and transformed into published knowledge; however, as Jim Endersby has shown, this oversimplifies their exchanges. Occasionally, Hooker and Colenso disagreed on how best to classify plants and Hooker often expected his status to assure him success. However, Colenso stressed the value of his local knowledge in a number of ways. First, he argued that his knowledge of living plants in their natural habitat gave him a significant advantage over anyone, such as Hooker, who was working from dried specimens. Second, his travels gave him detailed knowledge of the country's geography and therefore of botanical habitats. Thirdly, his contact with the Maori and familiarity with their language gave him unique access to indigenous knowledge. Ultimately, the disputes between Hooker and Colenso were predicated upon which kind of knowledge each valued and therefore who had the authority to classify and produce botanical knowledge. Hooker named thirty species or genera after Colenso and supported his application to the Linnean Society but remained unconvinced that Colenso's local expertise was a significant advantage. Both men had different kinds of botanical knowledge that the other valued and their correspondence developed on the basis of mutually beneficial exchanges, albeit with some tensions. Their association demonstrates that scientific knowledge can be traded commercially and through gift-exchange or trading favours. The relationship between Hooker and Colenso flourished through some elements of reciprocity, yet even this was routinely denied to indigenous peoples.

European science appropriated indigenous knowledge often without acknowledgement or recompense. Such contributions have been consistently erased both in the primary sources and by historians of science and have only received sustained attention in recent years. New histories of bioprospecting, for example, provide rich examples of such unequal exchanges and their importance for globalized science. Between the 1880s and 1920s, the European presence in West Africa created considerable imperial conflict. Abena Dove Osseo-Asare has shown that, in opposition to European incursions, the local 'Frafra' communities used their botanical expertise to produce poisoned arrows that killed quickly and effectively (Osseo-Asare). Local use of poison arrows had been known since the earliest voyages of the Portuguese in the fifteenth century and became synecdochally representative of the dangers of the African continent. As late as the 1880s, little was known of the exact plant species used to make the poison (although *Strophanthus* was believed to be the most likely), how the weapons were prepared or how to reverse the effects of poisoning. Fragmented intelligence had been collated from reports of wounded soldiers and the sporadic use of local antidotes. Yet, with the expansion of colonial occupation and warfare, poisoned arrows became of such concern for European doctors and colonial administrators that they were banned, thereby criminalizing certain kinds of African resistance, although their use continued until at least 1918. Meanwhile, following the 1884–85 Conference of Berlin, the governor of the Gold Coast set up a Commission for the Promotion of Agriculture and tasked it with investigating the use of *Strophanthus*. African healers had long used the genus to treat a variety of ailments but European interest in the plants escalated in the late 1880s. The physician Thomas Fraser produced purified strophanthin crystals following investigations of the stimulant effects of poisoned arrows. Fraser marketed his crystals as a more powerful drug than digitalin and suitable for treating weak hearts. By 1905, clinical trials were taking place in Germany and interest in the plants led to a failed export scheme after World War I. European interest in strophanthin systematically tried to suppress indigenous use of the plants by banning poisoned weapons or plundering seeds that were then sent to Kew for analysis. Attempts to create drugs from African plants exemplify how indigenous technologies and knowledges could become the basis for European medical treatments. In turn, the attempts to limit how Africans could use their botanical knowledge and, at times, coercively extract such expertise is a striking case of European appropriation through bioprospecting. Current and future histories of globalized science need to pay greater attention to these formerly neglected aspects of how global science developed. Without such consideration, we risk neglecting indigenous contributions and reiterating unhelpfully Eurocentric historiographies.

Conclusion

Scientific practice became increasingly international and global in the nineteenth century. Such changes brought significant opportunities for its practitioners and introduced new methodologies. The development of both international congresses and world fairs, for example, provided opportunities for collaboration, showcasing work and conducting national rivalries. Voyages of exploration and imperial reconnaissance brought new pathways for contributing to the sciences. Practitioners used voyages to conduct field research, publish original material and then claim authoritative expertise that ostensibly sedentary men of science could not challenge. As European empires expanded and consolidated themselves, colonial service provided new career paths and resources for would be scientists. Meanwhile, colonial infrastructure made colonies into important sites for fieldwork, earning them a reputation as vast laboratories (Tilley). The relationships between metropolitan centres and the so-called peripheries used to be characterized as one of the extraction of resources from abroad for research to be conducted in cities such as London or Paris and then made available to the broader scientific community. More recently, global interconnectedness has been imagined as networks of circulation and exchange that might operate independently of European and metropolitan centres entirely. This work has reconfigured the geography of how science became global and is a welcome intervention.

The sciences also made claims to globally relevant knowledge. In this chapter we have seen how scientific theories reshaped modern understandings of space and life and how this often depended on competing for authoritative knowledge with alternative epistemologies. The sciences contributed to the contraction of distance as ever greater speeds of travel were achieved. In doing so, the sciences contributed to shifts in globalization that led to greater similarities in the way people dressed, travelled, read new forms of print and consumed mass-produced goods. These knowledge claims highlight that the sciences are not inherently universal or international; rather, practitioners expended labour in trying to claim authority for their theories of how the natural world ought to be understood. The most dramatic examples of such rivalry can often be seen in histories of science that pay serious attention to local and indigenous knowledge. The sciences frequently appropriated local expertise without acknowledgement, remuneration or maintaining rights of use and access for local peoples, as is evident in bioprospecting. Yet, such connections were commonly erased both by nineteenth-century practitioners and by historians. Only more recently are the debts involved being retraced. Acknowledging the complicated history of how scientific practitioners sought to establish their authority provides a much more interesting and less Eurocentric perspective than proposing diffusionist models of scientific expansion. The challenge of how to embed these insights into new historiographies of science is ongoing and presents some of the most exciting ways to understand how and why science became global in the modern world.

Notes

1   This chapter is a longer version of one that will appear in P. Singaravélou and S. Venayre, Eds. *Histoire du Monde au XIXe Siècle*. Fayard, 2017. My thanks to Pierre Singaravélou and Fayard for permission to publish this version. My thanks to John Holmes and Sharon Ruston for their suggestions.
2   The great transregional history of how such collaboration was achieved across the sciences remains to be written. Alder's 'Scientific Conventions' presents the best introduction.

Bibliography

Primary texts

Darwin, Charles. *On the Origin of Species by Means of Natural Selection, or the Preservation of Favoured Races in the Struggle for Life.* London: John Murray, 1859.
Haddon, A. C. 'President's Address: Anthropology, its Position and Needs'. *Journal of the Anthropological Institute of Great Britain and Ireland* 33 (1903): 11–23.
Rivers, William H. R. 'Report on Anthropological Research Outside America'. *Reports on the Present Condition and Future Needs of the Science of Anthropology.* Ed. W. H. R. Rivers, A. E. Jenks, and S. G. Morley. Carnegie Publication 200.Washington, DC: Carnegie Institution, 1913.

Secondary texts

Alder, Ken. 'Scientific Conventions: International Assemblies and Technical Standards from the Republic of Letters to Global Science'. Eds. Mario Biagioli and Jessica Riskin. *Nature Engaged: Science in Practice from the Renaissance to the Present.* New York, NY: Palgrave Macmillan, 2012. 19–39.
Barth, Fredrik, Robert Parkin, Andre Gingrich and Sydel Silverman. *One Discipline, Four Ways: British, German, French and American Anthropology.* Chicago, IL: University of Chicago Press, 2005.
Bayly, Christopher A. *Empire and Information: Intelligence Gathering and Social Communication in India, 1780–1870.* Cambridge: Cambridge University Press, 1997.
Browne, Janet. *Charles Darwin: Voyaging: A Biography.* Princeton, NJ: Princeton University Press, 1996.
Desmond, Adrian. *The Politics of Evolution. Morphology, Medicine, and Reform in Radical London.* Chicago, IL: University of Chicago Press, 1989.
Drayton, Richard H. *Nature's Government: Science, Imperial Britain, and the 'Improvement' of the World.* New Haven, CT: Yale University Press, 2000.

Driver, Felix. 'Hidden History Is Made Visible? Reflections on Geographical Exhibition'. *Transactions of the Institute of British Geographers* 38 (2012): 420–35.

Edney, Matthew H. *Mapping an Empire: The Geographical Construction of British India, 1765–1843*. Chicago, IL: University of Chicago Press, 1997.

Elshakry, Marwa. *Reading Darwin in Arabic, 1860–1950*. Chicago, IL: University of Chicago Press, 2013.

Endersby, Jim. *Imperial Nature: Joseph Hooker and the Practices of Victorian Science*. Chicago, IL: University of Chicago Press, 2008.

Gillispie, Charles C. 'Scientific Aspects of the French Egyptian Expedition'. *Proceedings of the American Philosophical Society* 133 (1989): 447–74.

Hunt, Bruce J. 'Doing Science in a Global Empire: Cable Telegraphy and Electrical Physics in Victorian Britain'. *Victorian Science in Context*, 2nd edn. Ed. Bernard Lightman. Chicago, IL: University of Chicago Press, 1997. 312–33.

Kuklick, Henrika. *The Savage Within: The Social History of British Anthropology, 1885–1945*. Cambridge: Cambridge University Press, 1991.

Numbers, Ronald L. *The Creationists: From Scientific Creationism to Intelligent Design*, expanded edition. Cambridge, MA: Harvard University Press, 2006.

Osseo-Asare, Abena Dove. *Bitter Roots: The Search for Healing Plants in Africa*. Chicago, IL: The University of Chicago Press, 2014.

Pang, Alex Soojung-Kim. *Empire and the Sun: Victorian Solar Eclipse Expeditions*. Stanford, CA: Stanford University Press, 2002.

Qureshi, Sadiah. *Peoples on Parade: Exhibitions, Empire, and Anthropology in Nineteenth Century Britain*. Chicago, IL: University of Chicago Press, 2011.

Raj, Kapil. *Relocating Modern Science: Circulation and the Construction of Knowledge in South Asia and Europe, 1650–1900*. Basingstoke: Palgrave Macmillan, 2007.

Ratcliff, Jessica. *The Transit of Venus Enterprise in Victorian Britain*. London: Pickering & Chatto, 2008.

Rydell, Robert W. *All the World's a Fair: Visions of Empire at American International Expositions, 1876–1916*. Chicago, IL: University of Chicago Press, 1987.

Secord, James A. *Victorian Sensation: The Extraordinary Publication, Reception and Secret Authorship of 'Vestiges of the Natural History of Creation'*. Chicago, IL: University of Chicago Press, 2000.

Sera-Shriar, Efram. *The Making of British Anthropology, 1813–1871*. London: Pickering & Chatto, 2013.

Stocking, George W. *After Tylor: British Social Anthropology, 1888–1951*. London: Athlone, 1995.

Tilley, Helen. *Africa as a Living Laboratory: Empire, Development, and the Problem of Scientific Knowledge, 1870–1950*. Chicago, IL: University of Chicago Press, 2011.

Wenzlhuemer, Roland. *Connecting the Nineteenth-Century World: The Telegraph and Globalization*. Cambridge: Cambridge University Press, 2013.

Wheeler, Roxann. *The Complexion of Race: Categories of Difference in Eighteenth-Century British Culture*, New Cultural Studies. Philadelphia, PA: University of Pennsylvania Press, 2000.

Further reading

Bewell, Alan. *Romanticism and Colonial Disease*. Baltimore, MD; Johns Hopkins University Press, 1999.

Conklin, Alice L. *In the Museum of Man: Race, Anthropology, and Empire in France, 1850–1950*. Ithaca, NY: Cornell University Press, 2013.

Coombes. A. E. *Reinventing Africa: Museums, Material Culture and Popular Imagination in Late Victorian and Edwardian England*. New Haven, CT: Yale University Press, 1994.

Delbourgo, James, and Nicholas Dew (eds). *Science and Empire in the Atlantic World*. London: Routledge, 2008.

Drayton, Richard. 'Science, Medicine and the British Empire'. In R. W. Winks (ed), *Historiography*. The Oxford History of the British Empire V. Oxford: Oxford University Press, 1999: 264–75.

Edwards, Elizabeth. *Raw Histories: Photographs, Anthropology and Museums*. Oxford: Berg, 2001.

Fan, Fa-Ti. *British Naturalists in Qing China: Science, Empire and Cultural Encounter*. Cambridge, MA: Harvard University Press, 2004.

Felix Driver. *Geography Militant: Cultures of Exploration and Empire*. Oxford: Blackwell, 2001.

Fulford, Tim, Debbie Lee, and Peter J. Kitson. *Literature, Science and Exploration in the Romantic Era: Bodies of Knowledge*. Cambridge: Cambridge University Press, 2004.

Goodall, Jane. *Performance and Evolution in the Age of Darwin: Out of the Natural Order*. London: Routledge, 2002.

Harrison, Mark. 'Science and the British Empire', *Isis* 96 (2005): 56–63.

Jardine, Nicholas, James A. Secord, and Emma C. Spary (eds). *Cultures of Natural History*. Cambridge: Cambridge University Press, 1996.

Kitson, Peter. *Romantic Literature, Race and Colonial Encounter, 1760–1840*. New York; Basingstoke: Palgrave Macmillan, 2007.

Klaver, Jan Marten Ivo. *Scientific Expeditions to the Arab World* (1761–1881). Oxford: Arcadian Library, 2009.

MacLeod, Roy (ed.). Nature and Empire: Science and the Colonial Enterprise, *Osiris* 15 (2000).

Pratt, Mary Louise. *Imperial Eyes: Travel Writing and Transculturation*. London: Routledge, 1992.

Safier, Neil. *Measuring the New World: Enlightenment Science and South America*. Chicago, IL: University of Chicago Press, 2008.

Schaffer, Simon, Lissa Roberts, Kapil Raj and James Delbourgo, eds. *The Brokered World: Go-Betweens and Global Intelligence, 1770–1820*. Sagamore Beach, MA: Science History Publications, 2009.

Sivasundaram, Sujit, (ed.) 'Focus: Global Histories of Science'. *Isis*, 101 (2010): 95–158.

Spary, Emma C. *Utopia's Garden: French Natural History from Old Regime to Revolution*. Chicago, IL: University of Chicago Press, 2010.

Wulf, Andrea. *The Invention of Nature: Alexander Von Humboldt's New World*. New York, NY: Alfred A. Knopf, 2015.

2

SCIENTIFIC CULTURES AND INSTITUTIONS

Martin Willis

Cardiff University

The first half of the nineteenth century witnessed an explosion in the number of scientific societies, institutes and associations across Britain. This testified to the growing importance of structured organizations to the place of science in British society and by extension its many different cultural activities. In particular, major and minor figures from the worlds of literature and the arts engaged enthusiastically with these scientific organizations, which offered opportunities for access to both the latest scientific knowledge and the debates about science's value and influence. Indeed, interactions between new and existing scientific institutions and artists or writers were numerous. Their exchanges were influential on the practices of scientific institutions as well as on the imaginative production of poems, novels and art.

Scholarship investigating the place and function of scientific organizations across the nineteenth century had, until the early 2000s, focused attention on institutional histories: largely economic, political and experimental. The work of Morrell and Thackray (1981) on the British Association for the Advancement of Science is the best example of these preoccupations. A great deal was known about the type of scientific research conducted or supported by organizations, as well as how they functioned as discrete and autonomous economic entities with particular personnel. Less well known was their role in wider British society and their links within specifically cultural arenas remained under-investigated. But the scholarly work of the last fifteen years has made important advances so as to correct this imbalance in knowledge and in doing so has uncovered the richness of the connections between literary culture and institutionalized science.

In the early decades of the nineteenth century, many different scientific institutions were coming to understand their role as social as well as experimental; that is, as disseminators of knowledge as well as its producers. This epistemological shift was led in elite society by the chemical work being undertaken at the Royal Institution and in working-class communities by the rise of Mechanics' Institutes. Their increasing influence in what is commonly called public culture involved novelists, poets, and other key figures from a literary culture that already occupied a central place as knowledge-broker within the public sphere. As the century progressed, a greater number of diverse scientific institutions carved out a place within an increasingly fragmented public culture. Some institutions continued to exploit the cultural capital that their institutional status afforded them as science promoters, while others grew to become – or emerged for the first time as – commercialized sites of scientific education and entertainment. Poets, novelists and dramatists remained central to these institutions, either as communicators of scientific research or as models for the production of science-based creative performances. As the nineteenth century entered its final two decades, a significant number of scientific institutions became, or again were newly constituted as, professional scientific organizations. These organizations shifted their attention away from public culture onto their own institutional culture and the civic value of scientific research. The role that literary culture might play in these new organizations was less obvious, and the relationship between literature and science often became one of mutual and distrustful interrogation.

It is important to bear in mind, however, that cutting across these three periods of institutional evolution were a number of societies and institutes that did not fall into this kind of organizational hegemony. They either continued to work in the ways they had always done, or were early adopters of new modes that only reached a majority later. These institutions, and some of them are the most important of the period, rightly complicate and provide texture to the cultures of scientific institutions across the nineteenth century. While in this chapter I shall aim to provide an analysis of the evolution of scientific culture through the examples of these institutions, it should be understood as a map that provides useful boundaries and contours but which cannot reflect the complexity of the terrain that would emerge from a full analysis of the many different scientific organizations that were influential in the period.

One early boundary marker for nineteenth-century scientific institutions is the opening of the Royal Institution of Great Britain, which took place just before the turn of the century, in March 1799. The Royal Institution was Britain's most important scientific organization through the first three decades of the century. Although not as established as the Royal Society, it was, as Jenny Uglow says, 'ultra-respectable' (484) and therefore had access to Britain's powerful elite society. More than this, though, the Royal Institution was easily the most innovative of the new scientific organizations springing up all over the country, and was influential on the practice and principles of many others. Where the Royal Institution most made its mark was the ongoing series of lectures and later discourses (the Institution's title for more informal lectures) that attracted not only fellow scientists to its premises on Albemarle Street in London but also members of the public lucky enough to know a member and bag a ticket. Lectures had begun in 1800, became enormously popular after the appointment of the charismatic chemist Humphry Davy in 1802 and by the mid-1820s had diversified into both Christmas lectures, more focused on dramatic demonstration and short-form monologues, and the less-theatrical discourses, led by the research interests of members (Royal Institution). Although, as Sharon Ruston notes, the Royal Institution was 'part of the scientific establishment' (2005 38), this innate political conservatism did not halt its repositioning of science as part of public culture. For, as the lectures and discourses became an increasingly visible part of public life – to the extent that the artist James Gillray could make one of them the subject of a caricature (James 2000 6) – they began to inculcate scientific knowledge as a natural part of public knowledge.

The lecturers appointed by the Royal Institution, and the Institution's governing figures, consciously promoted the communication of scientific activity and discovery to new public audiences as a central part of the Institution's role. Unlike, for example, the Royal Society, which continued to focus on research and experiment, the Royal Institution gave over a considerable portion of its activities to engaging with audiences. Even the architecture of its first, and only, building suggests the primacy of its public role: the large lecture theatre was one of the first spaces to be designed after the Institution took up occupancy and it holds pride of place at the heart of the building, demanding that research and laboratory space fits in around it, rather than the reverse. The Institution's lectures did not earn a place in British high culture because of their scientific content alone. This was instead a result of both the form and content of the lectures, which delivered the latest scientific knowledge in dramatic formats that (sometimes) differed little from staged performance. The Institution's most prominent lecturer, Davy, was particularly adroit with this lecturing style and 'hugely popular' (Ruston 2013B 86) with the public, who attended to witness his particular brand of chemical dramaturgy. Davy, himself a poet, brought a theatrical edge to the dissemination of his scientific research that was undoubtedly drawn, in part at least, from witnessing performed dramas at theatres. At the same time, Davy and the other lecturers who followed him pursued the same techniques of practice, rehearsal and revision as professional actors did.

The impact of the Royal Institution's public dissemination of scientific knowledge developed, then, out of already existing techniques for dramatic performance, which had their fullest expression in the theatre. In turn, these lectures had a considerable impact on the literary figures who came to listen to them. Samuel Taylor Coleridge was one of those writers most obviously influenced by the Royal Institution's central place in early nineteenth-century public culture. As Nicholas Roe has shown, Coleridge's attendance at Davy's lectures on chemistry in the opening months of 1802 supported his poetry not only because of the philosophical ideas emerging from Davy's claims about chemistry but also because Coleridge was able to, in his own words, 'increase [my] stock of metaphors' (quoted in Roe 12) from the images and language Davy used to describe chemical reactions. Indeed Coleridge found inspiration in the linguistic diversity of chemistry, writing to Davy in May 1801 to say that 'as far as words go, I have become a formidable chemist' (quoted in Ruston 2013A 137). Nevertheless, the connections between scientific institution and writer were not always so direct. For example Percy Shelley's poetry, particularly that dealing with the nature of human vitality, and Mary Shelley's novel *Frankenstein* (1818) were part of a debate that also involved the lectures and subsequent books presented and written by Davy and others at the Royal Institution and elsewhere (Ruston 2005 1–23, Crouch 36). Shelley himself was not present, as Coleridge had been, but instead accessed their discoveries through the printed texts collected together and published after the lecture series. Although not directly the product of a scientific institution, these books could not have existed without organizational support and so form an important part of ongoing, and sustained, institutional influence on public (including literary) culture.

If it seems self-evident that new knowledge about chemical elements might prove both inspiring and epistemologically energizing to early nineteenth-century poets, then it should not seem surprising that science also inspired the poetic sensibilities of its practitioners and that this should be most marked within scientific organizations. Davy published a number of poems in his lifetime and was chosen as a proofreader by Wordsworth for the second edition of *Lyrical Ballads* in 1800, which persuaded him that 'chemistry was creative' (Ruston 2013B 79–80) in the same way as poetic composition. Davy asserted this view in the published version of his introductory course of lectures on chemistry, initially presented at the Royal Institution in 1802.

As scholars are discovering, the poetry written by nineteenth-century scientists can shed light on both the social and cultural meanings of scientific work and, of course, on scientific organizations (Brown; Willis 2011 68, 106). Poetry had, for example, a specific and vital role in the British Association for the Advancement of Science. This was a new scientific organization that had been founded in 1831, in part to provide greater promotional opportunities for the physical and natural sciences but also as a rival to the Royal Society, which was suffering a period of low standing in a range of influential scientific communities for its failures to support in adequate measure the work of the sciences across Britain. At the BAAS meeting in 1833, as Rosemary Ashton has shown, Coleridge found himself 'lionized' (404) by the young scientists in attendance and he also took part in a debate to discuss the collective name for a group of natural philosophers where the word 'scientist' was first suggested. Poetry also filled a more social role among the scientists themselves. From the late 1830s, as Daniel Brown has argued, leading BAAS figures such as Thomas Henry Huxley, Edward Forbes, John Tyndall and James Clerk Maxwell met to eat, drink and recite poetry together at each annual meeting. It would appear that the range of poetry ('pastiche, lampoon and doggerel' (Brown 90)) was matched only by the consumption of alcohol. This, the self-styled Red Lion Club was important for its sociability; for binding together the scientists in a shared cultural experience that sat alongside their shared scientific exploits. Indeed, these two were often merged in poetry evoking the scientific work being undertaken by the scientist-poets (just as Davy had written verse about his own experiments). Although a counterpoint to the official activity of the BAAS, these evenings of poetry and clubbability were spaces in which the possibilities, pitfalls and ethics of scientific work could be aired. They, as much as the visible workings of the Association's officers and members, defined what scientific activity should be and should not be (Brown 89–109).

The British Association for the Advancement of Science differed from the Royal Institution (and the Royal Society) less in its promotion of science, although its methods of promotion were very different, but far more in its efforts to engage the public across class, cultural, and geographical boundaries. By moving its large annual meetings to a different location each year, and by consciously choosing locations that were not only learned university towns but also commercial cities or provincial hubs, the BAAS aimed to connect with wider sections of the British public. The BAAS could reasonably claim, therefore, to be a scientific organization that was a part of British national civic culture. For the BAAS, the cultivation of a 'provincial knowledge culture', as Louise Miskell terms it, was an essential feature in their difference from elite London organizations such as the Royal Institution (2). One vital aspect of public connection that the BAAS expanded was the role played by women in scientific culture. While women had been present at the lectures at the Royal Institution, the BAAS meetings, especially from the middle of the nineteenth century onwards, engaged a large number of women in the circulation of scientific knowledge. As Charles Withers argues, women became 'crucial to the operation of the BAAS as its audience: they were participants in the social circuitry surrounding the civic engagement with science' (113; Higgitt and Withers).

The BAAS was only one example of the expanding role that science was playing across the civic lives of many British citizens of all classes. The BAAS itself had been founded, in collaboration with leading scientific figures, by the foremost members of the Yorkshire Literary and Philosophical Society (Orange). Many other 'Lit and Phils' existed in other towns and cities across Britain, providing largely genteel science to middle-class society. From the 1820s, however, the Mechanics' Institutes, which were founded in Edinburgh but soon spread across the United Kingdom, offered similar scientific sites for the (male) working class, including access to science's print culture through the Institutes' libraries. Although the Mechanics' Institutes were focused more on the practical sciences, or science in application – technology, in effect – they took their place in a fast-growing knowledge economy centred on scientific learning.

As the sciences infused themselves into British civic life, the organizations that promoted them became a focal point for literary and artistic culture. Scholars see the emergence of this intellectual culture of science as playing a key role in literary engagements with scientific ideas from the 1820s onwards, and most fervently in the Victorian period. As Alan Rauch has revealed, many writers, such as the Brontë sisters, whose brother Patrick Brontë was a central figure in the Keighley Mechanics' Institute, were connected to scientific organizations of one sort or another, and their work 'reflects an acute interest and engagement in the benefits and perils of science and technology' (59). The responses to scientific institutions could, however, be equivocal. Although it was before this explosion in scientific organizations that Mary Shelley had expressed her own reservations about the repercussions of scientific research in *Frankenstein* and focused the central scientific episodes around an institution of science, it was precisely during this period that Charles Dickens pointed up the absurdities of an organization like the BAAS. In two long articles for *Bentley's Miscellany* in 1837 and 1838, Dickens parodied the Association by retitling it 'The Mudfog Association for the Advancement of Everything'. Dickens satirizes both the scientists and the aims of the BAAS in a mock report of an annual meeting. In it he reports how arriving scientists are far more interested in superior hotel lodgings than in the scientific lectures on offer and notes that the only 'spread of science' is the excessively laden banqueting table. Humour aside, Dickens makes a point that he would return to in his fictions: his own distaste for the vacuity of those who would be popular rather than useful (Chaudry 111).

Unknown to Dickens in the 1830s, the tension between popularity and the dissemination of knowledge was to be a central feature of scientific institutions throughout the middle and later decades of the nineteenth century. As scientific organizations sought to disseminate scientific knowledge to ever-wider audiences, conflicting ideas on how to achieve this led to a split between, on the one hand, the value of teaching a lay audience about new scientific discoveries and, on the other hand, the desire to draw in that audience in the first place by offering something that appeared likely to be enjoyable. That slippage between education and entertainment, as scholars have generally seen the two sides of this debate, was at the forefront of organized science through the middle decades of the century. Indeed, the phrase 'rational recreation' (Kember, Plunkett and Sullivan 4), which was used to describe public scientific events, was supposed to capture the collision, or perhaps collaboration, of two fundamentally different epistemologies of knowledge dissemination. In fact, the contestation of these issues was part of a newly emerging aspect of scientific work that only in the second or third decade of the century gathered its now-commonplace designation as 'popular science' (Topham 135).

For the established scientific institutions, including many Mechanics' Institutes and Literary and Philosophical Societies, the Royal Society, the Royal Institution and, as we have seen, the British Association for the Advancement of Science, staging science as recreational performance ran counter to many members' views of proper scientific activity. This did not, however, halt them in continuing to provide public scientific demonstrations. The Royal Society and the Royal Institution, after all, had both laboratory spaces and lecture halls within their walls. But the more theatrical demonstrations of science – those lectures or discourses that seemed to diverge from sober and polite representations of scientific ideas in favour of entertaining the public for entertainment's sake alone – often came in for severe criticism. For example, the physicist John Tyndall, who was a lecturer at and then director of the Royal Institution (1853–87), a fellow of the Royal Society from 1852 and a leading member of the BAAS throughout the second half of the century, was often charged with introducing unnecessary theatricality into his lectures. James Clerk Maxwell, for one, was unimpressed by the 'morbid craving for excitement' (Brown 128) to which Tyndall played and the 'showmanship' (Brown 129) that characterized his lecture style in the 1860s and 1870s. At stake in these conflicts between institutional colleagues was the very nature of scientific work and how that might be communicated to those outside of scientific groups. For Tyndall, and many others, the popularizing of science was a tactic for achieving greater cultural capital for the sciences and for scientific institutions. To Maxwell and his supporters, the dramatic entertainment that Tyndall offered undermined the serious purposes of the sciences and made frivolous the philosophical sobriety with which it should be viewed.

At a huge number of other scientific institutions dramatic entertainment was itself a serious business. If concerns about the spectacular nature of the sciences continued in elite organizations, in the growing number of commercial institutions dedicated to presenting science to the public the dramatic was enthusiastically embraced. From just before the middle of the century, and continuing to its end, numerous commercial sites for scientific demonstration opened their doors to public audiences in all regions of the United Kingdom. The remit of these new organizations was to bring science closer to the public, to make it appear part of the lives of ordinary citizens rather than socially and intellectually exclusive. As Iwan Rhys Morus argues, dramatizing science and performing that drama were 'part and parcel of the business of making science and its products real to their audiences' (337).

The business was a highly competitive one and essentially placed science in the marketplace for the first time, as Aileen Fyfe and Bernard Lightman have revealed. London was a key site for these institutions: the Adelaide Gallery of Practical Science, the Panopticon of Science and Art and most importantly the Royal Polytechnic Institution on Regent Street vied with many other smaller, sometimes touring and often temporary places of scientific demonstration for ticket sales and public acclaim. Around Britain, too, other scientific institutions developed their own plans for scientific shows and spectacles, while organizations that had not previously offered much space for scientific events began to introduce such shows into their own schedules. John Plunkett and Jill A. Sullivan, in a study of provincial scientific shows taking place in Exeter and Plymouth, have shown, for example, how various literary societies, natural history clubs and other institutes hosted local and touring scientific displays and demonstrations (55–60). What is clear from this pattern of expansion is that the middle of the nineteenth century saw a change in the defining features of scientific institutions. Earlier in the century a scientific institution would have been defined by its ability to carry out research and offer occasional public events; it would have been led by reputable, active scientists and it would have reflected the values of a social and cultural elite. Indeed one of the key areas of concern about the future of the Royal Society in the early 1830s was the fact that it had elected a nobleman without scientific credentials as its President. Yet, by the 1850s, such organizations no longer dominated British scientific culture, although they remained individually the most powerful and influential. The new scientific organizations were led by impresarios and businessmen, fronted by showmen whose expertise lay in theatre and the circus, and they disseminated scientific discoveries made elsewhere rather than conducting or supporting basic research. Viewed from this perspective, it is clear that elite science no longer could claim cultural hegemony: science was increasingly available to the public on their own terms and responded to their interests and desires. The institutions of the second half of the century, then, reveal the strength and variety of citizen science.

The Royal Polytechnic Institution is one of the most fascinating examples of these new commercial scientific institutions. It has, for that reason, received considerable attention from historians of science and in recent years from scholars of literature and science. In its period of greatest success, through the 1860s, it was led by the self-styled 'Professor' Henry Pepper, who introduced to the public through a combination of lecture, demonstration, and exhibition a number of scientific and technological objects, discoveries and effects. These included the Great Induction Coil, which demonstrated new electrical knowledge, and 'Pepper's Ghost', which explained the nature of optics.[1] The Polytechnic's ability to show such objects and to produce such extraordinary displays arose from their commercial success. By attracting large metropolitan audiences, as Martin Hewitt has shown, the Polytechnic could 'develop and depend on a machinery of spectacle that was impossible to sustain elsewhere' (80). In certain respects the Polytechnic drew on earlier models of scientific institutions for the organization of its own rooms. It housed a demonstration space, a laboratory and exhibition galleries (Lightman 2013 38). In so deploying its spaces it followed elite organizations such as the Royal Institution. However, it also existed in 'the world of entertainment' (Lightman 2013 38) and so offered, too, musical acts, plays and magic lantern shows.

35

The hybrid nature of the Royal Polytechnic exemplified the double personality of many such commercial sites of scientific activity. But the function of the Royal Polytechnic extended beyond the binary suggested by its interior architecture. Like other, similar commercial scientific enterprises the Royal Polytechnic offered what Plunkett and Sullivan have called 'a potent mixture of instruction, amusement and spectacle' (41). At the heart of its public output was its culture of practised, modern demonstration: a culture of performance. The performances available at the Royal Polytechnic (of Professor Pepper, for example) owed a debt to literary and artistic culture, and perhaps especially to drama and the theatre, a point that Tiffany Watt Smith and Kirsten Shepherd-Barr also make, independently, about other institutional contexts. The Professor's most famous scientific demonstration, Pepper's Ghost, employed dramatizations of well-known literary fictions to stage its spectral illusion and so provide audiences with a spectacular new way of visualizing optical effects. Pepper's Ghost used dramatic moments from the work of various Victorian novelists and most successfully an adaptation of Charles Dickens's short story 'The Haunted Man', especially rewritten (and authorized by the author) for Pepper's scientific display.[2] Scientific organizations such as the Royal Polytechnic were drawing afresh a series of cultural connections that were normally absent, or played down, in other kinds of institutions. Not only did the performances reveal an indebtedness to literary and theatrical cultures but the demonstrations as a whole relied upon the kinds of professional showmanship that had been developed by successful circus acts such as P. T. Barnum's, and that continued into the twentieth century with the large-scale theatrical shows of magicians and conjurers (Morus 362, Willis 2013 176).

Commercial scientific institutions might have valued public entertainment and the profit this brought above the dissemination of scientific knowledge but this should not lead to an assumption that these organizations were either non-scientific or below other organizations in an accepted hierarchy of science. How, then, should they be regarded, and especially in what relation do they stand to elite organizations such as the Royal Society? One influential reading of the place and position of commercial organizations that popularized science is to see them as vital components in the transfer of knowledge from original research and experimentation to the wider public sphere. For James Secord, that is precisely the role played by institutions such as the Royal Polytechnic: purveyors of 'knowledge in transit' from discovery to dissemination, such organizations are part of a broad continuum of science and should be regarded as knowledge producers with equal value to sites designed solely for scientific experiment. Lightman notes, too, how popular scientific organizations offered a public keen to know more about the sciences access to scientific modes of thought that they understood as 'the glue holding together a new worldview far more relevant for living in an urban, industrialized, and middle-class society' (2007 4). Morus, however, recalls the differences between elite scientific organizations and popular institutions, and stresses the importance of recognizing that science was far from homogenous across the nineteenth century. Science was of varying 'epistemological or social status' (Morus 365) and involved a multitude of different activities and practices. Understanding different scientific institutions, therefore, is about recognizing their 'geographic (and therefore cultural) locations' (Morus 365) within particular kinds of networks, with different definitions of authority and different understandings of what science was.

As those new commercial organizations developed, so too did the most elite of all British scientific societies, the Royal Society. While the Royal Society had been mired in internal arguments over its future direction as well as in the relationship between scientific accomplishment and class privilege, it had been somewhat overtaken in cultural significance by other organizations and perhaps especially by the British Association for the Advancement of Science. As historian Marie Boas Hall noted some years ago in her work on the place of the Royal

Society in nineteenth-century society, when Davy become President in 1820 he tried to make a point of his dedication to science as well as to the aristocracy. But Davy's efforts to promote scientific accomplishment over the privilege of social rank within the Royal Society did little to bring to a halt the culture of an elite social club that had grown up within the Society in the late eighteenth century and against which many of the younger fellows railed. His presidency was, ultimately, a huge failure. This left the Royal Society rather without impetus during the period of great expansion in scientific institutions through the middle of the nineteenth century, despite initiating a number of key reforms from 1848 onwards.

However, in the 1860s another scientific organization, the X-Club, was initiated to give greater impetus to the reformist agenda within the Royal Society and especially to bring about change in its personnel and policies. The X-Club is perhaps best described as a club within a club, in that it existed unofficially within the Royal Society and its members were almost exclusively Royal Society fellows. It had been founded in November 1864 with just nine members, among whom were Huxley, Joseph Hooker and Tyndall.[3] Like the Red Lion Club, associated with the British Association for the Advancement of Science, the X-Club was a dining club, and it was at their monthly dinners that its policy and philosophy emerged. X-Club members were, as Ruth Barton has shown, against the aristocratic patronage that dominated Royal Society elections; they despised the increasingly commercial nature of British science and they were supportive of science as a subject for pure research, untainted by utilitarian limitations (53). However, they were also supportive of science as a cultural form; perhaps most especially, X-Club members realized the importance of placing science within broader conceptions of society and culture. To this end, one of the first nine members was the philosopher of evolution Herbert Spencer. From modest, if reasonably well-situated, beginnings the X-Club grew to be one of the most powerful organizations within the Royal Society and therefore in British science. By the 1870s, Barton argues, 'no other factional group [in the Royal Society] was large enough or organized enough to challenge the power of the X-Club' (59). Across the next two decades the X-Club was able successfully to promote its own candidates for senior office within the Royal Society and therefore deliver policies on science that had been restricted by previous conservative administrations. For example, under the presidency of X-Club member Hooker a fund for the relief of poor scientific men and their families was set up so as to allow ordinary (that is, those without independent wealth or means) members to join and take an active part in Royal Society research and affairs.

As an organization, the Royal Society was hugely altered by the influence of the X-Club: it shifted from being an institution reserved for the socially elite to one organized to support the scientific elite: its focus became research rather than amateur dilettantism. In turn, this also had an impact on how science was perceived in late nineteenth-century Britain. Since the Royal Society was, for many, the 'public representative of science' (Barton 72) as well as the image of science commonly seen by government, science became more about the search for knowledge and less about its influential place in the establishment. However, such a view must be tempered, first, by the recognition that scholarship on popular scientific institutions has revealed how science might mean different things to different constituencies and, second, by the relative paucity of research on some of the more significant scientific institutions. While the Royal Institution has been considered in detail by scholars of Romantic culture, neither the Royal Society nor the British Association for the Advancement of Science have had as much attention from outside the history of science. Rectifying this should be the aim of present and future scholars. What place these organizations will be seen to hold in nineteenth-century culture when this work has been done must be left for others to argue.

As the nineteenth century entered its final decades there were clearly two paths along which scientific institutions had travelled. One path had led to the growth of institutions of popular science. The other, as Lightman has argued, was towards ever greater professionalization (2007 494–6). Both paths had emerged from slow evolution rather than sudden revolution. Indeed, specialization – one key aspect of professionalization – had been increasingly apparent since the earliest years of the nineteenth century. While the Linnean Society, founded in 1788, was rather unusual at that time, specialist societies became almost commonplace from the first decade of the nineteenth century. The Geological Society had been formed in 1807, for example, the Astronomical Society in 1820, the Royal Society of Chemistry in 1841, the London Mathematical Society in 1865 and the Physical Society of London in 1874. These were only exemplary organizations for the most prominent sciences: numerous others emerged, grew and became defunct or flourished between 1800 and 1900.

Professional scientific institutions were, nevertheless, relatively new in the last few decades of the nineteenth century. These organizations introduced a different cultural paradigm: the corporate structure. Often funded either by government stipend or private finance or attached to larger research institutions such as universities, these new facilities (mostly laboratories and research institutes designed with very specific purposes) faced inwards rather than outwards. They had little contact with public audiences and were beholden instead to their owners and benefactors, as well as to their own regulatory systems and scientific aims. Often the objectives of these new scientific research facilities were, as the X-Club would have applauded, the investigation of pure scientific problems, but others had an essentially civic function. The British Institute of Preventive Medicine, for example, which was founded in London in the last decade of the century, was dedicated to the treatment, and ultimate eradication, of infectious diseases, with rabies as its first target following Louis Pasteur's successes in Paris.

For a range of public groups, however, this new kind of scientific organization gave rise to anxiety (Willis 2011 33–6). Unlike many of the other scientific institutions, the public was never admitted to their buildings, nor were their activities advertised or explained. What was clearly an attempt to maintain levels of safety (for, say, the British Institute) often appeared to the public as secrecy. Especially where scientific research had the potential to be dangerous – and the British Institute's use of infectious diseases meant it fitted that category – suspicion and a degree of fear were commonplace in British society. H. G. Wells captured this public concern in his 1896 novel *The Island of Doctor Moreau*. The explanation for Moreau's flight from London includes dangerous experimenting in secret and a public outcry at the immorality of his research. Wells clearly had the British Institute of Preventive Medicine in mind as his model.

Even where new corporate scientific institutions inspired interest and support rather than fear and distrust – at the Cavendish Laboratory for physical research in Cambridge, for example – it was clear that these later-century organizations felt little responsibility for their public constituencies. The architecture of their interior spaces did not allow for public lectures and were instead dominated by structures designed for management and control (office and administrative rooms alongside laboratories and research spaces). These institutions of science were becoming what the social theorist Max Weber would, in 1917, call bureaucratic, and it marked a new kind of corporate professionalism that was decidedly different in character from the knowledge-producing professional scientist for which the X-Club had lobbied.

Looking at the century as a whole from the standpoint of its final years it is clear that scientific cultures and institutions did not only evolve; they also multiplied. At the beginning of the century there were relatively few institutions and their defining features were reasonably homogeneous. By the century's end, scientific institutions could be numbered in the hundreds and their cultures varied enormously. Within the space of a few hours, a citizen of any one of Britain's larger cities could, if they wished, peer through the gates of a corporate scientific organization, join a fellow at an institutional discourse and pay the entrance fee to a popular scientific extravaganza.

What science might mean at each of these places and in those moments would differ, but none of them could claim a monopoly on science's meanings or its defining features.

Notes

1 See Iwan Rhys Morus, 'Staging Science', in this volume.
2 A contemporary restaging of this scientific demonstration can be seen online on YouTube under the title 'Pepper's Ghost Returns to 309 Regent Street'. This restaging took place in December 2013 in the Old Cinema at the University of Westminster, on the site where the original first took place in the 1860s. The restaging was devised, written and performed by Richard Hand and Geraint D'Arcy and produced by the author under the auspices of the Centre for the Study of Science and Imagination (SCIMAG). The film was produced and directed by Joost Hunningher.
3 Clubs associated with the Royal Society had only really taken root from the 1820s, after the long presidency of Joseph Banks (1778–1820), who actively discouraged them, came to an end.

Bibliography

Primary texts

Coleridge, Samuel Taylor. *The Complete Poems*. Ed. William Keach. London: Penguin, 1997.
Davy, Humphry. *A Discourse, Introductory to a Course of Lectures on Chemistry, Delivered in the Theatre of the Royal Institution on the 21st of January, 1802*. London: J. Johnson, 1802.
Dickens, Charles. 'Full Report of the First Meeting of the Mudfog Association for the Advancement of Everything' [1837]. *The Mudfog Papers*. London: Richard Bentley, 1880. 47–96.
Dickens, Charles. 'The Haunted Man and the Ghost's Bargain' [1848]. *The Haunted Man and the Haunted House*. Gloucester: Alan Sutton, 1985.
Shelley, Mary. *Frankenstein. The 1818 Text*. Ed. Marilyn Butler. Oxford: Oxford University Press, 1994.
Weber, Max. *Science as a Vocation* [1917]. London: Unwin Hyman, 1988.
Wells, H. G. *The Island of Doctor Moreau* [1896]. Ed. Patrick Parrinder. London: Penguin, 2005.

Secondary texts

Ashton, Rosemary. *The Life of Samuel Taylor Coleridge*. Oxford: Blackwell, 1996.
Barton, Ruth. '"An Influential Set of Chaps": The X-Club and Royal Society Politics 1864–85'. *British Journal for the History of Science* 23 (1990): 53–81.
Brown, Daniel. *The Poetry of Victorian Scientists*. Cambridge: Cambridge University Press, 2013.
Chaudry, G. A. 'The Mudfog Papers'. *The Dickensian* 70 (1974): 104–12.
Crouch, Laura E. 'Davy's *A Discourse, Introductory to a Course of Lectures on Chemistry*: A Possible Scientific Source of *Frankenstein*'. *Keats–Shelley Journal* 27 (1978): 35–44.
Fyfe, Aileen, and Bernard Lightman. 'Science in the Marketplace: An Introduction'. *Science in the Marketplace: Nineteenth-Century Sites and Experiences*. Eds. Aileen Fyfe and Bernard Lightman. Chicago, IL: University of Chicago Press, 2007. 1–19.
Hall, Marie Boas. *All Scientists Now: The Royal Society in the Nineteenth Century*. Cambridge: Cambridge University Press, 1984.
Hewitt, Martin. 'Beyond Scientific Spectacle: Image and Word in Nineteenth-Century Popular Lecturing'. *Popular Exhibitions, Science and Showmanship, 1840–1910*. Eds. Joe Kember, John Plunkett, and Jill A. Sullivan. London: Pickering and Chatto, 2013. 79–95.
Higgitt, Rebekah, and Charles W. J. Withers. 'Science and Sociability: Women and Audiences at the British Association for the Advancement of Science, 1831–1901'. *Isis* 99.1 (2008): 1–27.
James, Frank A. J. L. *Guides to the Royal Institution of Great Britain: 1 History*. London: Royal Institution, 2000. <www.rigb.org/docs/brief_history_of_ri_1.pdf>. Accessed 8 December 2014.
Kember, Joe, John Plunkett and Jill A. Sullivan. 'Introduction'. *Popular Exhibitions, Science and Showmanship, 1840–1910*. Eds. Joe Kember, John Plunkett and Jill A. Sullivan. London: Pickering and Chatto, 2013, 1–18.

Lightman, Bernard. *Victorian Popularizers of Science: Designing Nature for New Audiences*. Chicago, IL: University of Chicago Press, 2007.

Lightman, Bernard. 'Spectacle in Leicester Square: James Wyld's Great Globe, 1851–61'. *Popular Exhibitions, Science and Showmanship, 1840–1910*. Eds. Joe Kember, John Plunkett and Jill A. Sullivan. London: Pickering and Chatto, 2013, 19–39.

Miskell, Louise. *Meeting Places: Scientific Congresses and Urban Identity in Victorian Britain*. Farnham: Ashgate, 2013.

Morrell, Jack, and Arnold Thackray. *Gentlemen of Science: Early Years of the British Association for the Advancement of Science*. Oxford: Clarendon, 1981.

Morus, Iwan Rhys. '"More the Aspect of Magic Than Anything Natural": The Philosophy of Demonstration'. *Science in the Marketplace: Nineteenth-Century Sites and Experiences*. Eds. Aileen Fyfe and Bernard Lightman. Chicago, IL: University of Chicago Press, 2007, 336–70.

Orange, Derek. 'Science in Early Nineteenth-Century York: The Yorkshire Philosophical Society and the British Association'. *York, 1831–1981: 150 Years of Scientific Endeavour and Social Change*. Ed. Charles Feinstein. York: William Sessions, 1981, 1–29.

Plunkett, John, and Jill A. Sullivan. 'Fetes, Bazaars and Conversaziones: Science, Entertainment and Local Civic Elites'. *Popular Exhibitions, Science and Showmanship, 1840–1910*. Eds. Joe Kember, John Plunkett and Jill A. Sullivan. London: Pickering and Chatto, 2013, 41–60.

Rauch, Alan. *Useful Knowledge: The Victorians, Morality, and the March of Intellect*. Durham, NC: Duke University Press, 2001.

Roe, Nicholas. 'Introduction: Samuel Taylor Coleridge and the Sciences of Life'. *Samuel Taylor Coleridge and the Sciences of Life*. Ed. Nicholas Roe. Oxford: Oxford University Press, 2001, 1–21.

Royal Institution. 'RI Timeline'. <www.rigb.org/our-history/timeline-of-the-ri>. Accessed 19 September 2014.

Ruston, Sharon. *Shelley and Vitality*. Basingstoke: Palgrave Macmillan, 2005.

Ruston, Sharon. *Creating Romanticism: Case Studies in the Literature, Science and Medicine of the 1790s*. Basingstoke: Palgrave Macmillan, 2013 [2013A].

Ruston, Sharon. 'From "The Life of the Spinosist" to "Life": Humphry Davy, Chemist and Poet'. *Literature and Chemistry: Elective Affinities*. Eds. Margareth Hagen and Margery Vibe Skagen. Aarhus: Aarhus University Press, 2013, 77–97 [2013B].

Secord, James A. 'Knowledge in Transit'. *Isis* 95 (2004): 654–72.

Shepherd-Barr, Kirsten. *Science on Stage: From Doctor Faustus to Copenhagen*. Princeton, NJ: Princeton University Press, 2006.

Topham, Jonathan. 'Publishing "Popular Science" in Early Nineteenth-Century Britain'. *Science in the Marketplace: Nineteenth-Century Sites and Experiences*. Eds. Aileen Fyfe and Bernard Lightman. Chicago, IL: University of Chicago Press, 2007, 135–68.

Uglow, Jenny. *The Lunar Men: The Friends who Made the Future*. London: Faber, 2002.

Watt Smith, Tiffany. *On Flinching: Theatricality and Scientific Looking from Darwin to Shell-Shock*. Oxford: Oxford University Press, 2014.

Willis, Martin. *Vision, Science and Literature, 1870–1920: Ocular Horizons*. London: Pickering and Chatto, 2011.

Willis, Martin. 'On Wonder: Situating the Spectacle in Spiritualism and Performance Magic'. *Popular Exhibitions, Science and Showmanship, 1840–1910*. Eds. Joe Kember, John Plunkett, and Jill A. Sullivan. London: Pickering and Chatto, 2013, 167–82.

Withers, Charles W. J. 'Scale and the Geographies of Civic Science: Practice and Experience in the Meetings of the British Association for the Advancement of Science in Britain and Ireland, *c.* 1845–1900'. *Geographies of Nineteenth-Century Science*. Eds. David N. Livingstone and Charles W. J. Withers. Chicago, IL: University of Chicago Press, 2011, 99–122.

Further reading

Cunningham, Andrew, and Nicholas Jardine, eds. *Romanticism and the Sciences*. Cambridge: Cambridge University Press, 1990.

Golinski, Jan. *Science as Public Culture: Chemistry and Enlightenment in Britain, 1760–1820*. Cambridge: Cambridge University Press, 1992.

James, Frank A. J. L. *'The Common Purposes of Life': Science and Society at the Royal Institution of Great Britain*. Aldershot: Ashgate, 2002.

3

SCIENCE AND RELIGION

Paul White
University of Cambridge

The relationship between science and religion has a long and complex history. In the nineteenth century it was marked by the evangelical revival and the diffusion of natural theology, the rise of liberal Anglicanism and movements of secular or alternative religion, including positivism and agnosticism, spiritualism and theosophy. Key debates engaged the new sciences of geology, biology, physics and anthropology: the age of the earth, the nature of matter, the progressive development of life, the antiquity of 'man' and his place in creation, and the origins of mind and morality. Such debates were often entwined with programmes of reform or implicated in struggles over the production of knowledge, the leadership of social organizations, the curricula of universities and schools, authority in the periodical press and other popular forms of print and the role of science in public life.

Divergent approaches and frameworks have structured scholarship in this broad field. Influential early works characterized the nineteenth century as a period of gradual secularization, involving a 'crisis of faith' and the eventual dominion of a naturalistic world view (Chadwick; Himmelfarb; Houghton). Such accounts frequently drew upon conflicting models of science and religion that had been forged in polemical debates of the Victorian period. As Thomas Huxley wrote, 'Extinguished theologians lie about the cradle of every science as the strangled snakes beside the bed of Hercules' (556). Appearing in a review of *The Origin of Species* in 1860, these words advanced an aggressive secular version of Darwinism that became a fixture of later narratives (Draper, White 1896). Men of science stood opposed to biblical literalism and Christian apologetics. Scriptural geologists and clergyman-naturalists defended the Genesis story of creation and the flood, the beneficent design of nature and the special providence of 'man'. Like the Catholic officials who supposedly refused to look through Galileo's telescope, their success was only temporary. Geology triumphed over Genesis, apes over angels (Gillispie; Greene; Irvine).

In the 1970s and 1980s, as the history of science underwent a social turn, the conflict between science and religion was recast in terms of class struggle and professionalization. Science became an ideological weapon of radical artisans and industrial middle-class Dissenters as they vied with Anglican elites for political rights, social status and cultural authority (Desmond; Young). Clergymen contended with men of science who sought autonomy from religious traditions and institutions (Turner). At the beginning of the nineteenth century, the Anglican Church had a virtual monopoly in government, education and the most prestigious professions. Parliament and other political offices, together with degrees from the ancient universities, required subscription to the Thirty-Nine Articles of the church. Entry to the upper ranks of medicine and law was in turn largely restricted to university graduates. In the absence of paid positions in science, a number of practitioners were clergymen by profession. Cambridge and Oxford professors were ordained and some held clerical livings and gave parish sermons. Anglican control was not monolithic, but only in the 1820s and 1830s did opposing parties gain sufficient influence to implement changes. The admittance of Roman Catholics and Dissenters to government (1828–9, 1835) and the creation of the non-sectarian London University (1826) were among the first substantial reforms of Anglican hegemony.

The sciences played a role in challenges to the authority of the church and its divinely ordained hierarchies. Radical London publishers issued cheap reprints of materialist medical works, like William Lawrence's *Lectures on Physiology, Zoology, and the Natural History of Man* (1819), condemned as blasphemous by the Chancery Court for its dismissive remarks upon the soul and vital matter (Ruston). Comparative anatomy and zoology provided grounds for social transformism that threatened the order of ranks and the control of science and medicine by exclusive corporations such as the Royal Society and the Royal College of Physicians (Desmond). In the *Views of the Architecture of the Heavens* (1837), the Scottish astronomer and political economist John Pringle Nichol composed a cosmic narrative of progress, stretching from the origins of the universe in a gaseous nebula to the present state of society and beyond. Future generations would see the abolition of 'unnatural' laws that fixed grain prices in favour of idle landowners, and church rates that siphoned wealth from the poor and labouring classes (Schaffer).

Cosmic progress had to be secured against such radicalism. During this volatile era of reform, the sciences were enrolled in more moderate and conservative programmes, designed to stem the tide of democracy, popular enthusiasm and irreligion. Pringle's nebular hypothesis was condemned by Cambridge dons and Royal Society fellows, such as William Whewell and John Herschel. In his *A Preliminary Discourse on the Study of Natural Philosophy*, Herschel presented scientific theory as the work of rarefied genius, elevating the mind and spirit above the material (and political) world. Pedestrian platforms of improvement were mass-produced by steam press publishers, such as the Edinburgh firm of William and Robert Chambers (the later author of the evolutionary epic *Vestiges of the Natural History of Creation*). Their weekly *Journal* featured digests of politics, literature and the advancements of science. In 1835, the firm issued a people's edition of George Combe's *Constitution of Man*, an expansion of his earlier *Essays on Phrenology* (1819), a science of character based on the form and function of the brain. Though potentially radical, Combe's phrenology demonstrated God's design in the self-regulation of man, including a special organ of the brain for religious devotion. It was rapidly taken up by Christian readers as an aid to good living (Secord).

Central to this more conservative regime of science and religion was natural theology. The classic work in the field by William Paley depicted nature as God's handiwork, wrought with beautiful contrivances such as the lens of the eye, deep structures of organization and development and universal laws of matter and force. Despite the philosophical critiques of Hume and others in the previous century, natural theology endured because it drew together a wide range of increasingly technical research in newly formed disciplines of geology, zoology, physics and chemistry and set them within a common Christian framework, embedding the sciences in Anglican institutions of learning (Brooke 1991). Paley's book was reprinted many times, extracted and expanded with new scientific illustrations by authors like Henry Brougham, a Whig politician and education reformer, who envisioned natural theology as the crowning synthesis in a system of universal knowledge. A series of lavish treatises was funded by the Earl of Bridgewater, with leading experts in natural history, philosophy and medicine commissioned to detail the wisdom and beneficence of God as manifest in Creation (Topham).

Some evangelicals worried that natural theology weakened the authority of revealed religion and, especially in the aftermath of the Napoleonic wars, that it could be a mask for (French) deism. For a Christianity centred on fallen man, the malignancy of sin and salvation through Christ's atonement, Paley's beautifully ordered system could seem too complacent (Hilton). Natural theology had to be carefully presented so as to complement or reinforce revelation. In geological circles, Georges Cuvier's work on fossil mammals, with its arguments for serial extinction and the progressive development of life on earth, were presented as supportive of Genesis. William Buckland, appointed to the first readership in geology at Oxford, described creatures like *Megatherium* (the giant sloth) as an 'apparatus of colossal mechanism . . . calculated to be the vehicle of life and enjoyment to a gigantic race of quadrupeds . . . imperishable monuments of the consummate skill with which they were constructed' (1836 1:164). Buckland aligned the fossil record with biblical chronology by inserting millions of years of earth history before the Genesis story began. He used geological science as a tool of exegesis, deepening readers' understanding of biblical events such as the flood, emphasizing the sublimity of nature's forces.

Through dramatic descriptions of fossil forms and catastrophes of deep time, geology claimed a place beside other grand spectacles of science in the early Victorian period – astronomical shows, panoramas, chemical and electrical demonstrations, museum collections – all of which revealed the wonder and order of God's creation (O'Connor). Gentlemen practitioners such as Michael Faraday and Richard Owen demonstrated the power of science to manipulate matter and force, to determine the place of an organism in the system of creation, to divine nature's laws and unity of plan. Faraday combined public feats of electrical experimentation at the Royal Institution with private acts of virtue and charity, living according to the principles of Sandemanianism, a movement of primitive Christianity that had emerged from Scottish Presbyterianism in the eighteenth century (Cantor). Owen made his reputation as the British Cuvier by reconstructing the first dinosaurs and other giant creatures, such as the Moa, an extinct Australian land bird whose form he claimed to have deduced from a single bone.[1] Such performances rarely offered detailed proofs or evidence of God's design and attributes, but were delivered in an edifying and reverential tone. As a form of spectacle and rhetoric, natural theology helped to build broad, non-sectarian alliances, as well as common ground for different communities of belief and practice within the Anglican Church.

According to the older historiography, natural theology and the bonds it reinforced between science and religion in the first half-century were swept away by Darwin's *Origin of Species* and the secular campaigns conducted in its wake and often in its name. Huxley, by his own account and the generations of historians who accepted it, defended Darwinian theory against the haughty Bishop of Oxford, Samuel Wilberforce, at the 1860 British Association meeting, winning the battle of science against religious prejudice (Brooke 2001). Narratives of progress were rewritten, leaving out the guiding hand of Providence. God receded far into the background as a first cause or an unknowable force. Agnosticism became a watchword for scientific practitioners, a convenient veil for atheism. The reform of scientific organizations and education more generally accompanied this triumph of naturalism, and men of science ceased to be clergymen by profession and found careers in university teaching and research (Heyck).

Revisions to this picture began with closer attention to theological writing across the denominations, revealing a breadth of approaches to nature, the Bible and church doctrine. Scholars have now shown that many of the so-called science and religion controversies took place *within* religious circles. Disputes about Darwinian evolution were often continuations of earlier theological debates on the role of special providence, natural law, the authenticity of miracles and so forth (Moore). The clergyman–naturalist Charles Kingsley wrote to Darwin, describing how natural selection presented a 'noble conception of Deity' (Darwin 7:179–80). One of Darwin's leading supporters, the American botanist and ardent Presbyterian Asa Gray, argued that natural selection was not inconsistent with natural theology, for it left questions of first causes, the origin of life and the design of nature's laws quite open. Darwin tended to encourage such religious readings of his work, inserting a passage from Kingsley's letter into the second edition of *Origin of Species* and financing the republication of Gray's articles in pamphlet form in England.

Liberal clergymen and theologians accommodated Darwinian theory just as they took up historical and philological approaches to the Bible, adapting methods of scientific inquiry to sacred texts and traditions, advancing a view of God's revelation as a truth gradually unfolding in conjunction with scientific developments (*Essays and Reviews*). Reforming parties within the church struck alliances with men of science to challenge the authority of doctrine, especially subscription to the Anglican creed. Huxley himself worked closely with liberal clergymen in joint projects of reform that broadened the traditional classical curricula of schools and universities, introducing new science subjects, examinations and degrees (White 2003). In the periodical press and other forums of intellectual discussion such as the Metaphysical Society, potentially divisive issues such as the nature of the soul and its influence on matter were debated frankly and cordially. It was in the context of such debates that Huxley coined the term 'agnosticism', denoting an openness of conviction and an attitude of free enquiry to all matters of belief. This slender 'creed' derived in part from Protestant theological writing on the limits of knowledge, and could be foundational to both scientific and religious life (Lightman 1987).

In much popular science writing, natural knowledge continued to be placed within a Christian framework. Evangelical publishers such as the Society for the Promotion of Christian Knowledge and the Religious Tract Society commissioned works that targeted working and lower middle-class readers, emphasizing feelings of wonder and humility rather than rational grounds for belief (Fyfe). Popular authors such as Margaret Gatty, John George Wood and Arabella Buckley introduced the study of nature to the young, imparting reverence for the harmony and purpose of God's creation. The tradition of natural theology persisted in such writing, with the adaptive structures of form and function taking on a perfection that was more akin to Paley than Darwin. In *The Fairy-Land of Science*, Buckley evoked the 'unseen power' (25) behind the universal laws that reached from the heavens to each droplet of water, calling the reader to rise 'through nature to nature's God'. Even lengthy technical works of botany, entomology and microscopy might link the practice of science with spiritual discipline. In *The Microscope and its Revelations*, the Unitarian zoologist William Carpenter showed over the course of 700 pages how the wonders of creation, hidden to the naked eye, became manifest to the keen observer after patient and dedicated study. Close attention to detail and the self-control required in sound scientific practice became a means of shaping the moral and religious self.

From the beginning, most controversies over Darwinian theory were not centred on natural selection per se but on its application to 'man'. Many accepted the arguments for animal ancestry but retained belief in the uniqueness of humans, whose powers of mind and morality gave them a special place in nature. Liberal theologians who rejected the traditional foundations of faith in scripture and the church looked instead to inner feelings of obligation, conscience and reverence as marks of man's God-given nature. When the geologist Charles Lyell, a close friend and advocate of Darwin's, addressed the question in *Antiquity of Man* (1863), he refused to extend evolutionary theory to human mental and moral faculties, confessing privately that this removed 'much of the charm' (Darwin 11:218) from his speculations on the past. The Roman

Catholic zoologist St George Mivart granted the evolution of the human skeleton and brain by natural causes, but not man's abiding sense of sin, his abhorrence of cruelty and his esteem for charity. Scottish Presbyterianism remained important for leading physicists William Thomson, James Clerk Maxwell, Balfour Stewart and Peter Guthrie Tait, who regarded the second law of thermodynamics as evidence of the transitory nature of material existence. Energy, graced by God at the dawn of creation but gradually dissipating, was addressed as a moral problem through exhortations to good Christian work in the world and the belief that the ultimate power that had given life on earth would restore it after death (Smith). In *The Unseen Universe* (1871), Stewart and Tait speculated (anonymously) that energy flowed from the visible world to an invisible, spiritual realm, from whence it would ultimately return.

Insistence on the special providence and moral mission of man was not only important for individual believers, but fundamental to broader claims of European civilization and empire. In Britain's colonies, science was coupled with Christianity to demonstrate the inferiority of native peoples, and the study of nature was enlisted in their conversion from a heathen or barbarous state (Drayton; Sivasundaram). Ethnographers charted the advances in religion that accompanied those in knowledge, material technology and social institutions. Tracing the stages of culture from savagery to civilization, E. B. Tylor rooted primitive forms of religion, such as animism and fetishism, in the ignorance and fear that inspired tyrannical forms of government. Belief in the divine right of kings and other supernatural powers, he argued, were relics of primitive superstition. The anthropologist and liberal MP John Lubbock assured his readers that, while the details of savage beliefs and practices are often repugnant, 'the religious mind cannot but feel a peculiar satisfaction in tracing up the gradual evolution of more correct ideas and of nobler creeds' (114). Religion advanced with more enlightened forms of government, from animism to polytheism to the worship of one supreme being. Born by steam ships and steam presses, railroads and telegraphs, the message of Christianity was spread through Britain's imperial dominions.

The progress of religion did not always culminate in Christianity, however. In the *First Principles* of his synthetic philosophy, Herbert Spencer described the gradual decline of formal systems of belief through progressive stages of adaptation. As scientific knowledge advanced, the human mind and society would achieve perfect correspondence with nature and its laws, corrupt and restrictive institutions of government would dissolve and anthropomorphic deities would give way to a purer form of worship: the ultimate mystery of reality. Spencer insisted that this 'first principle' of philosophy, the 'Unknowable', was the shared basis of science and religion, an affirmation of the truth behind all appearances, and manifest in feelings of awe, wonder and reverence, indeed the same feelings that some liberal theologians had made the foundation of Christian faith. As Spencerian evolution gained in popularity over the second half of the century, more religious forms of agnosticism were taken up with evangelical fervour. Inspired by the success of the Religious Tract Society, the secularist publisher Charles Albert Watts founded the *Agnostic Journal* and recruited authors of books, pamphlets and articles to spread the doctrine of cosmic evolution and worship of the Unknowable (Lightman 1989).

In the closing decades of the century, other movements of secular and syncretic religion proliferated. The Oxford philosopher and physician Richard Congreve founded the London Positivist Society (1867) and later the Church of Humanity (1878) based upon the writings of the French theorist Auguste Comte. Comte's vision of progress, culminating in 'positive' knowledge and government by scientific principles, had been widely discussed among London radicals and critics, such as John Stuart Mill, George Henry Lewes and Harriet Martineau. It was also inspirational for a group of Oxford scholars in the middle of the century, who later moved away from Anglicanism and sought a religious foundation for reform outside of Christian orthodoxy (Kent). Instead of the idealized providence of Catholic, Protestant or deist, Comte declared, positivism reflected the 'real Providence in all departments–moral, intellectual, material' (1). The group, which included lawyers, civil servants, essayists and politicians, drew largely on a body of Comte's writings that gave secular (indeed technocratic) form to Catholic ritual and organization, including an elite priesthood of believers (the 'servants of humanity'), an elaborate system of worship, a positive 'catechism', a library and a calendar of scientific 'saints'. The positivist future was an altruistic utopia, in which selfish instincts and class conflict would be overcome. Guided by a small body of philosophers and educators, the great mass of workers would become a single family, ruled by social physics and bound by their common love for humanity (Dixon 2008A).

Analogous in its elitism, but drawing largely on eastern traditions, the theosophy movement originated in New York City, moving to London by way of India in the 1880s, and was composed mainly of urban artists, writers and intellectuals. Its charismatic founder, the Russian aristocrat Helena Blavatsky, described theosophy as a 'synthesis of science, religion, and philosophy' imparted to her by learned adepts in the course of her world travels. In *The Secret Doctrine* and other writings, she and her followers incorporated occult and mystical traditions (Zoroastrianism, Hermeticism, neo-Platonism), elements of physics, chemistry and psychology with a predominance of Hinduism, especially the idea of karma, yielding a superior moral system of just rewards. Alfred Sinnett, a leader of the London branch, worked to establish its credentials by experimental proofs of spiritual forces, while the socialist Annie Besant developed its evolutionary principles, from primitive matter through successive stages of reincarnation to the final ascent of man as a spiritual being. Blavatsky's *Secret Doctrine* engaged critically with evolutionary and atomic theory, energy physics and researches in magnetism, electricity and the ether, sifting modern scientific theories and discoveries and compounding them with 'ancient wisdom' while highlighting the limits of conventional science when it came to investigate the innermost secrets of nature.

Theosophy had originated out of the more popular movement of spiritualism by transposing communications with the dead that were the staple of Victorian seances into channels of arcane knowledge, unlocking the wisdom of past lives and higher beings. Spiritualism had gained currency in the middle decades of the century just as mesmerism was in decline. Both practices involved mediums who were able to harness hidden forces and act upon objects at a distance. Seances were typically held in private homes and grew popular among the middle and upper classes. Attendees sometimes paid a small fee for the chance of communicating with a departed loved one, a glimpse of the afterlife or amusement. Some claimed that spirit manifestations demonstrated the existence of life after death, the immortality of the soul and its ability to act on matter. Like natural theology, spiritualism could be harnessed in support of Christianity as a defence against atheism, materialism or agnosticism. But it could also pose a threat to traditional religious beliefs and institutions by allowing access from this world to the next and endowing professional mediums with authority usually reserved for the clergy.

Members of the scientific and medical communities were by no means united in opposition to the movement. Some critics explained the power of mediums and the material manifestations in closed rooms through the art of suggestion, and various unconscious operations of mind and body that were subject to manipulation by tricksters and showmen. But a number of leading practitioners became converts, convinced that the phenomena derived from purely natural and lawful, but unknown, causes. Alfred Russel Wallace campaigned for the scientific investigation of seances and mediums, having satisfied himself of their authenticity. The chemist William Crookes constructed special instruments to measure the power of mediums over inanimate objects, converting the seance into a kind of laboratory. In a series of articles in the *Popular Science Review*, of which he was the editor, he announced the discovery of a new 'psychic force'. Spiritual phenomena seemed analogous to electricity and magnetism – other invisible forces and fluids that had mystified investigators and thrilled audiences in earlier decades. Ongoing debates about subtle and diffused forms of matter, the existence of an electromagnetic ether and the power behind new technologies such as the telegraph all touched the borderland between nature and the supernatural and raised questions about the boundaries of science and its ability to probe the deepest mysteries of life and the universe (Noakes).

In 1882, Crookes and a team of philosophers, classicists, scholars of comparative religion, psychologists and experimental physicists came together to form the Society for Psychical Research, meeting regularly in London under the leadership of Henry Sidgwick, professor of moral philosophy at Cambridge. Special committees were formed to investigate thought transference, mediumship, apparitions and dreams with scientific rigour and impartiality, although some novel methods were developed to study the capricious phenomena and to evaluate evidence based largely on personal testimony, introspection and heightened emotional states. A series of experiments, conducted over two years on a clergyman's daughters, who exhibited the power to read one another's thoughts, was abandoned when the girls were detected signalling each other. Vast amounts of data were collected and published in regular proceedings. Two of the members, Edmund Gurney and Frederic Myers, devised a global census of hallucinations and evaluated over 700 cases of apparitions in their two-volume *Phantasms of the Living*. Other research by the Society on hypnotism and automatic writing seemed to demonstrate hidden levels of consciousness, manifestations of self outside the normal spectrum of awareness, like ultraviolet or infrared rays. As William James, a member of the American branch of the Society, remarked, 'there is nothing . . . that need hinder science from dealing successfully with a world in which personal forces are the starting-point of new effects' (327).

James often wrote with a view to broadening the scope of science, extending psychology to religious experience, such as mysticism, faith healing, demonic possessions and ecstasies. He also drew attention to the religious features of science itself, its peculiar faith, passions and enthusiasms. According to Victorian anthropologists, religion had been the science of savages, their primitive form of understanding and operating in a world whose true powers and causes were as yet unknown. Would science then be the religion of the future? Polymath and eugenicist Francis Galton (also Darwin's cousin) called upon English men of science to become a new priesthood, ministers of nature who would guide the human race towards perfection (260). 'We claim,' declared the physicist John Tyndall, 'and we shall wrest from theology, the entire domain of cosmological theory' (xcv). Such words, like those of Huxley in his defence of Darwinism, sound like a cry of victory in a war between science and religion. Huxley and his contemporaries knew the power of oppositional rhetoric and directed it towards political ends, especially the reform of Anglican institutions. But such polemical assertions belied a web of alliances between men of science and clergymen, intricate negotiations over cultural boundaries and expertise and enduring religious structures of belief about creation, human origins and providence. For better or worse, the history of science and religion remains a lively and controversial subject. The view of science and religion (or theology) as locked in perennial battle has proved so enduring in the public sphere that nearly every work by historians in recent years, including this one, begins and ends by denouncing it.

In conjunction with popular and specialist science writing, works of theology, moral philosophy and metaphysics, sermons, addresses and essays, nineteenth-century poetry and prose fiction helped to shape these complex and shifting relations of science and religion. Mary Shelley's *Frankenstein* (1818) drew on contemporary debates over the nature of electricity and the origins of life from inanimate matter, restaging the Promethean myth of godly invention through its twinned characters, the natural philosopher and polar explorer, shadowed by the monstrous creature of scientific hubris. The spark of life and generative power of imagination yield only hideous destruction when severed from moral sympathy and familial affections. The relationship between moral and material progress was reconfigured in the 'condition of England' novels of Charles Kingsley (*Yeast* and *Alton Locke*) and in a happier tale of human development, *The Water-Babies* (1863). In this popular children's book, the clergyman-naturalist expounded a version of Christian Darwinism, following a young chimney sweep through regressive stages of life, then forward, the boy becoming a heroic man of science and industrialist as a reward for his good deeds in the oceans of time. Natural history as an observational science and moral economy also informed George Eliot's early essays on literary realism and the 'organic' form of her novels *Romola* (1862–63) and *Middlemarch* (1871–72), in which evangelical piety, science and scholarship contend, their discipline and authority threatening to stifle or dissolve selfhood and human feeling until a complemental relation can be struck. A more radical solution was proposed by the late Victorian satirist Samuel Butler. At first enamoured by Darwinism as an antidote to Christianity, he grew critical of its sacred texts, such as *The Origin of Species*, that seemed to harden into orthodoxy, the 'man of science' being only 'the cleric in his latest development' (Butler 1913 179–80). Butler wrote scathingly about Victorian authorities and conventions in the semi-autobiographical *The Way of All Flesh* (1903), and in a series of speculative works beginning with *Life and Habit* (1878) he explored the unconscious forces of biological development and the cultural processes, especially authorship, that write, edit and erase the material of memory, reshaping the evolutionary future.

Despite the marked attention of scholars of literature and science to prose fiction, poetry was perhaps the most important literary form of engagement with science and religion in the nineteenth century. This was due in part to the enduring figure of the poet as a seer or prophet, epitomized by Wordsworth and Tennyson and their well-known ambivalence towards 'positive' or empirical science, but also because of the familiarity and intimacy of the poetic form, its common place in private life and communication between friends (or adversaries). Some correspondents of Darwin addressed their criticism in verse. Scientific practitioners often wrote poetry, not only in their youth or moments of 'idle fancy' and amusement, such as the comic rhymes and songs composed for annual meetings of the Red Lion Club at the British Association, but as serious speculative endeavour.[2] Charles Lyell's early geological writing took the form of Byronic verse, a model he extended to his *Principles of Geology*, converting the Earth into a Byronic hero (Buckland 2013 95–7, 127–30). For physicists and mathematicians, poetry opened a space of personal and metaphysical reflection that had been sealed off in science. James Clerk Maxwell wrote odes for colleagues and friends, first as a coping device for the mechanical exercises of the Cambridge examination system and later as a means of exploring foundational and methodological issues that had no scope in the official curriculum. He sustained his belief in Christian cosmology through personal hymns to the Creator, akin to the songs of praise in Scottish Presbyterian worship (Brown 62–3).

The most influential poetic reflection on religion and science of the second half-century was Tennyson's *In Memoriam*. Though published in 1850, it was composed over two decades, taking in the new science of geology, especially Lyell's *Principles*, the fossil evidence of large-scale extinction and the transformism of Lamarck. Beginning with the death of his friend Arthur Hallam, Tennyson's elegy extends to all of creation; personal loss swells to embrace earth history and the insignificance of human achievement ('the clock / Beats out the little lives of men' (Canto 2, ll. 7–8)) in the face of Nature, 'I care for nothing, all shall go' (Canto 56, l. 4). Grief and loss are not final, however, but stages in a larger process of development. Faith in a loving

God and trust in higher purposes falter, but return, reaching beyond the limits of knowledge to the promise of rebirth and revelation, when the book of nature is fully open to human understanding. Though written many years before *The Origin of Species*, *In Memoriam* would be evoked by later readers in response to Darwin's theory of descent by natural selection and the challenges it posed to a benevolently designed order (Holmes 2014).

In 1869, Tennyson delivered a poetic prologue to the first meeting of the London Metaphysical Society, an organization founded with the aim of discussing matters of common interest between men of science, clergymen and men of letters. The poem for the occasion, 'The Higher Pantheism', was characteristic of Tennyson's post-Darwinian works, more insistent on the separation of faith and knowledge and more assertive of the primacy of religious belief. Its quaint inversions would be parodied by Swinburne in 'The Higher Pantheism in a Nutshell': 'and God, who is not, we see . . . and diddle, we take it, is dee'. Such irreverence by a younger generation of poets was not shared by men of science, however, who began to forge Tennyson's reputation as the 'Poet of Nature' and 'Poet of Science' in the closing decades (Holmes 2012). Divergent readings of Tennyson reflected the wide range of conjunctions between science and religion across the nineteenth century: the enduring tradition of natural theology, the union of material and spiritual evolution, the reverence for scientific discipline and natural law, the poetic intuition of truth beyond appearances, and enchantment amid the harsh, impersonal forces and facts of nature.

Notes

1   See Gowan Dawson, 'Science in the Periodical Press', in this volume.
2   See Martin Willis, 'Scientific Cultures and Institutions', in this volume.

Bibliography

Primary texts

Blavatsky, H. P. *The Secret Doctrine: The Synthesis of Science, Religion and Philosophy*. 3 vols. London: Theosophical, 1888–97.
Buckland, William. *Geology and Mineralogy Considered with Reference to Natural Theology*. 2 vols. London: Pickering, 1836.
Buckley, Arabella. *The Fairy-Land of Science*. London: Stanford, 1879.
Butler, Samuel. *Life and Habit*. London: Trübner, 1877.
Butler, Samuel. *The Way of All Flesh*. London: Grant Richards, 1903.
Butler, Samuel. *The Humour of Homer and Other Essays*. London: A. C. Fifield, 1913.
Carpenter, William. *The Microscope and its Revelations*. London: John Churchill, 1856.
Combe, George. *The Constitution of Man*. People's edn. Edinburgh: Chambers, 1836.
Comte, August. *The Catechism of Positive Religion*. Transl. Richard Congreve. London: John Chapman, 1858.
Crookes, William. *Psychic Force and Modern Spiritualism*. London: Longmans, Green, 1871.
Darwin, Charles. *The Correspondence of Charles Darwin*. Ed. Frederick Burkhardt *et al.* 22 vols. Cambridge: Cambridge University Press, 1985.
Draper, John William. *History of the Conflict Between Religion and Science*. 2 vols. London: Henry King, 1875.
Eliot, George. 'The Natural History of German Life'. *Westminster Review* 66 (1856): 51–79.
Eliot, George. *Romola*. London: Smith and Elder, 1863.
Eliot, George. *Middlemarch: A Study of Provincial Life*. Edinburgh: William Blackwood, 1871–72.
Essays and Reviews. London: Parker, 1860.
Galton, Francis. *English Men of Science: Their Nature and Nurture*. London: Macmillan, 1874.

Gatty, Mrs Alfred [Margaret]. *Parables From Nature*. London: Bell and Daldy, 1855.

Gray, Asa. *Natural Selection Not Inconsistent with Natural Theology*. London: Trübner, 1861.

Gurney, Edmund, Frederic Myers, and Frank Podmore. *Phantasms of the Living*. 2 vols. London: Trübner, 1886.

Herschel, John. *A Preliminary Discourse on the Study of Natural Philosophy*. London: Longman, 1830.

Huxley, T. H. 'The Origin of Species'. *Westminster Review* 17 n.s. (1860): 541–70.

James, William. 'What Psychical Research Has Accomplished'. *The Will to Believe and Other Essays in Popular Philosophy*. London: Longmans, Green, 1897. 299–328.

Kingsley, Charles. *The Water-Babies: A Fairy Tale for a Land Baby*. London: Macmillan, 1863.

Lawrence, William. *Lectures on the Physiology, Zoology, and Natural History of Man*. London: Callow, 1819.

Lubbock, John. *On the Origin of Civilization and the Primitive Condition of Man*. London: Longmans, Green, 1870.

Lyell, Charles. *Principles of Geology*. 3 vols. London: John Murray, 1830–33.

Lyell, Charles. *The Geological Evidences of the Antiquity of Man with Remarks on the Theories of the Origin of Species by Variation*. London: John Murray, 1863.

Mivart, St George Jackson. *On the Genesis of Species*. London: Macmillan, 1871.

Nichol, John Pringle. *Views of the Architecture of the Heavens*. Edinburgh: Tait, 1837.

Paley, William. *Natural Theology, or Evidences of the Existence and Attributes of the Deity*. London: Faulder, 1802.

Paley, William. *Natural Theology, With Illustrative Notes by Henry Lord Brougham and Sir Charles Bell*. 4 vols. London: Knight, 1836–39.

Shelley, Mary. *Frankenstein, or, The Modern Prometheus*. Ed. Marilyn Butler. Oxford: Oxford University Press, 1994.

Spencer, Herbert. *First Principles*. London: Williams and Norgate, 1862.

Stewart, Balfour, and Peter Guthrie Tait. *The Unseen Universe, or Physical Speculations on a Future State*. London: Macmillan, 1875.

Swinburne, Algernon Charles. 'The Higher Pantheism in a Nutshell'. *Specimens of Modern Poets*. London: Chatto and Windus, 1880.

Tennyson, Alfred. *In Memoriam*. Ed. Erik Gray. New York, NY: Norton, 2004.

Tylor, Edward Burnett. *Primitive Culture: Researches into the Development of Mythology, Philosophy, Religion, Art, and Custom*. 2 vols. London: John Murray, 1871.

Tyndall, John. Presidential Address. *Report of the Forty-Fourth Meeting of the British Association for the Advancement of Science; Held at Belfast in August 1874*. London: John Murray, 1875, lxvi–xcvii.

White, Andrew Dickson. *A History of the Warfare of Science with Theology in Christendom*. 2 vols. New York, NY: Appleton, 1896.

Wood, John George. *The Common Objects of the Sea-Shore*. London: Routledge, 1857.

Secondary texts

Brooke, John Hedley. *Science and Religion: Some Historical Perspectives*. Cambridge: Cambridge University Press, 1991.

Brown, Daniel. *The Poetry of Victorian Scientists: Style, Science and Nonsense*. Cambridge: Cambridge University Press, 2013.

Buckland, Adelene. *Novel Science: Fiction and the Invention of Nineteenth-Century Geology*. Chicago, IL: University of Chicago Press, 2013.

Cantor, Geoffrey. *Michael Faraday: Sandemanian and Scientist*. London: Macmillan, 1991.

Chadwick, Owen. *The Secularization of the European Mind in the Nineteenth Century*. Cambridge: Cambridge University Press, 1975.

Desmond, Adrian. *The Politics of Evolution: Morphology, Medicine, and Reform in Radical London*. Chicago, IL: University of Chicago Press, 1989.

Dixon, Thomas. *The Invention of Altruism: Making Moral Meanings in Victorian Britain*. Oxford: Oxford University Press, 2008 [2008A].

Drayton, Richard. *Nature's Government: Science, Imperial Britain and the 'Improvement' of the World*. New Haven, CT: Yale University Press, 2000.

Fyfe, Aileen. *Science and Salvation: Evangelical Popular Science Publishing in Victorian Britain*. Chicago, IL: University of Chicago Press, 2004.

Gillispie, Charles Coulston. *Genesis and Geology*. Cambridge, MA: Harvard University Press, 1951.

Greene, John C. *The Death of Adam: Evolution and its Impact on Western Thought*. Ames, IA: Iowa State University Press, 1959.

Heyck, T. W. *The Transformation of Intellectual Life in Victorian England*. London: Croom Helm, 1982.

Hilton, Boyd. *The Age of Atonement: The Influence of Evangelicalism on Social and Political Thought, 1795–1865*. Oxford: Clarendon, 1988.

Himmelfarb, Gertrude. *Darwin and the Darwinian Revolution*. Garden City, NJ: Doubleday, 1959.

Holmes, John. '"The Poet of Science": How Scientists Read Their Tennyson'. *Victorian Studies* 54 (2012): 655–78.

Holmes, John. 'The Challenge of Evolution in Victorian Poetry'. *Evolution and Victorian Culture*. Eds. Bernard Lightman and Bennett Zon. Cambridge: Cambridge University Press, 2014, 39–63.

Houghton, Walter. *The Victorian Frame of Mind*. New Haven, CT: Yale University Press, 1957.

Irvine, William. *Apes, Angels, and Victorians: The Story of Darwin, Huxley, and Evolution*. New York, NY: McGraw-Hill, 1955.

Kent, Christopher. *Brains and Numbers: Elitism and Comtism in Mid-Victorian London*. Toronto: University of Toronto Press, 1978.

Lightman, Bernard. *The Origins of Agnosticism: Victorian Unbelief and the Limits of Knowledge*. Baltimore, MD: Johns Hopkins University Press, 1987.

Lightman, Bernard. 'Ideology, Evolution and Late-Victorian Agnostic Popularizers'. *History, Humanity, and Evolution: Essays for John C. Greene*. Ed. James Moore. Cambridge: Cambridge University Press, 1989. 285–310.

Moore, James. *The Post-Darwinian Controversies: A Study of the Protestant Struggle to Come to Terms with Darwin in Great Britain and America, 1870–1900*. Cambridge: Cambridge University Press, 1979.

Noakes, Richard. 'Spiritualism, Science and the Supernatural in Victorian Britain'. *The Victorian Supernatural*. Eds. Nicola Bowne, Caroline Burdett, and Pamela Thurschwell. Cambridge: Cambridge University Press, 2004. 23–43.

O'Connor, Ralph. *The Earth on Show: Fossils and the Poetics of Popular Science, 1802–1856*. Chicago, IL: University of Chicago Press, 2007.

Ruston, Sharon. *Shelley and Vitality*. Basingstoke: Palgrave Macmillan, 2005.

Schaffer, Simon. 'The Nebular Hypothesis and the Science of Progress'. *History, Humanity, and Evolution: Essays for John C. Greene*. Ed. James Moore. Cambridge: Cambridge University Press, 1989. 131–64.

Secord, James A. *Visions of Science: Books and Readers at the Dawn of the Victorian Age*. Oxford: Oxford University Press, 2004.

Sivasundaram, Sujit. *Nature and the Godly Empire: Science and Evangelical Mission in the Pacific, 1795–1850*. Cambridge: Cambridge University Press, 2005.

Smith, Crosbie. *The Science of Energy: A Cultural History of Energy Physics in Victorian Britain*. London: Athlone, 1998.

Topham, Jonathan R. 'Science, Religion, and the History of the Book'. *Science and Religion: New Historical Perspectives*. Eds. Thomas Dixon, Geoffrey Cantor and Stephen Pumfrey. Cambridge: Cambridge University Press, 2010. 221–44.

Turner, Frank M. *Contesting Cultural Authority: Essays in Victorian Intellectual Life*. Cambridge: Cambridge University Press, 1993.

White, Paul. *Thomas Huxley: Making the 'Man of Science'*. Cambridge: Cambridge University Press, 2003.

Young, Robert M. *Darwin's Metaphor: Nature's Place in Victorian Culture*. Cambridge: Cambridge University Press, 1985.

Further reading

Bowler, Peter J. *Monkey Trials and Gorilla Sermons: Evolution and Christianity From Darwin to Intelligent Design*. Cambridge, MA: Harvard University Press, 2007.

Brooke, John Hedley. 'The Wilberforce-Huxley Debate: Why did it Happen?'. *Science and Christian Belief* 13 (2001): 127–41.

Dawson, Gowan, and Bernard Lightman, eds. *Victorian Scientific Naturalism: Community, Identity, Continuity*. Chicago, IL: University of Chicago Press, 2014.

Dixon, Thomas. *Science and Religion: A Very Short Introduction*. Oxford: Oxford University Press, 2008.

England, Richard, and Jude V. Nixon, eds. *Victorian Science and Literature, Volume 3: Science, Religion, and Natural Theology*. Gen. eds Gowan Dawson and Bernard Lightman. London: Pickering and Chatto, 2011.

Fulweiler, Howard. 'Tennyson's *In Memoriam* and the Scientific Imagination'. *Thought* 59 (1984): 296–318.

Helmstadter, Richard, and Bernard Lightman, eds. *Victorian Faith in Crisis*. London: Macmillan, 1990.

Luckhurst, Roger, and Justin Sausman, eds. *Victorian Science and Literature, Volume 8: Marginal and Occult Sciences*. Gen. eds Gowan Dawson and Bernard Lightman. London: Pickering and Chatto, 2012.

Paradis, James. *Samuel Butler: Victorian Against the Grain*. Toronto: University of Toronto Press, 2007.

Wheeler-Barclay, Marjorie. *The Science of Religion in Britain, 1860–1915*. Charlottesville: University of Virginia Press, 2010.

4

WOMEN AND SCIENCE

Michelle Boswell

Thomas Jefferson High School for Science and Technology

In guidebooks and early versions of textbooks, in essays and conversations, in fairy tales and parables, in poetry and fiction, nineteenth-century British women employed a diverse range of genres to enter scientific conversations, despite the voices of many professional men of science who often sought to exclude them. For female practitioners of the biological and physical sciences in the long nineteenth century, these myriad genres were not only socially sanctioned but also pedagogically useful. Despite their marginalization from formal higher education and professional scientific societies for much of the century, female astronomers, botanists, chemists and naturalists found creative ways of entering scientific conversations: practising science as amateurs or hobbyists, popularizing it for a variety of public audiences – especially other women and children – and, finally, investigating contemporary topics as competent members of the scientific community and publishing their findings. Understanding the means by which these women addressed scientific questions not only provides insight into the debates themselves and the gendered notions of how women could enter them but also illuminates developments in nineteenth-century literary history and the evolution of its genres and forms.

Constrained by gendered assumptions of what women's proper roles in society should be, women throughout the era nevertheless found the means both to work within and to transgress these boundaries in their research and writing. Despite their limited opportunities for formal scientific study at the beginning of the nineteenth century, by the century's end women were finally able to enter women's colleges in London, Oxford and Cambridge, as well as 'red brick' institutions in Britain's industrial cities such as Birmingham, Bristol and Leeds. Women's entrance into formal higher education led to their wider – though not unproblematic – acceptance as practitioners in their own right, affording them the opportunity to communicate within the professional scientific community. As a consequence of their entry into formal scientific study, however, women adopted the same methods and rhetorical strategies of their male colleagues; thus, much of the generic diversity we can locate in women's scientific writing earlier in the century disappears by its end, a trajectory paralleling the shift towards disciplinary specialization taking place within late Victorian science more generally. But the genres women used to communicate science throughout the long nineteenth century deserve the attention of literary scholars and historians of science alike. These genres reveal how women compensated for their exclusion from the professional community by taking part in the biggest scientific conversations of their day via the popular realm. Using the genres available to them, women yet aimed to shape the debate.

With some notable exceptions, very few British women at the end of the eighteenth century had access to formal training in natural philosophy or natural history. Yet, by the middle of the nineteenth century many self-educated women had found publication niches within popularizations of science for children or guidebooks to wildflowers, ferns and aquatic plants and corals. Finally, by the late 1860s, young Victorian women gained access to higher education and began distinguishing themselves in their classes. Many excelled at the long-standing Mathematical tripos at Cambridge, including Charlotte Scott (1858–1931) of Girton College and Philippa Fawcett (1868–1948) of Newnham. Shouts of 'Scott of Girton!' drowned out the name of the male eighth 'wrangler' whose rank Scott unofficially equalled in 1880, and Fawcett's legendary examination results 10 years later placed her above the senior wrangler, as Mary Creese relates in her encyclopedic *Ladies in the Laboratory?* (1998). Though female students were able to sit the Oxbridge examinations, the universities refused to grant them degrees until the twentieth century, in part because the degree conferred a say in university governance (Sheffield 107).[1] Yet, one consequence of entry into the professional world of science was that women, too, became subject to the divisions among scientific disciplines. Rarely was it feasible for anyone, regardless of sex, to synthesize the most current scientific research as Mary Somerville (1780–1872) had done so concisely earlier in the century. Female students of science, like their male counterparts, began to focus on particular sciences for advanced study.

Given their exclusion from British institutions of higher learning until the last three decades of the nineteenth century, it was not formal education that allowed women to enter scientific conversations as knowledgeable members of the community. Rather, the larger context of the rapidly expanding British literary marketplace and the increasing dissemination of informative texts fostered women's pursuits in writing about science for women and children in the growing middle classes. The growth of the Society for the Diffusion of Useful Knowledge (SDUK) and the Society for the Promotion of Christian Knowledge (SPCK) both offered opportunities for women to publish on scientific topics. At the same time, of course, the idealization of the domestic, maternal woman – epitomized by Coventry Patmore in his poem *The Angel in the House* (1854) – likewise constrained scientific women into adopting particular genres during the first half of the century, such as the catechismal genre of the 'familiar format' of dialogues and conversations. Often composed of lessons presented by a female teacher and her young children or students, the familiar format followed in the footsteps of scientific, didactic works by men that had enjoyed success in the eighteenth century, for instance John Newbery's *The Newtonian System of Philosophy* (1761), featuring 'Tom Telescope' (Rauch 1989 14).[2]

The rhetorical effects of the dialogue genre deserve special mention. Writing about the dialogue form of women's popularizations, Greg Myers focuses on the hierarchical status implicit in this pedagogy. Though the dialogue form had a long history of discussing scientific topics in a way that could distance the writer from controversy (the dialogues of Galileo, for example), Myers demonstrates how later forms of dialogue divide public and scientific knowledge, creating a 'lower form of science' (173) for women and children. Instead of being dialectical and exploratory, debating two sides of an issue, the kinds of dialogues women wrote – coming by way of French catechismal dialogues – suggested a hierarchical relationship between knowledge and ignorance, setting readers up to be passive consumers of scientific knowledge. Yet, as science continued to pervade Victorian culture throughout the nineteenth century, women's choices in accommodating scientific theories and discoveries for their diverse audiences moved away from the familiar format and gradually expanded to include handbooks and illustrated guides – early versions of scientific textbooks – and fanciful tales for children that aimed to excite young readers' imaginations and curiosity. Still other women took their scientific knowledge and applied it in the literary realm, demonstrating how scientific ideas could serve as literary tropes, as George Eliot did so famously with her web-like structure in *Middlemarch*, or as fodder for social critique, as Constance Naden and other women poets suggested in their satirical verses about evolutionary theories, particularly sexual selection.

This chapter demonstrates ways in which women's marginalization from the increasingly professional realm of science was a productive constraint for their written work. Highlighting some of the most notable accomplishments and texts written by women specifically in the natural sciences, I have broken the discussion into rough categories familiar to early nineteenth-century readers: natural history and physical science (formerly natural philosophy). The following list of women writers is admittedly suggestive rather than comprehensive, and many more women's contributions to science over the course of the long nineteenth century may be found in full-length secondary works listed in this chapter's bibliography. A pleasantly surprising quantity of extant women's science writing is now accessible to researchers around the world thanks to digitization efforts in the past decade. The first section of this chapter describes women's efforts in botany, algology, zoology and palaeontology; the next on astronomy, chemistry and physics. A separate section follows on scientific texts for children, illustrating the styles in which women first became didactic voices on scientific topics – as in the dialogue or conversational form – and the uses to which women put the familiar literary genres of parable, fantasy, and fairy tale when accommodating science for young readers. Here the scientific topics vary even within a single volume, and the larger concern for today's reader focuses less on the details of the science being taught and more on the assumptions underscoring a particular approach to the subject and its style of delivery. Finally, this chapter concludes with a brief nod towards women who wrote about science strictly within the literary realm, not necessarily as practitioners but rather as interpreters and commentators. Because form and genre were especially important for women's communication and interpretation of science throughout the nineteenth century, considering those genres as interpretative tools – rather than as static categories – means that one can trace both how women entered scientific conversations and how they aimed to shape the reception of scientific ideas.

Natural history

Women's involvement in British science dates from a far earlier period than the nineteenth century. Londa Schiebinger has written extensively on the gendering of early modern science through the Enlightenment in *The Mind Has No Sex? Women in the Origins of Modern Science* (1989) and *Nature's Body: Gender in the Making of Modern Science* (1993). A helpfully concise discussion of European women's participation in science during the seventeenth and eighteenth centuries can also be found in Schiebinger's 'The Philosopher's Beard: Women and Gender in Science' (2003). In *Pandora's Breeches: Women, Science and Power in the Enlightenment* (2004), Patricia Fara revises a long-standing narrative of science's male heroes, demonstrating the extreme importance of the 'invisible' work that women did alongside their fathers and brothers. Part of an Enlightenment culture that valued scientific learning as one facet of general education for polite society, women were encouraged to become consumers of scientific knowledge in astronomy, physics, mathematics, chemistry and natural history as part of their spiritual and moral improvement (Shteir 1996 2). In the latter half of the eighteenth century, botany became a pursuit both fashionable and useful, capable of 'improving' the minds of its practitioners. Yet when Carl von Linné, better known as Linnaeus, decided to use plants' reproductive parts as the distinguishing feature of his taxonomic system, the sexual nature of the material at hand became fodder for debate about its suitability for female readers. In *Cultivating Women, Cultivating Science: Flora's Daughters and Botany in England, 1760–1860* (1996), Ann B. Shteir situates Linnaean taxonomic nomenclature within the era's broader debate about sexuality and morality (11–32). Sam George's *Botany, Sexuality, and Women's Writing, 1760–1830* (2007) focuses on the social dimensions of women's natural history study, examining how women's botanical studies seemed to threaten eighteenth-century female 'modesty'. Unsurprisingly, Mary Wollstonecraft scorned the notion of 'protecting' women's modesty by limiting their access to knowledge enjoyed by men in her *Vindication of the Rights of Woman* (1792), an inequity contrary to the 'enlightening principles' (298) of an equal social contract.

As women became visible presences within botanical studies, they drew their authority to teach science from their roles as mothers and teachers, often using the 'familiar format' of letters, dialogues and conversations as the mode of instruction (Shteir 1996 4; Gates and Shteir 8). Sarah Trimmer (1741–1810) was the literary predecessor of many women who continued using the familiar format into the nineteenth century. Trimmer's *Easy Introduction to the Knowledge of Nature* (1780) set a precedent in combining the study of natural history and morality – a practice common to adherents of natural theology – within conversations between a mother and her two children. Priscilla Wakefield (1751–1832) uses the epistolary form in *An Introduction to Botany, in a Series of Familiar Letters* (1796) and the conversational format in *Mental Improvement, or the Beauties and Wonders of Nature and Art* (1794–7). Wakefield's *Introduction to Botany* was the first systematic introduction to the science penned by a woman (Shteir 1996 83). A conversation between a mother and her son frames the botanical topics in Sarah Fitton's (*c.* 1796–1874) *Conversations on Botany* (1817) and demonstrates the utility of the form for both the juvenile reader and his mother or other guardian. The book went through nine editions between 1817 and 1840 (Torrens and Browne).

Other women earned reputations as botanical illustrators, including Elizabeth Twining (1805–89), whose family owned many tea plantations abroad and became renowned for its tea company. Twining travelled abroad and wrote comparative studies, including *Illustrations of the Natural Orders of Plants With Groups and Descriptions* (1849–55), a volume also based on her visits to the Royal Botanical Gardens at Kew (Gates 2002 654). Anne Pratt (1806–93) offers vibrant illustrations in her five volumes on British plants, *The Flowering Plants and Ferns of Great Britain* (1855), a thorough reference work published by the SPCK. Pratt follows the Linnaean taxonomic system but she uses English rather than Latin terms as part of her aim to assist readers new to the study of botany. These eighteenth- and early nineteenth-century botanical texts, especially those in the familiar format of conversations and letters, reveal how much notions of women's authority as teachers of natural science and their status as representatives of female modesty were bound up in choices of genre. Selecting formats that would conform to social expectations of modesty also brought similar expectations for those genres' uses in teaching moral lessons, such that by the middle of the 1800s scientific didacticism would merge neatly into 'natural theology', a theology based on observed facts rather than revelation and famously described in the nineteenth century by William Paley. In their content and their form, these texts would blend knowledge of the natural world with religious conviction for over a century.

Yet the rigidity and simplification required of the familiar form prompted a number of women science writers to seek alternatives to the constraints of dialogues. Influenced by Sarah Trimmer, poet Charlotte Smith (1749–1806) published *Rural Walks: In Dialogues Intended for the Use of Young Persons* (1795). Barbara Gates describes Smith's difficulties in rendering natural history in the dialogue form and emphasizes the ways Smith and other women often included disclaimers about the relative simplicity – or seeming insipidness – of accommodating the science for young readers (1998 40–41). Jane Haldimand Marcet (1769–1858), famous for her books of 'conversations', including *Conversations on Chemistry* (1806) – a book that notably inspired the young Michael Faraday – and *Conversations on Natural Philosophy* (1819), later authored a three-volume fictional travel and natural history narrative called *Bertha's Visit to Her Uncle in England* (1830). Voiced by a young woman who moves to England from Brazil after her father's death, the epistolary style of *Bertha's Visit* first describes Bertha's experiences aboard ship and then the lessons she learns with her cousins in various subjects, including natural history, arithmetic and scripture. Though Marcet figures prominently in histories of women's work in accommodating the physical sciences for juvenile audiences, *Bertha's Visit* has lingered in the shadows cast by the *Conversations* and merits further attention within studies of accommodated natural histories for children.

Jane Webb Loudon (1807–58) earned fame for her horticultural titles, including *The Ladies' Companion to the Flower Garden* (1841) and *Botany for Ladies* (1842), and her editing of the weekly *Ladies' Companion at Home and Abroad* (1849–51). But Loudon had originally hoped to earn a living as a novelist, anonymously publishing *The Mummy! A Tale of the Twenty-Second Century* ([1827] 1994), a speculative work of science fiction bringing together elements of the Gothic with political and technological commentary. Alan Rauch has recently abridged Loudon's lengthy triple-decker novel into a single volume and provides a detailed introduction to Loudon's life and work. Already in debt from her husband's earlier horticultural publications, Loudon wrote out of necessity when she was widowed in 1843 at the age of thirty-six (Shteir 1996 223–4). The retitling of *Botany for Ladies* as *Modern Botany* in 1851 speaks to the wider cultural questions about women's education and the Victorian public's appetite for scientific texts (Shteir 1996 225). When the craze for botany took hold by the mid-nineteenth century, inspiring such epithets as 'pteridomania' or 'fern fever', some scientific men like John Lindley strove to reclaim botany as a masculine pursuit. His *Ladies' Botany; or, A Familiar Introduction to the Study of the Natural System of Botany* (1865) may ostensibly have been part of the larger body of popular scientific literature, but it was likewise an act of disciplinary gatekeeping, reasserting a division between the kinds of botany men and women should pursue. David Elliston Allen's chapter on 'Exploring the Fringes' in *The Naturalist in Britain: A Social History* (1976) provides a useful context for the seeming explosion in amateur botanical explorations thanks to the period's emerging technologies, including rapid transit by means of Britain's new steam locomotives, the popular use of the compound microscope and 'Wardian cases', or miniature greenhouses and marine aquariums (108–25). Sarah Whittingham chronicles the Victorian fern craze in a marvellously illustrated and compellingly narrated volume, *Fern Fever* (2012).

The third quarter of the nineteenth century witnessed a veritable explosion of women's writings on botany, much of which drew connections to medicine and health. Into the famous mid-century 'fern fever', Phebe Lankester (1825–1900) wrote *A Plain and Easy Account of British Ferns* (1860), advocating the collection and cultivation of ferns, describing dozens of varieties and providing information on their habitats and medicinal uses (Shteir 2003). Lankester also authored *Wild Flowers Worth Notice* (1861) and a number of books aimed at a juvenile audience, including *Botany for Elementary Schools* (1875) and *Talks About Plants* (1879). Lankester also contributed to magazines throughout her career, writing for *Popular Science Review*, *The Queen*, *Chambers' Journal* and *Magazine for the Arts*. She also wrote about health in *Talks About Health* (1874) and *Domestic Economy for Young Girls* (1875), the latter a volume that also discussed food preparation, clothing and cleanliness, as did *The National Thrift Reader, With Directions for Possessing and Preserving Health* (1880). Texts such as these combined the scientific impulse with a subject likely perceived as more congenial to young Victorian women: the maintenance of a healthy household.

In the field of palaeontology, such connections to household economy were far less accessible, and, unsurprisingly, women's palaeontological writing and work tended more towards the value of discovery in adding knowledge to the scientific community. Mary Anning (1799–1847) made a name for herself not as a scientific writer but as a careful collector and preserver of palaeontological specimens, having begun combing the shores of Lyme Regis when she was a child. Anning found the first *Ichthyosaurus* fossil in 1811 (Gates 1998 68) and she went on to find the first complete *Plesiosaurus* fossil in 1823 and a *Pterodactyl* in 1828 (Torrens). Anning is the most famous woman of note in nineteenth-century geology; however, in *The Role of Women in the History of Geology* (2007), editors Cynthia V. Burek and Bettie Higgs offer a collection of essays demonstrating the geological work of many other women, including Eliza Maria Gordon Cumming (1798–1842), Mary Morland Buckland (1797–1857) and Elizabeth Anderson Gray (1831–1924). Other women naturalists include writer and illustrator Sarah Wallis Bowdich Lee (1791–1856), the first woman to collect plants systematically in tropical West Africa (Creese 129) and who wrote *The Fresh-Water Fishes of Great Britain* (1828), a biography of Cuvier (1833) and a number of popular books between 1841 and 1855 aimed at a juvenile audience (Creese 129, 132 n. 34).

Juvenile readers were just one part of the wider audience for whom popularizers of science directed their writing. Rosina Zornlin (1795–1859) authored a number of popular science books, including *Recreations in Geology* (1839), *The World of Waters; or, Recreations in Hydrology* (1843) and *Outlines of Geology* (1852). In *Recreations in Physical Geography; or, The Earth as It Is* (1840), Zornlin transitions from describing subterranean geology, which she describes as 'the earth in its former condition' to an illustration of 'the earth in its present condition' (iii) – its surface and waters, climate, people, animals and atmosphere. Zornlin studied and wrote about physical science as well, demonstrating her interests and abilities across a wide range of scientific topics and writing for various audiences. Her first books were aimed at a juvenile audience, following the popular dialogue format in books such as *What is a Comet, Papa?* (1835) and its sequel, *The Solar Eclipse* (1837). In *What is a Voltaic Battery?* (1842), Zornlin teaches the basics of batteries and electricity via a conversation between an uncle and his two nephews. Throughout her writing career, Zornlin demonstrates both her scientific acumen and her 'deeply held yet moderate faith', as James Secord describes in his introduction to the volume of Zornlin's works in Bernard Lightman's series *Science Writing by Women* (2004).

Many women science writers bridged the gap between purely scientific and popular works. Margaret Gatty (1809–73) may more often be remembered for her founding of *Aunt Judy's Magazine* in 1866, a publication for children she edited with her daughter, Juliana Horatia Ewing, than for her scientific texts. Yet as the daughter of one minister and the wife of another, Gatty combined her Christian faith with her botanical study to become a widely read author of moral children's stories on natural history and guidebooks for women about marine vegetation. For fourteen years living in the Ecclesfield parish of Yorkshire and raising a large family, Gatty collected algae and seaweeds along the British coastline, culminating in her publication of *British Sea-Weeds* (1863). Providing detailed descriptions and illustrations of seaweeds' colour, size, texture and habitats, Gatty also advises her female readers on the proper tools and attire for the activity. 'Let woollen be in the ascendant as much as possible,' Gatty counsels, continuing:

> [L]et the petticoats never come below the ankle. A ladies' yachting costume has come into fashion of late, which is, perhaps, as near perfection for shore-work as anything that could be devised . . . Cloaks and shawls, which necessarily hamper the arms, besides having long ends and corners which cannot fail to get soaked, are, of course, very inconvenient, and should be as much avoided as possible.
>
> *(viii–ix)*

Gatty's detailed, pragmatic suggestions for attire came out of her many years researching and writing her book, and likely her observations of fellow shore-hunters, too. William Dyce's

Figure 4.1 Pegwell Bay, Kent – a Recollection of October 5th 1858 by William Dyce
© Tate, London, 2015

painting *Pegwell Bay, Kent – A Recollection of October 5th 1858* may just be an idealized representation of Dyce's family on holiday, where the painter's wife and two sisters appear wearing long skirts, bonnets and shawls, but it might likewise suggest that many female seaside ramblers needed the practical advice that Gatty's volume would provide. Attuned to the needs of her diverse audience, Gatty understood how her handbook's novice readers might desire instruction not only in phycology itself but also in its methods and practice. Like the works of so many 'lady naturalists' in the nineteenth century, *British Sea-Weeds* contributes to the field of natural history while subtly revealing the gendered social expectations and limitations within which scientific women were working.

Physical science

Dubbed the 'queen of science' by the *Sunday Post* following her death, Mary Fairfax Grieg Somerville (1780–1872) was the foremost female expositor of science during the nineteenth century. Like Caroline Herschel (1750–1848), who became famous for her discovery or co-discovery with her brother William of at least eight comets, Somerville was an avid student of astronomy. But Somerville extended her studies into mathematics, chemistry, geography, microscopy, electricity and magnetism. An autodidact during most of her early years, she won a prize medal for her solution to a mathematical problem posed by Thomas Lebourn's *New Series of the Mathematical Repository* in 1811 and published her experiments seeking a relationship between light and magnetism in the *Philosophical Transactions* in 1826.[3] Somerville's scientific career flourished as she and her second husband, William Somerville, an army doctor, a fellow of the Royal Society and member of the Linnean Society, travelled in the most prominent scientific circles in Europe (Patterson 1983 1–17). She published a translation of Pierre-Simon Laplace's *Mécanique céleste* in 1831 as *The Mechanism of the Heavens*, a book used in university courses at Cambridge. This was followed by *On the Connexion of the Physical Sciences* in 1834, *Physical Geography* in 1848 and *On Molecular and Microscopic Science* in 1869. While Somerville's life has garnered scholarly interest among feminist historians of science, her wider influence in nineteenth-century British culture – especially among its literary elite – continues to warrant further examination. Anna Henchman attests to the richness of astronomical metaphors found within Somerville's *Connexion* for writers such as Alfred Tennyson in *The Starry Sky Within: Astronomy and the Reach of the Mind in Victorian Literature* (2014). But poetry likewise offered Somerville valuable means to teach these phenomena to her readers and to comment on their significance, as she does, for instance, in an allusive chapter on parallax in which she counters the despairing vertiginous perspective voiced by the eponymous speaker in Lord Byron's *Cain* (1821).

Agnes Clerke (1842–1907) began publishing on astronomy just after Somerville's death, with an 1877 article for the *Edinburgh Review*, 'Copernicus in Italy', and she went on to publish dozens more. However, because her articles were unsigned, she 'seemed to burst upon the scene out of nowhere' (Lightman 2007 471) in 1885 when she published her first book, *A Popular History of Astronomy During the Nineteenth Century*. Clerke went on to write five more books: *The System of the Stars* (1890), *The Herschels and Modern Astronomy* (1895), *Astronomy* (1898), *Problems in Astrophysics* (1903) and *Modern Cosmogonies* (1905). *A Popular History* became the standard book on astronomy, written for practitioners as well as non-specialists. Both Clerke and Alice Bodington (1840–97), who emigrated to Canada in 1887 and wrote *Studies in Evolution and Biology* (1890), can be read within a tradition of women writers beginning with Somerville, who sought to enter the professional realm of scientific study and wrote synoptic works for audiences of specialists. As the sciences fragmented towards increasing specialization, it became difficult for a specialist to know about relevant advances and techniques of use within his own field. Thus, as Lightman argues in *Victorian Popularizers of Science*, Clerke and Bodington saw the relevance of Somerville's emphasis on synthesis in their own era of greater specialization and aimed to fill this gap (461).

Science writing for children

Writing for children offered both male and female writers the most creative freedom in choosing useful styles of delivery in which to teach science, from Jane Marcet's didactic, familiar format to Mary Ward's conversational guides or from Margaret Gatty's fable-like stories to Arabella Buckley's lessons framed in fairy tales. In the early part of the nineteenth century, women drew upon the established authority of mothers to provide moral instructions to children, a role found in revolutionary texts of the genre, such as Anna Laetitia Barbauld's *Lessons for Children* (1778, 1779) and *Hymns in Prose for Children* (1781), books that finally considered the physical size and ability of a young child reader. Mitzi Myers identifies the late eighteenth century as the key period marking the establishment of children's books as a distinct genre (31). In the early nineteenth century, a child's moral education was likewise consistent with scientific training found in the tradition of Paleyian natural theology. Turning from this earlier maternal tradition, Lightman argues, women science writers in the middle of the century could redefine the source of their authority and their narrative voice, selecting different textual strategies (2007 97).

At the turn of the nineteenth century, a considerable number of writers engaged in writing texts providing 'useful' knowledge to children, including *Essays on Practical Education* (1798) by Maria Edgeworth and her father, Richard Lovell Edgeworth. The volumes of their treatise contained a host of 'conversations' on chemistry, natural philosophy, mineralogy and astronomy. A conversational delivery, the Edgeworths thought, resembled the manner in which children learnt and was more interesting than expository prose alone (Myers G. 175). The conversation format was common to various kinds of didactic juvenile literature, and in its scientific use, it was most famously employed by Jane Marcet in her *Conversations on Chemistry* and *Conversations on Natural Philosophy*, usually consisting of lessons given by a teacher to two students who could ask questions and converse about the topic. In form, these conversations resemble the question and answer format of catechism. Inspired, too, by the likes of Newbery's 'Tom Telescope', many of these texts included spiritual or religious digressions from the science lesson, so that, as a whole, this genre of didactic literature closely associated science and religion. (For a brief discussion of the close relationship between spiritual and scientific instruction, see Rauch 1989).

Margaret Gatty, whose *British Sea-Weeds* is discussed above, authored one of the most popular series of natural history books for children in the second half of the nineteenth century: *Parables From Nature* (Gates 1998 50). Penning five series of *Parables* between 1855 and 1871 for a total of thirty-seven stories, Gatty guided her juvenile readers – and their adult mentors as well – towards an understanding of plants and animals within a divinely ordered environment. Gatty devotes considerable narrative space to sermons on the merits of good behaviour alongside introductory natural history lessons. Her narrators admonish children's tendencies towards temper tantrums in stories like 'Kicking' – a tale about the taming of a young colt – and commend the proper behaviour of young women in an allegory of the destruction of flowers by a violent wind in 'Training and Restraining'. In this regard, Gatty's *Parables* demonstrate a decidedly conservative, rather than radical, attitude consistent with her ideals for science. Her third series, appearing in 1861, arrived in the midst of evolutionary debates. Gatty herself stood firmly against the secular implications of Darwinian notions of natural selection, and her strongest condemnations of evolutionary theory appear in 'Whereunto' and 'Inferior Animals'. The latter tale likely parodies the British Association for the Advancement of Science (BAAS), as a human narrator overhears the cacophonous cries of a parliament of rooks and imagines their debates. At the same time, Gatty's notes – appended to later editions of the *Parables* – describe the stories' scientific elements in expository prose, including plants' Latin nomenclature and references to contemporary experts. Combined with their technical notes, her tales offer both moral and technical lessons for the interested reader.

We can see Gatty's parables at work in promoting Britain's global reach in books by Mary (1817–93) and Elizabeth (1823–73) Kirby, two sisters who wrote books for young children about both natural history and the goods and technologies that Britain brought home from across its empire. *Things in the Forest* (1861) describes birds and mammals of tropical rainforests for very young children, while *Aunt Martha's Corner Cupboard* (1875) devotes different chapters to the cultivation and production of imported goods familiar to British families, including tea, china, coffee and sugar.[4] The Kirby sisters clearly take their cue from Gatty, turning butterfly metamorphosis into an allegory of resurrection in *Caterpillars, Butterflies, and Moths* (1861), just as Gatty had in her story 'A Lesson of Faith' (1855). The generic manipulations here are important, showing how forms of women's science writing themselves metamorphosed across the nineteenth century, incorporating and responding to Victorian Britain's key cultural trends, from religious belief and scepticism to imperial conceit.

Mary King Ward (1827–69) again strategically recycles genre in her two highly successful volumes for children, *A World of Wonders Revealed by the Microscope: A Book for Young Students* (1858) and *Telescope Teachings* (1859). In *A World of Wonders*, Ward adopts the epistolary style to introduce young readers to the various and minute details of nature. Illustrated in coloured plates throughout, *A World of Wonders* adapts the model of Robert Hooke's *Micrographia* (1665) for young readers, leading them through microscopic observations of insect wings and scales, the scales of fish, animal hairs, birds' feathers and even the lenses of fish eyes and insects' compound eyes, before moving on to examine flower petals, pollen grains and fern seeds. Like many of her female contemporaries, Ward too drew inspiration from a religious viewpoint, incorporating literary quotations ranging from the Bible to John Milton, all reinforcing the notion that a creator 'manifested His glory alike in things great and small' (1858 53). In *Telescope Teachings*, Ward instructs her readers in astronomy's stellar and lunar objects. The book contains fifteen vividly coloured plates illustrating the sun, lunar eclipses, maps of the moon's surface, planets of the solar system, comets and shooting stars, and drawings of fixed stars and nebulae. Ward's stated objective is to present astronomy 'to the young as an entertaining study' (1859 v) and to delight young readers with this 'sublime study' using 'simple words'. In so doing, she turns to yet another genre, lyric poetry, including for example Felicia Hemans's 'The Sunbeam' in the conclusion to her chapter on the sun (Ward 1859 38).

Perhaps the most striking example of applying literary genres to the teaching of science can be found in the work of Arabella Buckley (1840–1929). Managing geologist Charles Lyell's correspondence for eleven years, Buckley became well acquainted with the world of British science. Following Lyell's death in 1875, Buckley sought work as a popular science writer and lecturer. She published her first book, *A Short History of Natural Science*, in 1876, and she then edited the 1877 edition of Mary Somerville's *On the Connexion of the Physical Sciences*. Buckley aligned her scientific knowledge with her spiritualist beliefs, writing 'The Soul, and the Theory of Evolution', an anonymously published essay in the *University Magazine* in 1879 (Lightman 2007 244), the same year she wrote *The Fairy-Land of Science*, a set of lectures on physical science for children. An intangible spirit, power, or life force – as it appears in various iterations of her text – animates this series of lectures, even as she conveys the most up-to-date theories of physics as possible. Buckley stresses the immaterial element of physical sciences, invisible forces at work. Thus fairy tales offer the seeming magic of a world populated and influenced by fairies. Buckley's subsequent popularizations emphasized the parental fostering of offspring, not just the mechanisms of natural and sexual selection. Her book *Life and Her Children* (1881) reveals such a stance even in its title, a book that discusses six divisions of animal life, while *Winners in Life's Race* (1883) discusses vertebrates. In her correspondence with Darwin, Buckley kept her religious and spiritual beliefs to herself; she was more akin to Charles Lyell, who remained a Unitarian and accepted evolution as a mode of Providence, and she became friends with Alfred Russel Wallace, with whom she investigated spiritualism (Lightman 2007 242–3). Buckley was also a vocal proponent of teaching conceptual knowledge over examination skills, an argument she makes explicit in an article published in *The Sheffield and Rotherham Independent* on 30 January 1882, entitled 'Science in Elementary Schools'.

Margaret Gatty and Arabella Buckley share a similar understanding of how stories offer useful ways of entering scientific inquiry and naturalist study, and that such literary forms can help readers recall previous experience and awaken a receptive, enthusiastic frame of mind. They differ from their predecessors such as Jane Marcet and contemporaries such as Mary Ward in selecting these fanciful conceits to convey science. What the *Parables* and *Fairy-Land* share is the sense of joy in discovery, the sense that a student of natural history or physical science ought to be fully engaged in and excited by the inquiry at hand and that she should be welcomed into the study via creative means, asked questions and given the conceptual tools to apply in her everyday activities.

Beyond popularization

As science pervaded everyday life in nineteenth-century Britain, its audience grew beyond readers looking for popularizations or accommodations to include readers in the literary market as well. Novelists such as Mary Shelley, Charles Dickens, George Eliot, Elizabeth Gaskell, Robert Louis Stevenson and Thomas Hardy famously used scientific tropes productively in their works, while poets including Alfred Tennyson, Emily Pfeiffer, A. Mary F. Robinson, May Kendall, Mathilde Blind and Constance Naden often challenged the implications of science and technological progress in their verse.

Mary Shelley (1797–1851) figures predominantly in any list of Romantic- and Victorian-era women who used the space of a novel to interrogate and comment upon scientific inquiry and knowledge. If *Frankenstein* (1818) famously denounces its eponymous protagonist's scientific overreaching, *The Last Man* (1826) suggests science's impotence in an apocalyptic, twenty-first century future. Though Shelley's novels offer a predominantly cynical reading of science's ambitions and effects, the fiction of George Eliot lacks such outright scepticism as it considers science's methods in the context of Victorian realism and adopts discourses of Darwinian biology. Having read widely in philosophy and science – including Mary Somerville's *Connexion* – well before she became a novelist, Mary Ann Evans (1819–80) practices techniques of observation closely associated with natural history. Questions of knowledge and observation pervade Eliot's fiction, from *The Lifted Veil* (1859) to *Adam Bede* (1859) to *Middlemarch* (1871–2), just as *The Mill on the Floss* (1860) and *Daniel Deronda* (1876) also reflect on inheritance and descent. Elizabeth Gaskell (1810–65), too, makes scientific observation a focal point of her fiction, especially in the character of Roger Hamley in *Wives and Daughters* (1865), a scientific explorer evoking Charles Darwin or Alfred Russel Wallace. As Molly Gibson secretly falls in love with Roger, *Wives and Daughters* juxtaposes her own quiet observations of the Hamley family, including her knowledge of Osborne's secret marriage, with the false accusations of Hollingford's gossips (Gaskell 464 (Chapter 40)), who believe that Molly has formed an attachment to Mr Preston, a juxtaposition that adds a dimension of morality to the acts of observation and deduction.

During the same decades that women began attending Oxbridge, they also entered the various 'redbrick' civic institutions, located in Britain's most industrial cities, which frequently emphasized scientific and engineering studies. Constance Naden (1858–89), a poet, science student, philosopher and advocate for women's rights and charities, entered one such civic institution, Mason Science College in Birmingham (later the University of Birmingham) in 1881, one year after T. H. Huxley had given his famous address on science and culture at the college's opening. Here she excelled in science, earning first-class certificates in botany, organic chemistry, physiology, geology and physics (Holmes J. 2009 190). Previously, she attended botany and field classes at the Birmingham and Midland Institute (Hughes 17), where she composed the poems comprising the first of her two published volumes of poetry, *Songs and Sonnets of Springtime* (1881). Her second volume, *A Modern Apostle; The Elixir of Life; The Story of Clarice; and Other Poems*, appeared in 1887, containing a series of four poems under the heading 'Evolutional Erotics' that humorously take aim at the sexual politics implicit in Darwin's *Descent of Man*.

In taking contemporary science – especially evolutionary theory – as subject matter for poetry, Naden was not alone among her contemporaries. In *The Ascent of Man* (1888), Mathilde Blind (1841–96), for instance, sought a communion between the individual lyric voice and the community of living organisms united within Darwinian evolutionary theory, as Jason Rudy has argued in 'Rapturous Forms: Mathilde Blind's Darwinian Poetics' (2006). In Blind's poetics, her regular metre collides with isolated lyric 'spasms' (Rudy 444). A. Mary F. Robinson (1857–1944) offers a contemplative response to evolutionary theory in 'Darwinism' (1888), a poem written in regular quatrains of iambic tetrameter, and suggesting that evolution came in response to an 'unborn and aching thought', an unrest in the human soul (Leighton and Reynolds 547). In 'Lay of the Trilobite' (1885), May Kendall (1861–1943) imagines a fossil's rebuke of human hypocrisy in a wry but jaunty common metre. Her poem cautions against taking humanity's seemingly esteemed place in evolution too seriously, as John Holmes has argued (Holmes J. 2010). Emily Pfeiffer's (1827–90) poem 'Evolution' (1880) suggested that hunger – an insatiable longing – rather than an instinct for reproduction drove the changes in species over time, not unlike the 'unrest' in A. Mary F. Robinson's poem. Poetry by Naden, Blind, Robinson, Kendall and Pfeiffer remains important, both as historical instances of Darwinian evolution's reception and, as Holmes has argued in *Darwin's Bards*, as examples that reveal how little our own thinking about Darwinian biology has changed since then.

Conclusion

To discuss women's contributions to nineteenth-century science is to highlight a number of the contexts described in other chapters of this volume – not only the various disciplines of natural science themselves but also the varied approaches to writing for children, the relationship between science and religion and especially the genres of scientific writing. All of these related conversations implicitly bear witness to the ways in which women were largely marginalized from formal scientific study for much of the century and how they nevertheless found their voices within various scientific communities. Examining women and science requires revisiting a fact scholars in the field of literature and science have long recognized: the distinction we perceive today between these two disciplinary discourses would largely have been a specious division to many nineteenth-century practitioners and popularizers of science. Just as scientists employed allusions, rhetorical figures and theological or philosophical argumentative style to persuade their audiences to understand and believe new theories or discoveries, literary writers drew upon scientific knowledge or phenomena to consider questions of knowledge, perspective and objectivity. Yet nineteenth-century women practitioners and popularizers of science occupy a special case in this story of literature and science. The era's gendered expectations for women, including unsound assumptions among their male colleagues about the gentler sex's inferior aptitude and fitness for advanced education, meant that women's outlets for scientific inquiry and communication likewise became circumscribed by these gendered assumptions. Histories of nineteenth-century literature and science are complicated by women's contributions to science, not least because of these varied genres in which women communicated their work, a facet of science's literary history that merits wider appreciation among readers today.

Notes

1   Girton and Newnham Colleges at Cambridge were founded in 1869 and 1871, respectively; by 1874 women could sit for the degree exams. At Oxford, Lady Margaret Hall was founded in 1878, the same year that University of London began admitting women; Somerville College at Oxford was founded in 1879. Oxford began granting women degrees in 1921; Cambridge not until 1948. For a Victorian woman's perspective of women's education, see Davies (the author was a co-founder of Girton College, Cambridge). For a history of women's higher education in Britain, see, for example, Dyhouse. Also, Sheffield presents a history of British women's scientific education, including the inequality of resources and science equipment in girls' and boys' schools.

2   See Debbie Bark, 'Science for Children', in this volume.

3   Her early papers comprise one book of James Secord's nine-volume *Collected Works of Mary Somerville* (2004).

4   The Kirby sisters' works *Things in the Forest* and *Aunt Martha's Corner Cupboard* can be found in volume 5 of Bernard Lightman's series on nineteenth-century science writing, *Science Writing by Women* (2004).

Bibliography

Primary texts

Bowdich Lee, Sarah. *The Fresh-Water Fishes of Great Britain*. London: [no publisher], 1828.
Buckley, Arabella B. *Life and Her Children. Glimpses of Animal Life, From the Amœba to the Insects*. London: Edward Stanford, 1880.

Buckley, Arabella B. 'Science in Elementary Schools'. *The Sheffield and Rotherham Independent*, 30 January 1882.

Buckley, Arabella B. *The Fairy-Land of Science* [1879]. Ed. Barbara T. Gates. Bristol: Thoemmes, 2003.

Clerke, Agnes M. *A Popular History of Astronomy During the Nineteenth Century*. Edinburgh: Adam and Charles Black, 1885.

Clerke, Agnes M. *The Herschels and Modern Astronomy*. London: Cassell, 1895.

Clerke, Agnes M. *Problems in Astrophysics*. London: A. and C. Black, 1903.

Clerke, Agnes M. *Modern Cosmogonies*. London: A. and C. Black, 1905.

Clerke, Agnes M. *The System of the Stars* [1890]. London: Adam and Charles Black, 1905.

Clerke, Agnes M., A. Fowler, and J. Ellard Gore. *Astronomy*. New York, NY: D. Appleton, 1898.

Daniell, Madeline. 'Memoir'. In Constance Naden, *Induction and Deduction: A Historical and Critical Sketch of Successive Philosophical Conceptions Respecting the Relations Between Inductive and Deductive Thought and Other Essays*. Ed. Robert Lewins. London: Bickers, 1890. vii–xviii.

Dyce, William. *Pegwell Bay, Kent – A Recollection of October 5th 1858*. Oil paint on canvas. Tate Britain, London.

Edgeworth, Maria, and Richard Lovell Edgeworth. *Practical Education*. London: J. Johnson, 1798.

Fitton, Sarah Mary, and Elizabeth Fitton. *Conversations on Botany* [1817]. London: Longman, Orme, Brown, Green, and Longmans, 1840.

Gaskell, Elizabeth. *Wives and Daughters*. Ed. Angus Easson. Oxford: Oxford University Press, 2003.

Gatty, Margaret. *British Sea-Weeds: Drawn From Professor Harvey's "Phycologia Britannica". With Descriptions, an Amateur's Synopsis, Rules for Laying out Sea-Weeds, an Order for Arranging Them in the Herbarium, and an Appendix of New Species*. London: Bell and Daldy, 1863.

Gatty, Margaret. *Parables From Nature* [1888]. Facsim. edn. New York, NY: Garland, 1976.

Hughes, William R. *Constance Naden: A Memoir*. London: Bickers, 1890.

Kirby, Mary, and Elizabeth Kirby. *Things in the Forest; Sketches of Insect Life; and Aunt Martha's Corner Cupboard*. Repr. in vol. 5 of *Science Writing by Women*. Ed. Bernard Lightman. 7 vols. Bristol: Thoemmes, 2004.

Lankester, Phebe. *A Plain and Easy Account of the British Ferns: Together with Their Classification, Arrangement of Genera, Structure, and Functions; and a Glossary of Technical and Other Terms*. London: Hardwicke, 1860.

Loudon, Jane Webb. *The Ladies' Companion to the Flower Garden. Being an Alphabetical Arrangement of All the Ornamental Plants Usually Grown in Gardens and Shrubberies*. London: W. Smith, 1841.

Loudon, Jane Webb. *Botany for Ladies; or, A Popular Introduction to the Natural System of Plants, According to the Classification of De Candolle*. London: John Murray, 1842.

Loudon, Jane Webb. *The Mummy! A Tale of the Twenty-Second Century*. Ed. Alan Rauch. Ann Arbor, MI: University of Michigan Press, [1827] 1994.

Marcet, Jane. *Conversations on Chemistry in Which the Elements of That Science are Familiarly Explained* [1806]. Philadelphia, PA: M. Carey, 1818.

Marcet, Jane. *Bertha's Visit to Her Uncle in England*. 3 vols. London: John Murray, 1830.

Marcet, Jane. *Conversations on Natural Philosophy, in Which the Elements of That Science are Familiarly Explained and Illustrated by Experiments*. London: Hurst, Rees, and Orme, 1806.

Naden, Constance. *Songs and Sonnets of Springtime*. London: C. Kegan Paul, 1881.

Naden, Constance. *A Modern Apostle; The Elixir of Life; The Story of Clarice; and Other Poems*. London: Kegan Paul, Trench, 1887.

Paley, William, Henry Brougham, and Charles Bell. *Paley's Natural Theology*. 2 vols. London: C. Knight, 1836.

Pratt, Anne. *The Flowering Plants and Ferns of Great Britain*. 5 vols. London: Society for Promoting Christian Knowledge, 1855.

Smith, Charlotte. *Rural Walks: In Dialogues Intended for the Use of Young Persons*. 2 vols. London: Cadell and Davies, 1795.

Somerville, Mary. *On the Connexion of the Physical Sciences*. London: J. Murray, 1834.

Trimmer, Sarah. *An Easy Introduction to the Knowledge of Nature and Reading the Holy Scriptures. Adapted to the Capacities of Children*. London: Longman, Robinson, and Johnson, 1781.

Twining, Elizabeth. *Illustrations of the Natural Orders of Plants with Groups and Descriptions* [1849–55]. London: Sampson Low, Son, and Marston, 1868.

Wakefield, Priscilla. *An Introduction to Botany, in a Series of Familiar Letters* [1796]. Boston, MA: J. Belcher and J. W. Burditt, 1811.

Wakefield, Priscilla. *Mental Improvement, or the Beauties and Wonders of Nature and Art* [1794–7]. Ed. Ann B. Shteir. East Lansing, MI: Colleagues, 1995.

Ward, Mary. *A World of Wonders Revealed by the Microscope. A Book for Young Students*. London: Groombridge, 1858.

Ward, Mary. *Telescope Teachings: A Familiar Sketch of Astronomical Discovery Combining a Special Notice of Objects Coming Within the Range of Objects of a Small Telescope*. London: Groombridge, 1859.

Wollstonecraft, Mary. *A Vindication of the Rights of Woman: With Strictures on Political and Moral Subjects*. 2nd edn. London: J. Johnson, 1794.

Zornlin, Rosina. *Recreations in Physical Geography; or, The Earth as It Is*. London: John W. Parker, 1840.

Zornlin, Rosina. *The World of Waters; or, Recreations in Hydrology; What is a Voltaic Battery?; 'Observations of Shooting Stars on the Nights of the 10th and 11th of August, 1839'; 'On the Periodical Shooting Stars and on Shooting Stars in General'*. Repr. in vol. 4 of *Science Writing by Women*. Ed. Bernard Lightman. 7 vols. Bristol: Thoemmes, 2004.

Secondary texts

Allen, David Elliston. *The Naturalist in Britain: A Social History*. London: A. Lane, 1976.

Burek, Cynthia V., and Bettie Higgs, eds. *The Role of Women in the History of Geology*. London: Geological Society, 2007.

Creese, Mary R. S. *Ladies in the Laboratory? American and British Women in Science, 1800–1900: A Survey of Their Contributions to Research*. Lanham: Scarecrow, 1998.

Davies, Emily. *The Higher Education of Women*. London: Hambledon, 1988.

Dyhouse, Carol. *No Distinction of Sex?: Women in British Universities, 1870–1939*. London: UCL Press, 1995.

Fara, Patricia. *Pandora's Breeches: Women, Science and Power in the Enlightenment*. London: Pimlico, 2004.

Gates, Barbara T. 'Revisioning Darwin With Sympathy: Arabella Buckley'. *Natural Eloquence: Women Reinscribe Science*. Eds. Barbara T. Gates and Ann B. Shteir. Madison, WI: University of Wisconsin Press, 1997, 164–76.

Gates, Barbara T. *Kindred Nature: Victorian and Edwardian Women Embrace the Living World*. Chicago, IL: University of Chicago Press, 1998, 9.

Gates, Barbara T. *In Nature's Name: An Anthology of Women's Writing and Illustration, 1780–1930*. Chicago, IL: University of Chicago Press, 2002.

Gates, Barbara T., and Ann B. Shteir, eds. *Natural Eloquence: Women Reinscribe Science*. Madison, WI: University of Wisconsin Press, 1997.

George, Sam. *Botany, Sexuality, and Women's Writing, 1760–1830*. Manchester: Manchester University Press, 2007.

Holmes, John. *Darwin's Bards: British and American Poetry in the Age of Evolution*. Edinburgh: Edinburgh University Press, 2009.

Holmes, John. '"The Lay of the Trilobite": Rereading May Kendall'. *19: Interdisciplinary Studies in the Long Nineteenth Century* 11 (2010). <www.19.bbk.ac.uk/articles/10.16995/ntn.575>. Accessed 8 September 2014.

Leighton, Angela, and Margaret Reynolds, eds. *Victorian Women Poets: An Anthology*. Oxford: Blackwell, 1995.

Lightman, Bernard, ed. *Science Writing by Women*. 7 vols. Bristol: Thoemmes, 2004.

Lightman, Bernard, ed. *Victorian Popularizers of Science*. Chicago, IL: University of Chicago Press, 2007.

Myers, Greg. 'Science for Women and Children: The Dialogue of Popular Science in the Nineteenth Century'. *Nature Transfigured: Science and Literature, 1700–1900*. Eds John Christie and Sally Shuttleworth. Manchester: Manchester University Press, 1989. 171–200.

Myers, Mitzi. 'Impeccable Governesses, Rational Dames, and Moral Mothers: Mary Wollstonecraft and the Female Tradition in Georgian Children's Books'. *Children's Literature* 14 (1986): 31–59.

Patterson, Elizabeth Chambers. *Mary Somerville and the Cultivation of Science, 1815–1840*. Boston, MA: Martinus Nijhoff, 1983.

Rauch, Alan. 'A World of Faith on a Foundation of Science: Science and Religion in British Children's Literature: 1761–1878'. *Children's Literature Association Quarterly* 14.1 (1989): 13–19.

Rudy, Jason R. 'Rapturous Forms: Mathilde Blind's Darwinian Poetics'. *Victorian Literature and Culture* 34.2 (2006): 443–59.

Schiebinger, Londa. *The Mind Has No Sex? Women in the Origins of Modern Science*. Cambridge, MA: Harvard University Press, 1989.

Schiebinger, Londa. *Nature's Body: Gender in the Making of Modern Science*. New Brunswick, NJ: Rutgers University Press, 1993.

Schiebinger, Londa. 'The Philosopher's Beard: Women and Gender in Science'. *Eighteenth-Century Science*. Ed. Roy Porter. Cambridge: Cambridge University Press, 2003. 184–210.

Secord, James A. 'Introduction'. *Collected Works of Mary Somerville*. Ed. James Secord. 9 vols. Bristol: Thoemmes, 2004.

Secord, James A., ed. *Collected Works of Mary Somerville*. 9 vols. Bristol: Thoemmes, 2004.

Secord, James A. 'Introduction'. Rosina Zornlin, *The World of Waters; or, Recreations in Hydrology; What is a Voltaic Battery?; 'Observations of Shooting Stars on the Nights of the 10th and 11th of August, 1839'; 'On the Periodical Shooting Stars and on Shooting Stars in General'*. Repr. in vol. 4 of *Science Writing by Women*. Ed. Bernard Lightman. 7 vols. Bristol: Thoemmes, 2004.

Sheffield, Suzanne Le-May. *Women in Science: Social Impact and Interaction*. Santa Barbara: ABC-CLIO, 2004.

Shteir, Ann B. *Cultivating Women, Cultivating Science: Flora's Daughters and Botany in England, 1760–1860*. Baltimore, MD: Johns Hopkins University Press, 1996.

Secord, James A. 'Finding Phebe: A Literary History of Women's Science Writing'. *Women and Literary History: 'For There She Was'*. Eds. Katherine Binhammer and Jeanne Wood. Newark: University of Delaware Press, 2003, 152–66.

Torrens, H. S. 'Mary Anning'. *Oxford Dictionary of National Biography*. Oxford: Oxford University Press, 2004.

Torrens, H. S., and Janet Browne. 'Sarah Mary Fitton'. *Oxford Dictionary of National Biography*. Oxford: Oxford University Press, 2004.

Whittingham, Sarah. *Fem Fever: The Story of Pteridomania*. London: Frances Lincoln, 2012.

Further reading

Boswell, Michelle. 'Poetry and Parallax in Mary Somerville's *On the Connexion of the Physical Sciences*'. *Victorian Literature and Culture* 45.4 (2017).

Fyfe, Aileen, and Bernard V. Lightman, eds. *Science in the Marketplace: Nineteenth-Century Sites and Experiences*. Chicago, IL: University of Chicago Press, 2007.

Jacobus, Mary, Evelyn Fox Keller, and Sally Shuttleworth. *Body/Politics: Women and the Discourses of Science*. New York, NY: Routledge, 1990.

Jordanova, Ludmilla. *Sexual Visions: Images of Gender in Science and Medicine Between the Eighteenth and Twentieth Centuries*. Madison, WI: University of Wisconsin Press, 1989.

Merchant, Carolyn. *The Death of Nature: Women, Ecology, and the Scientific Revolution*. San Francisco, CA: Harper and Row, 1980.

Murphy, Patricia. *In Science's Shadow: Literary Constructions of Late Victorian Women*. Columbia, MO: University of Missouri Press, 2006.

Ogilvie, Marilyn Bailey, and Joy Dorothy Harvey, eds. *The Biographical Dictionary of Women in Science: Pioneering Lives from Ancient Times to the Mid-20th Century*. New York, NY: Routledge, 2000.

II

Genres

II

Genres

5

THE NOVEL AS OBSERVATION AND EXPERIMENT

Charlotte Sleigh
University of Kent

The realist novel is well established as a significant category of Victorian literature, and its name links it to a plethora of connected philosophical and natural philosophical (that is, scientific) debates stretching back into the eighteenth century and beyond. What things are real, and how can we know about them? Such are the questions to which science falls back in its methods of observation and experiment.

G. H. Lewes, the philosopher, critic, and partner of George Eliot, answers these questions in his account of an unexpected encounter. 'I am seated in my study,' he begins, 'and, on raising my head from a book, see a man slowly pass out of the room, cross the lawn, and seat himself on the garden wall' (Lewes 1904 40). His wife sees the man on the wall; his dog barks at him furiously. Finally, Lewes writes, 'I order my servant to fetch a policeman, and the policeman . . . carries off the struggling intruder'. Lewes's concern is not that the intruder might have trampled the flowers, or stolen the silverware, but a philosophical-cum-scientific one: was the intruder real? It is laughable, suggests Lewes, to suppose anything other than the intruder's reality. The corroboration of a wife, a dog and a policeman forces the realist belief of any sensible person, 'unless we declare all existence to be a dream'. Lewes's bourgeois tale relies upon a division of labour; the scholar in his study is relieved of certain foundational steps in logic by an animal, a woman, a servant, an arm of the state. The last of these is deputed to take the physical blows that might once have yielded empirical evidence for a natural philosopher. The fact that the story is – one presumes – fiction does not in the least hinder its teller's grip on reality.

In this episode, Lewes asserts a common-sense view of a material world, such as might underpin a faith in science – and yet uses fiction to do so. As this chapter will demonstrate, the Victorian novel may be understood as a space in which contemporary scientific claims were rehearsed and trialled. They were a space in which observation could be pursued and recorded and in which experiments could be essayed. For a full sense of this space of representation and intervention, one would wish to look beyond the text and into the full communication circuit of acquisition, reading and criticism. Only by looking beyond the text can one appreciate how it could itself be an experimental intervention. However, as these aspects of literature are covered elsewhere in this volume, the chapter is restricted mostly to the texts themselves.

The historian and critic David Masson, in a resonant phrase of 1859, suggested that the authors of novels were creating experimental systems that mimicked the real world:

[T]he novelist [is] . . . the creator of his mimic world . . . he makes the laws that govern it; he conducts the lines of events to their issue; he winds up all according to his judicial wisdom. It is possible, then, to see how far his laws of moral government are in accordance with those that rule the real course of things, and so, on the one hand, how deeply, and with what accuracy he has studied life, and, on the other, whether, after his study, he is a loyal member of the human commonwealth, or a rebel, or a cynic, or a son of the wilderness. In short, the measure of the value of any work of fiction, ultimately and on the whole, is the worth of the speculation, the philosophy, on which it rests, and which has entered into the conception of it.

(23)

Although there is a moral desideratum implicit in Masson's comments, there is also a natural one. There is faith that 'moral government' will be in accordance with the laws of nature. There is an obligation to study life carefully, and an assumption that doing so will place the author on the side of right. This conflation of the moral and the natural was arguably the outcome of the great early-century debates about knowing reality. During the politically unstable 1830s, science was a potential way of making sense of the world, of culture and government, and thereby controlling it – as James A. Secord has most recently described in his *Visions of Science*. Defining science was not just an interesting diversion but a political imperative. George Combe's *Constitution of Man* (1828), for example, attempted to establish science as a democratic art that would up-end inherited ranks of privilege.

The politics of epistemology (ways of knowing) during the 1830s included a critical questioning of imagination and its value. In the writings of Charles Lyell, a geologist and major scientific figure of the early Victorian period, the word 'imagination' is overwhelmingly used to indicate a source of error. It denoted primitive superstition, delusion and conversely (though more rarely) something whose limitations gave 'liability to error' (Lyell 66). For Lyell, the imagination had to be 'restored' (197) through the sober action of observation and inductive reasoning. William Whewell, who played a decisive role in disciplining visions of science, had a similarly dim view of the 'incredible and monstrous' (1837 1:180) scientific progeny of imagination. Whewell himself subscribed to a philosophy of 'inductivism', the notion that knowledge of nature was built up from repeated observations, gathered together, until laws could be inferred. That such laws existed was guaranteed by the existence of God, and their human intelligibility by the God-created harmonies between human intellect and the physical world (Chalmers). Despite the allegiance pledged to inductivism in his *History* and *Philosophy of the Inductive Sciences* (Whewell 1837; Whewell 1840), there was a strong idealist overtone to his science (Sleigh 2010B), a God-given dimension to the laws of nature that could not be inferred. By appeal to God, Whewell defused the danger that politically dangerous persons such as Combe might make observations, and that there was otherwise no way of saying whether their inductions were more valid than those of the clerisy. Whewell and his fellow men of science were caught up in a struggle for identity that sought, on the one hand, to take advantage of the allied interests in the useful arts of the new industrial elites, while maintaining on the other a safe, Anglican science that would not lend itself to revolutionary politics (Cannon; Inkster and Morrell; Morrell and Thackray; Yeo 1993).

This same period of instability was noted as a time in which the novel bloomed in number and quality. Masson calculated that about 3,000 new novels were published from the 1820s to the end of the 1850s (213). In quantity, if nothing else, he assessed his century as being the century of the novel. Masson's conflation of law as created order and moral imperative bears the imprint of Whewell's, or more likely Chalmers's, thought (Masson studied under Chalmers).

In reading novels of the 1820s–1840s, it is important to bear in mind that their status as a dominant form of literature was not yet settled, any more than expert, metropolitan science was settled as ascendant over industrial forms of knowledge (as it would eventually come to be). The novel was an intervention into an active debate – whose participants included scholars, industrialists, Romantics and radicals – about authoritative forms of knowledge and the politics of imagination, and demands to be read as such.

One text that encompasses the fluid forms of science and fiction in the earlier part of the century is Harriet Martineau's *How to Observe* (1838). This book constitutes advice to the traveller, telling her or him how to apply inductive science to the understanding of many areas of life, from soil to superstition and from popular song to technology. Martineau herself was on a journey into fiction at the time she wrote her guidance for travellers: her first novel, *Deerbrook*, was published the following year (1839). By making the right sort of observations, Martineau counsels, the reader will be equipped to reach legitimate conclusions. Martineau steers her readers between two possible sources of error. On the one hand, there is the error of over-confidence – that is, of too-easy generalization. Martineau is acutely aware that observation is in danger of revealing 'more of the mind of the observer than of the observed' (4). At the opposite extreme lies the temptation of false modesty – the unwillingness to risk induction. To avoid the latter of these sins one need only begin by observing in detail, and to mitigate the former one must have a panoramic picture of the whole in mind:

> [I]nstead of 'one of its faces', the moralist would see the whole of the earth in one contemplation; and . . . instead of a nebulous expanse here, and a brown or grey speck there, – continents, seas, or volcanoes, – he would look into the homes and social assemblies of all lands.
>
> *(Martineau 18)*

Geology – the continents, seas, and volcanoes of this passage – is a constant touchstone and comparison for Martineau in her account of human observation. One might say that the principles of geology are extended in this text to the understanding of humans. The moral sciences have become inductive. Indeed, in one particularly suggestive section, Martineau writes that a geological description of place directly relates to the people who live there, shaped in body and mind by their landscape (144–58; cf. Buckland). Such a reflection anticipates the novels of Thomas Hardy later in the century. As such, *How to Observe* is something like a primer for Martineau's mode of writing: the painstaking observation of life with a sympathy 'untrammelled and unreserved' (41). Later in this chapter, Eliot will be explored as an exemplar of this model.

History and heroes

Notwithstanding the development of inductive methods of novel writing, other traditions of creating and thinking about narrative literature flourished through the nineteenth century (Maitzen 2009) along with the kind of realism for which one can regard Martineau as a pioneer. Not least, one must remember that the practitioners of inductive writing and realism later in the century were raised on a diet of romanticism, and that the writing of these earlier authors remained very much alive in the later century. These traditions were rivers that divided and recombined many times over, crossing and re-crossing the fields of 'science' and 'literature'. Historicism and romanticism were two main flows; heroes and philology among the many streams one could identify for more specific study.

Martin Rudwick's magisterial histories of geology in the late eighteenth and early nineteenth centuries (Rudwick 2005; Rudwick 2008) are important for scholars of literature and science because of the way in which they identify literary techniques as a crucial tool in the construction of this nineteenth-century science. Rudwick draws our attention to the ways in which biblical hermeneutics yielded methods that were applied to the natural world, ultimately to great effect. Thus, Bishop Ussher, with his infamous dating of the earth to 4004BC, turns out not to be a fool but an interesting early example of someone attempting to apply concrete methods of calculation to the question of the earth's age. The historicity of the Christian religion (its plotted timeline from creation to last judgement) equipped natural philosophers such as Ussher and his successors to think about the precise chronology of the material world's narrative. Techniques developed in antiquarianism and historiography were applied to the earth to yield the results that Lyell built upon in his geology, so central to the Victorian age. Eliot, for one, began her career in historicist criticism. For her, this amounted to a common-sense rejection of miracles in a former age; it was a uniformitarianism of human history as of the geological eras. However, as Buckland reminds us, 'uniformitarianism' is a term of convenience and not an actor's category.

The river flowed the other way, too. Historians, in turn, began to cite science as the method that directed and justified their own efforts. Henry Thomas Buckle's *History of Civilisation in England* (1857–61) was pitched as a scientific account, modelled on the physical sciences and the statistical laws described by Auguste Comte (Hesketh). Meanwhile, another British school of historians followed the German Leopold von Ranke in an alternative school of 'scientific' history. Where Comteans eschewed a knowledge of deep-level realities, preferring the descriptive level of statistics, the von Rankeans claimed an inductive science that built from facts to deep-level laws. There was yet a third group of historians, as Hesketh explores, which privileged the narrative quality of their accounts: the sometime novelists Charles Kingsley and James Anthony Froude, to name two. The 'scientific' historians engaged in active 'boundary work' to distinguish their work from the realm of the novel. Or, to put it another way, Victorian historiography is a narrative pivot for both science and the novel.

Thus, it is relevant that Masson, who posited the novelist as 'creator of his mimic world' was first and foremost a historiographer. His appointment in 1852 as professor of English language and literature at University College, London, positioned him as an early academic specialist in the field, in which context he published his 1859 assessment of the novel. However, he came to public attention thanks to his historiographical books and essays. Masson's account of the development of the novel, and the qualities that he values within it, are closely related to history. Historians prefer novels to poems, he claims; they like the concreteness of the novel. In Masson's judgement, Walter Scott stood head and shoulders above all other novelists, on account of his historical sensibilities. Scott was, for Masson, defined by his passion for the antique and by his faculty of memory. It was Scott's power of creating characters that best exemplified his historical sensibilities. Here Masson was in agreement with Ruskin and Arnold, who in most other things opposed him. Scott is currently emerging as a key figure in research on literature and science (Buckland 46–55, Secord 78–89, O'Connor 205–6, 301, 441). He was an undisputed hero for Masson and Ruskin and had a central place in constructing Victorian literature through the desiderata of historiography, as well as in his own right.

One of the most interesting recent excavations of the complex relations between historical writing and the novel is Cregan-Reid's *Discovering Gilgamesh*, which like Rudwick's work relates history with deep time but in this case is refracted through the epic. Another nineteenth-century tradition of historical narrative that is relevant to science treated industrial progress and invention. In recent work, Courtney Salvey has convincingly argued that it is vital to understand the historical narrative framework for technology as a backdrop to later biological accounts. Darwin's *Origin* was, contrary to his intentions, received by many as a historical narrative of contrivance – of the upward ladder of the animal kingdom. Understanding the reading-frame of non-biological accounts of historical progress puts this interpretation in a fresh and telling context. Other than this, recent work on historiography as a genre in relation to the novel has frequently been done in relation to gender (e.g. Maitzen 1998 3–32).

One specific connection between historical narrative – Scott's in particular – and science, is the matter of dialect. Philology, sometimes (like theology) dubbed the 'queen of the sciences' (Holquist 77–9), can be considered as a model for nineteenth-century investigation (Turner 2), and it takes explicit form in Ruskin's first essay on 'Fiction, Fair and Foul' (1880–1). In this essay, Ruskin devotes a substantial final section to the dialect employed by Scott (201–8). For Ruskin, dialect is a form of language that is pure and naturally intelligent: moral and clean, not urban and corrupt. He links the question of language to fairy tales and their supposed autochthonic charm and validity. Eliot has a wry take on dialect in 'The Natural History of German Life' (1856), mocking those who 'mistake . . . an unintelligible dialect [for] a guarantee for ingenuousness' (1883 144). Sentiment aside, however, Eliot posits that 'the historical conditions of society may be compared with those of language' (1883 164). Like narratives of technical development, such histories of language as markers of civilization and human nature begin before Darwin. They also continue after him, in anthropology and child psychology as well as in critical studies (Dallas). Philology is a little like historiography in that it weaves between the realms of science and the humanities. Recent scholarship by Abberley, Ferguson and Radick has opened up critical studies of language as a route into the development of literature and science. It gives us, perhaps, a new way to think about realism as an intrinsically historicist account, its 'real' dialogue part of a trend that includes both biology and the novel, as opposed to these latter two being causally responsible for one another.

The uniting of natural history and history through language is also closely related, in Masson's and Ruskin's discussion of Scott, to the emergence of national character. Eliot too argues in her 'Natural History' that the peasant is a 'historical type' (1883 149), a remnant of ancient anthropology. Peasant physiognomy, dialect and popular culture are all inherited. Even names are amenable to natural-historical analysis, just as surely as the flora or fauna of a locale (Eliot 1883 151). Eliot's references to 'instinct' blur the cultural and the natural, since instinct was and is a term from animal as much as human science. History is, in short, 'incarnate' (Eliot 1883 166) in national character. Masson nods to the rightness and desirability of Eliot's frame when he calls for a 'Natural History of British life' (264).

National character in turn is closely related to the figure of the hero (since hardly anyone finds their national character anything other than wholly admirable). Novels generally have heroes, and if their main character does not live up to the qualities of heroism then that too is noteworthy; heroism is the frame within which she or he is judged. Thackeray's 'novel without a hero', *Vanity Fair* (1848), is the exception that proves the rule (Schor 331). In considering nineteenth-century notions of heroism, we are once again returned to history as mode, and to the admiration accorded to Scott for populating his historical romances with admirable and fully-characterized heroes. Thomas Carlyle's *On Heroes, Hero-Worship, and the Heroic in History*, a series of lectures given in 1840 and published the year after, helped theorize such positive critical reactions (Carlyle 2013). Carlyle gives a positive account of his chosen Great Men. Some are fictional (Odin), some counter-cultural (Muhammad) and others deliberately mis-categorized (Cromwell as king). The point is to distil the key qualities that (in Carlyle's view, rightly) evoke respect and adoration. However, this is not a pure exercise in idealism, as one might expect. The qualities are discovered as a set of deeper truths about history, truths which redound as opportunity and self-discovery in the present day. Carlyle's historicized account of the hero, yielding a harvest of opportunity to the ordinary reader, is a complement to the imagined histories produced by novelists.

Heroes may or may not be real, but hero-worship is. Moreover, so far as Carlyle is concerned, the latter is a morally sound and epistemologically trustworthy basis for true religion – religion being not a set of beliefs but 'the thing a man does practically lay to heart' (2013 5). It does not matter so much *what* one believes in so much as the *way* in which one believes – almost a pragmatist philosophy. The heroic pursuit of true religion, in Carlyle's faith, was the same spirit that produced true science:

> So much of truth, only under an ancient obsolete vesture, but the spirit of it still true, do I find in the Paganism of old nations . . . The primary characteristic of this old Northland Mythology I find to be Impersonation of the visible workings of Nature. Earnest simple recognition of the workings of Physical Nature, as a thing wholly miraculous, stupendous and divine. What we now lecture of as Science, they wondered at, and fell down in awe before.
>
> *(2013 21)*

In this passage, Carlyle does not pursue truth in the external world but investigates the virtue of truth-seeking in experience. It is a matter of faith, a postulate beyond investigation, that modern science and primitive religion are focused upon the same entities. The qualities of earnestness and simplicity that Carlyle attributes to mythology are positive ones and would be well carried over into the pursuit of science. The hero is the person who pursues truth in the right way.

It is a small but reasonable step on from all this to say that if Victorian readers could learn to appreciate heroes properly from novels then they would select the best ones in real life, including in the field of science. In this vein, Masson (267–74) gave an epitome of the good novel, recounting the education of the hero in the laws of nature, until he reaches 'a resting-ground of faith'. Readers who attended to Carlyle would not be sold the notion of science (or, rather, nature) as machine, and of science as merely the business of machine-tending (Carlyle 1829). Herbert Spencer derided Carlyle's account, stating that the man was the product of his society – the development of society being, moreover, a biological process. William James proposed a compromise, or perhaps a synthesis, in his lecture 'Great Men and Their Environment', in which he described a reciprocal process of shaping between the two (James). The fact that this debate continued through two generations demonstrates the ongoing significance of the hero figure.

Nationalism is one final theme to consider in relation to science and literature as mediated by history and the hero figure. 'National literature is national speech,' remarked Martineau, 'by this are its prevalent ideas and feelings uttered' (138). From Charles Babbage's demands for heroic medals in science (Babbage 1830) to the almost unprecedented interment of Darwin in Westminster Abbey, British science articulated itself through the literary vehicle of heroism. Of course, the hero is a more ancient construct than the novel, appearing in oral romance, hagiography and myth from far back in history. He (generally *sic*) was reworked during the nineteenth century, with both positive and negative representations. Smiles's heroes of engineering (Smiles 1861–62) received a critical reassessment in, for example, Dickens's *Hard Times* (not to mention Mill's *Autobiography*). Towards the century's end, ardently secularist hagiographies (Draper; White) pugnaciously established the scientist as hero in a manner that went on to provoke extreme reactions, whether restated as the scientist-hero of clichéd biography and fiction or rebutted via the equally clichéd antihero of horror. The hero, embedded in his history, had a long-lived and profound effect upon the nineteenth-century novel.

Reality and truth

Truth, so central to Carlyle's account, has also been co-opted as a key concept for a tradition of criticism that identifies later Victorian fiction, and writing about fiction, as realist. George Levine, author of several important books on realism (Levine 1981; Levine 1988; Levine 2002; and Levine 2008), judges that the search for truth was the great secular virtue common to both science and literature (2008 vi–viii). Examples additional to Levine's are not hard to find. Martineau's *How to Observe* opened by anchoring observation in the search for truth (1). G. H. Lewes shared these secular religious goals, commending the diligent study of natural history as productive of that 'inalienable value attendant upon all truth' (1860 412). Lewes's truth – sometimes italicized, sometimes capitalized (1858 494) – played a quasi-religious role in his otherwise secular outlook. When he discusses good art he reaches for a sacred exemplar, Raphael's *Madonna di San Sisto*. Although naturalistic, it has in its eyes 'an indefinable something . . . perfect *truth*' (Lewes 1858 494). Elsewhere I have described this sacralized goal of truth in the nineteenth century as a 'moral realism' (Sleigh 2010A 130–51).

The identification of realism as a historical movement within literature connects the novel to a political agenda associated with the 1848 uprisings across Europe (Lukàcs). This realism valued an account of real people, the 'us' of reading and writing communities, largely (but not entirely) the middle class (Lewes 1858 492). In Britain there was, of course, no 1848 revolution to focus this movement. Realism as political and aesthetic project evolved slowly, with less teleological impetus, and in a complex and sometimes paradoxical fashion (Teukolsky). Realist literature's valuing of truth is one of these paradoxes, at once a modern way of knowing the world, and a heroic – perhaps even Romantic – pursuit. Still, Eliot's 'Natural History of German Life' starts out by talking about the application of literature, or at least the application of the methods by which literature is made. A better grasp of 'the people' or 'the masses' is required if we are to legislate wisely in regard to them (Eliot 1883 142).

G. H. Lewes's 1858 essay 'Realism and Idealism' has been identified as a key intervention in the development of a realist aesthetic (Levine 2008 31). In it, Lewes argued that the opposition drawn by some critics between art and reality, or between idealism and realism, was a false one. The two were necessarily connected: 'Art always aims at the representation of Reality, i.e. of Truth' (Lewes 1858 493). 'Either give us true peasants, or leave them untouched,' he demanded; 'either keep your people silent or make them speak the idiom of their class.' Lewes's ability to find truth was underwritten by the 'objective method', a method that enabled the possibility of 'verification' (1904 9). It is the poet Samuel Taylor Coleridge that we must thank for the term 'objective', a term that is closely allied to realism and indispensable to later nineteenth-century conceptions of science. Coleridge established the modern English definition of objective qualities (real things inhering in the world) in opposition to subjective ones (those inhering in the self), a definition that was the very opposite of its previous (occasional) usage. The definitions fell on fertile soil; by 1840 the terms could be referred to as too obvious to require defence (Daston and Galison 30, 207; Lewes 1852 1:119). The historians Lorraine Daston and Peter Galison have explored the manifold meanings of 'objectivity' through the eighteenth and nineteenth centuries and conclude that around the mid-Victorian period – Lewes's period – it was being transformed into a new ideal of 'mechanical objectivity'.

Martineau, as it happens, concludes her advice in 1838 with an account of 'mechanical methods' of observation (196–202). Mechanical may seem a curious choice of adjective; by it she means simply the use of a journal (written up at the end of each day) and a notebook (carried around for constant jottings). The next twenty years saw a series of popular crazes in natural history (Allen), which generated further discussion on how to observe the natural world. One such account, concerning seaside life, was prepared by G. H. Lewes, in the midst of his writing on literature (Lewes 1860). Observation, 'cautious and patient' (Lewes 1860 40), was the basis for all knowledge of nature:

> He who would learn the exquisite delights Nature has for those who ardently pursue her, and would acquire a deep sense of reverence and piety in presence of the great and unfathomable mysteries which encompass Life, must give his mornings to laborious searchings on the rocks, his afternoons to patient labour with the Microscope.
>
> *(41)*

Like Martineau, Lewes sees observation as correcting a tendency to subjective speculation. However, to observation Lewes adds experimentation as a necessary process. Much of his seaside book is given over to the description of experiments he has conducted upon the creatures collected from rock pools and the conclusions he has been able to draw from them. Experiment is mediated for Lewes, above all, by the microscope. It enables a deeper knowledge than Martineau's traveller could ever have discovered:

> The observation of one detail is a step to the recognition of many. In this stage we resemble the traveller who has discovered the Alps to have valleys and habitations. If the Microscope be now placed in our hands, it brings us into the very homes and haunts of Life; and finally, the high creative combining faculty, moving amid these novel observations, reveals something of the great drama which is incessantly enacted in every drop of water, on every inch of earth.
>
> *(Lewes 1860 58)*

The microscope provides a qualitatively different way of observing:

> [T]he Microscope is not the mere extension of a faculty, it is a new sense . . . We all begin, where most of us end, with seeing things removed from us – kept distant by ignorance and the still more obscuring screen of familiarity.
>
> *(Lewes 1860 57)*

Eliot uses the qualitatively different reality of nature, revealed under the microscope, to describe Mrs Cadwallader's observations of female-kind:

> [W]hereas under a weak lens you may seem to see a creature exhibiting an active voracity into which other smaller creatures actively play as if they were so many animated tax-pennies, a stronger lens reveals to you certain tiniest hairlets which make vortices for these victims while the swallower waits passively at his receipt of custom.
>
> *(1994 59–60 (Chapter 6))*

What women are apparently doing, and what they are doing when truly scrutinized, may be two different things entirely. The microscope thus, for Lewes and Eliot, de-familiarizes human observation and makes it possible to get closer to the truth. It appears to approach what Daston and Galison have in mind; it is a machine that supposedly registers reality independent of the mind's subjective impositions.

According to Daston and Galison, the development of the mechanical-objective method was symbiotic with a disciplining of the scientific self: a deliberate suppression of the wilful, subjective element. Epistemology and the self-image of the scientist were developed in tandem:

> By *mechanical objectivity* we mean the insistent drive to repress the wilful intervention of the artist-author, and to put in its stead a set of procedures that would, as it were, move nature to the page through a strict protocol, if not automatically.
>
> (Daston and Galison 121)

This suppression of wilfulness recalls Levine's account of 'self-abnegation' in the Victorian era (2002 28), 'a willingness to repress the aspiring, desiring, emotion-ridden self and everything merely personal, contingent, historical, material that might get in the way of acquiring knowledge' (2). This new account of virtue cast a light on heroism that was rather different to Carlyle's:

> A hero must do battle with a worthy foe, and it was themselves whom the heroes of objectivity met on the field of honour. It was precisely because the man of science was portrayed as a man of action, rather than as a solitary contemplative, that the passive stance of the humble acolyte of nature . . . required a mighty effort of self-restraint.
>
> (Daston and Galison 231)

In *Middlemarch*, Lydgate's musings about his autobiography in relation to 'heroes of science' are ironically epitomized by Rosamund's failure 'to imagine much about the inward life of the hero' (Eliot 1994 166 [Chapter 16]). If we follow Daston and Galison's account, Lydgate's failure has been to conquer himself – to achieve 'self-abnegation', although, within Eliot's compass, this failure becomes manifest in the personal rather than the scientific realm. There was, perhaps, a danger that self-abnegation could make an amoral materialist of the man of science. This was problematic for the scientific naturalists – biologist T. H. Huxley *et al.* – who wanted to use their science as a moral platform for the direction of society. Huxley needed to protect himself from the charge of materialism, which even for Darwin's bulldog would be to court too much controversy. The question of vivisection was a pinch point where amoral objectivity in the scientist could appear culturally problematic. Animal suffering, to which the mechanically objective scientist could not admit any sympathy or even existence, was a step too far for some (Collins). Huxley was therefore obliged to temper his scientific objectivity and its materialist consequences, retaining a place for mind in his philosophy of nature. 'I am utterly incapable of conceiving the existence of matter if there is no mind in which to picture that existence' (Huxley 2011 245). Thus, although he claimed that scientists could discover objectively true facts, he did not make the leap to saying that they could know the real nature of the universe – that Lewes's intruder could be a fundamental feature of nature. Modern science, Huxley wrote, 'admits that there are two worlds to be considered, the one physical and the other psychical; and that though there is a most intimate relation and interconnection between the two, the bridge from one to the other has yet to be found' (1897 62). There were no grounds for accepting the doctrine of realism, in the sense that such things as colour or species had an absolute existence in nature.

Notwithstanding this caution on the part of some, certain writers pressed on to give a fully naturalized account of 'real' human life. Naturalism, as an approach to creating literature, shares a great deal with realism – particularly the scientifically literate realism of Eliot and Lewes – but adds to it an explicitly 'scientific' way of accounting for the events that unfold in its mimic worlds. Typically this science was drawn from statistics, from medical theories of body and environment and from theories of inheritance both before and after Darwin. Perhaps an over-neat definition, but one that gets us some way towards the state of things, would be to say that the realist novel contained observation, while the naturalist novel aspired towards experimentation – a text-lab in which life-like scenarios would play out just as they would in reality. Zola was the great practitioner of the form, as described in his tract 'Le Roman expérimental' ('The Experimental Novel' (1880); see Zola). In 'Les romanciers naturalistes' ('Naturalist Novelists' (1881)) he identified others, all French, whom he claimed to be writing in the same way. It is questionable whether such a strong tradition can be identified in British literature. Thomas Hardy is often read as examining the power of the environment, and of deep lines of inheritance (Radford), but he was not overt in claiming to write scientifically. His fatal predestination has something more ancient about it than mere Darwinism or physiology. To find a true British naturalist author one might, oddly enough, have to step into science fiction, such as the works of Grant Allen or H. G. Wells. Wells's *The Time Machine* (1895) offers a biologically plausible account of the future of humankind; *The War of the Worlds* (1898) suggests scientific explanation for how alien species may (unintentionally) be defeated by biological means.

There is perhaps one sense in which the British novel adopted a broadly naturalistic self-discipline, and that is in respect to psychology. Masson called for a 'consistency of incidents' (28), as opposed to the cheap appeal of 'thrilling interest' in 1859, and it was above all 'by his *characters* that a novelist [was] chiefly judged' in this respect. Lewes also drew attention to the proper priority that novelists gave to the psychological dimension or 'internal life' (1858 494) of their characters. As Sally Shuttleworth has explored in *The Mind of the Child*, psychology was one of the main ways in which nineteenth-century authors advanced (rather than merely reflected) contemporary science. The insights that they offered into them were transferrable into life, in turn demanding an authorial psychologization of the reader. Davis writes particularly succinctly of Eliot's psychological affinity for characters (384–92); others have done so in relation to the more forensic subjectivity of Henry James (Cameron). Conceivably, the sensation novel was distressing to its critics precisely because it took the real and material underpinnings of literature (nerves, bodies, electricity, communication networks) and showed what happened when they became disordered (Garrison).

The disciplining entailed in realism did not occur without resistance or without negotiation about its reach. For some, realism was simply a pointless and inartistic assemblage of detail (Lewes 1891 84; Pykett). In 'On the Modern Element in Literature' (1869), Arnold called upon literature to *deliver* its readers from the 'copious and complex' (305) present with its 'vast multitude of facts'. Ruskin in 'Fiction, Fair and Foul' went further, denouncing modern novels as precisely too scientific, too automatically registering, 'like specimens from Lyme Regis, impressions of skeletons in mud' (267–8). Masson, by contrast, was fully in favour of novel writing as a kind of natural history, and of the 'wholesome spirit of Realism' (257), with its 'painstaking accuracy', and renewed careful attention to facts. The most Masson would concede by way of criticism was the necessity for skill and style in arranging the facts and resemblances upon the page. Rather brilliantly, Masson turned John Ruskin upon his head, making him into a realist critic by highlighting his doubts about Scott's factual knowledge of the past – and then demolishing him by showing that the appeal of Scott's characters was what counted. Eliot and Lewes distinguished between realism that was the mere accumulation of detail and artistic realism, which consisted in knowing how to prioritize those details so as to portray the most important, and to arrange other details in a manner harmonious with the greater theme (cf. Daston and Galison's account of pre-mechanical truth-to-nature, 55–113). As such, realism was a rehearsal of the challenges of inductivism; the novel was the twin of the scientific text that boiled down copious observations into a coherent and plotted account of nature (Beer; Levine 1988).

So far, then, the conjoint writings of Eliot and Lewes appear to instantiate the development of realism in a context of, and in dialogue with, the development of mechanical-objective science. However, one must pause and note the philosophical inconsistencies at the heart of the realist project. Lewes's celebrated dictum that 'Art always aims at the representation of Reality, i.e. of Truth' (1858 493) is incoherent. At first sight, Lewes appears to agree with Arnold in his claim that the aim of both art and science is the same, namely 'to see the object as in itself it really is' (1914 9). However, as Daston and Galison put it, '[o]bjectivity' (which we might plausibly use as a gloss for good representation) '[is] a different, and distinctly epistemological goal – in contrast to the metaphysical aim of truth' (210). Lewes came to see this for himself. 'Truth', as he defined it, 'is the correspondence between the order of ideas and the order of phenomena, so that the one is a reflection of the other' (Lewes 1904 14); but he quickly conceded that it could only ever be relatively accurate due to the structure of the mind.

It is noteworthy that in his accounts of seaside life and landscape Lewes frequently protests that his powers of description are weak. Observation, and its verbal notation, is in some sense trumped by the second investigative stage of experimentation, and he seeks politely to efface his own writing – the very stuff of his professional identity. In *The Principles of Success in Literature*, Lewes makes explicit this previously implicit distinction; that is, he separates writing from the mental work that lies behind it. Skill in writing is separate from perceiving (as in, resemblances and relations); it consists in the selection and arrangement of symbols (Lewes 1891 144–6). To think of novel writing as scientific research with the addition of style, as one might, conceivably, from reading Lewes, is incoherent. It would amount to a stripping of aesthetic sensibility, followed by a mechanical-objective procedure of registering the world as it is, and finally a reintroduction of aesthetic sensibility.

Peter Garratt picks up on the complexities of Lewes's philosophy in his study *Victorian Empiricism* (102–26). Lewes, Garratt points out, resists 'the naïvely realist or common sense belief in the directness and simplicity of perception' (120). For Lewes, the retina is not the sole organ of sight; sight incorporates other senses; and perception is the reaction of the entire mind. Similarly:

> [I]n the hands of George Eliot the language of looking encodes not the neutrality of the gaze or the simple availability of a given 'natural' order of things, but rather the habitual problems of perception and human interpretation.
>
> *(Garratt 105)*

Lewes, then, that perfect candidate in whom to try to discover the relation between scientific and literary realism as understood by the Victorians, turns out to have a complex or perhaps even shaky philosophical grasp upon it. His distinction between writing and nature is a form of realism that inserts its own undoing, for that hiatus between word and world is a fatal glitch in the project of verisimilitude. Lewes commits to reality as an objectively knowable external world, but almost immediately unravels that possibility. Moreover he demonstrates that facts and theories are not strictly separable – the latter being generally just a fuller description of the former (Lewes 1904 18–19). In literature, he places his weight on what, in science, are divergent methods, namely clearness of vision (observation) and sincerity (idealism). Yet even in his description of science objectivity fails to bear his complete weight. Description, present at the heart of the matter, renders his distinction between scientific and non-scientific writing unconvincing.

Finding the fault lines in Lewes's thinking can act as a point of departure for discovering the stresses within late Victorian cultures of reading and writing, provoking a second look at both text and context. As Daston and Galison put it, 'for mid-nineteenth-century scientists, th[e] epistemological predicament [of telling the truth] was hopelessly entangled with an ethical one that was also cast in terms of the objective and subjective' (210). The incompleteness of the realists' programme makes space for an appreciation of the continuity of the century, for the strands of historiography and heroism, and for the loaded notion of truth. Religiously and culturally inflected notions of truth, and the search for truth, ran deep through the century as a whole, from Carlyle to Lewes and beyond. Novels knitted together the metaphysics and the epistemology of science despite literary claims to the equivalent of objectivity – that is, the discovery of the real.

Conclusion

The common-sense and concrete world of Eliot, Lewes and their circle – or perhaps one should say their cast-iron world – was upheld by a whole system of Victorian civilization. Idealism was even better refuted than by kicking a stone (Boswell 160 (6 August 1763)), by trying one's force against a steam machine. In that most practical of centuries, real steam machines had demonstrably real effects – even, as Carlyle lamented in his essay 'Signs of the Times', upon the human mind (Carlyle 1829). The novel had become a 'mimic world' in which readers and writers could observe and test reality – or at least their representations of it.

There was a basic expectation of truth, which not only implied a broadly objective-scientific outlook but also – without any precise logical connection to the business of science – entailed a set of moral obligations. Eliot's words on rationalism give a sense of the combination; science is the discipline that will produce the moral goals of impartiality and steadfastness:

> The great conception of universal regular sequence, without partiality and without caprice . . . could only grow out of that patient watching of external fact, and that silencing of preconceived notions, which are urged upon the mind by the problems of physical science.
>
> *(1883 271)*

Although many aspects of literature and science in the nineteenth century have been well worked through, a number of questions remain open for further study. Research is required to place science and the novel on common ground through more exploration of history and biography as linking genres. A big-picture account of philology through the eighteenth and nineteenth centuries would yield important insights into the understanding and place of literature in culture: a science of the word prior to the literary culture of the highbrow periodicals. In the terrain of the earlier nineteenth century, there is work to be done on what Masson calls North British literature. This may well complement Crosbie Smith's historical work on North British science (Smith). Carlyle, too, deserves elaboration in relation to the figure of the man of science, and to heroism across multiple genres including science. Such topics will contribute to an understanding of the making of the later nineteenth-century scientist, for if Daston and Galison are right there is a problem of chronology; scientists who disciplined their selves to create objective epistemologies in the 1860s and 1870s grew up in a literary landscape of romanticism. More work is needed on the literary shaping of the lives of scientists, in biography and autobiography; whom did they read? *Did* they read? Did they actively review and recast their literary influences (as Darwin seems to do in his *Autobiography*)?

Work remains to be done on the British reception of Émile Zola, and on Zola's hero Claude Bernard,[1] who is one of Daston and Galison's case studies in the history of the will-suppressing man of science. Extant work on Zola largely relates him to the specific topic of degeneration (McLean), but what did British critics, novelists and scientists make of Zola's overarching project to turn novel writing into a species of science? Huxley will likely be a promising place to begin.

Finally, the challenge of giving a post-Romantic account of imagination lies wide open. Is imagination, within realism, simply the element of novel writing that is not observation: the fizz that must be added to the real, proper ingredients? This hardly seems plausible. And can imagination, whatever it is, operate *through* science or must it operate around and outside it? As far back as 1859, Masson called for science to discipline the didactic purpose of novels: 'is it not to be wished that our novelists brought to their business a fair amount of scientific capital?' (300). The 1880s saw critical discussion of realism in relation to romances, which, as Scott Hames argues, 'can enrich and complicate monolithic notions of the novel's essential trueness to life' (61). Wells, as has been argued above, might be considered Britain's best example of naturalist fiction, and yet by tradition science fiction is commonly grouped with fantasy, in opposition to realism. The theme of the natural history of the imagination also runs through the very late nineteenth century in the work of George Macdonald. Understanding the post-realist conceptualization of imagination and its role in cultural politics will be another topic for future study.

Note

1 See Andrew Mangham, 'Medical Research', in this volume.

Bibliography

Primary texts

Arnold, Matthew. 'The Function of Criticism at the Present Time' [1865]. *Essays by Matthew Arnold*. London: Macmillan, 1914, 9–36.

Arnold, Matthew. 'On the Modern Element in Literature'. *Macmillan's Magazine* 19 (1869): 304–14.

Babbage, Charles. *Reflections on the Decline of Science in England: And on Some of its Causes*. London: Fellowes, 1830.

Boswell, James. *Boswell's Life of Johnson: Including Their Tour to the Hebrides*. Ed. John Wilson Croker. London: John Murray, 1848.

Carlyle, Thomas. 'Signs of the Times'. *Edinburgh Review* 59 (June 1829): 439–59.

Carlyle, Thomas. *On Heroes, Hero-Worship, and the Heroic in History* [1841]. Eds. David R. Sorensen and Brent E. Kinser. New Haven, CT: Yale University Press, 2013.

Chalmers, Thomas. *On the Power Wisdom and Goodness of God. As Manifested in the Adaptation of External Nature to the Moral and Intellectual Constitution of Man*. London: William Pickering, 1833.

Collins, Wilkie. *Heart and Science: A Story of the Present Time*. London: Chatto & Windus, 1883.

Dallas, E. S. [Eneas Sweetland]. *The Gay Science. On the Criticism of Literature and Art*. London: Chapman & Hall, 1866.

Draper, John W. *History of the Conflict Between Religion and Science*. New York, NY: D. Appleton, 1874.

Eliot, George. 'The Natural History of German Life' [1856] and 'The Influence of Rationalism' [1865]. *The Essays of George Eliot*. Ed. Nathan Sheppard. New York, NY: Funk & Wagnalls, 1883, 141–77 and 257–71.

Eliot, George. *Middlemarch* [1871–2]. Ed. Rosemary Ashton. London: Penguin, 1994.

Huxley, Thomas H. 'On the Hypothesis that Animals are Automata, and its History' [1874]. *Collected Essays. Vol. 1: Methods and Results*. Cambridge: Cambridge University Press, 2011, 199–250.

Huxley, Thomas H. 'Scientific and Pseudo-Scientific Realism' [1887]. *Science and Christian Tradition: Essays*. New York, NY: Appleton, 1897, 59–89.

James, William. 'Great Men and their Environment'. *Atlantic Monthly* 46 (October 1880): 441–9.

Lewes, George. *A Biographical History of Philosophy*. 2 vols. London: Cox, 1852.

Lewes, George. 'Realism in Art: Recent German Fiction'. *Westminster Review* 70 (October 1858): 488–518.

Lewes, George. *Sea-side Studies at Ilfracombe, Tenby, the Scilly Isles, and Jersey* [1858]. 2nd edn. Edinburgh: Blackwood, 1860.

Lewes, George. *The Principles of Success in Literature* [1865]. Ed. Fred K. Scott. Boston, MA: Allyn and Bacon, 1891.

Lewes, George. *Science and Speculation*. Reprint of 'Prolegomena' to *History of Philosophy*, 3rd edn 1871. London: Watts, 1904.

Lyell, Charles. *Principles of Geology; Or, The Modern Changes of the Earth and its Inhabitants Considered as Illustrative of Geology* [1830–33]. New edn. New York, NY: Appleton, 1854.

Martineau, Harriet. *How to Observe. Morals and Manners*. London: Charles Knight, 1838.

Masson, David. *British Novelists and Their Styles; Being a Critical Sketch of the History of British Prose Fiction*. Cambridge: Macmillan, 1859.

Ruskin, John. 'Fiction Fair and Foul' [1880–1] and 'Fairy Stories' [1868]. *On the Old Road: A Collection of Miscellaneous Essays, Pamphlets, etc. etc. Published 1834–1885. Vol. II: Literature, Economy, Theology etc*. London: George Allen, 1885, 175–289 and 290–97.

Smiles, Samuel. *Lives of the Engineers, With an Account of Their Principal Works: Comprising Also a History of Inland Communication in Britain*. 3 vols. London: J. Murray, 1861–62.

Whewell, William. *History of the Inductive Sciences, From the Earliest to the Present Times*. 3 vols. London: John W. Parker, 1837.

Whewell, William. *The Philosophy of the Inductive Sciences, Founded upon Their History*. 2 vols. London: John W. Parker, 1840.

White, Andrew. *A History of the Warfare of Science with Theology in Christendom*. London: MacMillan, 1896.

Zola, Émile. *'The Experimental Novel' and Other Essays*. Trans. Belle M. Sherman. New York, NY: Haskell House, 1964.

Secondary texts

Abberley, Will. *English Fiction and the Evolution of Language, 1850–1914*. Cambridge: Cambridge University Press, 2015.

Allen, David Elliston. *The Naturalist in Britain: A Social History*. London: Allen Lane, 1976.

Beer, Gillian. *Darwin's Plots: Evolutionary Narrative in Darwin, George Eliot and Nineteenth-Century Fiction*. 3rd edn. Cambridge: Cambridge University Press, 2009.

Buckland, Adelene. *Novel Science: Fiction and the Invention of Nineteenth-Century Geology*. Chicago, IL: University of Chicago Press, 2013.

Cameron, Sharon. *Thinking in Henry James*. Chicago, IL: University of Chicago Press, 1989.

Cannon, Susan F. *Science in Culture: The Early Victorian Period*. Folkestone: Dawson, 1978.

Cregan-Reid, Vybarr. *Discovering Gilgamesh: Geology, Narrative and the Historical Sublime in Victorian Culture*. Manchester: Manchester University Press, 2013.

Daston, Lorraine, and Peter Galison. *Objectivity*. New York, NY: Zone, 2008.

Davis, Philip. *The Victorians*. Oxford: Oxford University Press, 2002.

Ferguson, Christine. *Language, Science and Popular Fiction in the Victorian Fin-de-siècle: The Brutal Tongue*. Aldershot: Ashgate, 2006.

Garratt, Peter. *Victorian Empiricism: Self, Knowledge, and Reality in Ruskin, Bain, Lewes, Spencer, and George Eliot*. Madison, WI: Fairleigh Dickinson University Press, 2010.

Garrison, Laurie. *Science, Sexuality and Sensation Novels: Pleasures of the Senses*. Basingstoke: Palgrave, 2010.

Hames, Scott. 'Realism and Romance'. *Approaches to Teaching the Works of Robert Louis Stevenson*. Ed. Caroline McCracken-Flesher. New York, NY: MLA, 2013, 61–8.

Hesketh, Ian. *The Science of History in Victorian Britain: Making the Past Speak*. London: Pickering and Chatto, 2011.

Holquist, Michael. 'Why We Should Remember Philology'. *Profession* (2002): 72–9.

Inkster, Ian, and Jack Morrell, eds. *Metropolis and Province: Science in British Culture, 1780–1850*. London: Routledge, 1983.

Levine, George. *The Realistic Imagination: English Fiction from Frankenstein to Lady Chatterley*. Chicago, IL: University of Chicago Press, 1981.

Levine, George. *Darwin and the Novelists: Patterns of Science in Victorian Fiction*. Cambridge, MA: Harvard University Press, 1988.

Levine, George. *Dying to Know: Scientific Epistemology and Narrative in Victorian England*. Chicago, IL: University of Chicago Press, 2002.

Levine, George. *Realism, Ethics and Secularism: Essays on Victorian Literature and Science*. Cambridge: Cambridge University Press, 2008.

Lukàcs, Georg. *Realism in Our Time*. Trans. John Mander and Necke Mander. New York, NY: Harper, 1971.

Maitzen, Rohan Amanda. *Gender, Genre, and Victorian Historical Writing*. New York, NY: Taylor & Francis, 1998.

Maitzen, Rohan, ed. *The Victorian Art of Fiction: Nineteenth-Century Essays on the Novel*. Peterborough: Broadview, 2009.

McLean, Steven. '"The Golden Fly": Darwinism and Degeneration in Émile Zola's *Nana*'. *College Literature* 39.3 (2012): 61–83.

Morrell, Jack, and Arnold Thackray. *Gentlemen of Science. Early Years of the British Association for the Advancement of Science*. Oxford: Oxford University Press, 1981.

O'Connor, Ralph. *The Earth on Show: Fossils and the Poetics of Popular Science, 1802–1856*. Chicago, IL: University of Chicago Press, 2008.

Pykett, Lyn. 'The Real Versus the Ideal: Theories of Fiction in Periodicals, 1850–1870'. *Victorian Periodicals Review* 15.2 (1982): 63–74.

Radford, Andrew D. *Thomas Hardy and the Survivals of Time*. Aldershot: Ashgate, 2003.

Radick, Gregory. *The Simian Tongue: The Long Debate About Animal Language*. Chicago, IL: University of Chicago Press, 2007.

Rudwick, Martin J. S. *Bursting the Limits of Time: The Reconstruction of Geohistory in the Age of Revolution*. Chicago, IL: University of Chicago Press, 2005.

Rudwick, Martin J. S. *Worlds Before Adam: The Reconstruction of Geohistory in the Age of Reform*. Chicago, IL: University of Chicago Press, 2008.

Salvey, Courtney. 'Mechanism and Meaning: British Natural Theology and the Literature of Technology, 1820–1840'. PhD thesis, University of Kent, 2014.

Schor, Hilary. 'Fiction'. *A Companion to Victorian Literature and Culture*. Ed. Herbert F. Tucker. Oxford: Blackwell, 1999. 323–38.

Secord, James A. *Visions of Science: Books and Readers at the Dawn of the Victorian Age*. Chicago, IL: University of Chicago Press, 2014.

Sleigh, Charlotte. *Literature and Science*. Basingstoke: Palgrave, 2010 [2010A].

Sleigh, Charlotte. 'The Judgements of Regency Literature'. *Literature and History* 19.2 (2010): 1–17 [2010B].

Smith, Crosbie. *The Science of Energy: A Cultural History of Energy Physics in Victorian Britain*. Chicago, IL: University of Chicago Press, 1998.

Teukolsky, Rachel. 'Novels, Newspapers, and Global War: New Realisms in the 1850s'. *Novel* 45.1 (2012): 31–55.

Turner, James. *Philology: The Forgotten Origins of the Modern Humanities*. Princeton, NJ: Princeton University Press, 2014.

Yeo, Richard. *Defining Science: William Whewell, Natural Knowledge, and Public Debate in Early Victorian Britain*. Cambridge: Cambridge University Press, 1993.

Further reading

Barker-Benfield, Graham J. *The Culture of Sensibility: Sex and Society in Eighteenth-Century Britain*. Chicago, IL: University of Chicago Press, 1996.

Darwin, Charles. *On the Origin of Species by Means of Natural Selection*. London: John Murray, 1859.

Deane, Bradley. *Making of the Victorian Novelist: Anxieties of Authorship in the Mass Market*. London: Routledge, 2002.

Desmond, Adrian. *The Politics of Evolution: Morphology, Medicine, and Reform in Radical London.* Chicago, IL: University of Chicago Press, 1989.

Fyfe, Aileen, and Bernard Lightman, eds. *Science in the Marketplace: Nineteenth-Century Sites and Experiences.* Chicago, IL: University of Chicago Press, 2007.

Greenslade, William, and Terence Rodgers, eds. *Grant Allen: Literature and Cultural Politics at the Fin de Siècle.* Aldershot: Ashgate, 2005.

Jackson, Noel. *Science and Sensation in Romantic Poetry.* Cambridge: Cambridge University Press, 2008.

Lightman, Bernard, ed. *Victorian Science in Context.* Chicago, IL: University of Chicago Press, 2008.

Pittock, Murray. *Scottish and Irish Romanticism.* Oxford: Oxford University Press, 2008.

Ruston, Sharon. *Creating Romanticism: Case Studies in the Literature, Science and Medicine of the 1790s.* Basingstoke: Palgrave Macmillan, 2013.

Shaw, Harry E. *Narrating Reality: Austen, Scott, Eliot.* Ithaca, NY: Cornell University Press, 1999.

Shuttleworth, Sally. *George Eliot and Nineteenth-Century Science: The Make-Believe of a Beginning.* Cambridge: Cambridge University Press, 1984.

Shuttleworth, Sally. *The Mind of the Child: Child Development in Literature, Science, and Medicine 1840–1900.* Oxford: Oxford University Press, 2013.

Smith, G. G. 'Masson, David Mather (1822–1907)'. Rev. Sondra Miley Cooney. *Oxford Dictionary of National Biography.* First published 2004; online edn 2010. <www.oxforddnb.com/view/article/34924>. Accessed 30 April 2015.

Weeks, Mary Elvira. 'Some Scientific Friends of Sir Walter Scott'. *Journal of Chemical Education* 13.11 (1936): 503–7.

Whewell, William. 'On the Connexion of the Physical Sciences'. *Quarterly Review* 51 (March 1834): 54–68.

Yeo, Richard. 'Science and Intellectual Authority in Mid-Nineteenth-Century Britain: Robert Chambers and *Vestiges of the Natural History of Creation*'. *Victorian Studies* 28.1 (1984): 5–31.

6

FROM GOTHIC TO SCIENCE FICTION

Adam Roberts

Royal Holloway, University of London

The argument that recognizably modern science fiction (SF) developed out of Romantic Gothic writing has many adherents. Perhaps its most influential advocate has been writer and critic Brian Aldiss, who sees Shelley's *Frankenstein* as the first SF novel. 'Though the fantastical had been in vogue long before,' Aldiss argues,

> it was Mary Shelley, poised between the Enlightenment and Romanticism, who first wrote of life – that vital spark – being created not by divine intervention as hitherto, but by scientific means; by hard work and by research. That was new, and in a sense it remains new. The difference is impressive, persuasive, permanent.
>
> *(35)*

Joanna Russ is more hyperbolic: 'every robot, every android, every sentient computer (whether benevolent or malevolent), every nonbiological person . . . is a descendent' of the novel's central 'creature' (126).

Yet to set the early nineteenth-century vogue for Gothic – Radcliffe's *The Mysteries of Udolpho* (1794), Lewis's *The Monk* (1796), Shelley's *Frankenstein* (1818), Polidori's *The Vampyre* (1819), and Maturin's *Melmoth the Wanderer* (1820) – alongside the later-century developments in scientific romance and science fiction is to be struck by how little, relatively speaking, the genealogy has to do with 'science' and how much with the evolving valences of Gothic itself.

It ought not to surprise us that science and literature were mutually reinforcing. Not only did literature very often draw its subjects and values from science, but throughout the century scientists were also influenced by the themes and modes of successful novelists (see Otis 2002B). James Secord records the way that Robert Chambers, the author of the very influential work of cosmology and natural history *Vestiges of the Natural History of Creation* (1844), hoped to emulate Walter Scott. Chambers's 'self image' was 'dominated by Sir Walter Scott'. True, he insisted that 'the modern spirit of the age could best be realized in a scientific work' rather than in fiction; but Chambers's science took much from the Romantic novel. '*Vestiges* applied Scott's methods on a cosmic scale' is how Secord puts it: 'the historical novel of the early nineteenth century was characterized by its reconstruction of past worlds in relation to specific circumstances, so that history was seen as a shaping force in the most minute aspects of everyday life' (89). If Scott's historical writing provides one formal template for the metanarratives of science, Gothic provides another. Indeed, one way of looking at the Gothic mode is to see it as a special kind of haunted historical novel, shaped by buried secrets, tangled family trees and varied returns-of-the-repressed.

Gothic as a mode mediated and shaped nineteenth-century discourses of science in culturally discursive ways. Science was fed by Gothic fascinations with the scale and terror of the cosmos, with human insignificance and the sense of a 'haunted' universe – part of the broader Romantic resistance to the notion of a merely materialist or 'clockwork' world. This in turn was mediated back into 'science fiction' as a mode at the end of the century. This chapter sets out to trace some of the lineaments of this interaction and tries to touch on a process by which Gothic imaginative investments in scale, grotesqueness (savagery and monstrosity) and the dialectic of past and future pay out into culture more generally. It also argues for a process of 'interiorisation' of scientific enquiry prompted in part by a similar shift in the Gothic.

Gothic

The original vogue for Gothic fiction endured from Walpole's *Castle of Otranto* (1765) – usually pegged as the first 'Gothic' tale – through to the early years of the nineteenth century (Botting 1996 20–40). During this period, scores and perhaps hundreds of 'Gothic' novels were published (a complete tabulation has yet to be undertaken, although there is no shortage of critical studies and guides – see Hogle; Punter 2002; Punter 2012; Spooner). By the 1810s the mode had become something of a joke: all those outré tropes of haunted castles, subterranean passageways, ghostly midnight hauntings and wicked monks. Jane Austen's *Northanger Abbey* (composed around 1800, although not published until 1817) rather overplays its central joke: how foolish would a person be to actually believe all those Gothic conventions. Walter Scott's *The Black Dwarf* (1816) engages in a similar debunking of Gothic expectations.

The conventional critical narrative is that Gothic then becomes in effect submerged as the main literary currents of the nineteenth century turn towards domestic fiction and 'condition of England' social novels (see Kucich and Taylor). 'Gothic' informs some of the vigorous subcultures of Victorian publishing, such as 'penny dreadful' novels, 'Bloods' and the burgeoning culture of 'horror' epitomized by works such as Polidori's *The Vampyre* (1819) and James Malcolm Rymer's *Varney the Vampire; or, the Feast of Blood* (serialized 1845–7). But, though the Gothic is marginal to the work of Dickens, Thackeray, Eliot and the Brontës, it is not wholly absent. Rather it tends to erupt, as when an essentially Gothic individual such as Heathcliff disturbs the rather stifling conformity of the Earnshaws' and the Linton's family life in Emily Brontë's *Wuthering Heights* (1847) or the monstrous Quilp invades domestic London life in Dickens's *Old Curiosity Shop* (1841). Similar circumstances mark other key 'mainstream' Victorian novels (see Garrett).

In all this, Gothic figures as a kind of meta-trope for irrationality itself, the dark and buried forces repressed by the eighteenth century's commitment to a self-declared Age of Enlightenment. According to this critical line, Gothic stands at the opposite pole to 'science'. The latter is the action of reason and intellect; the former the irrational gush of intense emotion, skewed towards the pleasures of being terrified and horrified. Peter Garrett summarizes the conventional critical position (in order, in part, to establish his own slightly differently inflected reading of 'the Gothic') quoting Horace Walpole's insistence that he wrote *The Castle of Otranto* 'in spite of rules, critics and philosophers', in order to defy an age 'wanting only cold reason'. Garrett goes on:

> Much of the appeal of Gothic and its many offshoots from Walpole's time to our own has come from such resistance, from its promise of release from the limitations

of cold reason and the commonplace . . . Much of the recently increased critical attention to Gothic also dwells on its oppositional force, now given a more political valence. In these accounts its fantastic extremity opposes the middle-class ideology of the mainstream novel's domestic realism; its dark passions and nightmarish scenarios contest Enlightenment rationalism and optimize or destabilize the liberal humanist subject. Sometimes Gothic fiction is credited with deliberate subversion; sometimes it is read symptomatically for the ways its terrors betray cultural anxieties about sexuality and gender, the menace of alien races or the criminal classes – about whatever threatens the dominant social order or challenges its ideologies.

(1–2)

Garrett cites the most influential critical studies of the form (including Kilgour; Botting 1996; Watt), all of which develop this thesis.

According to this view, Gothic really ought to have nothing to do with nineteenth-century science, except perhaps to define it by opposition. But the situation is more nuanced than this. Quite apart from anything else, the widespread critical argument that Gothic informs the rise of science fiction seems to set this at odds. As Fred Botting puts it, 'despite so many Gothic science fiction mutations, it is strange the genres should cross at all' since Gothic 'conventionally deals in supernatural occurrences and figures' and eschews 'the enlightened reason and empirical technique so important in science fiction'. Nonetheless:

Realism and social reality recoils [*sic*] from strangeness and monstrosity. In contrast, science fiction embraces metaphorical possibilities to 'defamiliarize the familiar, and make familiar the new and the strange' (LeFanu 21). As a 'literature of cognitive estrangement' presenting 'a *novum*,' science fiction depends upon metaphorical creation (Suvin 4). But the 'novum' is simultaneously a 'monstrum,' the difference a matter of perspective . . . In making the familiar strange, the new can be seen as threat (monstrum) or promise (novum). Part of science fiction's success stems from the way it has 'diversified the Gothic tale of terror in such a way as to encompass those fears generated by change and technological advances which are the chief agents of change' (Aldiss 53).

(Botting 2008 111)

We can see this in the claim that Shelley's *Frankenstein* is foundational for science fiction. For what is most striking about the way 'science' is framed in this originary text is its Janus nature. Science for Shelley's novel is *both* a forward-looking process of rational enquiry and experimentation aimed at uncovering and advancing the natural world *and* simultaneously a backward-looking occult art, mastered in darkness and mystery and thoroughly supernatural in origin and effect. We can read Frankenstein himself either as a scientist or an alchemist, and we can read his 'creature', the unnamed monster at the novel's heart, either as a robot or a nightmare revenant or zombie. The text is very carefully balanced between these two interpretations.

This in turn has implications for the way 'science' – and in particular, the figure of the scientist – is troped in nineteenth- and twentieth-century culture. At the risk of perpetrating mere stereotypes, it is probably true to say that 'actual' scientists over the last two and a half centuries have tended to be scrupulous and methodical individuals, developing their knowledge in ways that can be replicated in order to open them to the potential of being falsified. They are therefore committed to a praxis of transparency. Some have surely been motivated by personal profit or gain, others by loftier ideals of public good or pure knowledge, but much actual science has been a community rather than an individual process. This is in rather striking contrast to the way the scientist figures in nineteenth-century literature (and in later literature as well). Here the scientist – always a *he* – is a loner, a solitary genius, even sometimes a madman. Surveying the complete range of scientists protracted in pre-twentieth-century science fiction, Brian Stableford and David Langford sketch the quasi-Gothic template for such representations. In nineteenth-century culture, scientists:

> [O]ften exhibited symptoms of social maladjustment, sometimes to the point of insanity; they were characteristically obsessive and antisocial. Some scientists were quasidiabolical figures, like Coppelius in E. T. A. Hoffmann's 'Der Sandmann' (1816) or Mary Shelley's eponymous *Frankenstein, or The Modern Prometheus* (1818) . . . In Honoré de Balzac's *La recherche de l'absolu* (1834) scientific research becomes an unholy addiction. Such stories make it clear that the scientist had inherited the mantle (and the public image) of medieval alchemists, astrologers and sorcerers, and certain aspects of this image proved extraordinarily persistent; its vestiges remain even today, with science-fictional alchemical romances still featuring in the work of authors like Charles L Harness. The founding fathers of SF, Jules Verne (Nemo and Robur) and H. G. Wells (Moreau, Griffin and Cavor), frequently represented scientists as eccentric and obsessive; Robert Louis Stevenson's Dr Jekyll is cast from the same anxious mould, as is Maurice Renard's *Dr Lerne*.
>
> *(Stableford and Langford)*

Again, where science has had many positive effects in the world (increasing agricultural yields, prolonging human life, fighting disease and facilitating transport and communication) the nature of dramatic interest has meant that the fictional scientist almost always trades in unintended consequences of a deleterious sort. The very name 'Frankenstein' has become a sort of shorthand for this: the idea that even with the best intentions scientists meddle and make the world worse – producing Frankenstein foods, or 'Frankenfreak' genetically engineered animals, or nurturing what Isaac Asimov (describing his 'robot' stories from the 1940s) calls 'the Frankenstein complex' (Freedman 24). Christa Knellwolf and Jane Goodall have contextualized the fraught interrelationship between the novel's 'romantic' imaginative excess and the developing cultures of science and discovery that also informed it. This is part of a recent push to challenge the notion 'that the novel is about masculinity and scientific hubris', a view 'that has led to an enduring use of the title as a byword for the dangerous potential of the scientific over-reacher'. Reinserting the work into the ideological contexts of its time produces a more nuanced understanding not just of this novel, but of the way 'science' as a whole is represented in the century:

> If we listen carefully for the contextual arguments into which the assessment of the benefits and dangers of a new discovery were embedded, we may have to relinquish the assumption . . . that this is a novel with an anti-Promethean message. In doing so, we can gain a more complex understanding of the cross-fertilisations between radical politics and the dramas of scientific exploration.
>
> *(Knellwolf and Goodall 1)*

'Of course,' they add, 'not every scientist subscribed to radical politics'. But a surprising number did; or, more to the point, there is a surprising interpenetration between the broader cultural ideas of science as such, and 'late eighteenth- and early nineteenth-century utopian thinking about the vast social benefits made possible by scientific innovation' (Knellwolf and Goodall 1).

This trio of elements – the Gothic, science and politics – are useful co-ordinates for understanding the complex interconnecting cultural and practical discourses of the century. An emerging theme in the scholarship is that the first of these terms provided a means for people in the nineteenth century to think about the last. Coleridge, himself the author (in 'The Rime of the Ancient Mariner' and 'Christabel') of two especially influential Gothic texts of the period, claimed that 'from my early reading of Faery Tales & Genii, &c &c – my mind had been habituated to *the Vast*' (1:354). This takes us back to the starting point for the Gothic vogue itself: Edmund Burke's *A Philosophical Enquiry into the Origin of Our Ideas of the Sublime and Beautiful* (1756), a direct motivation for Walpole to write *Otranto*. Burke's aesthetics finds a grandeur and terror in the consideration of enormousness, obscurity, darkness and limitlessness easily assimilable to the romance conventions of the Gothic novel (Morris). Coleridge's own fascination with these very conventions went hand in hand with a lifelong engagement with science. Trevor Levere has shown how, during what he calls 'Coleridge's productive middle years' (between the publication of the literary-critical *Biographia Literaria* in 1817 and the devotional *Aids to Reflection* in 1825), Coleridge read deeply in and responded to work by 'surgeons, chemists and animal chemists' in particular; but that he also had a lifelong, if theologically inflected, fascination with cosmology, general physics and theories of heat, light, colour and sound. 'Coleridge,' Levere says, 'recognising the creativity inherent in scientific discovery, saw in science a source of imaginative insight' (2).

There are a number of things to take away from this, in terms of the way a 'Gothic' sensibility mediated the relationship between science and literature in the nineteenth century. One is the question of scale – Coleridge's 'The Vast', itself an iteration of the same Burkean sublime that informed the original aesthetic of Gothicism itself. Another is the sense that this mode of engaging imaginatively with science was part of a reaction against mechanistic materialism – Nicholas Roe's 'Introduction' to his *Samuel Taylor Coleridge and the Sciences of Life* is very good on the way Romantic artists as a whole kicked back against this sense of the 'clockwork' universe (see also Smith).

Scale

To take the first of these first, nineteenth-century science was among other things a continual expansion or en-vastening of human perceptions of the cosmos. For example, Charles Lyell's *Principles of Geology* (1830–33) disseminated the idea that the Earth was vastly older than the few millennia sanctioned by the Bible. Charles Darwin's work opened similarly vertiginous perspectives on the history of life. The story of astronomy over this period was one of continually revising notions of how enormous the cosmos is, from the identification of the Milky Way as a galaxy (in Thomas Wright's *An Original Theory or New Hypothesis of the Universe* (1750)) through to the work of Harvard astronomer Henrietta Swan Leavitt in the 1890s and 1900s, which first allowed scientists accurately to measure interstellar and intergalactic distances. This is turn, in addition to providing a more (rationally) 'accurate' representation of the nature of the universe, was informed by a Gothic sense of the awe-inspiring and terrible sublimity of this new perspective. Wright's conclusion was that the size of the universe effectively swamped any pretions to significance on our own small globe:

> In this great Celestial Creation, the Catastrophy of a World, such as ours, or even the total Dissolution of a System of Worlds, may possibly be no more to the great Author

of Nature, than the most common Accident in Life with us, and in all Probability such final and general Doom-Days may be as frequent there, as even Birth-Days or Mortality with us upon this Earth.

(76)

Rather counter-intuitively, Wright goes on 'this Idea has something so chearful in it, that I own I can never look upon the Stars without wondering why the whole World does not become Astronomers' (76). Even infinity itself expanded under the impress of nineteenth-century science, with Georg Cantor's various publications (between 1874–97) on 'transfinite' infinites.

We can track this through the century with works of 'science fiction' that engage, more or less idiosyncratically, with science's new 'vastness'. Edgar Allan Poe's success as a writer of Gothic intensities, darkness and irrational horror was of a piece with his book-length attempt at a scientific cosmology, *Eureka: A Prose Poem* (1848). A strange work that wavers between gushing pseudo-science and more sure-footed science fiction – it begins, for instance, with a letter from the year 2048 that has been transported back in time – *Eureka*'s every page is dyed with Poe's wide-eyed wonder at the sheer scale of things. The word 'vast' tolls through the whole like a bell: the universe is 'a vast geometrical system' (Poe 47); in each star 'a vast quantity of nebulous matter has assumed globular form' (71); expansion has moved each star 'a vast distance' from its fellow (73); 'the light by which we recognize the nebulae now . . . left their surfaces a vast number of years ago' (91); the Milky Way is 'vast' (97), interstellar distances are 'unutterably vast' (116), and the more one looks the vaster it gets:

> [L]ooking back at the Divine plan . . . we shall be enabled easily to understand, and to credit, the existence of even far vaster disproportions in stellar size than any to which I have hitherto alluded. The largest orbs, of course, we must expect to find rolling through the widest vacancies of Space.
>
> *(112)*

Few take *Eureka* seriously as science, less because it eschews the scientific method in favour of 'intuition' and more because the cosmos as Poe describes it is so clearly a projection of his own remarkable ego onto the largest possible canvas. Poe's cosmos is alive, and shaped by a moral and spiritual destiny. In this he was channelling the Romantic hostility to a purely mechanistic, un-animistic universe – this too is, as it were, Gothic.

In France, Camille Flammarion (a practising astronomer as well as a writer of cosmos-spanning science fiction) explored the universe in less melodramatic mode. *Récits de l'infini* (1872) included 'Lumen', in which an extraterrestrial capable of travelling faster than light crosses and recrosses the galaxy finding various modes of life. In his bestselling *Astronomie populaire* (1880), Flammarion asked: 'N'est-il pas agréable d'exercer notre espirit dans la contemplation des grands spectacles de la nature? N'est-il pas utile de savoir au moins sur quoi nous marchons, quelle place nous occupons dans l'infini?' ('Is it not better to exercise our mind in the contemplation of Nature's great spectacles? Is there no benefit in at least knowing in which direction we are travelling, what place we have in all this infinity?') (2). Despite being one of France's most respected scientists, Flammarion used his science fiction to dramatize a fundamentally living cosmos, in which souls are reincarnated on multiple worlds in alien forms. Throughout his career Flammarion published extravagant and often Gothic science-fictional fables alongside his conventional popular science and astronomical researches. Brain Stableford convincingly argues for the direct influence of 'Lumen' on the later science fiction of Olaf Stapledon, William Hope

Un missionnaire du moyen âge raconte qu'il avait trouvé le point
où le ciel et la Terre se touchent...

Figure 6.1 Wood engraving by anonymous artist
Courtesy History of Science Collections, University of Oklahoma Libraries

Hodgson and Arthur Conan Doyle. Stapledon in particular shares Flammarion's fascination with cosmic scale and diversity, though not his optimism or geniality.

Also by Flammarion is the so-called 'Flammarion engraving'. This was long thought to be an actual woodcut from the German middle ages but no earlier appearance has been documented than Flammarion's *L'atmosphère: météorologie populaire* (1888), and the image is worked in a style not used before the 1800s. The case cannot be proven, but it is possible that Flammarion himself (who had been apprenticed to a Parisian engraver at the age of twelve) created the image. At any rate it has been widely disseminated as a visual rebus for the 'conceptual breakthrough' by which Kuhnian science, and science fiction more broadly, are defined.

What is especially fascinating about this is just how Gothic it is: medieval in cast and concerned with sublime mysteries and adyta; yet also playing games precisely with the opposition between a 'mechanistic' or clockwork explanation of the cosmos – those wheels resembling giant cogs on the far side of the sky – and older, more rooted 'human' and spiritual explanations. The image's sole figure, after all, is dressed as a pilgrim.

Monsters

By the end of the century, Wells cannily inverts this trope of the observing eye, thereby only intensifying the sense of human insignificance and weakness in the larger scheme of things. The opening paragraph of *The War of the Worlds* (1897) sets the tone for a story of Martian invasion that draws equally on Gothic monstrosity and body-horror as it does the discourses of astronomy and microscopy:

No one would have believed, in the last years of the nineteenth century, that human affairs were being watched keenly and closely by intelligences greater than man's and yet as mortal as his own; that as men busied themselves about their affairs they were scrutinized and studied, perhaps almost as narrowly as a man with a microscope might scrutinize the transient creatures that swarm and multiply in a drop of water. With infinite complacency men went to and fro over this globe about their little affairs, serene in their assurance of their empire over matter. It is possible that the infusoria under the microscope do the same . . . Yet, across the gulf of space, minds that are to our minds as ours are to those of the beasts that perish, intellects vast and cool and unsympathetic, regarded this earth with envious eyes, and slowly and surely drew their plans against us.

One implication – that Gothic monsters already exist, hidden from us only by their miniature size until the invention of the microscope – is neatly inflected to orient us with the 'infusioria'. The invocation of interplanetary scales has the equal-and-opposite effect of effectively shrinking us down to microscopic significance by comparison. Wells's *The Time Machine* works its effects via a similar Gothicized 'expansion' of temporal perspectives, as the unnamed protagonist uses his machine first to visit the future of the year 802,701 and then further on to the 'terminal beach' of the world's eventual death, where evolution has worked a Gothic monstrosity into the human form, and our descendants – for we have become huge crustaceans:

Can you imagine a crab as large as yonder table, with its many legs moving slowly and uncertainly, its big claws swaying, its long antennae, like carters' whips, waving and feeling, and its stalked eyes gleaming at you on either side of its metallic front? Its back was corrugated and ornamented with ungainly bosses, and a greenish incrustation blotched it here and there. I could see the many palps of its complicated mouth flickering and feeling as it moved.

(89)

Perhaps this describes a far-future in which humanity has become extinct and the descendants of crabs rule the world. I prefer to read this description, with its suggestion of 'metallic' as well as organic components, as balancing a 'forward'-looking aesthetic we might associate with a cyborg science-fictional speculation against an atavistic terror, evolution-as-regression, and pursuit by terrifying Gothic monstrosity. In this, it recapitulates the formal or interpretive logic of Shelley's *Frankenstein*.

This *'longue durée'* conception of time was a direct reaction to the new temporal vistas opened by geology. Adelene Buckland frames a compelling argument that the developing sciences of nineteenth-century geology provided not a simple 'authoritative plot pattern on which to hang a fiction, as many critics have suggested' (27), but rather 'offered the Victorians a language for the breakdown that might secure narrative authenticity'. But it is significant that her examples (Scott, George Eliot, Dickens) are not drawn from SF. She is in part working in dialogue with Gillian Beer's older but still important *Darwin's Plots: Evolutionary Narrative in Darwin, George Eliot and Nineteenth-Century Fiction* (1983) – another work that does not discuss SF. What this difference of emphasis highlights, among other things, is the way science fiction towards the end of the century became less interested in the newly 'objective' flavour of science.

Recently there has been an increased focus of interest on the ways Darwin and Darwinism articulated an ideologically freighted discourse of monstrosity and domination, of a present haunted by its past, and of a future – of transcendence or decay, depending on one's view – prophetically revealed. This is, in Brian Baker's phrase, 'Darwin's Gothic' (199; see also Creed; Feldman; and, for a perspective from the hard sciences, Chase *et al.*). This in turn is connected to a shift in the cultural meaning of Gothic through the second half of the century, along with a (related) shift in the valences of literary mediation of the prevailing scientific culture. Gothic became more than a set of novelistic contrivances for the creation of a readerly mood of terror and awe. While never entirely leaving this behind, Gothic as a cultural logic broadened in significance as the century aged. Wells's semi-autobiographical novel *The History of Mr Polly* (1910) records how the young Alfred Polly came under the influence of John Ruskin and learnt to love the Gothic architecture of Canterbury Cathedral ('There was,' the narrator notes, slightly mockingly, 'a blood affinity between Mr. Polly and the Gothic'). Ruskin's mid-century analysis of the cultural significance of Gothic, 'The Nature of Gothic' (1853), is a vital way station in the evolution of the concept. Ruskin's essay sets out to trace 'this grey, shadowy, many-pinnacled image of the Gothic spirit within us' via six quintessentially Gothic elements, and ranked 'in order of importance': 1. Savageness; 2. Changefulness; 3. Naturalism; 4. Grotesqueness; 5. Rigidity; 6. Redundance. This schema has, as Ruskin applies it, a clear applicability not just to architecture but to the visual arts more broadly. What is less well explored, and could stand to be more thoroughly theorized, is the way Ruskin's categories speak to the broader logic of science. Only a few years after the publication of this essay, Darwin's long-gestated *On the Origin of Species* (1859) reconfigured our sense of the natural world from being a finished and supremely accomplished divine artwork, to being a project of immense timescales marked precisely by its savage changefulness, the grotesque redundancy of its naturalism. And this 'Darwinian Gothic' fed into the late-century boom in science fiction romances, often (as in Wells's *The Island of Doctor Moreau*, Rider Haggard's *She* or Conan Doyle's *The Lost World*) directly about savagery and naturalism and the past, but even when not then still informed by an aesthetic commitment to the grotesque that played the rigidity of generic forms against the fecundity of unfettered imagination.

Interiority

One element in particular is worth stressing, not least because it is perhaps counter-intuitive: the internalization of this Ruskinian Gothicism. The savagery (as in *Jekyll and Hyde*), the grotesqueness and the changefulness are uncovered as residing *inside* the individual. This connects with one of the most critically influential readings of this shift, Roger Luckhurst's argument that the resurgence of Gothic at the end of the century owed less to the conventional setting and props of the late eighteenth-century mode (haunted castles, subterranean passages and so on) and more to developing discourses of the psyche:

> The fin-de-siècle Gothic was fascinated by forms of psychic splitting, trance states and telepathic intimacies. It adopted the language of the psychical researcher and relocated spooky phenomena within the quotidian spaces of English modernity; rather than the pre-modern landscapes of its eighteenth-century antecedent.
>
> *(182)*

Perhaps counter-intuitively, this interiorization of the Gothic was in part a reaction to what Martin Willis has identified as the shift to a more exterior, Newtonian mode of science. Willis's reading encompasses texts across the whole of the span of the century from Hoffmann's 'The Sandman' (1816) to Wells's *The Island of Doctor Moreau* (1896) in order to show that 'science and science fiction changed dramatically' over that period. 'Romantic science gave way to

Newtonian methodologies of theory and experiment; scientific disciplines began to rise from the more general category of natural philosophy, and professionalization and institutionalization marked the end of uncoordinated amateurism' (Willis 2006 2).

Willis notes how often nineteenth-century science used 'literary language' ('the language of literature . . . was a vital part of the scientists armoury in imagining the principles of the natural world' (7)). The *Origin of Species* is, for instance, both elegantly and readably written. But, as Willis also says, after the 1870s – broadly speaking – this began to change. Science became professionalized, specialized and elite; and the scientific idiom began to grow into today's intractable thicket of jargon-strewn prose. The 'professional' obscurity was matched by a resurgence of the idiom of Gothic 'obscurity' and 'darkness'. Where Edmund Burke had identified immensity and obscurity in the vastness of the cosmos, late-century writers found it in the soul. The more late nineteenth-century science stressed the functioning of laws of physics and chemistry, the more late Gothic and science fiction dramatized a moral, animist universe.

Arguably, the key sciences at the end of the century, as far as this is concerned, were the developing ones of 'neurology' and 'psychology', as well as 'parapsychological' pseudo-sciences such as spiritualism and telepathy, nowadays discredited but still regarded as viable avenues of scientific enquiry in the nineteenth century. In *Popular Fiction and Brain Science in the Late Nineteenth-Century* (2013), Anne Stiles draws interesting connections between the development of 'neurology' as a science and certain specific romance and Gothic narrative forms.[1] Gothic, in particular (she suggests), 'presented certain advantages in grappling with fin-de-siècle neurology' (Stiles 17). Gothic fiction:

> [S]imultaneously complements and critiques classical, post-Jacksonian neurology by focusing on subjective experience rather than objective data . . . in contrast to the oversimplifying linearity of classical neuroscience, Gothic prose is snarled by multiple narrators, embedded texts, instances of doubling and mistaken identity and numerous indications of narrative instability and unreliability.
>
> *(Stiles 16)*

Stiles pegs the beginnings of the 'late-Victorian romance revival' to the 1880s 'as works like Stevenson's *Treasure Island* (1883) and Haggard's *King Solomon's Mines* (1885) signalled a new direction in popular fiction' (17).

This direction was dominated by what Patrick Brantlinger has called 'Imperial Gothic' – and so politics comes again to the fore. Discussing Stevenson's *Strange Case of Dr Jekyll and Mr Hyde* (1886) alongside Wells's science fiction from the 1890s and Bram Stoker's *Dracula* (1897), Brantlinger suggests that 'Imperial Gothic is related to several other forms of romance writing which flourished between 1880 and 1914', tracing the 'subterranean links between late Victorian imperialism, the resurrection of Gothic romance formulas, and the conversion of Gothic into science fiction' (233). He makes the point that British imperialism was based as much on the justification of moral superiority as anything else, asking the crucial question 'If an empire based on a morality declines, what are the implications?' (Brantlinger 232). But to insist that this end-of-century fascination with disease and decline (commonly but rather misleadingly tagged as 'decadence') was a wholly *moral* anxiety is to miss the larger context. The frame was political, not just in the sense that a project like 'imperialism' can hardly help be ideological and political but also in the sense that science itself increasingly came to inform political thought. Here is Laura Otis:

> The growth of industrialism and colonialism in the nineteenth century greatly increased contact between nations and cultures. If diseases were contagious, commercially

crippling quarantines had to be established, so political beliefs inevitably affected people's opinions as to whether disease spread through foul air or human contact. Anticontagionist reformers, largely liberals and radicals, fought for scientific, commercial and individual freedom simultaneously, regarding the three as inseparable. Contagionism consequently attracted conservatives, officers and bureaucrats who thought that centralized power structures not individual citizens or local authorities, should control policies affecting public health . . . Politics thus shaped medical opinion, but simultaneously, scientific discoveries influenced political thinking.

(2002A 11)

Cultural representations of disease still tended to frame contagion in terms of aerial miasma for decades after John Snow had taken away the handle of the Broad Street pump, and even for a few years after Koch had conclusively demonstrated the microorganisms responsible for major diseases (see Willis 2008; Willis 2011). One reason for this is that the textual prop cupboard of the Gothic, with its invisible spirits, strange gusts and spectres, continued providing the symbolic terms for conceptualizing science. Mary Shelley's *The Last Man* (1826) describes a plague that annihilates almost the entire human race. This contagion is an aerial miasma ('infection depended upon the air, the air was subject to infection') but one that becomes operative only because the victims, in some sense, unconsciously collaborate with it:

But how are we to judge of airs, and pronounce – in such a city plague will die unproductive; in such another, nature has provided for it a plentiful harvest? . . . [B]odies are sometimes in a state to reject the infection of malady, and at others, thirsty to imbibe it.

(Shelley 1826 1:203)

This inevitably invokes a moral dimension to the conceptualization of contagion. Disease is not a merely material infelicity; it is, for Shelley, the articulation of some personal failing – because, precisely, the Gothic mode of writing about such a subject is predicated upon the sense that the universe is (precisely) not clockwork and impersonal. In Poe's 'The Mask of the Red Death' (1842), Prince Prospero and his revellers are targeted by a spectral embodiment of contagion itself, precisely because of the immorality of their debauched lives. By the 1890s a sense that disease was caused by germs rather than miasmas had percolated into popular culture; as, for instance, in H. G. Wells's ominous short story 'The Stolen Bacillus' (1895). But this tended to reinforce rather than anything else the political dimension. Stoker's *Dracula* is a case in point. Scholars have, for some time, been exploring the extent to which this novel dramatizes a late imperial anxiety about what Stephen D. Arata calls 'reverse colonisation' (621). Dracula himself arrives in the country like a contagion, spreading his blood-borne evil and corrupting true Englishmen and women. Significantly, he arrives both as a 'germ' – his ship carries quantities of dirt from his own land into England – and as a Gothicized 'miasma', a spectre capable of moving through the air. More than this, the novel stages the larger debate implied by a Gothic mediation of 'science'. To quote Carol Senf, *Dracula*

[E]mphasizes the conflict between people who believe that the world is systematic and subject to both reason and human control, and individuals whose very existence embodies mystery and the total lack of human control over a powerful and over-whelming universe.

(19)

This is the key: by stressing the supra-human scale of the cosmos, Gothic 'sublime' speaks to something important about the way science itself is conceived. Hilary Grimes puts it well:

> [S]cientists of the fin de siècle were deeply conflicted between a desire to police the boundaries of science ... and conversely, to experience the obscure thrill of the 'Unknown'. Although elements of the unknown like telepathy, spiritualism and spirits, mesmerism and extrasensory perception threatened to compromise their rational borderlines, they were also intoxicating and inspiring, both dangerous and delightful.
>
> *(1)*

The path of the Gothic, from a set of generic conventions and story elements through to a broader cultural logic, shadows that of nineteenth-century science itself: from a disparate set of often amateur interests and topics to a unified and professionalized discourse. Many key fascinations of Victorian scientific enquiry are specifically Gothicized: fascinations with size and scale, cosmic and ultra-*longue durée* timescales; with the intimate connections of past to future; with the savage and irrational underpinning of apparently rigorous laws and rules. We might say, with Julian Wolfreys, that 'the Gothic becomes truly haunting' (11) where science is concerned.

Note

1 See Suzy Anger, 'Sciences of the Mind', in this volume.

Bibliography

Primary texts

Coleridge, Samuel Taylor. *The Collected Letters of Samuel Taylor Coleridge*. Ed. Earl Leslie Griggs. 6 vols. Oxford: Clarendon, 1956–71.

Flammarion, Camille. *Astronomie populaire*. Paris: Marpon/Flammarion, 1880.

Poe, Edgar Allan. *Eureka: A Prose Poem*. New York, NY: Putnam, 1848.

Shelley, Mary. *Frankenstein*. 3 vols. London: Lackington, Hughes, Harding, Mavor & Jones, 1818.

Shelley, Mary. *The Last Man*. 3 vols. London: Henry Colburn, 1826.

Wells, H. G. *The Time Machine*. London: Heinemann, 1895.

Wells, H. G. *The War of the Worlds*. London: Heinemann, 1898.

Wright, Thomas. *An Original Theory or New Hypothesis of the Universe, Founded on the Laws of Nature and Solving by Mathematical Principles the General Phaenomena of the Visible Creation; and Particularly the Via Lactea*. London: Henry Chapelle, 1750.

Secondary texts

Aldiss, Brian. *Billion Year Spree: The History of Science Fiction*. London: Weidenfeld and Nicolson, 1973 (Later updated, with David Wingrove, as *Trillion Year Spree: The History of Science Fiction*. London: Victor Gollancz, 1986).

Arata, Stephen D. 'The Occidental Tourist: *Dracula* and the Anxiety of Reverse Colonization'. *Victorian Studies* 33.4 (1990): 621–45.

Baker, Brian. 'Darwin's Gothic: Science and Literature in the Late Nineteenth Century'. *Literature and Science: Social Impact and Interaction*. Eds. John H. Cartwright and Brian Baker. Santa Barbara: ABC-CLIO, 2005, 199–222.

Beer, Gillian. *Darwin's Plots: Evolutionary Narrative in Darwin, George Eliot and Nineteenth-Century Fiction*. London: Routledge, 1983.

Botting, Fred. *Gothic*. London: Routledge, 1996. 2nd edn 2014.

Botting, Fred. '"Monsters of the Imagination": Gothic, Science, Fiction'. *A Companion to Science Fiction*. Ed. David Seed. Oxford: Blackwell, 2008, 111–26.

Brantlinger, Patrick. *Rule of Darkness: British Literature and Imperialism, 1830–1914*. Ithaca, NY: Cornell University Press, 1988.

Buckland, Adelene. *Novel Science: Fiction and the Invention of Nineteenth-Century Geology*. Chicago, IL: University of Chicago Press, 2013.

Chase, Mark W., Maarten J. M. Christenhusz, Dawn Saunders and Michael F Fay. 'Murderous Plants: Victorian Gothic, Darwin and Modern Insights Into Vegetable Carnivory'. *Botanical Journal of the Linnean Society* 161.4 (2009): 329–56.

Creed, Barbara. 'Darwin, Early Cinema and the Origins of Uncanny Narrative Forms'. *Darwin's Screens: Evolutionary Aesthetics, Time and Sexual Display in the Cinema*. Carlton: Melbourne University Press, 2009. 1–18.

Feldman, Mark B. 'Love in the Age of Darwinian Reproduction'. *Darwin in Atlantic Cultures: Evolutionary Visions of Race, Gender, and Sexuality*. Eds. Jeanette Eileen Jones and Patrick B. Sharp. London: Routledge, 2010, 73–89.

Freedman, Carl Howard, ed. *Conversations with Isaac Asimov*. Oxford, MS: University Press of Mississippi, 2005.

Garrett, Peter. *Gothic Reflections: Narrative Force in Nineteenth-Century Fiction*. Ithaca, NY: Cornell University Press, 2003.

Grimes, Hilary. *The Late Victorian Gothic: Mental Science, the Uncanny and Scenes of Writing*. Farnham: Ashgate, 2011.

Hogle, Jerrold E., ed. *The Cambridge Companion to Gothic Fiction*. Cambridge: Cambridge University Press, 2002.

Kilgour, Maggie. *The Rise of the Gothic Novel*. London: Routledge, 1995.

Knellwolf, Christa, and Jane Goodall, eds. *Frankenstein's Science: Experimentation and Discovery in Romantic Culture, 1780–1830*. Farnham: Ashgate, 2008.

Kucich, John, and Jenny Bourne Taylor, eds. *The Oxford History of the Novel in English: Volume 3: The Nineteenth-Century*. Oxford: Oxford University Press, 2012.

LeFanu, Sarah. *In the Chinks of the World Machine: Feminism and Science Fiction*. London: Women's Press, 1988.

Levere, Trevor Harvey. *Poetry Realized in Nature: Samuel Taylor Coleridge and Early Nineteenth-Century Science*. Cambridge: Cambridge University Press, 1981.

Luckhurst, Roger. *The Invention of Telepathy, 1870–1901*. Oxford: Oxford University Press, 2002.

Morris, David B. 'Gothic Sublimity'. *New Literary History* 16 (1985): 299–319.

Otis, Laura. *Membranes: Metaphors of Invasion in Nineteenth-Century Literature, Science and Politics*. Baltimore, MD: Johns Hopkins University Press, 2002 [2002A].

Otis, Laura, ed. *Literature and Science in the Nineteenth Century: An Anthology*. Oxford: Oxford University Press, 2002 [2002B].

Punter, David, ed. *A Companion to the Gothic*. Oxford: Blackwell, 2002.

Punter, David, ed. *A New Companion to the Gothic*. Oxford: Blackwell, 2012.

Roe, Nicholas, ed. *Samuel Taylor Coleridge and the Sciences of Life*. Oxford: Oxford University Press, 2001.

Russ, Joanna. *To Write Like a Woman: Essays in Feminism and Science Fiction*. Bloomington, IN: Indiana University Press, 1995.

Secord, James A. *Victorian Sensation: The Extraordinary Publication, Reception, and Secret Authorship of 'Vestiges of the Natural History of Creation'*. Chicago, IL: University of Chicago Press, 2000.

Senf, Carol A. *Science and Social Science in Bram Stoker's Fiction*. Westport, CT: Greenwood, 2002.

Smith, Jonathan. *Fact and Feeling: Baconian Science and the Nineteenth-Century Literary Imagination*. Madison, WI: University of Wisconsin Press, 1994.

Spooner, Catherine, ed. *The Routledge Companion to Gothic*. London: Routledge, 2007.

Stableford, Brian, ed. and transl. *Camille Flammarion, Lumen*. Middletown, CT: Wesleyan University Press, 2000.

Stableford, Brian, and David Langford. 'Scientists'. *The Encyclopedia of Science Fiction*. Eds. John Clute, Peter Nicholls, and David Langford. 3rd edn. Gollancz, 2012. <www.sf-encyclopedia.com>. Accessed 15 September 2015.

Stiles, Anne. *Popular Fiction and Brain Science in the Late Nineteenth Century*. Cambridge: Cambridge University Press, 2013.

Suvin, Darko. *Metamorphoses of Science Fiction: On the Poetics and History of a Literary Genre.* New Haven, CT: Yale University Press, 1979.

Watt, James. *Contesting the Gothic: Fiction, Genre and Cultural Conflict 1764–1832.* Cambridge: Cambridge University Press, 1999.

Willis, Martin. *Mesmerists, Monsters and Machines: Science Fiction and the Cultures of Science in the Nineteenth-Century.* Kent, OH: Kent State University Press, 2006.

Willis, Martin. 'Le Fanu's "Carmilla", Ireland, and Diseased Vision'. *Literature and Science. Essays and Studies.* Ed. Sharon Ruston. Cambridge: Brewer, 2008. 111–30.

Willis, Martin. *Vision, Science and Literature, 1870–1920: Ocular Horizons.* London: Pickering and Chatto, 2011.

Wolfreys, Julian. *Victorian Hauntings: Spectrality, Gothic, the Uncanny and Literature.* Basingstoke: Palgrave, 2002.

Further reading

DeLamotte, Eugenia C. *Perils of the Night: A Feminist Study of Nineteenth-Century Gothic.* Oxford: Oxford University Press, 1990.

Elbert, Monika, and Bridget M. Marshall, eds. *Transnational Gothic: Literary and Social Exchanges in the Long Nineteenth Century.* Farnham: Ashgate, 2013.

Evans, Arthur B. 'Nineteenth Century SF'. *The Routledge Companion to Science Fiction.* Eds. Mark Bould et al. London: Routledge, 2012.

Höglund, Johan. *The American Imperial Gothic: Popular Culture, Empire, Violence.* Farnham: Ashgate, 2014.

Halberstam, Judith. *Skin Shows: Gothic Horror and the Technology of Monsters.* Durham, NC: Duke University Press, 1995.

Hughes, William. 'The Gothic'. *The Oxford Handbook of Science Fiction.* Ed. Rob Latham. Oxford: Oxford University Press, 2014. 463–74.

Malchow, Howard L. *Gothic Images of Race in Nineteenth-Century Britain.* Redwood, CA: Stanford University Press, 1996.

Page, Michael R. *The Literary Imagination from Erasmus Darwin to H. G. Wells: Science, Evolution, Ecology.* Farnham: Ashgate, 2012.

Schmitt, Cannon. *Alien Nation: Nineteenth-Century Gothic Fictions and English Nationality.* Philadelphia: University of Pennsylvania Press, 1997.

7

POETRY AND SCIENCE

Gregory Tate

University of St Andrews

In 1884 the poet George Barlow published his *Poems Real and Ideal*. One of these poems, 'Poetry and Science', directly addresses the binary distinction set out in the volume's title, identifying scientific knowledge with the 'real' and poetic imagination with the 'ideal', and asserting that the two are irreconcilably opposed:

> Give me the days of faith, and not the days of Science,
> Where fancy is concerned. Each new exact appliance
>   Leaves still less room to dream.
> The fairies, like the wolves that haunted forest-marches,
> Are disappearing fast. No white robe thrills the larches:
>   Titania travels not the moony gleam!
>
> *(Barlow 250–1) (ll. 55–60)*

This is not Victorian poetry at its best. The language of Barlow's stanza is deliberately and therefore embarrassingly archaic: 'fairies', 'forest-marches', 'moony gleam'. The argument which that language articulates is simplistic and repetitive, insisting rather than demonstrating that there is a straightforward antipathy between a reductive scientific rationalism and a poetry characterized vaguely by faith and by dreams. The growing influence of scientific thinking within modern culture, according to this poem, is inexorably damaging to poetic 'fancy'; Barlow's view of the rigid opposition between them is encapsulated in the stanza's rhyme scheme, which shifts from the rhyme of 'science' and 'appliance' to the self-conscious poeticisms of 'dream' and 'gleam', 'marches' and 'larches'. The poem's stance, then, is unsophisticated and schematic, but this is the source of its interest to researchers of poetry and science, because it communicates a conventional assumption that was pervasive and influential throughout the nineteenth century: poetry, more than other literary forms, was consistently defined in opposition to scientific practice. Examining and interrogating this opposition is one of the foundational aims of research on nineteenth-century poetry and science.

As Gowan Dawson and Sally Shuttleworth have pointed out, the widespread nineteenth-century view of the antagonism between science and poetry enjoyed a surprisingly long afterlife. Much twentieth-century criticism on literature and science focused exclusively on the novel, implying through omission that Romantic and Victorian poetry was somehow removed from the intellectual and cultural debates that surrounded science and that the novel as a form sought to document and study. This approach tacitly perpetuated the 'rigid and ahistorical contrast between abstract poetic idealism and empirical scientific positivism' (Dawson and Shuttleworth 1) endorsed in the nineteenth century by Barlow and many others. Recent research, however, has set about complicating this contrast, showing that science was a topic of discussion in every area of poetry and poetics: in conventional verse such as Barlow's; in the more complex and influential writings of other poets; and in theorizations of poetry. This research has also revealed how poets borrowed, interpreted and misinterpreted concepts from each of the proliferating disciplines and specializations of nineteenth-century science, from the human sciences of medicine and psychology to physical sciences such as geology and astronomy. Poets appropriated scientific concepts in order to rethink both the content and the formal design of their verse, and scientific writers similarly made use of poetry to structure their arguments and to articulate their theories. This chapter will discuss the various ways in which science and poetry intersected in the nineteenth century, focusing on three key issues: the supposed epistemological opposition between poetic subjectivity and scientific objectivity; the quotation and the composition of poetry by scientific researchers; and the influence of science on the language and the generic conventions of nineteenth-century verse. The chapter will also, in the process, consider some possible directions for future research in this field.

Subjectivity and objectivity

In nineteenth-century debates about literature and science, the word 'poetry' was frequently employed to signify not just verse but artistic expression more generally: the phrase 'poetry and science' summarized a whole argument about the essential distinction between art's dependence on the subjective personality of the artist and science's factual objectivity and empirical method. Building on the philosophy and aesthetics of Immanuel Kant, Samuel Taylor Coleridge articulates this key argument of romanticism in *The Friend* (1818), differentiating the method of the arts from that of the sciences by asserting that 'in all, that truly merits the name of *Poetry* in its most comprehensive sense, there is a necessary pre-dominance of the Ideas (i.e. of that which originates in the artist himself) and a comparative indifference of the materials' (1:464; author's italics). The exemplary statement of this position, however, is William Wordsworth's account of the 'contradistinction' between 'Poetry and Matter of fact, or Science' in his preface to *Lyrical Ballads* (1802). Science and poetry, according to Wordsworth, are manifestations of two essentially different kinds of knowledge:

> Aristotle, I have been told, hath said, that Poetry is the most philosophic of all writing: it is so: its object is truth, not individual and local, but general, and operative; not standing upon external testimony, but carried alive into the heart by passion; truth which is its own testimony.
>
> *(1992 750–1)*

The 'truth' of poetry, Wordsworth asserts, is not dependent on 'external testimony', the examination and observation of factual data that are central to scientific practice. Instead, poetic expression is self-validating, based on the feelings of the poet, and it speaks to and elicits a corresponding 'passion' in readers. This argument constitutes part of Wordsworth's attempt in the preface to promote an active role for poetry in nineteenth-century Britain: concerned about the social effects of the industrialization of the British economy, he presents poetry as the repository and transmitter of the subjective but shared feelings that guarantee community and social cohesion. The science that underpins industrial technology, conversely, is divorced from feeling and defined as the straightforwardly objective study of matters of fact.

Poets returned to Wordsworth's contradistinction throughout the century, but, while Barlow simply reiterated Wordsworth's argument, other writers contested and interrogated the opposition between the subjective imagination of poetry and the factual objectivity of science. In 1864, for instance, Gerard Manley Hopkins wrote this short poem:

> It was a hard thing to undo this knot.
> The rainbow shines, but only in the thought
> Of him that looks. Yet not in that alone,
> For who makes rainbows by invention?
> And many standing round a waterfall
> See one bow each, yet not the same to all,
> But each a hand's breadth further than the next.
> The sun on falling waters writes the text
> Which yet is in the eye or in the thought.
> It was a hard thing to undo this knot.
>
> *(31)*

John Keats, in his 1820 poem 'Lamia', worried that scientific rationalism had the dangerous potential to 'unweave a rainbow'; the scientific theories of Isaac Newton threatened to rob the rainbow of its magic by explaining it as the product of sunlight refracted through water in the atmosphere (473) (Part 2, l. 237). Hopkins's poem, however, is less certain that science is at odds with imaginative responses to nature, and its formal structure, with its half rhymes and its circular return to the opening expression of knotty and unresolved difficulty, enacts the ambiguities of its argument. The poem at first promotes a subjective reading of natural phenomena: the rainbow, seen from different perspectives by its various spectators, is constructed 'only in the thought'. As a devout Christian, Hopkins was often sceptical of scientific positivism, but these lines also highlight a similarity between scientific method and Romantic and post-Romantic poetry: their shared reliance on the observation and description of nature. Gillian Beer comments in an essay on Hopkins that 'the urgency of his writing is coiled upon contesting forces' and 'the irresolvable energies of counter-explanation' (244), and this is evident in the negotiation between subjectivity and objectivity in this poem. The rainbow shines in part 'by invention', but it is also an objective fact that exists prior to perception, a 'text' written in the sky through the quantifiable process of the refraction of sunlight. Hopkins's poem emphasizes the discrepancy between subjective experience and the scientific explanation of nature, while also indicating that both perspectives are epistemologically valid. Nineteenth-century poets frequently placed scientific concepts alongside other philosophical or religious discourses in their verse. Their goal was often not to draw a straightforward distinction between these discourses but to reflect on the complexities of their competing models of explanation.

The examples of Barlow and Hopkins show that Victorian responses to Romantic assessments of science were diverse and often ambivalent. The study of literature and science complicates the division of nineteenth-century poetry into discrete Romantic and Victorian periods by showing that debates about the relation between poetry and science, and between subjectivity and objectivity in particular, played a pivotal role in understandings of poetry throughout the century. Consistently, some writers asserted that poetic and scientific practices were mutually exclusive, while others argued for their close connection to each other. This is not to say that the priorities, practices and dominant concerns of poetry (and of the sciences) did not shift and change over the course of the century, but there was no abrupt or easily definable switch from Romantic to Victorian attitudes towards science. Research on nineteenth-century poetry and science enables a rethinking of the separation of Romantic and Victorian poetry, inviting instead a more fine-grained historicist approach that considers the various ways in which the terms 'poetry' and 'science' were understood across the century. Noah Heringman has set out a historicist model of criticism specifically targeted towards the study of literature and science; calling it 'aesthetic materialism', he defines it as an approach that responds to theoretical models of 'historical and cultural materialism by articulating a historically specific conception of materiality' (21). Heringman argues that scientific and literary accounts of material, physical nature (for example, in geology or in nineteenth-century poetry) are historically contingent, determined in part by cultural contexts. Even when a writer such as Barlow argues that poetic imagination is removed from the objective and factual concerns of scientific knowledge, his stance is shaped by what Heringman calls 'material and social conditions' (22).

These contextual conditions include the formal conventions of poetic and scientific writing and the modes of publication (such as poetry volumes, scientific treatises and the periodical press) through which that writing was disseminated. Research on poetry and science, therefore, also benefits from asking how poets and scientists worked within these modes and conventions. Recent scholarship on poetry has been characterized by a renewed interest in poetic form. This scholarship, variously identified as 'new formalism' or 'historical poetics', does not consider the formal elements of verse in exclusively aesthetic terms, cut off from the poem's historical contexts. Instead, it analyses, in Marjorie Levinson's words, 'the processes and structures of mediation through which particular discourses and whole classes of discourse (literary genres, for example)' (561) are constructed and communicated in specific historical circumstances. The value of this approach to the study of poetry and science is that it can also be applied to the processes and structures of scientific discourse in the nineteenth century. The formal and generic features of poetic and scientific texts can be read as valuable evidence for changing historical understandings of poetry and science and of the connections between them.

The relation between subjectivity and objectivity, for instance, was an issue not just in nineteenth-century debates about the content and epistemological stance of poetry but also in discussions of poetic form. These discussions were not limited to poets and literary critics; many scientific writers contributed too, using the theory and practice of verse as evidence for their arguments, particularly about psychology and physiology. John Wyatt notes that the historian and philosopher of science William Whewell 'was passionately interested in metre and rhythm in poetry and music' (96). In *The Philosophy of the Inductive Sciences*, published in 1840, Whewell presents prosodic analyses of lines of Latin and English verse to support his argument, derived from Kant, that time exists as an a priori idea in the mind. He identifies poetic metre as the clearest example of 'the apprehension of *rhythm*', the structuring pattern of duration and repetition, which is subjectively imposed on experience: 'All the forms of versification and the *measures* of melodies are the creations of man, who thus realizes in words and sounds the forms of recurrence which rise within his own mind' (Whewell 1840 1:134; author's italics). Poetic forms, according to Whewell, are demonstrations of 'the forms of recurrence' that the mind employs to organize and regulate time and that constitute one of the foundations of perception and knowledge and therefore of scientific investigation.

While Whewell describes metre as an abstract and subjective idea rather than a material and objective fact, other writers put forward physiological definitions of metre as an expression of quantifiable embodied rhythms. Jason David Hall notes that 'the complex interplay between metrical abstraction and embodiment can, in fact, be seen as central to much verse theory' (180) in the nineteenth century. Hall argues that, as the science of psychology developed a more firmly empirical methodology, scientific approaches to poetry became more materialist in orientation. Scientists devised experimental protocols and instruments for measuring and recording the bodily rhythms, modulations of voice and aural sensations involved in the reciting of verse. This approach also contributed to poets' own discussions of their work. Even in the early years of the century, John Thelwall claimed to have developed a theory of rhythm founded on 'the facts and principles of physiological science' (xii). By 1880, the American poet Sidney Lanier was arguing, in *The Science of English Verse*, that the essence of poetry resided not in its semantic meaning but in sensory 'perceptions of sound which come to exist in the mind' (22), either 'by virtue of actual vibratory impact upon the tympanum' or 'by virtue of indirect causes (such as the characters of print and of writing)'. Emphasizing the scientific basis of his account, Lanier identifies the sound of verse as just one example of the physical transmission of sound waves, pointing out that 'what we call "sound," the physicist only recognizes as "vibrations"' (25). As the work of Hall and Matthew Rubery has shown, scientific concepts and new technologies made important contributions to nineteenth-century poetics. Further research is needed, though, on the ways in which physiology, psychology and the scientific study of the senses informed theorizations of poetic form.

Scientific theory, poetic quotation and the writing of verse

Poetry made a contribution to nineteenth-century definitions of science, most frequently through the medium of poetic quotation. Allusions to and quotations from the work of poets were utilized in almost every form of scientific discourse, from public lectures to specialist treatises, and for various purposes. Science was engaged in an ongoing process of professionalization and legitimization in the nineteenth century; its status within British culture was insecure and poetic quotation represented a means of borrowing some of the established cultural prestige of poetry. Quotations might also be deployed in support of specific scientific arguments: because the methodological norms of repeatable observation and experimental verification were not as rigorously defined or as firmly entrenched as they would become in the twentieth century, scientific writers could, with a degree of legitimacy, quote poets' statements about the mind or nature as evidence for their claims. They could also alter and reinterpret the meaning of a quotation by isolating it from its original context and repositioning it in a different argument and a different textual form. The third volume of Whewell's *History of the Inductive Sciences* (1837), for instance, opens with an epigraph from Wordsworth's 1814 poem *The Excursion*:

> 'Go, demand
> Of mighty Nature, if 'twas ever meant
> That we should pry far off yet be unraised;
> That we should pore, and dwindle as we pore,
> Viewing all objects unremittingly
> In disconnection dead and spiritless;
> And still dividing, and dividing still,
> Break down all grandeur, still unsatisfied
> With the perverse attempt, while littleness
> May yet become more little; waging thus
> An impious warfare with the very life
> Of our own Souls!'

(2007 156–7) (Book 4, ll. 953–64)

These lines are spoken in *The Excursion* by the Wanderer, a figure who exemplifies, for Wordsworth, the virtues of traditional rural society. These virtues are demonstrated here in the Wanderer's scepticism towards modern scientific practices; his critique centres on science's 'disconnection', by which he means both its analytical methods and its promotion of intellectual specialization. The chiasmus of 'still dividing, and dividing still' conveys his concern that science fragments nature, and the knowledge of nature, into ever more rigidly demarcated classifications and disciplines. This is contrary to 'the very life / Of our own Souls', which, in the view of the Wanderer and of Wordsworth, is defined by sympathetic connection rather than analytical division. These lines are not an attack on science itself, but on certain aspects of scientific practice; nonetheless, they make a surprising epigraph to a volume of scientific writing.

Whewell's reasons for using this quotation become clearer when the lines are read in the context of his argument in the *History of the Inductive Sciences*. He claims that the history of science is characterized by the incorporation of specific theories and discoveries into ever more comprehensive and accurate understandings of nature: 'the earlier truths are not expelled but absorbed, not contradicted but extended; and the history of each science, which may thus appear like a succession of revolutions, is, in reality, a series of developements [*sic*]' (Whewell 1837 1:10). This cumulative model of the history of science maps a progression towards a unity of knowledge, and the lines from *The Excursion* support Whewell's argument inasmuch as they criticize those scientific practices that tend, conversely, towards analysis and the division of knowledge into separate disciplines. By appropriating the elevated language, moral earnestness, and cultural authority of Wordsworth's attack on scientific 'disconnection', Whewell tacitly enlists the poet in support of his own, unifying, theory of science. In a discussion of scientists' allusions to the poetry of Alfred Tennyson, John Holmes suggests that 'scholarship on literature and science', which 'tends to focus on the interrelationship between literary texts and authors on the one hand and scientific discourses and ideas on the other' (2012 675), would benefit from paying closer attention to the other side of this exchange, to the ways in which scientific authors utilize literary forms. Whewell's quotation of Wordsworth, for example, shows that poetry was employed not just in the communication of specific scientific concepts but also in the construction of philosophies of science. The volume and variety of poetic quotations in nineteenth-century scientific texts means that more work is needed on this important link between science and poetry.

Many practitioners of science wrote verse as well as quoting it. As Daniel Brown has shown, this poetry was not merely a kind of recreation; instead, it offered an alternative means of examining hypotheses and commenting on scientific debates. Brown argues that the formal and linguistic capacities of verse, particularly 'the unruly play of the pun, the tense relations of analogy, and the variegated repetition of rhyme' (2013 261), helped Victorian scientists to rethink and interrogate the experimental comparisons and theoretical complexities that formed the basis of their scientific research. Other connections between the practices of science and of poetry are traceable in the poems of Humphry Davy. A pioneering chemist, experimentalist and inventor, and a charismatic lecturer at the Royal Institution from 1801, Davy was an influential figure in the growing professionalization and public profile of science in the early nineteenth century. He also wrote poetry throughout his life. A friend of Coleridge, Wordsworth and Robert Southey, he published several poems in Southey's *Annual Anthology*. Anthologies of verse such as this, which were popular in the first half of the century and which featured contributions from a wide range of (often non-professional) poets, are an example of a specific mode of publication in which science and poetry might intersect. In Davy's 'Lines Descriptive of Feelings', printed in the *Annual Anthology* for 1800, a visit to his childhood home in Cornwall prompts reflections on his early intellectual development:

Here first I woo'd thee Nature, in the forms
Of majesty and freedom, and thy charms
Soft mingling with the sports of infancy
Its rising social passions and its wants
Intense and craving, kindled into one
Supreme emotion.

(294) (ll. 19–24)

The dynamic of these lines is strikingly similar to that of Wordsworth's poems of retrospect such as 'Lines Written a Few Miles Above Tintern Abbey' (1798): a return to a place of emotional significance triggers memories of youth and of the process of psychological growth, a process which is guided first and foremost by nature but also by emotional sympathy with other people ('social passions'). Written in the artfully unstructured blank verse frequently employed by Wordsworth and Coleridge in their lyrics of memory, these lines are an assured if conventional expression of a key theme of Romantic poetry: the importance of nature's beauty and sublimity, and of childhood emotion, to the formation of personal identity.

The significance of this poem, however, is that it merges this poetic convention with a recognizably scientific approach to the observation of nature. Sharon Ruston has shown that the Romantic notion of the sublime, of nature's 'majesty', is just as important to Davy's scientific writing as it is to his verse (2013 132–74). This poem demonstrates that the language and concepts of science reciprocally informed his poetry. Recalling his childhood responses to the landscapes of Cornwall, Davy writes:

Here first my serious spirit learnt to trace
The mystic laws, from whose high energy
The moving atoms in eternal change
Still rise to animation.

(294) (ll. 33–6)

The 'serious spirit' with which the young Davy studied the laws of nature points towards the rigorous objectivity that would emerge during the nineteenth century as a methodological norm of science. Davy's description of 'The moving atoms in eternal change' suggests that his particular understanding of chemistry, as the study of the 'laws' and powers of nature as they are manifested in the changing states of matter, is founded on his youthful admiration of Cornwall's natural scenery. This poem about memory, in other words, is also about intellectual and professional development. This is not unusual in Romantic verse, which often combines reflections on the formation of personality with considerations of how that formation prepared the way for a poetic vocation; as Brian Goldberg comments, 'the growth of a poet's mind is a process in which credentials are earned' (220). The difference here is that the credentials are scientific rather than poetic; Davy reworks Romantic accounts of the link between nature and poetic creativity to suggest that science too is informed by a love of nature. To some extent, the professionalization of work, and especially of scientific work, in the nineteenth century established and enforced a strict disciplinary division between science and literature; Nicholas Roe, for example, argues that Keats, a medical student, was forced to choose between a career in medicine and his work as a poet (160–81). Davy's poem, however, written early in his career, suggests that the generic conventions of poetry might be used to support and contribute to the construction of a professional scientific persona.

Poetic genres and the language of science

The growing cultural influence of science, and of scientific practices, fundamentally altered the conventions of more than one poetic genre. The eighteenth-century discourse of human science was founded on the application of the empirical scientific method to the study of the mind; together with the literary preoccupation with introspection that shaped romanticism, this discourse participated in a wider cultural reorientation through which human character was reimagined as a legitimate object of observation and classification. This had significant consequences for lyric verse: as Noel Jackson points out, human science and Romantic lyric poetry both 'turn human subjects into experimental objects' (105) while also promoting a 'self-experimental approach' based on the detailed examination of the practitioner's own subjective interiority. This analytical and objectifying approach to the study of the self and other people became more firmly established in science over the course of the nineteenth century, with the incorporation of scientific methods into medicine and the treatment of mental illness, and the emergence of discrete disciplines such as psychology and sociology within human science. The same approach underpinned the dramatic monologue, the influential Victorian poetic genre that severed the link between the lyric voice and the poet's identity, refiguring lyric poetry as the analytical assessment of the mind of a dramatic character.

In Tennyson's 'monodrama' *Maud* (1855), for example, the poet's sustained if ambivalent interest in science is filtered through the pathological perspective of the poem's speaker, a misanthropic young man who rails unhappily at what he considers to be the social and political inequities of nineteenth-century Britain:

> A monstrous eft was of old the Lord and Master of Earth,
> For him did his high sun flame, and his river billowing ran,
> And he felt himself in his force to be Nature's crowning race.
> As nine months go to the shaping an infant ripe for his birth,
> So many a million of ages have gone to the making of man:
> He now is first, but is he the last? is he not too base?
>
> *(2:530) (Part 1, ll. 132–7)*

Published four years before Charles Darwin's *On the Origin of Species*, these lines reflect the widespread cultural influence of pre-Darwinian models of evolution. The speaker's dismay at the 'baseness' of modern society is sharpened by his awareness of the expanses of evolutionary time over which humanity has developed (or failed to develop sufficiently), and by the comparison with the 'monstrous eft', the dinosaur, which once thrived but is now extinct. *Maud* questions the teleological optimism of pre-Darwinian evolutionary theories, the majority of which equated evolution with progressive improvement, by suggesting that there is no such thing as a 'crowning race', or that if there is humanity deserves that title no more than the prehistoric eft. The immediate and incongruous juxtaposition of the references to evolution and the speaker's jaundiced social criticism demonstrates how poets employed scientific notions to comment on a range of (often non-scientific) issues. It also, however, contributes to Tennyson's depiction of his speaker's individual character and mindset. There is no clear argumentative link between the speaker's discussion of evolution and his bitterness towards nineteenth-century 'man'. For another person, knowledge of the vast timescale of evolution and the comparative brevity of human existence might help to put feelings of resentment and hurt pride into some perspective. In *Maud*, conversely, the scientific concept is assimilated into the speaker's prevailingly cynical stance, and so becomes supporting evidence for Tennyson's representation of his morbid misanthropy.

Science was also central to developments in other poetic genres. Some of the most high-profile sciences in nineteenth-century Britain were essentially historical disciplines: evolutionary theory, geology, and astronomy used evidence from the present to construct narratives of the pre-human history of the earth that radically subverted religious accounts of creation. These scientific perspectives were registered in epic poetry, the genre of verse most closely concerned with historical narrative and the relation between past and present. Mathilde Blind's *The Ascent of Man* (1889) can be read as a kind of scientific epic, which seeks to incorporate human history into an even more comprehensive narrative of the history of the universe. The poem begins with a description of the formation of the earth:

> Struck out of dim fluctuant forces and shock of electrical vapour,
> Repelled and attracted the atoms flashed mingling in union primeval,
> And over the face of the waters far heaving in limitless twilight
> Auroral pulsations thrilled faintly, and, striking the blank heaving surface,
> The measureless speed of their motion now leaped into light on the waters.
> And lo, from the womb of the waters, upheaved in volcanic convulsion,
> Ribbed and ravaged and rent there rose bald peaks and the rocky
> Heights of confederate mountains compelling the fugitive vapours
> To take a form as they passed them and float as clouds through the azure.
>
> *(Blind 7–8) (Part 1, ll. 1–9)*

As its title suggests, *The Ascent of Man* is ostensibly a response to Darwinian evolutionary theory, as set out in *On the Origin of Species* (1859) and *The Descent of Man* (1871). Blind aims to reshape Darwinism into an optimistic vision of human and universal development. Her poem, however, says relatively little about the scientific details of Darwinian theory. As these lines suggest, *The Ascent of Man* is less concerned with discussing specific theories than with incorporating a range of sciences, including geology and cosmology, into its totalizing account of history. In this respect it takes as its model pre-Darwinian scientific texts such as Robert Chambers's influential *Vestiges of the Natural History of Creation* (1844), which sought to synthesize diverse scientific findings into a comprehensive explanation of the physical universe. In *The Ascent of Man*, the branches of science are linked together and are communicated in poetry, through what Jason R. Rudy calls 'the universality of rhythmic experience' (156). For Blind, the rhythms of verse (such as the insistent anapaestic metre of these lines) enact the repetitive 'pulsations' and cycles that structure all physical phenomena and that have been observed and quantified through scientific investigation. Yet, while Blind's account of the earth's formation relies on scientific approaches and terminology ('electrical', 'atoms', 'volcanic'), it also borrows the language of the biblical creation narrative of Genesis, in which God moves 'over the face of the waters'. As Charles LaPorte has shown, Blind's writing offers a form of 'atheist prophecy', in which the moral authority of poetry is rooted in the 'evidence of scientific truths' (439) while also retaining elements of religious discourse. Nineteenth-century poetry constituted a medium in which scientific and spiritual modes of explanation could be compared, examined and to some extent reconciled.

Poetic representations of science often hinged on particular arrangements of language, and the use of scientific terminology in nineteenth-century poetry is a topic that would benefit from further research. Several questions remain to be answered: to what extent does the use of scientific terms in a poem signify a sustained engagement with, as opposed to a passing reference to, science? How is the jargon of science incorporated into the formal structures of metre and rhyme? And how did reviewers and readers respond to the introduction of specialist vocabularies into poetry? Many scientists were acutely aware of the complex function of language in their own work; in 1838, the influential geologist Charles Lyell argued that his discipline was characterized by an approach to language directly opposed to that of poetry. Lyell quotes the seventeenth-century poet Edmund Waller on the ephemeral nature of poetic styles and trends: 'We write in sand, our language grows, / And, like the tide, our work o'erflows' (264). He then comments:

> But the reverse is true in geology; for here it is our work which continually outgrows the language. The tide of observation advances with such speed, that improvements in theory outrun the changes of nomenclature; and the attempt to inculcate new truths by words invented to express a different or opposite opinion, tends constantly, by the force of association, to perpetuate error.

Lyell sets out a conventional distinction here: literature is inherently textual and science essentially factual. Poetry, the quotation from Waller suggests, is defined by language, and so alterations in language over time impinge on the meaning of poems, possibly rendering them irrelevant or even unintelligible. Geology, conversely, consists of theories founded on the 'observation' of concrete data; 'changes of nomenclature' are too cumbersome to keep pace with the rapid and efficient progress of the scientific method. Lyell's argument, however, points to a different conclusion from the one he intends: by recognizing that scientific 'truths' must be communicated through words, and that these words sometimes inevitably 'perpetuate error', he demonstrates that science is just as textual and linguistic a discourse as poetry. The development of language, and its role in defining their work, was a concern shared by poets and scientists.

The relations between the languages of geology and of poetry are explored in Tennyson's poem *The Princess* (1847). Tennyson owned many of Lyell's writings on geology, and his poetry frequently employs geological theories, concepts and terms. The characters of *The Princess* engage, at one point, in some geological investigation of a kind that was practised in the nineteenth century not just by professional scientists but also by amateur enthusiasts. The group travels to some mountains to collect geological samples:

> Many a little hand
> Glanced like a touch of sunshine on the rocks,
> Many a light foot shone like a jewel set
> In the dark crag: and then we turned, we wound
> About the cliffs, the copses, out and in,
> Hammering and clinking, chattering stony names
> Of shale and hornblende, rag and trap and tuff,
> Amygdaloid and trachyte, till the Sun
> Grew broader toward his death and fell, and all
> The rosy heights came out above the lawns.
>
> *(Tennyson 2:230) (Canto 3, ll. 338–47)*

These lines begin and end with some fairly unremarkable nineteenth-century poetic diction: the male speaker starts by making romanticized (and belittling) references to his female companions ('little hand', 'light foot') and concludes by evoking the visual spectacle of a sunset. In between, though, the register shifts away from the conventionally lyrical, with the introduction of geological terminology: 'shale and hornblende, rag and trap and tuff, / Amygdaloid and trachyte'. The geological terms are located strictly within the iambic metre of the lines and are linked together through the alliterative patterning of 'a' and 't' sounds, but this seamless integration of scientific terminology into the poem's blank verse is ambiguous in its effects. Is Tennyson's use of these terms an assertion of the inclusive potential of poetic writing, its capacity to appropriate and assimilate the languages of science? Or is it a comic strategy, mocking the intellectual pretensions and claims to scientific knowledge of these amateur geologists?

There is evidence for both interpretations. Dennis R. Dean, in his pioneering study *Tennyson and Geology*, argues that this geological outing, with its 'hammering and clinking' and 'chattering', descends 'into congenial triviality' (16). This reading is supported, to some extent, by a consideration of the particular 'stony names' voiced here, which designate a mix of igneous and sedimentary rocks and minerals with no clear geological or geographic connection to each other. The characters' chatter, a rote recycling of jargon rather than a demonstration of substantive knowledge, seems to reinforce a distinction between scientific and poetic discourse. Yet these geological terms also highlight the problems with scientific nomenclature discussed by Lyell. These names are not scientific neologisms, formulated to identify specific geological phenomena. Instead, many of them carry broad and ambiguous meanings, having evolved from a variety of vernacular languages: 'rag' from Middle English, 'trap' from Swedish and 'tuff' from Italian. In this respect, the names suggest that there is no straightforward difference between scientific terminology and other forms of language, and that poetry, therefore, can legitimately make use of technical terms for purposes other than ridicule. Such a reading is offered by Virginia Zimmerman, who argues that *The Princess* takes its geological terminology seriously, and that the scientific conversation between men and women in these lines makes an important contribution to the poem's wider debates about gender relations and social change (65–95).

Comedy and social commentary, however, were not mutually exclusive: some poets utilized scientific language expressly for the purpose of satire. This approach did not limit itself to mocking science and scientists: for example, Constance Naden's 'Evolutionary Erotics' (1887) wittily explores some surprising connections between science, gender relations and the traditions of love poetry. Discussing this series of four poems, Holmes observes that Naden's comic verse is designed both 'as a piece of broad comedy mocking the supposed social ineptitude of scientists and their unflinching commitment to science, and as a witty critique of scientific practice aimed at scientists themselves and their readers' (2009 194). Naden's humour also depends on a self-conscious awareness of the complex interaction of scientific concepts with poetic conventions. In 'Scientific Wooing' a male student draws on a plethora of scientific images in an effort to express his love:

> I'll sing a deep Darwinian lay
> Of little birds with plumage gay,
>> Who solved by courtship Life's enigma;
> I'll teach her how the wild-flowers love,
> And why the trembling stamens move,
>> And how the anthers kiss the stigma.
>
> Or Mathematically true
> With rigorous Logic will I woo,
>> And not a word I'll say at random;
> Till urged by Syllogistic stress,
> She falter forth a tearful 'Yes,'
>> A sweet '*Quod erat demonstrandum!*'

> (Naden 138) (ll. 73–84)

The humour here is directed partly at the speaker: his rapid jumps from one scientific model to another suggest that his knowledge of the sciences may not be particularly deep, and some of his suggestions ('With rigorous Logic will I woo') are laughably incongruous and inept. The poem also, though, satirizes the genre of love poetry itself, and specifically its dependence on elaborate metaphors. The conventional comparison of human relationships with nature is enhanced but also subverted by Naden's allusion to Darwin's discussions of the evolved characteristics used by birds in 'courtship' displays. On the one hand, the Darwinian reference suggests that the metaphorical link between human and animal behaviours may be firmer than previous poets supposed. On the other hand, it highlights the discrepancy between the biological 'courtship' of birds and the arbitrary social rituals of human relationships. Throughout these stanzas, the tension between the perspectives of science and love poetry is enacted in the poem's language and form: in the juxtaposition of 'Darwinian' with the archaic poeticism of 'lay', for example, and in the deliberately obtrusive and ridiculous rhyme of 'random' and '*Quod erat demonstrandum*'. Naden's satirical and poetically self-conscious deployment of technical terminology represents just one of the ways in which such terminology was incorporated into verse, and more research is needed on how scientific language contributed to poetic writing.

The notion that poetry and scientific vocabulary were incompatible was itself, at times, a target of satire in the nineteenth century. In 1888, Algernon Charles Swinburne published an article in the periodical *The Nineteenth Century*, purporting to be 'a contribution to the Tennyson-Darwin controversy'. Mocking recent speculations that Shakespeare's plays were in fact written by the philosopher Francis Bacon, Swinburne satirically presents evidence for the argument that Tennyson's poetry was composed by Darwin. The real authorship is revealed, he claims, by the evolutionary speculations of *Maud*'s 'monstrous eft' stanza and by the geological terminology of *The Princess*: in these poems 'the borrowed plumes of peacock poetry have fallen from the inner kernel of the scientific lecturer's pulpit' (Swinburne 129). Swinburne's satire ridicules the view that science and scientific discourse have no place in poetry. His article also shows that the periodical press was an important forum for the exchange of views about the connections between poetry and science. Inherently intertextual and dialogic, periodicals frequently brought together scientific articles, poetry, and reviews of scientific and poetic books, making the press one of the most significant forms of publication through which relations between poetry and science could be set out and debated.[1] More detailed research is needed on the role of the press in fostering and disseminating discussions about science and poetry in nineteenth-century Britain. Another priority for future research is to consider how these discussions resonate after 1900, for example in the verse of Thomas Hardy, which frequently responds to Victorian scientific theories, and in the writings of later generations of scientists and poets whose views and practices were shaped by the debates of the nineteenth century.

Note

1 See Gowan Dawson, 'Science and the Periodical Press', in this volume.

Bibliography

Primary texts

Barlow, George. *Poems Real and Ideal*. London: Remington, 1884.
Blind, Mathilde. *The Ascent of Man*. London: Chatto and Windus, 1889.
Coleridge, Samuel Taylor. *The Friend*. 2 vols. Ed. Barbara E. Rooke. Princeton, NJ: Princeton University Press, 1969.

Davy, Humphry. 'Lines Descriptive of Feelings Produced by a Visit to the Place Where the First Nineteen Years of My Life Were Spent, in a Stormy Day, After an Absence of Thirteen Months'. *The Annual Anthology*. Ed. Robert Southey. London: Longman and Rees, 1800. 293–6.

Hopkins, Gerard Manley. *The Poetical Works of Gerard Manley Hopkins*. Ed. Norman H. Mackenzie. Oxford: Clarendon, 1990.

Keats, John. *The Poems of John Keats*. Ed. Jack Stillinger. London: Heinemann, 1978.

Lanier, Sidney. *The Science of English Verse*. New York, NY: Charles Scribner, 1880.

Lyell, Charles. *Elements of Geology*. London: John Murray, 1838.

Naden, Constance. *A Modern Apostle, The Elixir of Life, The Story of Clarice, and Other Poems*. London: Kegan Paul and Trench, 1887.

Swinburne, Algernon Charles. 'Dethroning Tennyson: A Contribution to the Tennyson-Darwin Controversy'. *The Nineteenth Century* 23 (1888): 127–9.

Tennyson, Alfred. *The Poems of Tennyson*. Ed. Christopher Ricks. 2nd edn. 3 vols. Harlow: Longman, 1987.

Thelwall, John. *Illustrations of English Rhythmus*. London: J. McReery, 1812.

Whewell, William. *History of the Inductive Sciences*. 3 vols. London: John W. Parker, 1837.

Whewell, William. *The Philosophy of the Inductive Sciences*. 2 vols. London: John W. Parker, 1840.

Wordsworth, William. Preface to *Lyrical Ballads*. *Lyrical Ballads and Other Poems, 1797–1800*. Eds. James Butler and Karen Green. Ithaca, NY: Cornell University Press, 1992, 741–60.

Wordsworth, William. *The Excursion*. Eds. Sally Bushell, James A. Butler, and Michael C. Jaye. Ithaca, NY: Cornell University Press, 2007.

Secondary texts

Beer, Gillian. 'Helmholtz, Tyndall, Gerard Manley Hopkins: Leaps of the Prepared Imagination'. *Open Fields: Science in Cultural Encounter*. Oxford: Oxford University Press, 1996, 242–72.

Brown, Daniel. *The Poetry of Victorian Scientists: Style, Science and Nonsense*. Cambridge: Cambridge University Press, 2013.

Dawson, Gowan, and Sally Shuttleworth. 'Introduction: Science and Victorian Poetry'. *Victorian Poetry* 41 (2003): 1–10.

Dean, Dennis R. *Tennyson and Geology*. Lincoln: Tennyson Society, 1985.

Goldberg, Brian. *The Lake Poets and Professional Identity*. Cambridge: Cambridge University Press, 2007.

Hall, Jason David. 'Materializing Meter: Physiology, Psychology, Prosody'. *Victorian Poetry* 49 (2011): 179–97.

Heringman, Noah. *Romantic Rocks, Aesthetic Geology*. Ithaca, NY: Cornell University Press, 2004.

Holmes, John. *Darwin's Bards: British and American Poetry in the Age of Evolution*. Edinburgh: Edinburgh University Press, 2009.

Holmes, John. '"The Poet of Science": How Scientists Read Their Tennyson'. *Victorian Studies* 54 (2012): 655–78.

Jackson, Noel. *Science and Sensation in Romantic Poetry*. Cambridge: Cambridge University Press, 2008.

LaPorte, Charles. 'Atheist Prophecy: Mathilde Blind, Constance Naden, and the Victorian Poetess'. *Victorian Literature and Culture* 34 (2006): 427–41.

Levinson, Marjorie. 'What is New Formalism?'. *PMLA* 122 (2007): 558–69.

Roe, Nicholas. *John Keats and the Culture of Dissent*. Oxford: Clarendon, 1997.

Rubery, Matthew. 'Thomas Edison's Poetry Machine'. *19: Interdisciplinary Studies in the Long Nineteenth Century* 18 (2014). <www.19.bbk.ac.uk/articles/10.16995/ntn.678>. Accessed 27 July 2015.

Rudy, Jason R. *Electric Meters: Victorian Physiological Poetics*. Athens, OH: Ohio University Press, 2009.

Ruston, Sharon. *Creating Romanticism: Case Studies in the Literature, Science and Medicine of the 1790s*. Basingstoke: Palgrave Macmillan, 2013.

Wyatt, John. *Wordsworth and the Geologists*. Cambridge: Cambridge University Press, 1995.

Zimmerman, Virginia. *Excavating Victorians*. Albany, NY: State University of New York Press, 2008.

Further reading

Allard, James Robert. *Romanticism, Medicine, and the Poet's Body*. Aldershot: Ashgate, 2007.

Armstrong, Isobel. *Victorian Poetry: Poetry, Poetics and Politics*. London: Routledge, 1993.

Blair, Kirstie. *Victorian Poetry and the Culture of the Heart*. Oxford: Clarendon, 2006.

Brown, Daniel. *Hopkins' Idealism: Philosophy, Physics, Poetry*. Oxford: Clarendon, 1997.

Cohen, William A. *Embodied: Victorian Literature and the Senses*. Minneapolis, MN: University of Minnesota Press, 2009.

Faas, Ekbert. *Retreat into the Mind: Victorian Poetry and the Rise of Psychiatry*. Princeton, NJ: Princeton University Press, 1988.

Gold, Barri J. *ThermoPoetics: Energy in Victorian Literature and Science*. Cambridge, MA: MIT Press, 2010.

Henchman, Anna. *The Starry Sky Within: Astronomy and the Reach of the Mind in Victorian Literature*. Oxford: Oxford University Press, 2014.

Kaston Tange, Andrea. 'Constance Naden and the Erotics of Evolution: Mating the Woman of Letters With the Man of Science'. *Nineteenth-Century Literature* 61 (2006): 200–40.

O'Neill, Patricia. '*Paracelsus* and the Authority of Science in Browning's Career'. *Victorian Literature and Culture* 20 (1992): 293–310.

Richardson, Alan. *British Romanticism and the Science of the Mind*. Cambridge: Cambridge University Press, 2001.

Ross, Catherine E. '"Twin Labourers and Heirs of the Same Hopes": The Professional Rivalry of Humphry Davy and William Wordsworth'. *Romantic Science: The Literary Forms of Natural History*. Ed. Noah Heringman. Albany, NY: State University of New York Press, 2003. 23–52.

Ruston, Sharon. *Shelley and Vitality*. Basingstoke: Palgrave Macmillan, 2005.

Tate, Gregory. *The Poet's Mind: The Psychology of Victorian Poetry 1830–1870*. Oxford: Oxford University Press, 2012.

8

SCIENTIFIC LITERARY CRITICISM

Peter Garratt

Durham University

Literary critics in the nineteenth century became interested in establishing the view that science was a legitimate source of the *ethos* of criticism. This way of thinking has been more or less erased from the common idea of criticism's historical development, even by current movements such as literary Darwinism that pursue a scientific (or perhaps scientistic) agenda without being sufficiently aware of Victorian evolutionary criticism and its 'false steps' (Holmes 102). Yet it was during the nineteenth century that it became possible to consider criticism in relation to scientific ideas, concepts and methods, and even to fashion criticism itself as a science. Writers of the eighteenth century would have been perplexed by this formulation, since 'criticism' then designated an operation of taste and appreciation rather than the pursuit of empirical knowledge. Between the Augustan age and the twentieth-century formation of modern literary studies, however, there was an influential strain of scientific thought mixed in with mainstream literary theory and criticism. What obscures it today is almost certainly a dominant story of the development of aesthetic criticism traced by a line running typically from the Romantic period through Matthew Arnold to the likes of Walter Pater and Henry James – an intellectual tradition rekindled in the Leavisite era and inflected by its own two-cultures perspective.

Science played a more central role than this story allows, as the present chapter explores. Undeniably, there are important nineteenth-century literary critics with only minor relevance to this account, and figures such as Leigh Hunt, Edward Bulwer-Lytton and Walter Bagehot are omitted from the discussion in favour of others with a more direct bearing on the emergence of scientific criticism. The sciences of psychology and biology receive particular emphasis since they had the greatest influence on literary thinking, as well as being distinctive in the period in a more general sense. What follows is broadly chronological in nature but three discernible themes recur. One might be termed the Romantic critical legacy, concerning the entanglement of aesthetic and scientific theories in critical prose at the start of the century and its influence on later writers. Another is the importance of investigating pleasure, an emotional, physiological and literary state framed by the period's moral attitudes. Finally, as criticism embraced scientific knowledge or sought to become scientific by, say, pursuing fixed laws of literature, it opened onto another anxious issue that was a feature of the massive growth of nineteenth-century print and periodical culture: the professionalization of the literary critic.

Science and the spirit of the age

Reason, or an excess of it, has been presented as an enemy of aesthetic experience since the Romantic period. A complaint that literary criticism relates parasitically, even enviously, to the truly primary aesthetic work, composing itself at a distance as a discourse of reason, stretches back even further into the cultural past and may be associated with the legacy of Plato's philosophy. 'Unlike any other discipline, literary criticism arose in hostility to the object of its study', Seán Burke has suggested, with reference to the 'ethical denunciations' (25) of literature that would see poets banished from Plato's rationally conceived *polis*. Echoes of this inaugurating banishment undoubtedly reverberate in romanticism's central statements of artistic purpose and in its essays on poetic theory, especially Percy Shelley's *A Defence of Poetry* (written in 1821 though not published until 1840), which responds to it by reclaiming Plato himself as 'essentially a poet' (679) who never wrote in verse, and by casting Shakespeare, Dante and Milton as 'philosophers of the very loftiest powers', in a reversal of conventional categories. Shelley's immediate critical interlocutors in the *Defence* appear to be Philip Sidney and Thomas Love Peacock, but implicitly he has in mind a much longer sense of tradition encompassing Plato and Platonism, albeit in a spirit of profound ambivalence to both of them (see Ware). This can be seen at the end of the Victorian period, too, in Oscar Wilde's Platonic dialogues, notably *The Critic as Artist* (1890), which in its subversive Paterian fashion lends literary artifice to the performance of criticism itself. At this decadent extreme, all that can be said to matter is the creation of exquisite verbal structures, without the taint or imposition of discursive hierarchies. If Wilde's critical decadence expresses one form of Romantic inheritance, this is because it appears to extend and exaggerate a critique of detached reason and artless abstraction already audible in Wordsworth's ironic complaint in 'The Tables Turned' (1798) that the 'meddling intellect' (81–2) (l. 26) distorts and disfigures what it seeks to know. Variations of this Romantic protest against Enlightenment standards of knowledge run through the critical writings of Coleridge, Hazlitt and Shelley, amid arguments embracing intensity of feeling, spontaneity and imagination. What Hazlitt called the 'spirit of the age' in 1825 has since become associated to a significant degree with these dissenting anti-scientific postures.

Even if a grain of truth dwells in every cliché, this now-familiar picture of romanticism's hostility to reason, empiricism and science needs serious qualification, as several important studies have shown (see, for example, Smith; Roe; Ruston 2005). One source of the problem is a distinction that Wordsworth draws between poetic sensibility and the 'Man of Science' in his preface to *Lyrical Ballads* (105–7), an area of discussion that was expanded in his revised version of 1802. This aspect of the preface has widely been taken to licence a trenchant split between science and poetry (poetry being for Wordsworth and other Romantic critics a larger category than metrical language), and solid evidence for the way that Wordsworth and Coleridge 'both distinguish poetry (whose object is pleasure) from science (whose object is truth)' (Esterhammer 145). Yet, in principle, according to the preface, pleasure underwrites all forms of knowing – indeed 'the Chemist and Mathematician . . . know and feel this' (Wordsworth and Coleridge 105–6), just as the poet does – and, moreover, poetry is acknowledged to be purposeful in ways that reach beyond the solicitation of pleasure, for 'its object is truth'. While this truth may be derived internally, by means of the feelings, and while it may be considered distinct from the kind of truth that science aims at, the pleasure of the aesthetic deepens understanding and crystallizes as purposive knowledge.

One possible explanation for this move in Wordsworth's thinking is that his theory wishes tactically to extend poetry's natural perimeter such that it encompasses the territory of the sciences, a kind of definitional imperialism that Shelley then repeats. 'Poetry is the first and last of all knowledge' (Wordsworth and Coleridge 103), insists the preface, memorably; *A Defence of Poetry* claims more stridently still that it is 'at once the centre and circumference of knowledge; it is that which comprehends all science, and that to which all sciences must be referred' (Shelley 696). Two points reveal this view to be a simplification, however, or only one part of the story. First, as a number of commentators have pointed out (for example, Jackson; Page; Mitchell; Ruston 2013), the preface declares that the poems in *Lyrical Ballads* have amounted to an 'experiment' (Wordsworth and Coleridge 95), recycling a keyword from the volume's original 1798 advertisement. This carefully chosen term, *experiment*, reinforces the sense that *Lyrical Ballads*, in design and execution, can be correlated with the 'experimental paradigm of science' (Page 48). One could go further in linking the two, as Robert Mitchell has done, by suggesting that the poetic transposition of the term reveals Wordsworth and Coleridge to have been experimenting with prevailing assumptions of what it meant to conduct experiments at all, for they 'did not simply *apply* an existing scientific concept of experiment to their poetry but rather *experimented* with the concept and practice of experimentation itself' (34). At the very least, *experiment* conveys a prospective and anticipatory mood arising from having deliberately set up structured conditions for the arrival of newness, the sense not merely of leaving behind past conventions (in this case, Augustan poetic form and diction) but, more urgently, of establishing a structured context in which new effects, interactions or sets of relations might emerge. Wordsworth's language at the start of the preface recreates such a mood and situation, initially by recalling how *Lyrical Ballads* was angled openly towards the future – 'I had formed no very inaccurate estimate of the probable effects of these Poems' (95) – and then by setting forth the precise terms of its experimentation:

> Low and rustic life was generally chosen, because in that condition, the essential passions of the heart find a better soil in which they can attain their maturity, are less under restraint, and speak a plainer and more emphatic language; because in that condition of life our elementary feelings co-exist in a state of greater simplicity, and, consequently, may be more accurately contemplated, and more forcibly communicated; because the manners of rural life germinate from those elementary feelings; and, from the necessary character of rural occupations, are more easily comprehended, and are more durable; and lastly, because in that condition the passions of men are incorporated with the beautiful and permanent forms of nature.
>
> *(97)*

What this describes is a finely calibrated experimental apparatus. These details are choices designed to maximize the accessibility of the targeted phenomena, namely excited states of feeling. It supplies an account of finding the best conditions in which such feeling can be 'contemplated', 'communicated' and 'comprehended' – in short, realized and observed. In this respect, the poems in *Lyrical Ballads* form an experiment in human feelings, which cross inside and outside the domain of the aesthetic, as well as an experiment in transfiguring common or rustic life into a valued poetic object. The feelings in question arise from the stimulus of poetry's own language, which by Wordsworth's admission may include 'strangeness and aukwardness' (96) caused by an adoption of the 'language really used by men' (97). They emanate expressively from the poetic self in the form of the 'spontaneous overflow of powerful feelings' (Wordsworth and Coleridge 111) and 'emotion recollected in tranquillity'. That this experiment should be traceable to the 'primary laws of our nature' (97) is a sign of the influence of David Hartley's associationist psychology on *Lyrical Ballads* and of Wordsworth's reliance on Hartley's science of mind to develop an account of this porous affective system of author, work and reader.

A second point to register, briefly, is that its concern with futurity leads the preface to imagine a moment when the findings of natural science will be integrated with the aims of poetry. 'The remotest discoveries of the Chemist, the Botanist, or Mineralogist,' it predicts, 'will be as proper objects of the Poet's art as any upon which it can be employed, if the time should ever come when these things shall be familiar to us' (Wordsworth and Coleridge 105–6). This vision of poeticized science, for Wordsworth, expresses how all fields of learning fundamentally require the verve and authority of the creative imagination. Crucially, poet, critic and the man of science have yet to be segregated by specialization or by the professionalization of knowledge, even if the preface regards the scientist as a producer of relatively isolated knowledge. In exemplary fashion, Wordsworth's revisions to his poetic theory in the 1802 version of the preface bear the very probable imprint of his friendship with the chemist Humphry Davy. As Sharon Ruston has illustrated, the text establishes a subtle dialogue with a surprisingly wide range of sources from science, medicine, political theory and philosophy, though the connection with Davy is especially lively (2013 7–27). This may well explain 'textual echoes' (Ruston 2013 21) in their writing. According to Roger Sharrock, the first critic to consider this relationship in detail, Wordsworth's additions to the preface (which depart from his earlier theme of poetry's expressive power by focusing on the figure of the poet) most likely reflect his awareness of Davy's inaugural lecture for the Royal Institution in January 1802, in which he delivered a popular argument for the constitutive role of imagination in chemistry and for the dynamic interdependence of the arts and sciences generally. In the revised preface that came after it, Wordsworth accepted 'Davy's vision of a future transformed by the chemist and his colleagues' (Sharrock 74). But the important wider point illustrated here is that Wordsworth's poetic theory, arguably the keystone of Romantic critical writing, opens up the prospect of entangling scientific ideas and values with the vitalizing power of the imagination.

If the spirit of the age, for Hazlitt, could be signalled by attitudinal terms such as 'gusto', then he also felt that literary criticism had to break free from the barren and stultifying habits of earlier neoclassical models, especially the prose of Dryden, while rejecting just as firmly the 'modern or metaphysical system of criticism' (2:123). This pointed reference to modern metaphysics, from his essay 'On Criticism' (1824), was intended, of course, for Coleridge, who was perceived to have absorbed excessive amounts of German philosophy from around the turn of the nineteenth century while trying to develop a serious literary-critical method. Coleridge may indeed have 'found his desire for the marriage of nature and spirit sympathetically voiced by Kant' (Ashton 49–50), and furthermore by Fichte and Schelling, especially when seeking to define the imagination as healing the rift between the mind and matter, subject and object, in *Biographia Literaria* (1817). There, dryly, he admits that metaphysics and psychology have long been his 'hobby-horse' (Coleridge 204). Even so, the metaphysical drift of Coleridge's critical project can be overstated: he remained engaged with natural philosophy and scientists such as Davy; he took chemistry to be a principle of material interconnection that had appeal as a metaphor for life's – and art's – deep unity; and the theory of the association of ideas, taken from Hartley and later famously disowned for being mechanistic and passive, continued to influence his critical lexicon.

Organicism, psychology and positivism

Coleridge's thought has two important legacies for post-Romantic scientific literary criticism. The first of these is the notion of organicism; the second is the language of mental association, which many mid-Victorian writers, both critical and literary, continued to use freely. Under this broad head, it should be noted that the first sustained attempts to formulate a science of criticism were able to cast themselves in this role because they identified consciously with the nascent discipline of psychology.[1] To begin by addressing the former issue, Coleridge presents his aesthetic theory of organicism by the time of the *Biographia*, a work that shares *The Prelude*'s interest in individual life, that is, autobiographical growth, while managing to be formally anything but continuous and holistic (rather, patchy and miscellaneous). Organic form, though, refers to wholes and to the achievement and apprehension of wholeness. Even if the term *organic* 'is less figurative for Coleridge than is commonly supposed' (Jones 104), it does represent a borrowing from biology, most likely plant botany, and implies a strong correspondence with physical dynamics in nature, especially the idea of an organism's unique instantiation of self-furthering life. In *Aids to Reflection* (1825), Coleridge uses 'organic' at one moment in a narrowly biological way: 'The fairest part of the most beautiful body will appear deformed and monstrous, if dissevered from its place in the organic whole' (281). Gathering these resonances, it becomes a critical term naming the realized structural unity of the verbal artwork, including the way that a poem or text originates its own laws rather than obeying pre-set or mechanically applied criteria. As has been discussed widely, organicism is at once liberating, conservative, autonomous and figurative, capturing the spirit of the genuinely creative imagination as opposed to merely mechanical fancy (Abrams 168–9).

Organicist rhetoric was taken up by a wider Romantic critique of mechanical or merely materialist philosophies of the self and nature, one of its targets being Bentham's utilitarian pleasure calculus, which would resonate with late Romantic early Victorians like Carlyle and Dickens. Shelley's *A Defence of Poetry*, which makes use of organic metaphors of growth and vitality, as in the image of the poet's thoughts as 'the germs of the flower and the fruit of latest time' (677), also launches an attack on the narrow economic concept of utility. Much like John Stuart Mill's later distinction between higher and lower types of pleasure, Shelley wishes to show that love, friendship and reading literature exemplify a complex and superior form of affective gratification that resists being reduced to quantification, and precisely therefore troubles its authority. 'The production and assurance of pleasure in this highest sense is true utility,' Shelley says (695). Mill himself made an important contribution to literary criticism with two essays that appeared in the *Monthly Repository* in 1833, titled 'What is Poetry?' and 'The Two Kinds of Poetry'. Alerting Carlyle by letter to the publication of the first, Mill worried that his correspondent would find his work of criticism 'too much infected by mechanical theories of the mind: yet you will probably in this, as in many other cases, be glad to see that out of my mechanical premises I elicit *dynamical* conclusions' (1910 1:67). Organicism hangs in the air here. In a related way, Mill criticizes utilitarianism in his well-known essay 'Bentham' (1838) by identifying Bentham's 'deficiency of Imagination':

> In many of the most natural and strongest feelings of human nature he had no sympathy; from many of its graver experiences he was altogether cut off; and the faculty by which one mind understands a mind different from itself, and throws itself into the feelings of that other mind, was denied him by his deficiency of Imagination.
>
> *(1963–91 10:91)*

Mill's liberalism, in contrast to Bentham's utilitarianism, required a model of selfhood that could be harnessed to the advantages of poetic sympathy. According to his 1833 essays, it was poetry that provided a foundational model for accessing the minds of others. Earlier understandings of sympathy had garnered scientific or quasi-scientific associations, in particular with physiology, which informed ideas of embodied emotional communication in the second half of the eighteenth century, though conceptual links with folk wisdom and 'quack science' were also retained (Fairclough 22). Mill's use of *sympathy* reflects a shift that is at once Romantic and post-Romantic: his criticisms of Bentham's philosophy turn on the observation that the purely deductive method neglects how to cultivate social minds effectively (Snyder 15–18), a purpose aptly served by the phenomenon of artistic – and specifically poetic – pleasure. For Mill, poetry is special in that it discloses states of feeling rather than external events, and in doing so directly offers its reader 'the delineation of the deeper and more secret workings of human emotion' (1833 62). This truth helps to explain why infant minds relish narrative forms before they can appreciate a great poem, he claims. Poetic eloquence then puts minds in a communicable relation: 'Eloquence is feeling pouring itself out to other minds, courting their sympathy, or endeavouring to influence their belief, or move them to passion or to action' (Mill 1833 64–5).

If ideas of organic form were never separable from economic and political questions of social structure and reform, then discoveries in physiology in the middle of the nineteenth century brought organicism into contact with mid-Victorian science. The consequences directly shaped the possibilities of literary criticism, especially of the novel, in an era much influenced by Comtean positivism. A key figure at the centre of these developments was George Eliot, whose intellectual world view was first shaped by an intense familiarity with the poetry of Wordsworth and Romantic ideas of organic form. In Eliot's mature view, her novels were to be regarded as 'experiments in life' (1954–78 6:216), as she wrote to the publisher John Blackwood in 1876, distantly recalling Wordsworth's experimental purpose in *Lyrical Ballads*. But Eliot was increasingly familiar with biology and physiology, too, and her partner was the influential polymath George Henry Lewes. An important critical essay by Eliot, 'Notes on Form in Art' (1868), illustrates how her views negotiate finely between an inherited Romantic discourse of formal organic unity and more recent scientific models of mind and matter that pictured these phenomena as vital distributed processes. A key idea Eliot explores here is 'complexity', in relation to both art and the form of biological organisms:

> Poetic Form was not begotten by thinking it out or framing it as a shell which should hold emotional expression, any more than the shell of an animal arises before the living creature; but emotion, by its tendency to repetition, i.e., rhythmic persistence in proportion as diversifying thought is absent, creates a form by the recurrence of its elements in adjustment with certain given conditions of sound, language, action, or environment. Just as the beautiful expanding curves of a bivalve shell are not first made for the reception of the unstable inhabitant, but grow and are limited by the simple rhythmic conditions of its growing life.
>
> *(2000 358–9)*

As well as being a refutation of abstract a priori ideals, Eliot's thinking about form here is deeply interested in relational interdependence and in the dynamic reciprocal relationship between an organism and its environment. However clearly marked the passage is by her Romantic antecedents, its organicism is no longer merely a metaphorical borrowing from biology but a committed turn to biological science as a way of modelling epistemology and aesthetics.

Lewes, himself a critic, dramatist, novelist, scientist, psychologist and philosophical historian, was one of the pioneers of a scientific method in literary criticism, and Eliot's critical vocabulary owes something of a debt to him, as well as to other broadly positivist writers such as Herbert Spencer who pursued an anti-metaphysical systemization of final and binding knowledge. As Charlotte Sleigh points out,[2] Lewes significantly shaped the mid-century understanding of realism as a goal of novelistic representation, especially via his essay 'Realism and Idealism' (1858), while at the same time exemplifying commensurate observational practices in scientific works such as *Seaside Studies* (1860). Nevertheless, Lewes has an understated role in the usual narrative of the history of criticism. His talents and writing are regarded as somewhat diffuse. But he is in fact central to the development of critical models bridging the worlds of literature and science, as several studies have explored at length (such as Rylance; Dames). His work drives at establishing laws of literature and systemic methods for understanding mind, body and literary form, integrating criticism with a wider array of concerns in psychology, physiology, and biology. One of the important features of his critical output is its serious engagement with a flourishing, if still relatively recent, literary form – the novel. Well before Henry James would declare in 'The Art of Fiction' (1874) that the novel was under-theorized, Lewes set about devising a theory of the novel and its relation to criticism.

While not entirely successful in its task, 'The Principles of Success in Literature', a series of essays initially serialized in the *Fortnightly Review* in 1865, gives a clear sense of Lewes's mission as a critic and theorist. 'All Literature,' it quickly pronounces, 'is founded upon psychological laws, and involves principles which are true for all peoples and for all times' (Lewes 1865A 87). It makes an ally of the positivist method: one can immediately detect an important shift to principled and systemized aesthetic understanding in a way that may seem quaintly neo-Aristotelian but in fact signals the professionalization of knowledge. Already in 1847, in a piece for *Fraser's Magazine*, Lewes had remarked that 'Literature has become a profession' (285), a comment bringing to light not just the commercial advantages to literary reviewers of a buoyant literary marketplace but arguably also a form of existential anxiety about the figure of the critic, as an explicator and mediator, which ran in parallel with concerns in some circles over the status of the man of science in a post-Romantic age of institutionalization and professional authority (DeWitt 9–11). What kinds of expertise did the critic and scientist possess? The aim of Lewes's 'Principles of Success in Literature' is to bolster the intellectual prominence and repute of criticism by staging an appeal to its underlying laws and principles, while at the same time contributing to the cultural development of the novel as a literary genre by raising the standards of novel criticism. As he points out in 'Criticism in Relation to Novels', again published in 1865 while serving as editor of the *Fortnightly*, 'the vast increase of novels, mostly worthless, is a serious danger to public culture' (Lewes 1865B 354), and so a sharper, more scientifically alert practice will presumably help to assuage this effect.

Something of the impact of this agenda can be gauged from the example of Lewes's own scientifically informed practical criticism, illustrated aptly by an important review of *Jane Eyre* (1847), again in *Fraser's*, and by his later well-known appraisal of Dickens's life and writing in 1872, effectively blending the genres of literary criticism and obituary. To dwell briefly on the latter, 'Dickens in Relation to Criticism' is a striking performance in technique and tone, in which Lewes hones in on the gap between the author's popularity with general readers and the lukewarm reviews of some professional critics. In other words, the spectres of expertise, authority and professionalization loom once more. Lewes presents Dickens as an importantly thorny case for criticism to address – 'Dickens has proved his power by a popularity almost unexampled, embracing all classes. Surely it is a task for criticism to exhibit the sources of that power?' (1872 143) – and in effect he thereby clears a space for scientific critical analysis aided and abetted by psychology and physiology. One insight that Lewes offers draws on his understanding of the processes of delusion and hallucination. These were not signs of mental collapse in Dickens but, startlingly, features of his healthy imaginative processes: 'to him . . . *created* images have the coercive force of realities' (Lewes 1872 145). But, elaborating on his genius, Lewes then weighs in with a devastatingly ambivalent analysis of Dickens's literary style that turns on the vivid scientific image of a dissected frog. His characters' rootedness in habitual modes of behaviour, in fixed gestural and linguistic tics, recalls for Lewes 'the frogs whose brains have been taken out for physiological purposes, and whose actions henceforth want the distinctive peculiarity of organic action, that of fluctuating spontaneity' (1872 149). A brainless frog is merely a machine; alive, it has dynamism. Dickens fails with the latter. What his technical characterization cannot grasp is, once again, *complexity*, in the scientific sense of that term registered by Eliot. As Lewes concludes, 'It is this complexity of the [uninjured] organism which Dickens fails wholly to conceive' (1872 149). The use of a laboratory image accesses an experimental domain that Lewes was directly familiar with: frogs as experimental organisms are discussed in his *Physiology of Common Life* (1859). The more significant point is that this moment aspires to integrate criticism, physiology and theories of organicism.

Empirical aesthetics, evolution and the *fin de siècle*

Lewes was not alone in pursuing a scientific model of criticism. With a different emphasis, for example, the French critic and historian of English literature, Hippolyte Taine, devised a critical method of scientific literary historicism. On this view, literature became a form of empirical evidence of its own, to be read and interpreted for scientific ends. A position closer to Lewes's was taken by Eneas Sweetland Dallas, a journalist and critic who had benefited from the mentorship of the Scottish metaphysician Sir William Hamilton, and who absorbed a good deal of early scientific psychology and sought to extend these findings to poetry and art. Like Lewes, Dallas confidently predicted that such a model of criticism was possible. In his two main published works, one an essay dedicated to Hamilton called *Poetics* (1852) and the other a longer study titled *The Gay Science* (1866), Dallas marshals complex arguments that cover topics such as the nature of pleasure and pain, the imagination, unconscious mental activity, dreaming and so forth. These contributions have been mostly forgotten, perhaps in part because the author and his ideas had their roots in the less intellectually respected arena of newspaper journalism, and most recent synoptic surveys of the history of literary criticism do not discuss Dallas much or at all. Looking back on the nineteenth century in 1911, George Saintsbury, one of the early professors of English literature, took a different view, however. Saintsbury's *A History of English Criticism* remembers Dallas as 'a remarkable critic' (464) who had a strong scientific vision, if perhaps less talent as a reader of individual literary works, someone who is 'on the whole a Pre-Arnoldian type . . . to whom justice, I think, has not unusually been done'. His first long essay on poetry, from the 1850s, was to be especially admired as an adventurous prose study.

In *Poetics: An Essay on Poetry*, Dallas has in view two main themes, both of which connect his approach to that of Lewes and to the Romantic critical legacy. One is the notion of pleasure. Dallas takes literary and aesthetic pleasure seriously, indeed scientifically. Second, he is interested in the possibility of establishing an approach to literature that has procedural principles and clarity. An early reference to Aristotle and Bacon as 'two of the world's greatest thinkers' (Dallas 6) amplifies his intention to apply the scientific method to the aesthetic realm of literature, wedded to an articulation of definitive positivist laws that demarcate and classify analytical phenomena. Dallas cites three such laws of poetics: the law of imaginative activity (Wordsworth was wrong to separate fancy and imagination since all cognitive effort contains imaginative power); second, the law of harmonious power (echoing Bacon, the imagination serves a connective purpose linking sense and spirit); and lastly the law of unconscious might (poetry is linked closely to the mind's non-volitional activity). In these respects, Dallas sets off in empirical pursuit of the aesthetic. Yet simultaneously his project feels surprisingly Shelleyan. Just as *A Defence of Poetry* turns artfully around its anti-utilitarian premise that 'pleasure in this highest sense is true utility' (Shelley 696), so Dallas earnestly defends poetic pleasure. He extends a point running through the critical prose of Wordsworth, Shelley and Mill. Dallas also has a vivid sense of image, as in this delicately sustained scientific analogy:

> Imagination mingles with every pleasure, but the largest share goes to poetry. This is true; but it is not the less true that a difference of degree will often in a certain sense constitute a difference of kind. Add a little warmth, and ice will become water; a little more heat will turn the water into steam. Thus will a mere increase of imagination thaw the most stubborn reason, melt the hardest prose, and make it flow forth in song; and thus, too, might we speak of genius as different from talent . . . And every pleasure, too, has a degree of its own at which it becomes poetry, just as ice, glass, and iron have each a degree at which they melt.
>
> *(47)*

For all of its confident precision, and its moments of poised conceptualization such as this one, Dallas cannot entirely shed a sceptical and relativistic sense of the unknowable nature of whatever his analysis tries to uncover. Here, Hamilton's influence on him as a Kantian in the Scottish tradition of common-sense philosophy may be decisive. 'We do not understand Keats; we do not understand comets; perhaps we never will', he concedes (Dallas 77). But Dallas made an undoubted mark with *Poetics*. The Scottish literary critic, historian and university professor David Masson discusses him seriously in a study of poetic theories in *Essays Biographical and Critical* (1856), putting Dallas in the august company of Goethe, Wordsworth and Coleridge. Masson writes glowingly of his 'relish for the many lucid and deep remarks which drop from his pen' (413) in *Poetics*. A weakness comes from Dallas's lack of scientific interest in imagination in comparison with pleasure, Masson notes, without making any principled objection to his scientific literary criticism per se.

Herbert Spencer, among others, also recognized the appeal of an empirical approach to literary texture. His essay 'The Philosophy of Style' (1852) expounds a theoretical and practical analysis of linguistic presentation, reaching for instances of poetic and literary form to illustrate his mostly associationist account of the mind in the act of reading. Spencer's views bristle at times with eccentric conviction but remain largely consistent with his own systematic psychology. One of his overriding concerns is with embodied comprehension, in keeping with his interactionist model of mind and body. Language's materiality, its size and shape in the mouth, matters to its denotative context. So, for example, a 'voluminous, mouth-filling epithet is, by its very size, suggestive of largeness or strength, as is shown by the pomposity of sesquipedalian verbiage; and when great power or intensity has to be suggested, this association of ideas aids the effect' (Spencer 1996 337–8).

Another matter of concern for Spencer is the order in which words follow one another, a formal feature of syntactical expression related to the mental expenditure involved in comprehension. Approved style seems to minimize readerly effort. Compression, in effect, typifies 'direct style' (Spencer 1996 347) and it may be no accident that his accompanying case studies are drawn from poetry. Examples, all metrical, from Shakespeare, Milton, Coleridge, Shelley and Tennyson illustrate ideal sentence construction, corresponding to a 'law' of composition based on associationist notions of sequence and relation. For example, it is more effective in English to say *black horse* than *horse black*. In the former case, the adjective 'black' prepares the mind abstractly to receive a subsequent object, whereas, if the noun comes first, the mind begins by summoning an image of a horse, most likely one of brownish colour, and then is forced to do a kind of colour correction with the late arrival of the adjective, from brown to the requisite black. The better syntax is the one that aligns with smooth and natural chains of association. On these quasi-empirical grounds, Spencer advocates a form of stylistic minimalism (in his native language, at least). The 'aim must be to convey the greatest quantity of thoughts with the smallest quantity of words' (Spencer 1996 356), he concludes, in almost a proto-Imagist vein.

Spencer's entire intellectual system was shaped by evolutionary theory, though 'The Philosophy of Style' is not especially inflected by those concerns. Spencer does though tackle the question of literature and evolution elsewhere in his large body of writing, notably in *First Principles* (1862) and in a related essay for the *Westminster Review* titled 'Progress: Its Law and Cause' (1857). Here Spencer presents an account of progress – meaning, in effect, the inexorable superiority of European society and civilization as it has formed through time – in what may seem to be the domineering high-Victorian fashion beloved by caricaturists of the period. Underpinning it is his scientific view of evolution, a hypothetical dynamic according to which things move from a state of homogeneity to one of increasing heterogeneity. Literature, like any aspect of nature and culture, can be understood by considering its development in these Spencerian master terms. Language itself emerged as a diffusion pattern, resulting in many different local languages with a common original form; then came pictorial representation, hieroglyphs, and the invention of writing, all gradual evolutionary stages of human culture. In a similar way, poetry eventually became differentiated from earlier primitive chants, dances and religious rituals. Over time, it then acquired greater internal differentiation as a genre of its own, with subtly different forms and styles as its particular family of diverse subcategories. When retraced to its ancient cultural roots, literature can be found mixed in with 'theology, cosmogony, history, biography, civil law, ethics, poetry' (Spencer 1857 464), whereas in its present heterogeneous state literature's 'divisions and subdivisions are so numerous and varied as to defy complete classification'. Moreover, there may have been a distant development moment when literature, art, and science were not yet distinguished, an era in which science was 'in union with Art, the handmaid of Religion' (Spencer 1857 465). However problematic it may be, this view allows Spencer to develop an evolutionary logic that will help justify his synthetic philosophy and its objective of structurally reintegrating all areas of knowledge and understanding.

Other critics and theorists in the last third of the nineteenth century were drawn to post-Darwinian accounts of literary and aesthetic pleasure, such as Vernon Lee and Grant Allen (for a detailed discussion of these critics, including others such as A. T. W. Borsdorf and Constance Naden, see Holmes). Allen's *Physiological Aesthetics* (1877) turned once again to the embodied feelings to scrutinize the origins of aesthetic experience, establishing that 'pleasures of imagination can be finally affiliated on our general principle of pleasure and pain' (248), but in a way that differed from, say, Lewes's related investigations. Allen's evolutionary biology brings him to contrasting views of literature and aesthetics, one in search of evolutionary explanations for the distinction of artistic genius and cultural achievement in a larger perspective. 'Savage' and 'civilization' are important terms in the argument of *Physiological Aesthetics*, and of course ideologically also very laden ones. Its convictions can feel shaky. Allen worries about how empirically valid it can be to investigate the nature of human aesthetic taste by referring to art from advanced European culture as if it were universal:

> We cannot expect a child or a savage to admire the poetry of Wordsworth, the landscapes of Turner, the sonatas of Beethoven, the duomo of Milan. Yet even here we see the community of feeling, that what is at once simple and beautiful is pleasing to the highest and lowest intelligences alike:—instance a daisy or a butterfly, an old English melody, or a Homeric ballad.
>
> *(47–8)*

In the space of a paragraph we can witness the relativist fighting with the cultural imperialist in Allen's mind. As John Holmes observes, such arguments are also beset by a logical flaw: for Allen, advanced aesthetic taste and physiological refinement serve as each other's proof, in a circular fashion (106). But in its racial assumptions and in its struggle to reconcile absolute and relative standards in human experience, Allen's work epitomizes a late Victorian anthropological moment.

One could also mention here a number of other critics and theorists of literary aesthetics who were linked by a set of cultural anxieties that can be recognized as signatures of the *fin de siècle*, including post-Darwinian pessimism, degeneration and insanity. These would have to include the alienist Henry Maudsley, the biologist and eugenicist Francis Galton, and Max Nordau. Maudsley's psychiatric writings make many frequent allusions to poets and dramatists, including Shakespeare, Blake and Dickens, in a way that does not so much add to our understanding of mainstream critical tradition but rather demonstrates how literature took on illustrative or evidential significance in crucial ways as new disciplines were formed at the end of the nineteenth century. James Sully, another psychologist of the late Victorian period, who pioneered child development, made instinctive use of literary character and situation in a similar fashion in his extensive scientific writings. Galton, too, was much taken by the scientific value of literature to his attempt at establishing the empirical study of genius and heredity.[3] Yet it would be a mistake either to describe these developments as a linear historical extension of what came before, or to insist at all on a well-defined British tradition of scientific literary criticism in the nineteenth century. The picture is less tidy and more varied. Nonetheless, a body of criticism appeared over the course of the nineteenth century written by figures who crossed, like Lewes, between the worlds of science and literature, which drew extensively on knowledge of biological life and psychological processes in order to forge models of literary evaluation that might either exemplify or help disclose something close to definitive, or at least dependable, laws of literature. This critical tradition may be partially occluded today, but it was highly visible in the period. Many histories of criticism, focusing perhaps on Coleridge, Arnold and Pater, omit this vibrant experimental side of Romantic and Victorian critical writing. Scientific perspectives on the interpretation of poetry and fiction emerged much earlier than I. A. Richards, F. R. Leavis or Northrop Frye, and they did so in relation to similar contextual pressures (specialized knowledge, disciplinary identity, cultural authority, professionalization) that arose in the twentieth century and continue to exist, if in different ways, today.

Notes

1   See Suzy Anger, 'Sciences of the Mind', in this volume.
2   See Charlotte Sleigh, 'The Novel as Observation and Experiment', in this volume.
3   See David Amigoni, 'Writing the Scientist: Biography and Autobiography', in this volume.

Bibliography

Primary texts

Allen, Grant. *Physiological Aesthetics*. London: Henry S. King, 1877.
Coleridge, Samuel Taylor. *The Major Works*. Ed. H. J. Jackson. New York, NY: Oxford University Press, 2008.
Dallas, Eneas Sweetland. *Poetics: An Essay on Poetry*. London: Smith, Elder, 1852.
Eliot, George. *The George Eliot Letters*. Ed. Gordon S. Haight. 9 vols. New Haven, CT: Yale University Press, 1954–78.

Eliot, George. *Selected Critical Writings*. Ed. Rosemary Ashton. Oxford: Oxford University Press, 2000.

Hazlitt, William. *Table Talk; Or, Original Essays on Men and Manners*. 2nd edn. 2 vols. London: Henry Colburn, 1824.

Lewes, George Henry. 'The Condition of Authors in England, Germany and France'. *Fraser's Magazine* 35 (1847): 285–95.

Lewes, George Henry. 'Principles of Success in Literature'. *Fortnightly Review* (15 May 1865): 85–95 [1865A].

Lewes, George Henry. 'Criticism in Relation to Novels'. *Fortnightly Review* (15 December 1865): 352–61 [1865B].

Lewes, George Henry. 'Dickens in Relation to Criticism'. *Fortnightly Review* (1 February 1872): 141–54.

Masson, David. *Essays Biographical and Critical: Chiefly on English Poets*. Cambridge: Macmillan, 1856.

Mill, John Stuart. 'What is Poetry?'. *Monthly Repository* 7 n.s. (1833): 60–70.

Mill, John Stuart. *The Letters of John Stuart Mill*. Ed. Hugh S. R. Elliott. 2 vols. London: Longmans, Green, 1910.

Mill, John Stuart. *The Collected Works of John Stuart Mill*. Ed. John M. Robson. 33 vols. Toronto: University of Toronto Press, 1963–91.

Saintsbury, George. *A History of English Criticism*. Edinburgh: William Blackwood, 1911.

Shelley, Percy Bysshe. *The Major Works*. Eds. Zachary Leader and Michael O'Neill. Oxford: Oxford University Press, 2003.

Spencer, Herbert. 'Progress: Its Law and Cause'. *Westminster Review* 67 (1857): 445–85.

Spencer, Herbert. 'The Philosophy of Style'. *Essays: Scientific, Political and Speculative*. London: Routledge/Thoemmes, 1996, 333–69.

Wordsworth, William, and Samuel Taylor Coleridge. *Lyrical Ballads: 1798 and 1802*. Ed. Fiona Stafford. Oxford: Oxford University Press, 2013.

Secondary texts

Abrams, M. H. *The Mirror and the Lamp: Romantic Theory and the Critical Tradition*. Oxford: Oxford University Press, 1971.

Ashton, Rosemary. *The German Idea: Four English Writers and the Reception of German Thought, 1800–1860*. Cambridge: Cambridge University Press, 1980.

Burke, Seán. *The Ethics of Writing: Authorship and Legacy in Plato and Nietzsche*. Edinburgh: Edinburgh University Press, 2008.

Dames, Nicholas. *The Physiology of the Novel: Reading, Neural Science and the Form of Victorian Fiction*. New York, NY: Oxford University Press, 2007.

DeWitt, Anne. *Moral Authority, Men of Science, and the Victorian Novel*. Cambridge: Cambridge University Press, 2013.

Esterhammer, Angela. 'The Critic'. *The Cambridge Companion to Coleridge*. Ed. Lucy Newlyn. Cambridge: Cambridge University Press, 2002. 142–55.

Fairclough, Mary. *The Romantic Crowd: Sympathy, Controversy and Print Culture*. Cambridge: Cambridge University Press, 2013.

Holmes, John. 'Victorian Evolutionary Criticism and the Pitfalls of Consilience'. *The Evolution of Literature: Legacies of Darwin in European Cultures*. Eds. Nick Saul and Simon J. James. Amsterdam: Rodopi, 2011. 101–12.

Jackson, Noel. *Science and Sensation in Romantic Poetry*. Cambridge: Cambridge University Press, 2008.

Jones, Ewan. *Coleridge and the Philosophy of Poetic Form*. Cambridge: Cambridge University Press, 2014.

Mitchell, Robert. *Experimental Life: Vitalism in Romantic Science and Literature*. Baltimore, MD: Johns Hopkins University Press, 2013.

Page, Michael. *The Literary Imagination from Erasmus Darwin to H. G. Wells: Science, Evolution and Ecology*. Farnham: Ashgate, 2012.

Roe, Nicholas. *Samuel Taylor Coleridge and the Sciences of Life*. Oxford: Oxford University Press, 2001.

Ruston, Sharon. *Shelley and Vitality*. Basingstoke: Palgrave Macmillan, 2005.

Ruston, Sharon. *Creating Romanticism: Case Studies in the Literature, Science and Medicine of the 1790s*. Basingstoke: Palgrave Macmillan, 2013.

Rylance, Rick. *Victorian Psychology and British Culture, 1850–1880*. Oxford: Oxford University Press, 2000.

Sharrock, Roger. 'The Chemist and the Poet: Sir Humphry Davy and the Preface to *Lyrical Ballads*'. *Notes and Records of the Royal Society of London* 17:1 (1962): 57–76.

Smith, Jonathan. *Fact and Feeling: Baconian Science and the Nineteenth-Century Literary Imagination*. Madison, WI: University of Wisconsin Press, 1994.

Snyder, Laura J. *Reforming Philosophy: A Victorian Debate on Science and Society*. Chicago, IL: University of Chicago Press, 2010.

Ware, Tracy. 'Shelley's Platonism in *A Defence of Poetry*'. *Studies in English Literature, 1500–1900* 23.4 (1983): 549–66.

9

WRITING THE SCIENTIST
Biography and autobiography

David Amigoni
Keele University

Scientific biography and autobiography is one of the most obvious, yet paradoxically under-researched, genres for exploring the relationship between literature and science in the nineteenth century (Söderqvist 1). Biography and autobiography are of course *literary* genres in which scientific endeavours, pursued throughout the life of an individual, are narrated. It is clear from Liba Taub's work on extant *bioi* from the ancient world that there were generic continuities between the literary contours of the ancient life philosophical – ancestry, childhood, education, great deeds, virtues, death and consequences – and the narrative shape of the life scientific in the nineteenth century (24). But there were also many distinctive features that marked nineteenth-century life writing. Anna Seward, the biographer of Dr Erasmus Darwin at the very beginning of the nineteenth century (1804), remarked that 'Generosity, wit, and science, were his household gods' (58). In the idea of the 'household god' Seward laid bare a hybrid complex of ancient and modern generic ingredients: ancient ancestral veneration coupled with an exploration of modern domesticity and manners, and the modern knowledge formations born of scientific reason, observation, and inventiveness. Seward's emphasis on 'wit', or a style of expressiveness, highlights a strand of distinctively *Romantic* thought that started to shape the nineteenth-century life writing landscape at the end of the eighteenth century (Cantor 219). Samuel Johnson was an eighteenth-century figure, but in *The Life of Johnson* (1791) his biographer James Boswell's emphasis on Johnson's unique subjectivity and the traces of individual inventiveness, inscribed in 'private' letters, notebooks and recorded conversations, established deep Romantic markers on nineteenth-century life writing. Rather in the way that eighteenth- and nineteenth-century literary figures left behind manuscripts and letters that informed later generations about the creative process, we can see now that this provided knowledge of how the nineteenth-century scientist thought, observed and reasoned.

The written life and the forms of publication from which it emerged were vehicles that shaped science's cultural authority in the nineteenth century. This chapter explores the multiplicities of *literature* as not only the narrative vehicle but also as contested sources of expressive and inventive notation, ranging from formal poetical composition to notebooks and other manuscripts in which the unfolding of thought can be traced. These contributed varied and distinctive accents to the task of 'writing the scientist' in the nineteenth century – accents that shaped narratives about the making of the individual person, their scientific thought and its contribution to scientific legacy, as well as their personal and even domestic relationships: all of these factors had important implications for the status of science itself as a series of increasingly differentiating and institutionally defined fields of inquiry, contributing to the progress of the nation. These fields included chemistry in the early nineteenth century (Humphry Davy), mathematics and mechanics (Charles Babbage) and evolutionary biology (Charles Darwin and T. H. Huxley).

I shall begin with the Darwin family and the publication of a biography to record the life of Erasmus Darwin at the opening of the nineteenth century and return to the Darwin family in the early years of the twentieth as the Darwinian evolutionary legacy reached a high watermark. I take my selected texts to be noteworthy because of the way in which they were organized by topics that were constitutive of science's claim to cultural authority: thinking as discovery; social mobility and progress; the place of family, religion and secularity. As I shall show, key institutions of scientific organization, such as the Royal Society, the Royal Institution and the British Association for the Advancement of Science, were contexts that coloured the writing of lives that were offered to populate science's cultural authority.[1] That authority could not be taken for granted and the topics and contexts figure as sources of literary contestation in the shaping of life narratives that were often replies to rival narratives delineating another version of the life in science.

The question of genre and voice is important: my selection of texts interrogates the generic differences between biography and autobiography, acknowledging that it is in no sense simple, while paying close attention to literary textures of voice through which *literariness* itself comes to contest authority. To that end, I argue for a particular critical role for narrative reverie through the autobiography of Charles Babbage, the literary consequences of which are further explored in the more managed autobiographical writings of Charles Darwin and T. H. Huxley. The chapter concludes with a focus on the self-consciously authoritative and field-shaping aspirations of *The Life, Letters and Labours of Francis Galton*, written by his disciple Karl Pearson over a long period (1914–30). This work about the birth of eugenics marked, arguably, the terminus for the fashioning of scientific biography during the long nineteenth century, yet Galton's autobiography could, I argue, still be an occasion for literary reverie that underlined a critical role for the ludicrous in literature and science.

It is instructive to begin with Anna Seward's controversial *Life of Erasmus Darwin* precisely because, in gesturing to the centrality of science to its subject's life, it then went on to sidestep that very topic. The biography appeared in 1804, following Darwin's death in 1802. While Erasmus Darwin's contribution to the history of science and medicine has come to be recorded in depth in work by Desmond King-Hele and others, his 'household gods' of science and medicine were originally presented in less than luminous detail by Seward's biography. Robert Waring Darwin, Erasmus's son and Charles's father, asked the Lichfield poet Seward to supply anecdotes about the Lichfield life of the most eminent physician in Britain. As Darwin's friend, Seward had observed at close hand the capacious intellect of the inventor, speculator and poet. Seward went much further and turned her collection of anecdotes into an uncommissioned, indeed unflattering, life, which deeply displeased the Darwin family, not only because the anecdotes reflected little on Erasmus Darwin's role as elite physician but chiefly because of Seward's allegation that he was the perpetrator against her of the literary crime of plagiarism. As Teresa Barnard's recent study of the writer suggests, Seward's approach to writing the life of a 'great man' was in a state of tension with her own somewhat brittle sense of a hard-won reputation and identity, achieved and defended through her meticulous self-archiving of her letters and poetry (5).

Seward had come to occupy a central intellectual position in the powerful literary and Anglican circle of the Midlands city of Lichfield. In fact, the idea of the intellectual circle played an important, indeed slightly eccentric, role in Seward's biography, given the less than flattering attention devoted to Erasmus Darwin's friends, especially the philosopher Thomas Day and his proprietorial predilection for young, foundling women. Seward's critical ire was also directed at alleged improprieties in Erasmus Darwin's own literary achievements, embodied most famously in *The Botanic Garden* (1789, 1792). The controversy that Seward's biography initiated had at its heart a 'tangled and sequestered scene' (108) (the 'wild, umbrageous valley' on the outskirts of Lichfield purchased by Darwin), which led to an intellectual entanglement between literary and scientific aspirations. According to Seward's account, the idea for *The Botanic Garden* came from her own poetical impulses, inspired by Dr Darwin's enclosed section of the Derwent valley (but not the physician himself, who on the day in question was attending to 'a medical summons' (109)). On reading Seward's fragment on 'these HALLOWED VALES' and its humanized flora, Darwin conceived the idea of the exploration of the Linnaean system through 'metamorphoses of the Ovidian kind' (111), inviting her to write the verse, for him to provide the scientific notes. According to Seward, Darwin 'objected the professional danger of coming forward an acknowledged poet . . . all risque of injury by reputation flowing in upon him from a new source'. Yet, while Darwin ultimately set aside these professional anxieties to undertake the entirety of *The Botanic Garden* by himself, Seward charged Darwin with publishing her fragment of verse as the exordium of the first part of the poem: 'No acknowledgment,' she protested, 'was made that those verses were the work of another pen'.

If literary circles and coteries were sites for collaborative work on manuscripts (it is unlikely that Erasmus Darwin would have *recognized* himself as a plagiarist), Seward was inclined to claim the status of the Romantic, individual author. Following from this, there is also a sense in which Seward's biographical criticism presents Darwin's poetical achievement as running counter to the most vibrant emerging tendencies in poetical expression in the later eighteenth century through her distinction between 'vivid poetry which does *not* excite sensation, and vivid poetry which *does* excite it' (131): thus an autumnal mood from *The Loves of the Plants* is compared unfavourably to a sonnet about a November scene by Charles Lloyd, published in the second edition of S. T. Coleridge's first volume of *Poems* (1797), which is said to 'thrill our nerves' (132) in ways that Darwin's more measured verse does not. Of course, Seward's predominantly neoclassical frame of critical reference could not deliver precisely the story about 'the poetry of sensation' as the harbinger of the modern age, which Arthur Henry Hallam would tell through the early poetry of Tennyson in the 1830s. It is striking, above all, that Seward's focus on the Lichfield life of Erasmus Darwin undertook to tell no particular story about his significance for the nation – in contrast to James Boswell, who, in the advertisement to the second edition of his biography of Samuel Johnson (1793), claimed to have '*Johnsonised* the land' (8). If science as a source of the nation's story sought to reveal its stepping stones to progress through biography, it needed to look beyond Seward's assertively anti-heroic model.

The relationship between science and its contribution to the nation as reflected in the scientific life was formulated in popular, but contested, schemes of publication between the 1830s and 1850s as a range of organizations, from the Society for the Diffusion of Useful Knowledge (SDUK) to the Religious Tract Society (RTS), used biography for improving and educational purposes. Religion was a central fault line in this contest; and religion was, of course, an enduring source of contestation in nineteenth-century writing about the scientific life. As the research of Aileen Fyfe has shown, competing secular and Christian versions of knowledge classification used biography as a way of establishing a place for science in the hierarchies of knowledge comprising the national culture. Isaac Newton was *the* canonical figure, representing the birth of scientific reason and a vision of a mechanical universe respectfully subordinate to the truths of revelation (Fyfe 74–9). In Rebekah Higgitt's work on the transmission of Newton's manuscript sources –

dealing sometimes with alchemical speculation – from which biographies of the natural philosopher were produced in these new publication contexts, she records the staunchly Anglican-Tory biographer of Newton, S. P. Rigaud, informing William Whewell of Trinity College, Cambridge, in private correspondence in 1836 of the need for editorial and curatorial care in this enterprise for 'if Newton's character is lowered, the character of England is lowered and the cause of religion is injured' (163). The 1830s and 1840s were a key period of institutional formation for science in which its relationship to the established Church had also to be defended: this was also the moment, as Geoffrey Cantor points out, when Thomas Carlyle's model of heroic biography, along with narratives of self-helping labour and perseverance, became discourses for shaping stories of national morality and progress (219). It was the formidable Whewell, through his position as ordained Cambridge divine and savant, his philosophies and histories of science, as well as his organizational work for the British Association for the Advancement of Science, who coined the new term 'scientist' in 1833. There remained the question of what one ought to embody in order to qualify for this office – embodiment that was both represented and contested through biographies about subjects who advanced both natural philosophy and their own social position in the nineteenth century.

Biographies of Sir Humphry Davy, who made great strides in the fields of electricity, chemistry and the technological application of science in the century's first decades, and who died in 1829, constitute an important case study. Davy crowned his career as President of the Royal Society in 1820, the most eminent office in British science, a position which had been held since the late 1770s by the aristocratic Sir Joseph Banks. Davy, the son of a wood-carver from Penzance, Cornwall, came from origins that made his story one of social, as well as scientific, advancement. These progresses were first inscribed into biographical narrative in 1831 by John Ayrton Paris, fellow of the Royal Society, a London physician (a member of the Royal College) with professional connections to Cornwall, who had attended the charismatic Davy's lectures at the Royal Institution. Paris was able to appeal to the full institutional authority of the Royal Society to validate his narrative before the public, dedicating it to H. R. H. Prince Augustus Frederick, Duke of Sussex, President of the Royal Society from 1830–38, a validation through which one can glimpse something of the professional rivalry between the two leading, and most nationally prestigious, scientific institutions. Davy's presidency of the Royal Society was not a happy one; he was much criticized. Paris used the biography of Davy to subtly promote the virtues of Dr William Wollaston – edged out of the presidency of the Royal Society by Davy after a matter of months in the role – through comparisons between the temperaments of the two men that worked to the disadvantage of Davy (1:296–7).

Paris's narrative authority was, however, explicitly 'corrected' by John Davy in 1836. A military physician and Humphry Davy's younger brother, the new biographer made a sustained case against Paris's alleged carelessness and inaccuracy, while there was clearly also a broader resentment at the condescending tone adopted by Paris. Thus, Chapter VII of John Davy's work records that 1807 was an important year in the history of chemical science, when Humphry Davy successfully used his researches in electricity to decompose fixed alkalis through his experiments with potash. John Davy points out that Paris's biographical account of the process of discovery (1:274–5) used only the Royal Institution laboratory register to note a final sequence of experiments, concluding that Davy commenced his work on 11 October and obtained his 'great result' on 19 October (Davy 1:384). Paris represented Davy's power as vested in 'rapidity', tending to an actual recklessness in moving between experiments to 'grab' data (Paris 1:144–5). In contrast, John Davy argued, based on access to his brother's notebooks, that the researcher's thinking on this theme went back to 1800 at least (1:378): in other words, while only a very few might have first-hand experience of 'the operations of the laboratory and the exciting nature of original research', these valued 'private' sources of expression could give a valued insight into the true nature of the scientist's 'previous views, and the analogies by which he was guided' (1:383). These notebooks revealed the workings of 'a young mind with avidity for knowledge and glory commensurate' (Davy 1:384). Thus, access to valued 'private' sources tended to assist family members in securing authoritative roles in the nineteenth-century business of writing the scientist – we shall see more of this tendency when we come to look at the biographies of Charles Darwin and T. H. Huxley, written at the end of the century by their sons.

John Davy crafted a biographical portrait in which every occurrence contributed to the continued elevation of Humphry Davy's national fame and esteem: even the near-fatal illness which afflicted him following his successes of 1807 underlined how much his country was indebted to him, following as he did in a line including Bacon and Boyle (1:389). In promoting Humphry Davy's contribution to the nation, John Davy subtly revised the image of Humphry Davy 'fixed' by Paris: as a man forever associated with the wild but backward-looking regional fringe of Cornwall. Thus, in his representation of Davy during his life-threatening illness of 1807, Paris claimed that 'no Swiss peasant ever sighed more deeply for his native mountain than did Davy for the scenes of his early years' (Paris 1:290). For John Davy, not only was there no evidence for this Romantic account of his brother's illness, but there was no justification for allusions to the Radcliffian Gothic that Paris explicitly invoked in his account of Humphry Davy's Cornish childhood (Paris 1:10). John Davy was careful to emphasize a Cornwall that had kept pace with a modernizing nation through intellectual and transport networks, but Paris equated the chemist's reputation with a Romantic literariness that, when taken in connection with accounts of a chaotic pre-marital domestic life and a lecturing style that tended to oratory and vanity (Paris 1:136–7), hardly bolstered the scientific reputation. Of course, Humphry Davy composed poetry in the 1790s, knowing both Wordsworth and Coleridge: Paris states early in his biography that his subject would have become the 'first' of poets had he not become a chemist (1:40); a sentiment that may echo Coleridge's own assessment of Davy in *Sibylline Leaves* (89–90). It is notable that John Davy mentions this only very late on in the concluding 'character encomium' section of the second volume (2:391–4), in which letters about the affections of friendship to Coleridge, passed to John Davy by Wordsworth, are included. If John Davy's biography played down his brother's Romantic literary affiliations to counter the effect of Paris's hints at Romantic excess, Davy's epigraph ('The affections are their own justification') was from his friend Wordsworth's first 'Essay upon Epitaphs': Davy's biographical narrative was, self-consciously, designed to be a dignified yet affectionate epitaph to his brother's contribution to science *and* human life.

The self-written scientific life necessarily avoided epitaphic finality: for this very reason, scientific autobiography could be more engagingly uneven in tone and form. This was the case with Charles Babbage's *Passages From the Life of a Philosopher* (1864), which, in the words of its author, did 'not aspire to the name of an autobiography. It [the work] relates a variety of isolated circumstances in which I have taken part' (1994 vii). The disconnected 'passages' that the narrative recounts relate to Babbage's varied career as a mathematician, theorist of the factory system, statistician of life assurance and inventor of the so-called 'difference engine', usually represented as the first computer. As such, he was a savant who moved originally, and influentially, in the world of scientific and philosophical institutions led by Humphry Davy's Royal Society, to which he wrote an open letter in 1822 extolling the virtues of the difference engine for the business of producing error-free tables for the Board of Longitude. To this extent, Babbage was committed to scientific analysis that improved the efficiencies of an industrializing and trading Britain: he was the author of *On the Economy of Machinery and Manufactures* (1832) and he claimed, in *Passages*, that in adapting Adam Smith's principle of the division of labour he was the first to fully grasp the idea that its successful implementation cut the *cost* of production (1994 328). However, Babbage's refusal to claim the coherence of autobiography for his text may be seen in large part as a consequence of his failing to achieve continuity of office and output in the changing world of nineteenth-century science. His penultimate chapter (XXXV) records a long list of failures to secure office, bearing witness instead to appointments made on the basis of patronage. His failure to be appointed to the mastership of the mint in 1849 – an office that, as he was painfully aware, had been occupied by Isaac Newton – was because he 'had no political interest' (Babbage 1994 358).

In some sense, appeals to literary form and allusiveness had, for the duration of his career, been a paradoxical part of Babbage's armoury, defending his 'practical science' of mechanical philosophy and mathematics against the condescension of politically powerful ordained science. His *The Ninth Bridgewater* (1837), a kind of riposte to William Whewell's series of endowed treatises advancing the claims of natural theology, known as 'the Bridgewater Treatises', was subtitled *A Fragment* in ways that echoed literary romanticism's dialectic of blindness and insight. The fragmentary arguments on which it was based (which included gestures towards evolutionary speculations) were provocatively framed by, and in oppositional 'reply' to, an epigraph from Whewell's first *Bridgewater Treatise* denying that 'mechanical philosophers' had any insight into the profound, revealed truths organizing the universe.

Such literary playfulness also performed an oppositional role in Babbage's *Passages*; for example, his account of his 'Experience Amongst Workmen', a narrative about his visit to manufacturing districts in Leeds and Bradford (Chapter XVII), records his meeting with an informed, working-class mechanic who holds out his 'brawny hand' (1994 170) in delight at meeting the author of *On the Economy of Machinery and Manufactures*. The worker leads him to inspect a large iron-works, wherein he could enter 'a large tunnel through a rock which had originally been intended for a canal: but that . . . was now used as an air-chamber, to equalize the supply of the blast furnaces' (Babbage 1994 171). Babbage's journey into the great chamber of intensely hot, pumped air, contrasting sublimely with the cool air of the night outside, becomes increasingly mythic and allusive as he ventures into 'this Temple of Aeolus' as though he was the Greek Cynic philosopher Diogenes. The 'truth' of the cavern, which is intense in the heat of 'tons of air . . . driven without cessation by the untiring fiery horse', leads him to think of the Old Testament story of 'Shadrach, Meshach and Abed-nego' (Book of Daniel, Chapters 1–3), protected from the heat of Nebuchadnezzar's furnace by their belief in one true God, or, as Babbage will have it implicitly, science.

This becomes an occasion for an allusive foray into the heat and chill of public, yet ultimately transient, theological and parliamentary controversy among those who, for Babbage, were implicit producers of hot air during the 1850s; and the extent to which reputations might be consumed by the 'heat' of publicity or frozen out of view before reaching the attention of posterity. Referencing the Anglican heresy pronounced against the Christian Socialist cleric F. D. Maurice, and the (in 1864) unfulfilled career of the Conservative statesman Benjamin Disraeli, 'refrigerated by his *friends*', Babbage self-consciously uses this 'reverie' to claim that the writing of his life will 'give each of them his last chance of celebrity preserved in the modest amber of my own simple prose' (1994 171–2). All of this is framed by the self-promoting irony through which it is Babbage's name that is recognized by the brawny-handed working mechanic who has read *On the Economy of Machinery and Manufactures*.

In episode after episode, Babbage demonstrates the authorial and inventive act as a *calculated* mastery over nature's extremes, exemplified in his measured descent into the crater of Vesuvius (1994 160–65). These are, in a sense, allegories of science's capacity to harness those extremes in technology. Babbage styles his own inheritance and enduring identity as that of an inventor of tools (1994 2). Nonetheless, the centres of power of nineteenth-century science were being realigned by 1864 towards the increasingly authoritative biological sciences of evolution. In talking about his ancestral tool-makers at the beginning of his text, Babbage indicated that he would not have recourse to the 'philosophic, but unromantic, views of our origin taken by Darwin' (1994 1). Yet, it was those 'unromantic views' of Darwin's on the origins of life that would, precisely, help to shape a paradoxically *humane* and naturalistic image of the scientific life inscribed in Darwin's biography.

Between May and August 1876, Charles Robert Darwin, author of the revolutionary *On the Origin of Species* (1859), broke from his ongoing scientific labours (on self- and cross-fertilization in plants) for an hour each afternoon to write 'Recollections of the Development of My Mind and Character', a text that has come to be known as 'The Autobiography'. While the bulk of the text, reflecting on early recollections, the *Beagle* voyage and an ensuing life in science, was written in 1876, Darwin continued to write additions to his text in the six years that remained of his life, including a sketch of the character of his father, Robert Waring Darwin. Familial legacy became increasingly important: in 1879, Darwin wrote a life of his grandfather, Erasmus Darwin, which aimed to correct the rather negative and eccentric portrait left by Anna Seward in 1804 with which I began this chapter. Charles Darwin embarked on this autobiographical and biographical activity chiefly for his family: he conceived that it would be of interest to his children and their children, given his regret at the absence of an autobiographical articulation of his own grandfather's life (Barlow 21).

In fact, the autobiography that Darwin wrote would become a public document when it became the prefatory frame of Darwin's posthumous biography, published by his son, Francis, in 1887; at which point the life of the scientist was increasingly embedded within a public industry of life writing that characterized the 1880s which saw developments such as the inauguration of the massive *Dictionary of National Biography* (Francis wrote the entry on his father for Leslie Stephen's monumental project). The late Victorian biographical industry sought to satisfy a burgeoning public appetite for the lives of noteworthy authors, politicians and, increasingly, scientists as celebrities. It produced tension within the Darwin family, where there was profound disagreement about the wisdom of publishing the autobiography: Darwin's vehement rejection of Christianity (he condemned eternal punishment for non-belief as 'a damnable doctrine') inevitably generated family anxieties and conflicts (Barlow 12). On the other hand, the print infrastructure serving public interest in the life and views of the scientist, which had grown exponentially since the time of the Davys, gave rise to a more urgent role for late nineteenth century family members in the management of the biographical, public image. Francis Darwin, having been his father's scientific assistant for eight years, became his father's biographer, while Leonard Huxley, son of Thomas Henry Huxley, the great public leader of the evolutionary controversy, wrote the biography of his own father.

Francis Darwin took it upon himself to manage the autobiographical image bequeathed by his father. This included the publication of numerous letters and notebooks. Extracts from Darwin's now-famous 'Transmutation Notebooks', from 1837, made their first appearance as transcription and facsimile in the second volume of the biography (Darwin 2:4–5). In time, drafts of scientific manuscripts were published, such as the early case for evolution by natural selection set out in the so-called 'Essay of 1842'. Inaugurating this 'trail' of scientific speculations, Francis published Charles's 'Autobiographical Chapter' in the first volume of the *Life and Letters*. He appeased his mother and sister (Henrietta) by editing out the reference to 'damnable' Christian doctrine. Nonetheless, this did not detract from Darwin's 'Recollections' as a powerfully coherent narrative, organized around the central life event of the *Beagle* voyage, which produced a complex image of human motivation at the same time as the very origin of human exceptionalism was scrutinized by the speculations. For Darwin was simultaneously able to present the pleasures of, and desire to pursue, scientific investigation that would 'add a few facts to the great mass of facts in Natural Science' (Darwin 1:65), thus satisfying 'an ambition to take a fair place among scientific men' while also projecting intense humility. Darwin can form 'no opinion' whether he is 'more ambitious or less so than most of my fellow-workers'. Moreover, the scientific allure of the *Beagle* voyage had to be offset against the feeling of being 'out of spirits at the thought of leaving family and friends' (Darwin 1:64) for so long. For Darwin's image is also domestically resonant: marriage and the arrival of children follow the *Beagle* voyage, and the acquisition of a family home in a carefully chosen venue of quiet rusticity (Downe, Kent).

Darwin's autobiography, in Victorian fashion, tells a story of gains and losses. The most 'lamentable' loss is that of the 'higher' aesthetic sense for painting, music and poetry in later life. Towards the end of the text, recalling an earlier love for the poetry of Milton and Wordsworth, Darwin claims now to be 'nauseated' (1:101) by Shakespeare, having become 'a kind of machine for grinding general laws out of large collections of fact'. What is notable is Darwin's paradoxical resort to metaphor to style this scientific, anti-aesthetic self that became, in the long view, a challenge to find the 'poetry' in Darwin's writing, despite this self-perception. It is also a metaphor that perhaps projects a sense of literature as a land of lost content, and as such also perhaps constitutes an elision of the role of literary expression in more embattled cultural and intellectual contexts. Darwin's humane and sympathetic image, tinged with literary (but not religious) regrets, has been an enduring one (Levine), but it still required vigilant management as Francis saw off rival biographers and fugitive religious statements that seeped out of his father's uncensored correspondence into febrile public contexts (Amigoni).

While Charles Darwin pursued his evolutionary speculations in rural isolation, protected by his family, Thomas Henry Huxley was the public controversialist who asserted and defended the new evolutionary ethos in the service of new scientific, governmental and educational institutions before his death in 1895. Huxley was the crucial figure bringing the authority of scientific knowledge to bear on an increasingly complex 'social organism' and its rules of governance and opportunities for education. Leonard Huxley, his son and a school master at Charterhouse, took on the responsibility for recording the public life, though 'impersonally' as he put it in a preface that played down the family investment. The 'impersonal' imperative derived from the fact that the son had occasionally to search for proportion and consistency in his father's self-image when curating his writings into an authoritative 'life and letters'. Huxley was a reluctant autobiographer whose final published self-testimony had been extracted from him by the late Victorian vogue for the celebrity life story. It first appeared in the unlikely setting of Louis Engel's predominantly musical and artistic *From Handel to Hallé: Biographical Sketches* (1890) and in it Huxley expressed a strong distaste for and scepticism about the authority of autobiographical expression. Leonard recalled this scepticism as a context for discovering in the disordered mass of his father's papers, 'an [earlier] entirely different sketch of his early life, half-a-dozen sheets describing the time he spent in the East End [of London, during his medical training] with an almost Carlylese sense of the horrible disproportions of life' (Huxley 2:231). In borrowing the literary, protesting and prophetical idioms of Thomas Carlyle's voice, the young T. H. Huxley fostered an autobiographical experiment in an accent of grotesque 'disproportion' that the public scientist had suppressed, though Leonard published extracts at the beginning of the biography (Huxley 1:15–16).

For Huxley, a highly skilled rhetorician, literary inventiveness provided a code for interpreting 'the horrible disproportions of life'. Huxley opened his later life autobiographical narrative with a tale about the denial to him at birth of 'mellifluous eloquence', though it is more socially combative than Darwin's lamentation at the loss of an aesthetic sense:

I am not aware that any portents preceded my arrival in this world, but, in my child-hood, I remember hearing a traditional account of the manner in which I lost the chance of an endowment of great practical value. The windows of my mother's room were open, in consequence of the unusual warmth of the weather. For the same reason, probably, a neighbouring beehive had swarmed, and the new colony, pitching on the window-sill, was making its way into the room when the horrified nurse shut down the sash. If that well-meaning woman had only abstained from her ill-timed interference, the swarm might have landed on my lips, and I should have been endowed with that mellifluous eloquence which, in this country, leads far more surely than worth, capacity, or honest work, to the highest places in the Church and State. But the opportunity was lost, and I have been obliged to content myself through life with saying what I mean in the plainest of plain language, than which, I suppose, there is no habit more ruinous to a man's prospect of advancement.

(Engel 122–3, Huxley 1:3)

What characterizes this birth myth is its hybrid status: Huxley describes it as a 'traditional account' of loss, rather than a completely fabricated event, so there is no reason to doubt the reported presence of the bees (they are, in the language of the naturalist, making their advance as a colony) or of the nurse's swift action in denying them access to the room and the baby Huxley's lips. But it is the bees' designs on the baby's lips, of course, that renders the story traditional and mythic. The story alludes to a myth of Greek antiquity that lips touched by bees were sweetened by honey and would, thereby, be the source of 'mellifluous' poetry. The English 'mellifluous' derives from the Latin compound for 'flowing like honey'. Thus, Huxley's story, written in 1889–90, is one of the lesser-known contributions to what Marion Thain, in her study of the poetry of Michael Field, has described as 'apian aesthetics', or those multiple *fin de siècle* contexts in which decadent associations between bees, honey and sensuous, eloquent poetry were woven. Thain points out that the 'resonance achieved a much more precise intellectual currency around the middle of the nineteenth century with Matthew Arnold's popularization of the phrase "sweetness and light" in *Culture and Anarchy*' (141). Huxley concludes that the denial of 'mellifluous eloquence' initially blocked his access to influential positions in the Church and the state; it is the inheritance of 'literary' eloquence that has been blocked at the same as its status as cultural capital is critiqued. Huxley aligns himself with Leslie Stephen's school of rationalist 'plain speaking and freethinking', and almost certainly against the former Oxford professor of poetry, champion of Hellenic eloquence and state functionary (inspector of schools), Matthew Arnold. Huxley's rhetoric returns 'sweetness and light' to its original, satiric, Swiftian context, while putting it into battle against Anglican privilege and the Classics-based model of literacy that continued to validate a route into governance.

Huxley's self-conscious playfulness as an autobiographical rhetorician constituted an opening to popular biographers, a category of life writing explored by Bernard Lightman (2007) in his work on late Victorian scientific popularizers. Figures such as Grant Allen (who wrote a brief biography of Darwin in 1885, obtaining the cooperation of Francis Darwin to do so) re-energized at the very end of the century the popular trend of widening access to the life of the scientist that had first emerged between the 1830s and 1850s. The scientific journalist and popularizer Edward Clodd wrote a brief life of Huxley for the series entitled 'Modern English Writers' in 1902, notably placing Huxley among writers rather than scientists. Clodd had worked with Huxley on educational causes for many years and was a rationalist and agnostic who used the brief, condensed biographical model to energetically extend his mentor's combative rhetorical legacy in ways that Leonard Huxley's more sober and exhaustively detailed authorized performance could not. In this context, Clodd wrote what members of Huxley's own family recognized and approved as a 'handbook to evolution and the agnostic position' (Collier). In authorizing this position, Clodd appealed to a literary frame of reference, invoking Arnold: under the title *Thomas Henry Huxley*, Clodd added the epigraph 'Who saw life steadily, and saw it whole', applying it to his subject. This epigraph derives, of course, from Matthew Arnold – a quotation from his poem of *c.* 1848, 'To A Friend' (Arnold 105). This was one of a number of self-conscious moves that Clodd made to realign Thomas Henry Huxley with an Arnoldian inheritance. Huxley's patchy early education was treated in both Leonard Huxley's and Edward Clodd's biographies. However, Clodd added an Arnoldian supplement stating that Huxley's:

'boyhood was a cheerless time. Reversing Matthew Arnold's sunnier memories:

No rigorous teachers seized his youth,
And purged its faith, and tried its fire,
Shewed him the high, white star of truth,
There bade him gaze, and there aspire'.

(2)

Thus, Clodd uses Arnold's 'sunny' eloquence from 'Stanzas From the Grand Chartreuse' in order to 'reverse' the poem's memory of the rigorous education experienced by aspiring youth. A literary appropriation enables Clodd to develop his own eloquent rendering of the biological 'curriculum' (7) experienced on the exploratory sea voyages undertaken by Darwin, Huxley and Hooker, which had 'schooled' these scientists in nature's essential medium of water: 'life had its origin in water, and therein the biologist finds his most suggestive material. Darwin and Joseph Hooker had passed through a like curriculum'.

Thus, at the very beginning of the twentieth century at the point where the Darwinian revolution had reached a high watermark, popular biographies of evolution's champions could articulate the foundations of biological knowledge and the spaces of nature where one might be schooled in it. Literariness could function as an inventive source of expression, but it could also be used to authorize, as in the case of Clodd's use of the Arnoldian 'seeing life steadily and whole' epithet, the 'wholeness' of a research paradigm and its fitness to direct social thinking.

Perhaps the terminal point of the Victorian life and letters tradition for writing the life of the scientist was a monumental attempt to consolidate the authority of the new field of biological population management, or eugenics. In writing *The Life, Letters and Labours of Francis Galton*, Karl Pearson blended many aspects of the tradition outlined in this chapter. Compiled by Pearson in the role of impersonal scientific heir to Galton, his benefactor, as the director of the Galton Laboratory at University College, London, it was nonetheless full of filial respect and indeed love for a man who had come to be seen by Pearson as a father. Punctuated by the cataclysms of twentieth-century European war and its preparations (the first volume appeared in 1914, the second in 1924 and the final volumes in 1930), the biography painstakingly recorded a life of scientific achievement that blended Darwin's evolutionism (Galton was Darwin's cousin; they shared a grandfather in Erasmus Darwin) with Babbage's conviction about the centrality of computation and statistics in designing social policy. If it was now population, mate selection and the weeding out of the 'unfit' instead of the economy of manufacture that was this science's new object of national salvation, a variety of literary language and metaphor still had a legitimating role to play. As Pearson observed in outlining the mission of the Galton Laboratory in a quasi-religious, visionary discourse,

> The garden of humanity is very full of weeds, nurture will never transform them into flowers; the eugenist calls upon the rulers of mankind to see that there shall be space in the garden, freed of weeds, for individuals and races of finer growth to develop the full bloom possible of their species.
>
> *(3B:220)*

There is something unsettling about a metaphorical advocacy of the weeding of races, published in 1930, that reaches back for its source into Galton's life, recorded in a Victorian biographical monument.

However, as I have argued throughout this chapter, Victorian scientific biography and auto-biography were inscribed in multiple literary accents. In *Memories of My Life*, the autobiography published by Francis Galton just three years before his death in 1911, Galton's memories of being a student at Cambridge in the 1840s recall a scene involving the 'formidable' (60) Dr Whewell, master of Trinity, the man who gave birth to the office and nomenclature of the 'scientist', as he recited Milton's *Paradise Lost* to the woman who would become his first wife. It constitutes another, quite different perspective on the mechanics of mate selection and the preservation of the fit:

> All male animals, including men, when they are in love, are apt to behave in ways that seem ludicrous to bystanders. . . . I fancied I could almost hear the rustling of his stiffened feathers, and did overhear these sonorous lines of Milton rolled out to the lady *à propos* of I know not what, "cycle and epicycle, orb and orb", with hollow o's and prolonged trills on the r's.

Galton's voice here rediscovers sources of youthful humour which satirized the scientific standing of Whewell and other professors. This was humour that he and many Cambridge undergraduates, as Janet Browne has demonstrated, would have consumed from student magazines (174–5, 183–7). In Book VIII of *Paradise Lost*, which Whewell 'performs' and which Galton interprets as a bizarre mating ritual, Raphael addresses Adam, urging him to overcome his doubts and to look for evidence to God's great book of Nature: he can see its revealed workings in the movements of the planets, 'cycle and epicycle, orb and orb'. Trilling his r's, Whewell imagines himself to be impressing his lady admirer by drawing on Milton to write an orthodox Bridgewater Treatise, sonority underlining God's epic, whereas, for Galton, Whewell seems to be performing an early draft of Darwin's theory of sexual selection *avant* the *Origin of Species*. In mixing these sources, Galton's accent makes the scene 'ludicrous': a *Punch* cartoon rather than a chapter from nature's epic. There are many layers to Galton's autobiographical voice here: the Miltonic grand style; youthful humour from the 1830s and 1840s; Darwin's defining statement of evolutionary scientific naturalism from 1859; and the satiric humour of the magazine culture that greeted and filtered it during the 1860s and beyond. In any event, it is a more complex voice than Pearson's priestly biographical sermon, speaking on behalf of Galton's project and the institutions that sought to shape public policy in its name.

As I have argued in this chapter, if the Romantic emphasis on subjectivity and voice marked all nineteenth-century life writing, including the writing of scientific lives, then researchers in the field of literature and science studies need to pay closer attention to the multiple ways in which voices work in these still somewhat neglected texts. This is not simply because such a method gives us more complex, multifaceted images of individual scientific lives – though undoubtedly it can do that. More importantly, the voices give us insight into the way in which the sources comprising 'the life' interacted with the scientific institutions and discourses of public value that enabled science to claim its authority. Literature played a crucial role in the construction of this authority, but it could also, through forms of reverie and humour, play a role in critically exposing the foundations on which it was built.

Note

1   See Martin Willis, 'Scientific Cultures and Institutions', in this volume.

Bibliography

Primary texts

Arnold, Matthew. *The Poems of Matthew Arnold* (Longmans Annotated English Poets). Ed. Kenneth Allott. London: Longmans, 1965.
Babbage, Charles. *The Ninth Bridgewater: A Fragment*. London: John Murray, 1837.
Babbage, Charles. *Passages From the Life of a Philosopher* [1864]. Ed. Martin Campbell-Kelly. London: William Pickering, 1994.
Barlow, Nora. *Autobiography of Charles Darwin*. London: Collins, 1958.
Boswell, James. *Life of Johnson* [1791]. Ed. R. W. Chapman. Oxford: Oxford University Press, 1980.
Clodd, Edward. *Thomas Henry Huxley*. New York, NY: Dodd, Mead, 1902.
Coleridge, Samuel Taylor. *Sibylline Leaves: A Collection of Poems*. London: Rest Fenner, 1817.
Collier, Ethel G. Letter to Edward Clodd, 9 March 1902. BC 19c Clodd/Hux/C: Clodd Collection, Brotherton Library, University of Leeds.
Darwin, Francis. *Life and Letters of Charles Darwin*. 3 vols. London: John Murray, 1887.
Davy, John. *Memoirs of the Life of Sir Humphry Davy*. 2 vols. London: Longman, Rees, Orme, Brown, Green, and Longman, 1836.
Engel, Louis. *From Handel to Hallé: Biographical Sketches*. London: Swan Sonnenschein, 1890.
Galton, Francis. *Memories of My Life*. London: Methuen, 1908.
Huxley, Leonard. *Life and Letters of Thomas Henry Huxley*. 3 vols. London: Macmillan, 1901.
Paris, John Ayrton. *The Life of Sir Humphry Davy*. 2 vols. London: Henry Colburn and Richard Bentley, 1831.
Pearson, Karl. *The Life, Letters and Labours of Francis Galton*. 4 vols [1–3B]. Cambridge: Cambridge University Press, 1914–30.
Seward, Anna. *Life of Erasmus Darwin* [1804]. Ed. Philip K. Wilson, Elizabeth A. Dolan, and Malcolm Dick. Studley: Brewin, 2010.

Secondary texts

Amigoni, David. 'Between Medicine and Evolutionary Theory: Sympathy and Other Emotional Investments in Life Writings by and About Charles Darwin'. *After Darwin: Animals, Emotions and the Mind*. Ed. Angelique Richardson. Amsterdam and New York, NY: Rodopi, 2013. 172–92.
Barnard, Teresa. *Anna Seward: A Constructed Life*. Farnham: Ashgate, 2009.
Browne, Janet. 'Squibs and Snobs: Science in Humorous British Undergraduate Magazines Around 1830'. *History of Science* 30 (1992): 165–97.
Cantor, Geoffrey. 'Scientific Biography in the Periodical Press'. *Science in the Nineteenth-Century Periodical*. Eds. Geoffrey Cantor *et al.* Cambridge: Cambridge University Press, 2004. 216–37.
Fyfe, Aileen. *Science and Salvation: Evangelical Popular Science Publishing in Victorian Britain*. Chicago, IL: University of Chicago Press, 2004.
Higgitt, Rebekah. 'Discriminating Days? Partiality and Impartiality in Nineteenth-Century Biographies of Newton'. *The History and Poetics of Scientific Biography*. Ed. Thomas Söderqvist. Aldershot: Ashgate, 2007. 155–72.
Levine, George. *Darwin Loves You: Natural Selection and the Re-enchantment of the World*. Princeton, NJ: Princeton University Press, 2006.
Lightman, Bernard. *Victorian Popularizers of Science: Designing Nature for New Audiences*. Chicago, IL: University of Chicago Press, 2007.

Taub, Liba. 'Presenting a "Life" as a Guide to Living: Ancient Accounts of the Life of Pythagoras'. *The History and Poetics of Scientific Biography*. Ed. Thomas Söderqvist. Aldershot: Ashgate, 2007. 17–36.

Thain, Marian. *'Michael Field' (1880–1914): Poetry, Aestheticism, and the Fin de Siècle*. Cambridge: Cambridge University Press, 2007.

Further reading

Gagnier, Regenia. *Subjectivities: A History of Self-Representation in Britain, 1832–1920*. New York, NY: Oxford University Press, 1991.

Humphry Davy and His Circle [Davy Letters Project]. <www.davy-letters.org.uk>. Accessed 17 October 2015.

Knight, David M. 'From Science to Wisdom: Humphry Davy's Life'. *Science in the Romantic Era*. Aldershot: Ashgate, 1998, 283–94.

Lightman, Bernard. 'The Many Lives of Charles Darwin: Early Biographies and the Definitive Evolutionist'. *Notes and Records of the Royal Society* 64.4 (2010): 339–58.

Morrell, Jack, and Arnold Thackray. *Gentlemen of Science: Early Years of the British Association for the Advancement of Science*. New York, NY: Oxford University Press, 1981.

Rosenberg, John D. 'Mr Darwin Collects Himself'. *Nineteenth-Century Lives: Essays Presented to Jerome Hamilton Buckley*. Eds. Laurence S. Lockridge *et al.* Cambridge: Cambridge University Press, 1989, 82–111.

White, Paul. *T. H. Huxley: Making the Man of Science*. Cambridge: Cambridge University Press, 2003.

10

WRITING SCIENCE
Scientific prose

Jonathan Smith
University of Michigan–Dearborn

In *The Queen of the Air* (1869), his study of Greek myths of clouds and storms, John Ruskin took one of his many swipes at the science of Thomas Henry Huxley. Just a few months earlier, Huxley had popularized the term 'protoplasm' in his controversial essay 'On the Physical Basis of Life'. Huxley's claim that life at root was a physical substance in the cells of all organisms struck many contemporaries as rank materialism and atheism. Ruskin largely shared that view, but his response wittily focused on the shortcomings of scientific terminology and Huxley's prose:

> Now, on heat and force, life is inseparably dependent; and I believe, also, on a form of substance, which the philosophers call 'protoplasm'. I wish they would use English instead of Greek words. When I want to know why a leaf is green, they tell me it is coloured by 'chlorophyll', which at first sounds very instructive; but if they would only say plainly that a leaf is coloured green by a thing which is called 'green leaf', we should see more precisely how far we had got. However, it is a curious fact that life is connected with a cellular structure called protoplasm, or, in English, 'first stuck together': whence conceivably through deutoroplasms, or second stickings, and tritoplasms, or third stickings, we reach the highest plastic phase in the human pottery, which differs from common china-ware, primarily, by a measurable degree of heat, developed in breathing.
>
> *(355)*

Ruskin, the great art and social critic, widely regarded as one of the period's premier prose stylists, here satirized an opponent in Huxley celebrated for his own satirical brilliance and rhetorical acumen. Huxley's science, Ruskin implies, is jargon-riddled; its concepts and explanations are empty. Worse, in Ruskin's eyes, Huxley's language obscures the materialism that reduces human beings to the level of mass-produced dinnerware, and fails to distinguish between the heat produced by a living body and that produced by a kiln. Ruskin had a low opinion of much contemporary science, which in his view failed to attend to the habits and appearance of living beings. He frequently lamented science's inability to answer the simplest but most consequential questions, like why a leaf is green or how a bird flies. He repeatedly excoriated scientists for their imprecise descriptions and inaccurate illustrations.

141

Few of Ruskin's contemporaries shared his opinion that the nineteenth century was an era of wrong-headed science badly presented. Indeed, the prominence of science was a given, and works of science aimed at or accessible to an audience not comprised exclusively of practitioners were expected to be engaging and well-written, to instruct and delight. In numerous magazines and newspapers, articles and reviews of recent science often appeared cheek by jowl with – and in implicit or explicit conversation with – articles and reviews about politics, history, religion and literature, or even with the latest instalment of a serialized novel.[1] Readers were used to seeing the prose of scientists and scientific popularizers evaluated and assessed, often favourably. From the earliest decades of the century, works of natural science employed language and techniques more commonly associated with works of literature. Generally speaking, these features were more prominent in texts aimed at a general audience. Increasingly, though, historians and literary scholars have come to see these borrowings, even in prose written for professional or well-educated readers, as ubiquitous and constitutive of knowledge-making, rather than as isolated instances of mere phrase-making, dashes of eloquence or poetic fancy, or bits of story designed to add variety to the presentation.

Perhaps the first point to be made, however, is that there was no such thing – or at least no single, monolithic thing – as 'scientific prose' in the nineteenth century, any more than there was such a thing as 'novelistic prose'. Scientific prose, like its novelistic counterpart, varied with the writer and the situation. Audience, venue and genre mattered enormously when it came to writing about science. What were the writer's credentials and identity? Who was the audience for the work? Where would the work appear? Huxley's 'On the Physical Basis of Life' began as a Sunday evening address in an Edinburgh lecture series devoted to non-theological topics initiated by James Cranbrook, the controversial former pastor of the Congregationalist Albany Street Chapel. One of Huxley's 'lay sermons', it stands as a classic example of his use of that form to subvert religious orthodoxy. It gained wider fame and notoriety when it was published in the liberal *Fortnightly Review*, creating what the editor declared the greatest sensation initiated by a single article in a generation (Desmond 367). Parts of Ruskin's *Queen of the Air* also began as public lectures, and his satire of Huxley's celebration of protoplasm must be seen in the tradition of skewering the hubristic obtuseness of science found in Book 3 of *Gulliver's Travels* and in Charles Dickens's caricature of the British Association for the Advancement of Science as the Mudfog Association for the Advancement of Everything in *Bentley's Miscellany* in 1837–38.

Satire of science and scientific prose represented only one strand of the complex relationship between literary authors and science during the nineteenth century, though one that paid backhanded tribute to science's growing influence.[2] By defining itself as hostile to unbridled speculation and imagination, science implicitly and sometimes explicitly claimed for itself a path to truth different from and superior to that of the literary and poetic. Too much emphasis on the factual, however, left science open to complaints about its dryness and its lack of human interest and passion. Embracing literary language and techniques helped to counter this and to fuel the oft-invoked position that science was just as poetic, just as fanciful, as anything generated by the literary imagination – and yet was also true. When the physicist John Tyndall, friend and ally of Huxley and another *bête noire* of Ruskin, spoke in 1870 of 'the scientific use of the imagination', the term struck a cultural chord and became widely adopted. While Tyndall sought both to defend the role of the imagination in science and to delineate its limits, the term was almost invariably alluded to or appropriated in support of the former.

The long nineteenth century in Britain has long been regarded by literary scholars as a golden age of non-fiction prose. Burke, Wollstonecraft and Coleridge. Hazlitt, Lamb and De Quincey. Carlyle, Newman and Macaulay. Eliot, Arnold and Mill. Ruskin, Pater and Wilde. A place at that canonical table has always been occupied by two scientists: Charles Darwin and Huxley, his self-proclaimed 'bulldog'. Indeed, as far back as 1938, in their magisterial anthology *English Prose of the Victorian Era*, Charles Frederick Harrold and William D. Templeman devoted a section to Huxley's essays and included excerpts from Darwin in their appendix of 'Passages Illustrative of Nineteenth-Century Conceptions of Growth, Development, Evolution'. Despite radical changes in the study of nineteenth-century scientific prose over the last three-quarters of a century, and particularly over the last thirty years, little has changed in its presentation in most current literary anthologies. The *Norton Anthology of English Literature* also includes selections from Huxley's essays and a section on evolution with excerpts from Darwin, while the competing *Longman Anthology of British Literature* reverses that, giving Darwin his own section and including Huxley only in its religion and science sections. Even the more specialized *Victorian Prose: An Anthology* (1999) adds to its selections from Darwin and Huxley only an excerpt from Robert Chambers's anonymous 1844 bestseller *Vestiges of the Natural History of Creation*. Anthologies of British romanticism, including the romanticism sections of the *Norton* and *Longman*, tend to give no special attention to natural science, or to include writings by natural scientists, presumably because science is regarded as less central to and characteristic of British culture in the first few decades of the century than it became in the Victorian era. The romanticism volume of the 2010 *Broadview Anthology of British Literature* is a bit better, having sections on 'The Place of Humans and Non-Human Animals in Nature' and 'Steam Power and the Machine Age'. The former, however, is without any excerpts from works of natural science and the latter, which does include an excerpt from Humphry Davy's 'Introductory Discourse' to his 1802 lectures on chemistry at the Royal Institution, is only available on the *Anthology*'s electronic resources site.

Huxley's presence in anthologies, moreover, stems more from his writings as a controversialist than as a scientist per se. Huxley's joustings with Matthew Arnold over the place of science in a liberal education and with John Henry Newman and other religious figures over agnosticism and the extent of science's voice in religious and moral questions are the texts reproduced most frequently. 'In Victorian controversies over religion and education,' write the editors of the *Norton Anthology*, Huxley was 'one of the most distinctive participants', a scientist whose 'clear, readable, and very persuasive English prose' drew him out of the laboratory and 'onto the platforms of public debate' (Greenblatt *et al.* 1449). True enough, and yet Huxley wrote some of the century's best- and steadiest-selling works of science for general audiences, and his widely read *Evidence as to Man's Place in Nature* (1863) was the first major book to apply Darwinian principles to human evolution, doing the very thing that Darwin had avoided four years earlier in *On the Origin of Species* and would not do for another eight years with *The Descent of Man*.[3]

In short, then, scientific prose holds almost no formal place in the major anthologies of nineteenth-century British literature, even for the Victorians. As indices of the canonical, anthologies are generally slow to register scholarly changes, but the venerable *Norton* has made surprisingly few revisions to its selections and treatment of Victorian science over the course of half a century, even as other topic clusters have been changed substantially or new ones have been added. The recent *Broadview Anthology of Victorian Prose, 1832–1901* (2012) is thus particularly welcome. Organized thematically rather than biographically, it contains an entire section devoted to science, introduced by historian Bernard Lightman. Of that section's fourteen selections, two are from Darwin (the *Origin* and the *Descent*) and one is by Huxley (a late essay on 'The Struggle for Human Existence'). Charles Lyell's *Principles of Geology* (1830–33) is excerpted, as are *Vestiges* and Herbert Spencer's *Social Statics* (1851). But also included are excerpts from Charles Bell's Bridgewater Treatise on *The Hand* (1833), Francis Galton's *Hereditary Genius* (1869) and works of popular science by both women (Mary Somerville, Mary Ward, Lydia Becker) and men (Philip Henry Gosse and Richard Proctor). While still somewhat weighted towards scientific naturalists and evolutionary thinkers and texts, these selections are a much more appropriate reflection of science's place in Victorian prose, and of the kinds of scientific prose to which nineteenth-century readers were exposed.

As the above confirms, Charles Darwin stands out as the nineteenth-century scientist whose scientific prose has been most extensively studied by literary scholars. An early spur for this came with the centenary celebration of the *Origin* in 1959 and the subsequent – and ongoing – publication of the voluminous correspondence and other archival materials in the Darwin Papers at the Cambridge University Library. The initial focus of this work on Darwin's prose had to do with determining Darwinism's literary form. For Jacques Barzun in *Darwin, Marx, Wagner* (1942, 2nd edn 1958) and Stanley Edgar Hyman in *The Tangled Bank: Darwin, Marx, Frazer and Freud as Imaginative Writers* (1962), the characteristic literary form of Darwinism was tragedy: his narratives of nature were tales of struggle and death, and the moral of evolution's story for humanity is that we, too, are animals, our significance shaped only by ourselves rather than being divinely and transcendently ordained. For A. Dwight Culler, however, the characteristic Darwinian form was comedy, with the Darwinian revolution enunciated in the *Origin* offering a reversal of orthodox thinking best likened to subversive strands and movements in Victorian literary culture: Aestheticism, nonsense literature and especially the satire of Samuel Butler. Culler's shift from examining *Darwinism's* relation to *literary form* to considering *Darwin's* relation to *Victorian literature* provided a bridge to the groundbreaking work on Darwin by Gillian Beer and George Levine in the 1980s. Beer's *Darwin's Plots* (1983) and Levine's *Darwin and the Novelists* (1988) have been most influential for their examinations of relationships between Darwin and Victorian fiction, particularly the realism of figures like Eliot, Trollope and Hardy, but also the novelistic worlds of Dickens. It was Beer who first fully drew attention to the similarities between Darwin's mapping of the cumulative effects of small forces acting over extended periods of time, his portrait of a dynamic, interconnected, competition-laden natural world, and the societies of Eliot's Middlemarch and Hardy's Wessex. But both Beer and Levine took as initial and central questions the challenges faced by Darwin in his prose. How could Darwin hope to explain natural selection to a culture saturated with the assumptions of natural theology, in which an omnipotent and benevolent god designed living things expressly for their places in nature? Indeed, how could Darwin use language, which is fundamentally about human agency, to describe the agency-less, non-anthropocentric process that is natural selection? Darwin had to deploy literary tools and techniques to convey his theory, Beer argued, but he could not control them. He relied on an analogy between artificial selection by humans in the breeding of domesticated animals and the operation of natural selection, but he thereby implied that nature is an intelligent, selecting entity. He deployed simile and metaphor for discussing relationships but wanted his readers to accept them as literally true, contending that similarities among organisms were the result of descent from a common ancestor. He drew on common cultural narratives of transformation, progressive improvement and of course fall, but in doing so he invited associations and appropriations at odds with or at least tangential to his theory. To speak of the 'tree of life' and of life having been originally 'breathed into' the first organisms (as Darwin did in Chapter 4 of the *Origin* and in its final paragraph, respectively) was to align one's text with Genesis in ways that different readers could find comforting, disingenuous or heretical.

The work of Beer and Levine for a time seemed to be the last word on Darwin's prose, at least in literary studies. Apart from James Krasner's *The Entangled Eye* (1992), which focused on the role of visual perception in the *Origin*, it has only been in the last decade that new, full-length studies of Darwin's writing have emerged. These have moved to different cultural aspects and implications of Darwin's work. In *Colonies, Cults, and Evolution* (2007) David Amigoni has argued that the modern conception of 'culture' is rooted in a nineteenth-century interdisciplinary nexus with Darwin's theories at their centre. Cannon Schmitt, in *Darwin and the Memory of the Human* (2009), traces the role of 'savages' in Darwin's life and writings, examining evolutionary theory's shaping of the imperial self. Gowan Dawson's 2007 study, *Darwin, Literature and Victorian Respectability*, demonstrates the ways that Darwin's treatment of sex and morality in *The Descent of Man* entered charged cultural debates about avant-garde literary movements and progressive social movements. Darwin and his allies, Dawson shows, fiercely defended Darwin's reputation and sought to limit or repudiate efforts to associate his theories with anything that smacked of sexual or intellectual impropriety. My own *Charles Darwin and Victorian Visual Culture* (2006), meanwhile, argued that Darwin's writings sought to advance an evolutionary understanding of xsex and beauty that challenged prevailing aesthetic theories and connected Darwin to 'lower' forms of visual culture.

No other nineteenth-century British scientist has received as much attention from literary scholars as Darwin. In many ways that is – or was – understandable. Darwin looms extraordinarily large over nineteenth-century British culture, not just over its science, and historical scholarship on Darwin still dwarfs that on other nineteenth-century figures. Surprising as it is to say, however, considerable room exists for further examination of Darwin's prose. Almost all of the attention to date has been concentrated on the *Origin* and the *Journal of Researches* and, to a lesser but growing extent, *The Descent of Man* and *The Expression of the Emotions in Man and Animals* (1872); the *Variation of Animals and Plants Under Domestication* (1868) and the botanical writings so prominent in the last two decades of Darwin's life have been given short shrift. The neglect of the *Variation* by literary scholars, given its focus on inheritance, breeding, domestication and cultivation, is particularly surprising. And Darwin's botanical works, concerned with plant sexuality, movement and the behaviour of insectivorous species, point to connections not with realism but rather with sensation fiction, that hybrid form as much a part of the Darwin family's evening novel-reading as the works of Eliot and Trollope.

But Darwin was not the typical nineteenth-century scientist and his prose cannot stand in for scientific prose generally. Nor can Huxley's. Indeed, literary scholars have only fairly recently begun to appreciate that Darwin and Huxley were part of a relatively small (but highly influential) group of scientists that historians (drawing on Huxley) refer to as the Victorian 'scientific naturalists'. As Frank Turner has succinctly put it, the scientific naturalists 'sought to expand the influence of scientific ideas for the purpose of secularizing society rather than for the goal of advancing science internally. Secularization was their goal; science, their weapon' (1974 16). Along with Tyndall, the botanist Joseph Hooker, the polymathic John Lubbock and others, Huxley led a largely successful campaign in the latter decades of the century to establish scientists as the pre-eminent authorities on the natural world and to a considerable extent on the human one, finally freed from clerical interference and the need to avoid ruffling religious feathers.[4] In another widely adopted concept and turn of phrase, Turner has called the scientific naturalists' efforts a quest for 'cultural authority'. Like other historical victors, Huxley and his allies and acolytes got to write the history of their triumph, so that triumph has tended to be seen as the inevitable outcome of the struggle between noble truth-tellers like Darwin and Huxley and powerful religious bigots and obscurantists.

The long-standing focus on Darwin and Huxley has also meant that major scientific figures of the nineteenth century and even many of the leading scientific naturalists have remained under-studied by literary scholars. Humphry Davy, Michael Faraday, Charles Lyell, Richard Owen and Alfred Russel Wallace have begun to receive attention, but all merit much fuller examination, something approaching the level that Darwin has received.[5] Tyndall's name was mentioned in almost every Victorian breath with Huxley's, yet this successor of Faraday at the Royal Institution, author of numerous successful works of popular science and deliverer of the notorious

'Belfast Address' to the British Association for the Advancement of Science has received comparatively limited attention from historians of science and is known to literary scholars largely through Ruskin's hostility. With his correspondence now being published and a major biography finally in the offing, that should change.[6]

As literary scholars have taken up nineteenth-century scientific prose armed with the work of historians of science, an emergent theme has been the ubiquity and significance of 'literary' language and techniques in that prose. Earlier scholarship tended to be drawn to passages in science writing that seemed self-consciously literary, the bits of purple prose or sprinkled allusions to Shakespeare or Milton. Focusing on such passages, however, assumed or implied that the literary and the scientific were separate and separable. More recently, literary scholars have demonstrated the rhetorical work being done by such passages, and have looked beyond them to subtler examples, finding that the literary and the scientific are often seamlessly intertwined. In his more recent work on Darwin, *Darwin Loves You* (2006) and especially *Darwin the Writer* (2011), George Levine has complained that cultural scholarship has lost sight of Darwin's persona and imaginative art. Emphasizing the joy and exuberance often to be found in Darwin's scientific prose, and the imaginative and literary quality of the *Origin*, Levine argues that the 'literary' elements of these works are inseparable from their factual descriptions and theoretical pronouncements. Yet much cultural work has stressed a comparable point, that the 'literary' is not something inserted into scientific prose at particular moments but rather is part of the very fabric of scientific prose, part of the knowledge-making process. As I once emphasized in an essay about the geological imagination in Darwin's *Journal of Researches*, the passages most frequently singled out by literary scholars for their poetic and imaginative qualities were also ones deeply imbued with Darwin's theoretical commitments to Charles Lyell's geological approach (Smith 1994 (Chapter 3)). Ralph O'Connor in *The Earth on Show* (2007), his study of the representation of fossils in the first half of the nineteenth century, similarly contends that 'literary' techniques and 'poetic' passages are not restricted to popular science – though they are often more visible there – and cannot be cordoned off from the 'scientific' content; rather, these devices help to produce scientific knowledge and are elements in constituting it.

Other recent examples of literary scholarship informed by deep familiarity with work in the history of nineteenth-century science have also borne this out. In *Romantic Rocks, Aesthetic Geology* (2004), for example, Noah Heringman has traced the formative influence of landscape aesthetics on both romanticism and geology from 1770 to 1820. Gowan Dawson has recently explored relationships between palaeontology, particularly as practised by Richard Owen, and the reading and writing of Victorian serial fiction (Dawson 2010; Dawson 2012). Adelene Buckland, in *Novel Science* (2013), has drawn on the work of historians of geology to argue that what nineteenth-century geological writing offered novelists – and what Walter Scott's historical romances offered to geologists – was not *plot* or *story* but a model for representing *structure* and *chronology*. Nineteenth-century novelists adopted from geology not the triumphant narratives of uniformitarian geology and Darwinian evolution, Buckland contends, but the challenges of making sense of complex and ambiguous sequences and a suspicion of stories that are linear and neat. In his account of energy physics in *Victorian Literature, Energy, and the Ecological Imagination* (2014), Allen MacDuffie extends the work of Beer, Levine and others on this topic by examining nineteenth-century thermodynamic narratives, what MacDuffie calls 'stories about the way energy travels through social, natural, and cosmic systems' (13), for the convergences and especially divergences of literary and scientific imaginaries. In all of these works, the two-way nature of the traffic between literature and science in the nineteenth century is vividly demonstrated, not by the 'influence' of scientific ideas on literature, but by the ways that scientists and novelists, poets and essayists deploy similar tools – of narrative, of visualization and description, of structure and publishing format – in their work and in their writing.

The major shift in the history of nineteenth-century science, and a move with extensive implications for the literary study of the period's science, has come with regard to popularization. Popularization had long been seen by historians and scientists alike through a 'diffusionist' lens.[7] That is, popular science writing was seen to spread the work of professional scientists to a broader, more general audience. The work of popularization was regarded as a sort of act of translation that involved replacing technical terms with more everyday language and explaining concepts with homespun analogies; popularizers were assumed to remain as true as possible to the original work, to stick to the facts, to take themselves out of the picture. But we now know that this is rarely how popularization works, and it certainly wasn't how it worked in the nineteenth century. The line between professional and popular science is a blurry one. Popularizers are active shapers of knowledge, constituters of it, even, rather than passive transmitters. The mere context in which their work appears, whether a religious periodical or a series of science books for children or a provincial lecture for working people, shapes the work's reception and perception. Popularizers also frequently comment on and critique the science they popularize. They can extend and add to it, combine it with similar or conflicting work, package it in ways that change or distort its meaning. As Bernard Lightman has shown, for example, most people in the nineteenth century learnt about Darwin's theories not from Darwin's books but from the numerous popular books and periodical articles on them, and even when these were faithful to Darwin's ideas they often incorporated them into explicitly religious frameworks that certainly departed from Darwin's own emphasis (Chapters 2 and 8).

This new approach to popularization has led to an awareness of how deeply both popular and professional science depended on the 'literary' in the nineteenth century. Several prominent strands of literary borrowings have become evident. One is the language of visual spectacle. Iwan Rhys Morus's *Frankenstein's Children* (1998) demonstrated the importance of popular public experiment and exhibition of electricity in early nineteenth-century London, and Bernard Lightman and Ralph O'Connor have traced the importance of the language of visual display and spectacle to all of nineteenth-century popular science.[8] A second, one in which women played a key role, involved what Barbara Gates and Ann Shteir have called 'the narrative of natural history' (11). While female popularizers of the late eighteenth and early nineteenth centuries were more likely to cast their work as educational conversations in a domestic setting between a mother or mother figure and children, later work by women and men alike tended overwhelmingly to be cast as diverting, non-theoretical, anecdotal narratives, focused on encounters with nature rather than the activity of the scientist. Often, these narratives of natural history were framed as rambles in the country or strolls at the shore. Since many popularizers were also writing fiction, and certainly were reading it, they were keenly aware of the possibilities of story for creating and maintaining the reader's interest. A third influential literary form for popular science writing was the evolutionary epic, a term used by James Secord to describe the form of *Vestiges*. Unlike the rambling, peripatetic structure of the narrative of natural history and its series of Wordsworthian discoveries of the extraordinariness of common things, the evolutionary epic started from a daisy or a walnut to construct vast narratives of global and even cosmic change. The evolutionary epic could be used to tell gripping stories of the universe's formation, the earth's geological history, the development of life – or, as in *Vestiges*, all of these together. While the evolutionary epic was a narrative form particularly congenial to writers promoting evolutionary theories, Lightman shows that it could be used by popularizers who sought to subvert the secular goals of Huxley and the other scientific naturalists.

Throughout the nineteenth century, then, scientific prose for popular audiences deployed 'literary' language and techniques of many different types for a variety of purposes. In the first half of the century in particular, when men of science frequently wrote for an educated elite as well as for other men of science, their prose also drew heavily on the literary. Thus Adelene Buckland can present Lyell's *Principles of Geology* as a mock epic (Chapter 3), and Ralph O'Connor can trace the presence of Byron in a range of geological writings (345–55). As the century progressed and science increasingly professionalized, professional and popular science writing began to diverge, particularly, as Alice Jenkins has noted, in areas of the physical sciences that adopted the language of mathematics (2007 17). Nonetheless, many important works of science from the second half of the nineteenth century were written for the educated reader as well as the professional, and professional scientists often wrote for popular audiences. As public exposure to science writing grew and science's cultural authority expanded, literary writers increasingly framed their own work in relation to writing about natural science.

One of the century's earliest such reflections came in William Wordsworth's preface to the 1800 edition of *Lyrical Ballads*, in which Wordsworth famously considered the relationship between the poet and the 'man of science'. As Sharon Ruston has noted, while this comparison was long seen by literary scholars as critical of science, in recent decades it has been interpreted as being more favourable towards science, if only in the future (25–6). That Wordsworth, in defining the identity and role of the poet, felt compelled to differentiate the poet from the man of science speaks volumes about the growing cultural authority of science. In Wordsworth's rendering, although both figures seek truth, the poet is an emphatically social being who joins with all humans in celebrating truth as a 'visible friend and hourly companion' (1:xxxvii), while the man of science seeks truth 'as a remote and unknown benefactor' (1:xxxvi). The poet's search for truth leads to 'habitual and direct sympathy connecting us with our fellow-beings', whereas the man of science is a kind of intellectual hermit or miser who cherishes truth 'in his solitude' and regards it as 'a personal and individual acquisition'. Nonetheless, says Wordsworth,

> [i]f the time should ever come when . . . Science . . . shall be ready to put on, as it were, a form of flesh and blood, the Poet will lend his divine spirit to aid the transfiguration, and will welcome the Being thus produced, as a dear and genuine inmate of the household of man.
>
> *(1:xxxvii–xxxix)*

Wordsworth's reference to 'transfiguration' evokes the transfiguration of Jesus, a New Testament episode in which Jesus appears in a radiant glory and is identified by God as his son. That reference makes *both* poetry and science divine, but it makes the poet Christ-like, having already undergone the transfiguration, a 'divine spirit' in 'a form of flesh and blood' (1:xxxix). Over the course of the ensuing century, as science arguably did create what Wordsworth called a 'material revolution . . . in our condition' (1:xxxviii), and as its discoveries became 'manifestly and palpably material to us as enjoying and suffering beings', scientific discoveries became increasingly, as Wordsworth predicted, 'proper objects of the Poet's art'. That 'material revolution', however, garnered ever more attention and influence for science from all quarters.

Wordsworth's distinction between poetry and science set something of a pattern for literary responses to science's truth claims: acknowledge the validity of those claims, but make poetry, and literature more generally, also a pathway to truth, and then elevate the truth of literature over that of science. In romanticism, this was part of the corrective effort to rescue the imagination from the Enlightenment's elevation of the reason, but as the nineteenth century progressed these moves were more specifically responses to science in its emerging modern sense rather than its more traditional sense as 'knowledge'. In his discussion of what he termed 'the literature of knowledge' and 'the literature of power', for example, Thomas De Quincey elevates the latter over the former. In words that echo Wordsworth, he argues that even '[t]he commonest novel, by moving in alliance with human fears and hopes, with human instincts of wrong and right, sustains and quickens those affections' (De Quincey 303). By contrast, even the greatest examples of the literature of knowledge are instantly superseded the moment they are partially revised or expanded. The literature of power is permanent, universal, and moral. Similarly, Matthew Arnold, in his debate with Huxley during the 1880s about the place of science in education, contrasts knowledge of the diameter of the moon with the inability to paraphrase a line from *Macbeth*. 'I think I would rather have a young person ignorant about the moon's diameter', remarks Arnold, drily, than unable to gloss Shakespeare well, the implication being that scientific knowledge is a mere recitation of facts (127). Even when he cites a more momentous example, Darwin's statement in *The Descent of Man* that humans have probably evolved from a hairy, tree-dwelling quadruped with pointed ears and a tail, his point is that this is mere knowledge, 'knowledge not put for us into relation with our sense for conduct, our sense for beauty' (Arnold 112). (Arnold is surely being disingenuous here, as the *Descent* provided an evolutionary account of how our moral and aesthetic senses arose – it is the nature of Darwin's account, not Darwin's lack of attention to conduct and beauty, that bothers Arnold.) At least for the purposes of this public debate, Arnold concedes science's truth value but hollows it of almost all significance.

Another major text from the Romantic era offers what would become another common pattern of literary response to science's truth claims. In his anonymous preface to his wife Mary's *Frankenstein* (1818), Percy Shelley justified the novel's major plot device – the reanimation of a corpse – as being, according to some contemporary physiologists, 'not of impossible occurrence' (5). The double negative suggests reluctance, and, indeed, in the very next sentence Shelley denies 'according the remotest degree of serious faith to such an imagination', and two sentences after that declares that bringing the dead back to life by physical means *is* impossible. Nonetheless, by basing the story on the reanimation that some scientific writers have decreed 'not impossible', Shelley differentiates this work from other tales of Gothic horror, for the scientific basis 'affords a point of view to the imagination for the delineating of human passions more comprehensive and commanding than any which the ordinary relations of existing events can yield'. Shelley's tortured and contradictory statements are revealing. A basis in science, however slim, provides the grounds for presenting *Frankenstein* not, like other Gothic fiction, as a tissue of 'supernatural terrors' or as 'a mere tale of spectres or enchantment'. At the same time, Shelley clearly wants to avoid being seen as endorsing this piece of scientific speculation, a reflection, perhaps, of his awareness of the hostility in England towards Continental science with perceived radical implications, politically or religiously. While unsympathetic with this hostility, Shelley uses the occasion of the preface to immunize the novel from conservative critiques about this and other aspects of its content. Thus, too, Shelley claims that it is the literary imagination that channels and brings value to the potentially unbridled scientific one, leading to a delineation of human truths of a type with those to be found in Homer, Greek tragedy, Shakespeare and Milton. Unlike Wordsworth, who equates science with affectless 'fact', Shelley sees science as imaginative, and even capable of being overly imaginative, but he shares with Wordsworth the sense that the literary imagination is what can identify and display the moral and emotional significance of scientific knowledge.

Science writers increasingly embraced what Shelley implied: that science could be just as poetic and imaginative as literature. To the notion that science offered 'mere' facts or knowledge, popular and professional science alike highlighted the creativity of scientists and their 'amazing but true' accounts of the natural world. In the same year that De Quincey published his distinction between the literatures of knowledge and power, for example, Robert Hunt, who worked in the physical and geological sciences, published his *The Poetry of Science* (1848). Subtitled *Studies in the Physical Phenomena of Nature*, Hunt's book was a primer on the current state of knowledge in the natural sciences, but its prose was particularly vivid and its expressions of wonder frequent and lofty. 'The phenomena of Reality are more startling than the phantoms of the Ideal', wrote Hunt (xxiii). Charles Dickens, reviewing the book for *The Examiner*, concurred: 'To show that the facts of science are at least as full of poetry, as the most poetical fancies' (787) was not only a worthy purpose but a salutary one in a utilitarian age. Despite this sentiment, Dickens's critique of utilitarian science in *Hard Times* (1854) just a few years later had room only for poetical fancies, not scientific ones, with the exemplar of scientific knowledge in this case being the definition of a horse as a graminivorous quadruped (Chapter 2).

The reduction of scientific knowledge to 'mere' knowledge speaks, however, to science's (and especially, later in the century, scientific naturalism's) increasing claims to comment on issues previously dominated by literary figures, religious authorities, and politicians, and to invoke and deploy its work as deeply imaginative rather than hostile to the imagination. Tyndall's involvement in what came to be known as the 'prayer-gauge debate' – he proposed an experiment to test the efficacy of prayers for the sick after claims that national prayers for healing had saved a dangerously ill Prince of Wales in 1871 – can stand as a mark of the former, his 'scientific use of the imagination' for the latter. Delivered to the British Association for the Advancement of Science in September 1870, Tyndall's defence of the imagination's role in science sought to cast the imagination as essential, particularly when dealing with phenomena unobservable to the senses. 'There are tories even in science who regard imagination as a faculty to be feared and avoided rather than employed,' Tyndall complained (1870 16). In fact, however, the imagination, guided and constrained by the reason, is the scientist's mightiest tool, without which 'our knowledge of nature would be a mere tabulation of coexistences and sequences'. Newton relied on it in formulating the theory of universal gravitation, Tyndall argued, and scientists of the present use it when speaking of atoms, the luminiferous ether, the nebular hypothesis, protoplasm and evolution. And in one of the address's most famous and widely quoted passages, if the nebular and evolution hypotheses are correct, then 'at the present moment all our philosophy, all our poetry, all our science, and all our art – Plato, Shakespeare, Newton, and Raphael – are potential in the fires of the sun' (Tyndall 1870 47).

For someone like Ruskin, such passages were precisely what made Tyndall both alluring and dangerous. If Huxley's lay sermons cast the scientist in the role of a secularized Protestant moralist, Tyndall's expressed a scientist's version of the mystical beauty and sublimity that the Romantics, and Ruskin himself, had experienced in Alpine landscapes. Tyndall's scientific imagination unleashed Hunt's poetry of science, but it did so in the service of advancing scientific naturalism and the theories it embraced, many of which seemed to Ruskin and other critics to lead straight to materialism and atheism, whatever Tyndall might say to the contrary. And Ruskin was certainly not alone among literary figures in worrying about and resisting scientific naturalism's claims about the imagination. As George Levine has argued in *Dying to Know* (2002), the Victorians made scientific objectivity a moral ideal, with the death of the (scientific) self a guarantee of truth.[9] This in turn helped enable the scientific naturalists to assert not just cultural but *moral* authority for scientific investigation, as Anne DeWitt has shown. In so doing, the scientific naturalists began to tread on the turf of the novelist, whose province, as Shelley had stated in the preface to *Frankenstein*, was 'to preserve the truth of the elementary principles of human nature' (5). Thus it is, argues DeWitt, that even late-century novelists sympathetic to science and scientific naturalism such as George Eliot and H. G. Wells filled their novels with deeply flawed, immoral scientist figures as a way to maintain the novel's own cultural status as a venue for working out questions of personal and public morality.

The scholarly work of the 1980s that launched the study of literature and science sought to challenge or overturn C. P. Snow's 'Two Cultures' thesis. Collections with titles like George Levine's *One Culture* (1987) and Judith Yaross Lee's and Joseph Slade's *Beyond the Two Cultures* (1990) signalled what their contents made explicit: that literature and science were parts of a single, unified culture. Recently, that 'one culture' thesis has been challenged by such scholars as Dawson, Buckland and DeWitt, who argue that the 'one culture' pendulum swung too far, moving from a corrective to a dogma of its own, blinding critics to the genuine tensions and conflicts between literature and science in the nineteenth century. That shift has come, ironically, as historians of science have turned more and more to the tools of literary analysis to understand how scientific prose does its work, and to argue increasingly for the essential contributions of the literary to the writing of science. James Secord's *Visions of Science* (2015), which examines seven books from the 1830s, among them Humphry Davy's *Consolations in Travel* (1830), Lyell's *Principles of Geology* and Thomas Carlyle's *Sartor Resartus* (serialized 1833–34), epitomizes those moves by offering extended close readings of the seven texts that chart their intertwined philosophical, imaginative, and practical aims within the rich cultural and publishing contexts of a tumultuous decade. Yet, in reasserting the existence of tensions and points of resistance in nineteenth-century literature and science, literary scholars do so this time as a result of careful historical analysis rather than anachronistic assumption. And in that sense, historians and literary scholars studying the prose of nineteenth-century science enact an array of relationships – from sceptical appraisal to wary embrace to enthusiastic coupling – similar to those displayed by their objects of study.

Notes

1 See Gowan Dawson, 'Science in the Periodical Press', in this volume.
2 On satire and Victorian science, see Paradis.
3 For data on sales of scientific works in the period, I rely on Lightman 490–93.
4 For recent reappraisals of scientific naturalism, particularly in light of work by historians that has drastically broadened our sense of nineteenth-century science, see the volumes edited by Dawson and Lightman and Lightman and Reidy.
5 On Davy, see Ruston (Chapter 4) and Jenkins 1998; on Faraday, see Jenkins 2008 and Jenkins 2007 (Chapter 6); on Lyell, see Buckland (Chapter 3); on Owen, see Dawson 2010 and Dawson 2012; on Wallace, see Schmitt (Chapter 2).
6 On Tyndall and Ruskin, see O'Gorman, Sawyer and Smith 1994 (Chapter 5). On the Tyndall correspondence, see Elwick, Lightman and Reidy. Lightman is at work on a Tyndall biography.
7 See Ralph O'Connor, 'Science for the General Reader', in this volume.
8 See Iwan Rhys Morus, 'Staging Science', in this volume.
9 For an influential account of the conception of scientific objectivity in the nineteenth century parallel to Levine's but focusing on scientific atlases, see Daston and Galison.

Bibliography

Primary texts

Arnold, Matthew. *Discourses in America*. London: Macmillan, 1885.
Chambers, Robert. *Vestiges of the Natural History of Creation*. Ed. James A. Secord. Chicago, IL: University of Chicago Press, 1994.
Darwin, Charles. *Journal of Researches into the Geology and Natural History of the Various Countries Visited by H. M. S. Beagle*. London: Colburn, 1839.

Darwin, Charles. *On the Origin of Species*. London: Murray, 1859.
Darwin, Charles. *The Variation of Animals and Plants Under Domestication*. London: Murray, 1868.
Darwin, Charles. *The Descent of Man, and Selection in Relation to Sex*. London: Murray, 1871.
Darwin, Charles. *The Expression of the Emotions in Man and Animals*. London: Murray, 1872.
De Quincey, Thomas. Review of *The Works of Alexander Pope, Esquire*, by W. Roscoe. *North British Review* 9 (1848): 299–333.
Dickens, Charles. Review of *The Poetry of Science*, by Robert Hunt. *Examiner*, 9 December 1848: 787–8.
Dickens, Charles. *Hard Times*. Ed. Fred Kaplan. 3rd edn. New York, NY: Norton, 2000.
Hunt, Robert. *The Poetry of Science: Studies in the Physical Phenomena of Nature*. London: Reeve, 1848.
Huxley, Thomas H. *Evidence as to Man's Place in Nature*. New York, NY: Appleton, 1863.
Huxley, Thomas H. 'On the Physical Basis of Life'. *Fortnightly Review* 5 n.s. (1869): 129–45.
Ruskin, John. *The Queen of the Air*. *The Library Edition of the Works of John Ruskin*. Ed. E. T. Cook and Alexander Wedderburn. 39 vols. London: George Allen, 1903–12. 19:282–423.
Shelley, Mary. *Frankenstein, or, The Modern Prometheus*. Ed. J. Paul Hunter. 2nd edn. New York, NY: Norton, 2012.
Tyndall, John. 'Scientific Use of the Imagination'. *Essays on the Use and Limit of the Imagination in Science*. London: Longmans, Green, 1870, 13–51.
Tyndall, John. *Address Delivered Before the British Association Assembled at Belfast*. New York, NY: Appleton, 1874.
Wordsworth, William. *Lyrical Ballads*. 2 vols. London: Longman and Rees, 1802.

Secondary texts

Amigoni, David. *Colonies, Cults, and Evolution: Literature, Science and Culture in Nineteenth-Century Writing*. Cambridge: Cambridge University Press, 2007.
Barzun, Jacques. *Darwin, Marx, Wagner: Critique of a Heritage*. London: Secker and Warburg, 1942.
Beer, Gillian. *Darwin's Plots: Evolutionary Narrative in Darwin, George Eliot, and Nineteenth-Century Fiction*. London: Routledge, 1983.
Black, Joseph, Leonard Connoly, Kate Flint, Isobel Grundy, Lloyd Riuzza, Jerome J. McGann, Anne Lake Prescott, Barry V. Qualls and Claire Waters, eds. *The Broadview Anthology of British Literature: The Age of Romanticism*. 2nd edn. Peterborough: Broadview, 2010.
Buckland, Adelene. *Novel Science: Fiction and the Invention of Nineteenth-Century Geology*. Chicago, IL: University of Chicago Press, 2013.
Culler, A. Dwight. 'The Darwinian Revolution and Literary Form'. *The Art of Victorian Prose*. Ed. George Levine and William Madden. New York, NY: Oxford University Press, 1968, 224–46.
Damrosch, David, Kevin J. H. Dettmar, Christopher Baswell, Clare Carroll, Andrew Hadfield, Heather Henderson, Peter J. Manning, Anne Howland Schotter, William Sharpe, Stuart Sherman and Susan J. Wolfson, eds. *The Longman Anthology of British Literature*. Vol. 2. 5th edn. London: Longman, 2013.
Daston, Lorraine, and Peter Galison. *Objectivity*. New York, NY: Zone, 2007.
Dawson, Gowan. *Darwin, Literature and Victorian Respectability*. Cambridge: Cambridge University Press, 2007.
Dawson, Gowan. '"By a Comparison of Incidents and Dialogue": Richard Owen, Comparative Anatomy and Victorian Serial Fiction'. *19: Interdisciplinary Studies in the Long Nineteenth Century* 11 (2010). <www.19.bbk.ac.uk/articles/10.16995/ntn.577>. Accessed 12 June 2015.
Dawson, Gowan. 'Paleontology in Parts: Richard Owen, William John Broderip, and the Serialization of Science in Early Victorian Britain'. *Isis* 103.4 (2012): 637–67.
Dawson, Gowan, and Bernard Lightman, eds. *Victorian Scientific Naturalism: Community, Identity, Continuity*. Chicago, IL: University of Chicago Press, 2014.
Desmond, Adrian. *Huxley: The Devil's Disciple*. London: Michael Joseph, 1994.
DeWitt, Anne. *Moral Authority, Men of Science, and the Victorian Novel*. Cambridge: Cambridge University Press, 2013.
Elwick, James, Bernard Lightman, and Michael S. Reidy, eds. *The Correspondence of John Tyndall*. London: Pickering & Chatto, 2015.
Gates, Barbara T., and Ann B. Shteir, eds. *Natural Eloquence: Women Reinscribe Science*. Madison, WI: University of Wisconsin Press, 1997.

Greenblatt, Stephen, Carol T. Christ, Alfred David, Barbara Lewalski, Lawrence Lipking, George M. Logan, Deidre Shauna Lynch, Katharine Eisaman Maus, James Noggle, Jahan Ramazani, Catherine Robson, James Simpson, Jon Stallworthy, Jack Stillinger and M. H. Abrams, eds. *The Norton Anthology of English Literature.* Vol. 2. 9th edn. New York, NY: Norton, 2012.

Harrold, Charles Frederick, and William D. Templeman. *English Prose of the Victorian Era.* New York, NY: Oxford University Press, 1938.

Heringman, Noah. *Romantic Rocks, Aesthetic Geology.* Ithaca, NY: Cornell University Press, 2004.

Hyman, Stanley Edgar. *The Tangled Bank: Darwin, Marx, Frazer and Freud as Imaginative Writers.* New York, NY: Atheneum, 1962.

Jenkins, Alice. 'Humphry Davy: Poetry, Science and the Love of Light'. *1798: The Year of the 'Lyrical Ballads'.* Ed. Richard Cronin. Houndmills: Macmillan, 1998. 133–50.

Jenkins, Alice. *Space and the 'March of Mind': Literature and the Physical Sciences in Britain 1815–1850.* Oxford: Oxford University Press, 2007.

Jenkins, Alice, ed. *Michael Faraday's Mental Exercises: An Artisan Essay-Circle in Regency London.* Liverpool: Liverpool University Press, 2008.

Krasner, James. *The Entangled Eye: Visual Perception and the Representation of Nature in Post-Darwinian Narrative.* New York, NY: Oxford University Press, 1992.

Lee, Judith Yaross, and Joseph W. Slade. *Beyond the Two Cultures: Essays on Science, Technology, and Literature.* Ames, IA: Iowa State University Press, 1990.

Leighton, Mary Elizabeth, and Lisa Surridge, eds. *The Broadview Anthology of Victorian Prose, 1832–1901.* Peterborough: Broadview, 2012.

Levine, George, ed. *One Culture: Essays in Science and Literature.* Madison, WI: University of Wisconsin Press, 1987.

Levine, George. *Darwin and the Novelists: Patterns of Science in Victorian Fiction.* Cambridge, MA: Harvard University Press, 1988.

Levine, George. *Dying to Know: Scientific Epistemology and Narrative in Victorian England.* Chicago, IL: University of Chicago Press, 2002.

Levine, George. *Darwin Loves You: Natural Selection and the Re-enchantment of the World.* Princeton, NJ: Princeton University Press, 2006.

Levine, George. *Darwin the Writer.* New York, NY: Oxford University Press, 2011.

Lightman, Bernard. *Victorian Popularizers of Science: Designing Nature for New Audiences.* Chicago, IL: University of Chicago Press, 2007.

Lightman, Bernard, and Michael S. Reidy, eds. *The Age of Scientific Naturalism: John Tyndall and His Contemporaries.* London: Pickering & Chatto, 2014.

MacDuffie, Allen. *Victorian Literature, Energy, and the Ecological Imagination.* Cambridge: Cambridge University Press, 2014.

Morus, Iwan Rhys. *Frankenstein's Children: Electricity, Exhibition, and Experiment in Early-Nineteenth-Century London.* Princeton, NJ: Princeton University Press, 1998.

Mundhenk, Rosemary J., and LuAnn McCracken Fletcher, eds. *Victorian Prose: An Anthology.* New York, NY: Columbia University Press, 1999.

O'Connor, Ralph. *The Earth on Show: Fossils and the Poetics of Popular Science, 1802–1856.* Chicago, IL: University of Chicago Press, 2007.

O'Gorman, Francis. '"The Eagle and the Whale?" Ruskin's Argument with John Tyndall'. *Time and Tide: Ruskin and Science.* London: Pilkington, 1996, 45–64.

O'Gorman, Francis. '"The Mightiest Evangel of the Alpine Club": Masculinity and Agnosticism in the Alpine Writing of John Tyndall'. *Masculinity and Spirituality in Victorian Culture.* Eds. Andrew Bradstock, et al. New York, NY: St Martin's, 2000, 134–48.

Paradis, James G. 'Satire and Science in Victorian Culture'. *Victorian Science in Context.* Ed. Bernard Lightman. Chicago, IL: University of Chicago Press, 1997, 143–75.

Ruston, Sharon. *Creating Romanticism: Case Studies in the Literature, Science and Medicine of the 1790s.* Houndmills: Palgrave Macmillan, 2013.

Sawyer, Paul. 'Ruskin and Tyndall: The Poetry of Matter and the Poetry of Spirit'. *Victorian Science and Victorian Values: Literary Perspectives.* Eds. James Paradis and Thomas Postlewait. New Brunswick, NJ: Rutgers University Press, 1985, 217–46.

Schmitt, Cannon. *Darwin and the Memory of the Human: Evolution, Savages, and South America.* Cambridge: Cambridge University Press, 2009.

Secord, James A. *Visions of Science: Books and Readers at the Dawn of the Victorian Age*. Chicago, IL: University of Chicago Press, 2015.

Smith, Jonathan. *Fact and Feeling: Baconian Science and the Nineteenth-Century Literary Imagination*. Madison, WI: University of Wisconsin Press, 1994.

Smith, Jonathan. *Charles Darwin and Victorian Visual Culture*. Cambridge: Cambridge University Press, 2006.

Snow, C. P. *The Two Cultures* [1959]. Ed. Stefan Collini. Cambridge: Cambridge University Press, 2012.

Turner, Frank Miller. *Between Science and Religion: The Reaction to Scientific Naturalism in Late Victorian England*. New Haven, CT: Yale University Press, 1974.

Turner, Frank Miller. *Contesting Cultural Authority: Essays in Victorian Intellectual Life* [1993]. Cambridge: Cambridge University Press, 2008.

11

SCIENCE FOR THE GENERAL READER

Ralph O'Connor
University of Aberdeen

Literary scholars have become increasingly aware of the ways in which nineteenth-century authors' encounters with science were shaped by science writings aimed at non-specialist readerships. Such texts form an important part of the cultural background to literature and science studies today. However, they are rarely treated as literary texts worthy of study in their own right. Of the hundreds of books and thousands of essays and articles, only a handful have ever been illuminated through literary analysis or mentioned in literary histories. They are usually (and misleadingly) all lumped together as 'popular science', as if they belonged to a single genre. Like science writing more generally, however, they comprise a wide range of genres, most of which have never been identified as genres, let alone described.

Since the 1980s, some important steps along this path have been taken by historians of science, a field now well into its 'cultural turn' and thus occupying a fertile borderland with literary history. It is these studies that now constitute most of the basic essential reading for literature students and scholars interested in science writings for general readers. Those writings still occupy a blank space at the edge of the literary-historical map, however, on the rare occasions that they are included at all. In order to set out recent and future directions for possible research, this chapter will give some account of the terrain itself and set out the case for treating much of it as literature, as well as outlining the important insights provided by historians of science. I will conclude by suggesting some of the benefits that could result from bringing these texts fully into the orbit of literary studies, even for those whose primary interest is the analysis of fiction, poetry, drama and life writing.

The emergence of 'popular science'

During the nineteenth century, many texts were published in which an important aim was to communicate and reflect on scientific knowledge for general readers: that is, for reader-ships outside specialized scientific communities.[1] In earlier periods, the boundary between scientific practitioners and the (small, upper-class) literate public had been much more permeable. By 1800 literacy was fast spreading to social groups outside the intellectual and political elites. As historians of science demonstrated in the 1980s and 1990s, the authority to make knowledge was increasingly claimed by new specialist disciplines directed by gentlemanly elites, instead of being the property of an enlightened literate public (Schaffer; Morrell and Thackray;

Rudwick; Yeo). As existing sciences withdrew from general learned culture, and literacy and print became available to more of the middle and working classes (McKitterick), more of the literate public found themselves excluded or marginalized from these elites, although the scientifically active among them sometimes maintained their own traditions of knowledge-making that did not recognize the absolute authority of the elites (Secord A. 1994). By the end of the century, the professionalization of science and the secularization of science education had opened up the scientific elites to participation from a much wider social spectrum, but such participation was rigorously disciplined. The new elites were even more determined than their predecessors to emphasize the inability of outsiders (including the emerging *literary* elites) to contribute meaningfully to science.[2]

Nineteenth-century science writing for the general reader embodied three different relationships between scientific practitioners and the public:

1   As in the late eighteenth century, many contributions to elite science – including most of the famous scientific books of the Victorian era – were written for a dual audience of educated general readers as well as specialists in the field of inquiry.

2   Knowledge produced *outside* those emerging scientific elites also continued to be offered to the educated general reader in literary form, perpetuating older Enlightenment notions of the public sphere as the producer and arbiter of knowledge.

3   The perceived gap between emerging scientific elites and a scientifically uninformed public was met and reaffirmed by texts that (from the 1820s onwards) were increasingly labelled 'popular science' or 'popularization'. These texts were designed to make science attractive to general readers and often presupposed little or no prior knowledge. They were written both by well-known men and women of science, and by authors with no claim to membership of a scientific community. In the early decades of the century, this field was dominated by books for children in affluent families, often in dialogue form,[3] but it subsequently broadened to target adult readerships of both genders and various social classes using a wider range of literary forms and styles.

The nature and boundaries of scientific elites were continually shifting and contested, varying depending on the science, geographical space and social context. Consequently, these three science –public relationships were far from stable and often overlapped. I will begin with the third group, since it contains by far the largest number of texts. Also, of our three groups, its readerships spanned the widest range of social class and education, thus approximating best to modern notions of the 'general reader'. But it is important to recognize that this category, while numerically dominant, constitutes neither the totality of the literature of science for the general reader, nor some of its best-known texts.

The term 'popular science' did not appear until the late 1820s, but the practice of self-consciously disseminating science to the uninitiated had been going on since the mid-eighteenth century in specific contexts. As Debbie Bark notes in her contribution to this volume, scientific dialogues remained a prolific genre for home schooling among well-heeled families throughout the early decades of the nineteenth century, with Jane Marcet's bestselling *Conversations on Chemistry* (1806) rapidly achieving classic status. Upper-class adult audiences in the 1800s and 1810s were initiated into the wonders of science more frequently in *viva voce* settings than through publications: Humphry Davy's flamboyant lectures on chemistry and geology at the Royal Institution at once educated his audiences and helped construct an image of the man of science as authoritative, benevolent and forward-looking (Golinski).

The aim of 'diffusing' scientific knowledge via publications among a wider class base took off in the 1820s, as part of the mass education movement spearheaded by Whig political reformers (Secord J. 2014 10–19). Here the boundary between expert and non-expert became more sharply defined. As Jonathan Topham has shown (Topham 2004; Topham 2007; Topham 2009), the new label 'popular science' emerged in cheap periodicals of the late 1820s to denote science *for* the people, not science *by* the people, and it is in this sense that I use the term in this chapter. As in the present day, it implied a hierarchy of knowledge between the makers and the receivers of science. Like other commercial categories used by booksellers and magazine editors (such as humour or classics), 'popular science' is not and was not a genre but a vague, nebulous indicator of content and purpose that covers many different genres. It was not used consistently or used to label all the works discussed below. However, these works can all be described as popularizations because they share the underlying stance of introducing uninitiated readers to new knowledge associated with scientific elites – even if, in practice, this stance can mask different underlying purposes and even subversions.

There has never been any consensus on what genres constituted 'popular science', although the label has remained in wide use since the 1820s. Individual works were often defined *against* aspects of recognized genres, for example the impenetrable technical jargon of a high-level scientific treatise, the systematic and disciplined approach of a textbook or the comprehensiveness of an encyclopedia. This did not prevent considerable overlap with such genres,[4] but it does suggest the construction of an implied readership who did not know much about science and who needed to be courted rather than crammed. They could include men, women and children: as Bark observes in her chapter in this volume, many of these books were targeted at the young but also read by their elders (Rauch 1997). In such works, disciplining in scientific methods and providing facts were expected to take place within a context of pleasure and recreation (Secord A. 2002).

Consequently, numerous book titles advertised their wares using words connoting the elementary (*Introduction, First Lessons, For Beginners, Simplified, Outline*), the easily accessible (*Popular, Plain, Short, Easy*), the miscellaneous taster (*Sketches, Glimpses, Vignettes, Notes, Jottings, Thoughts*), the sensational or fantastic (*Wonders, Marvels, Wonderful, Strange, Fairyland, Romance*), amusement (*Recreations, Entertaining*), narrative appeal (*Romance, Story, Fairy Tales*), and the familiarity of everyday experience (*Common, Familiar*).[5] Awakening and exercising the imagination was vital, and many of these texts were designed to evoke or connect with more immediate sites of scientific encounter: museums, galleries, outdoor field trips, lecture halls, Sunday schools. Some books were intended or used as guidebooks to exhibitions, collections or sites of scientific interest (Fyfe 2011). Others were based on public lectures or informal talks (Lightman 2007 406–8, Brock xvii–xviii, xx). Printed images became increasingly widespread in the second half of the century, but with or without such visual aids, a rhetoric of spectacular display was built into many works of science popularization, as signalled in titles such as Thomas Milner's 1846 compendium *The Gallery of Nature: A Pictorial and Descriptive Tour Through Creation* (Secord J. 2000 439–41, O'Connor 2007A, Fyfe 2007).

The juxtaposition of different discourses was central to the literature of 'popular science'. Much popularization took place within periodicals: some existed solely to disseminate scientific knowledge, such as the *Popular Science Review* and *Science Gossip*. More frequently, entertaining science articles appeared in general-interest newspapers and magazines alongside fiction, poetry, politics and reportage (Dawson, Noakes and Topham).[6] Many popularizing books reunited individual articles from this hybrid context under a single scientific theme, but the effect was often that of a miscellany, as in W. J. Broderip's *Zoological Recreations* (1847) and Frank Buckland's *Curiosities of Natural History* (1857–72). The label 'popular science' was in fact first used in the 1820s to denote bite-sized extracts from existing scientific writings incorporated into cheap miscellanies (Topham 2004; Topham 2007). Compilatory procedures remained central to the literature of science popularization in periodical and book formats (O'Connor 2007A 236–41, 325–55). Indeed, the word 'compiler' was sometimes used by gentlemen of science to refer to popularizers in a derogatory sense, connoting a watered-down, commercially driven rearrangement of facts discovered – and expressed much better – by the experts (Buckland 61). There was, to be sure, plenty of artless cutting and pasting going on; but many of the most successful works incorporated quotations in purposeful, creative and dialogic ways that lent new meaning both to the framing discourse and to the material quoted.

Further adding to these texts' generic hybridity, they adapted several existing literary traditions: travel writing and topographical description in Philip Henry Gosse's *A Naturalist's Rambles on the Devonshire Coast* (1853), personal journal and animal biography in Eliza Brightwen's *Wild Nature Won by Kindness* (1890), biography and intellectual history in Edward Clodd's *Pioneers of Evolution* (1897), devotional reflection in Ebenezer Cobham Brewer's *Theology in Science* (1860), practical guides to observation in Thomas Webb's *Celestial Objects for Common Telescopes* (1859), fable in Margaret Gatty's *Parables From Nature* (1855–61), animal autobiography and dream-vision in John Mill's *The Fossil Spirit: A Boy's Dream of Geology* (1854), philosophical dialogue and educational narrative in Maria Hack's Harry Beaufoy stories (1820s–1830s) and cosmic narrative and natural history description in Arabella Buckley's *Winners in Life's Race* (1883). At the level of rhetoric and narrative technique, some texts stick closely to the conventions of their host genres, such as Gatty's fables and Brewer's catechisms. Others refuse to settle in a single mode, instead moving freely between past-tense narration, vivid description, familiar metaphor, literary allusion, fantastic thought experiment, imaginary journey, explanation and reflection, drawing on established genres where appropriate. A strong narratorial voice ('I', 'we' or both) often ties the whole together, with a range of subsidiary narrators deployed for specific purposes. Texts of this kind – Charles Kingsley's *Glaucus; or, the Wonders of the Shore* (1855), Gosse's *Romance of Natural History* (1860–61), Buckley's *Fairy-Land of Science* (1878), various geological bestsellers by the Scottish stonemason, geologist and newspaperman Hugh Miller – challenge any casual attempt to classify them in traditional literary-historical categories.[7]

A closer look at one such book, which has received more literary and historical attention than most (and stayed in print longer than most), will show how all these aspects and attributes come together in a single work. Miller's first book on fossils was *The Old Red Sandstone, or New Walks in an Old Field* (1841), its subtitle conveying both novelty and familiarity, fresh air and informal conversation.[8] Among its purposes was to enthuse and inform general readers about geology and palaeontology. It originated in a series of articles that Miller contributed to the *Witness*, the Edinburgh-based biweekly newspaper which he edited, but the book does not simply reprint the articles. Miller recomposed and reordered much of their content and added a great deal more, at once clarifying the overall argument and enriching it poetically. The result was a substantial, engaging literary work about the Devonian fossil fish of Scotland's Old Red Sandstone formations and their place in the history of life. Several of the fossils are illustrated in full-page lithographs bound into the book. But the imaginative appeal of *The Old Red Sandstone* for the non-specialist reader rests chiefly on Miller's writing, especially his skill at bringing the fossils and scenery of ancient Scotland to life before the reader's mind's eye.

This feat was achieved by Miller's narrative voice moving freely, sometimes disconcertingly, between modes, genres and fields of allusion. Fossils are thus brought to life using homely or picturesque comparisons: the 'enamelled scales and plates' of the *Cheirolepis* 'glitter with minute ridges, that show like thorns in a December morning varnished with ice' (Miller 97), while the *Cephalaspis* resembles 'a saddler's cutting-knife' (138). Comparisons of fossils and rock formations with sculptures, buildings, and Egyptian antiquities run throughout the book, giving life to the familiar metaphor of 'nature's monuments' and infusing the text with human interest and 'epic energy' as James Paradis has put it (135). Narrative and theatrical modes also have their place. In the later parts of the book, Miller sometimes places his readers in the midst of an imagined prehistoric scene described with present-tense immediacy, staging himself as a friendly guide on a journey to the centre of the earth (Merrill 238–51). Geological Virgil to the reader's Dante, he calls up these visions of former times in order to hint at a vast, obscure and ongoing narrative of cosmic history and divine Providence, which he occasionally slips into narrating in the past tense (O'Connor 2007A 361–433). More than one reviewer felt that Miller's power to evoke the past so vividly and intimately gave *The Old Red Sandstone* 'pre-eminence as a book either of science or of imagination' ([Anon.] 6), and its (prose) author the status of a 'didactic poet' superior to Erasmus Darwin.

Miller's overarching vision of earth history emerges in full colour, as it were, only in the final chapters of his book. But its direct and immediate impact on the reader has been prepared from the very start by Miller's adoption of an engagingly personal narrative voice identified with himself (Paradis 143–4, Taylor 109–10). The book begins with a chapter of autobiography, telling the story of how Miller's eyes were opened to the wonders of geology as a young mason working in a quarry (1–17). This prepares the reader to expect the whole book to be a personal narrative rather than a work of impersonal scientific exposition. The autobiographical mode remains a strong generic current throughout the book, enlivening Miller's descriptions with personal anecdotes and providing the framework for two of the book's central chapters, in which Miller resumes the story of his geological awakening (109–33). Miller thus ensures that the emotions impelling his study of the rocks – wonder, suspense, passionate interest – remain legible even through the drier moments of technical description and argument. Many of his descriptions of present-day scenery, too, are framed as if to suggest that Miller and his reader are walking along the beaches and bays of the Black Isle together. The rapport established between narrator and reader ensures that we are more willing to suspend disbelief when Miller spectacularly exceeds the normal generic limitations of early Victorian topographical writing or life writing and takes us on a fantastic time-travelling journey into a Carboniferous swamp – one of the many virtuoso descriptive set-pieces with which the book is punctuated (269–72).

The Old Red Sandstone was almost immediately hailed as a literary classic and remained in print well into the twentieth century. But, if Miller was perceived as unusually eloquent, the generic fluidity and narrative agility seen in his writing was characteristic of many works of science for the general reader. Each book had its own literary procedures, which generated meanings from its scientific reference-points in different ways (religious, philosophical, personal or political). Books like Miller's may all share the purpose of introducing science to the uninitiated in attractive ways – hence they have often been described as 'popular science' or 'popularization' – but this does not mean that they all belong to the same genre. To understand how they work as texts, we need to dig deeper.

Looking beyond 'popularization'

The concept of science popularization has limited generic value, but it also poses problems from a historical perspective. It has a basic usefulness as an umbrella category that was in use by booksellers, authors, and readers from the 1820s onwards. However, its frequent connotations of watering down 'real' knowledge for a uniformly passive, ignorant and grateful audience have been recognized by recent historians and sociologists of science as deeply problematic (Myers 2003; Topham 2009). Such scholars have employed new perspectives from the history of popular culture and the sociology of knowledge to challenge the dominant 'deficit' or 'diffusion' models of science's relation to public culture. One of the most-cited reference-points for such approaches is Roger Cooter's and Stephen Pumfrey's critical survey 'Separate Spheres and Public Places', published in a special issue of *History of Science* in 1994 alongside important research articles on non-elite science by Anne Secord ('Science in the Pub') and Alison Winter ('Mesmerism and Popular Culture').

Most recent historical studies of nineteenth-century science popularization reject assumptions of the top-down diffusion of knowledge from scientific elites through popularizers to passive audiences. Many of them attend instead to the ways in which authors, publishers, printers and readers represented, transformed and deployed scientific knowledge for various purposes (commercial, religious, political, aesthetic) that did not necessarily match the aims and ideologies of scientific elites and that sometimes actively subverted them. Influential and accessible examples, drawing on the history of the book and the history of reading, include Bernard Lightman's and Aileen Fyfe's studies of the persistence of religious teachings in late Victorian science popularization, Topham's and James A. Secord's studies of the diverse reading communities encountering science writing, and the work of all four scholars on scientific and educational publishing and authorship (Lightman 1997; Lightman 2007; Fyfe 2004; Fyfe 2012; Topham 1998; Secord J. 2000; Secord J. 2014).[9] In particular, Secord's groundbreaking *Victorian Sensation* (2000), his recent book *Visions of Science* (2014), and Lightman's wide-ranging study *Victorian Popularizers of Science* (2007) are essential reference-points for anyone interested in nineteenth-century science writing and reading. Periodicals, too, have been well served, as Dawson discusses in the present volume. Complementing such investigations has been a renewed attention to non-textual media and sites of public scientific engagement, such as museums, shows, lecture halls and the home, sometimes including close readings of texts produced or used in association with such sites (Gooday 2008; Qureshi; Finnegan; Fyfe and Lightman).

Cooter and Pumfrey, however, wanted historians to jettison the whole concept of popularization, not just its connotations of passive diffusion, and instead to study science as *made* in the public realm as much as in specialist communities. This perspective remains controversial, but it has been consistently pursued by James A. Secord. For Secord, the enrolment of those outside scientific communities is part of what has come to *constitute* science 'proper', not a subsidiary by-product of knowledge-making. He thus advocates removing the distinction between the making and the communication (or popularization) of scientific knowledge and thus reuniting the study of science popularization with the history of science *tout court* (Secord J. 2004; see also Topham 2009 16–20). This, in turn, has opened up new approaches to novels on scientific themes, seeing them as helping to *make* science, not just to comment on it (Buckland).

In practice, most historians continue to use the nineteenth-century actors' categories 'popularization' and 'popular science' as convenient ways of dividing up their material. But Secord's work shows how vital it is not to swallow nineteenth-century rhetoric about the supposed gulf between scientific communities and general readers, or indeed its present-day equivalent. Parcelling up science writing between texts producing new knowledge for scientists and texts communicating that knowledge for non-scientists may work for some texts, such as papers in specialist journals versus children's books about nature's wonders. But hundreds of nineteenth-century writings on science either did both of these things or were produced in contexts where this dichotomy made no sense.

Consequently, when surveying literary manifestations of science for the general reader, we cannot limit ourselves to works labelled as popularization alone but need to take in many writings that claimed to offer new knowledge, whether produced within or outside the new professionalizing scientific communities. Not all sciences acquired elites and orthodoxies at an equal rate, so some – such as antiquarianism and sciences of the mind – remained open to contributions from wider constituencies for longer than others. Early nineteenth-century writings on prehistoric monuments were produced, not within a specialized discipline but along Enlightenment lines, within a socially elevated and well-educated general readership of fellow-scholars and philosophers (Heringman). This was not popularization: the general readers targeted *were* the relevant scientific community.

Even for new sciences with rigorously policed boundaries, contributions to knowledge continued to be offered by authors who had no recognized position in the relevant scientific community but who wrote as if the Enlightenment public sphere still operated on something like the old terms – which, in many contexts, it did. The focus of Secord's *Victorian Sensation*, an anonymous evolutionary treatise entitled *Vestiges of the Natural History of Creation* (1844), is a good example. The Scottish journalist and editor Robert Chambers drew on his scientific reading and general education to join several existing sciences together into a controversial new argument for species transmutation. Though aimed at a very wide readership, this book was not popularization in the usual sense. It claimed to contribute new knowledge by drawing hitherto unseen links between discrete areas of study. Its primary purpose was not to persuade non-scientific readers to engage with science but to invite general readers and specialists alike to participate in a natural philosophical discussion. In this case, both readerships took plenty of notice. Most works of this kind were ignored by scientific elites but were taken seriously by other knowledge-communities, such as the many erudite books which offered new cosmological hypotheses by blending biblical exegesis, natural philosophy and natural history including Isabelle Duncan's bestselling 1860 account of prehistoric angelology, *Pre-Adamite Man*, or contributions to other investigations deemed unphilosophical by the elites, including mesmerism, occultism and psychical research (Snobelen; Winter 1998; Luckhurst and Sausman; O'Connor 2007B).

Marginal sciences aside, a dual audience of scientific specialists and educated general readers was also courted by books produced by (or in alliance with) those who had secure positions within the new scientific establishments. Few would dream of calling Darwin's *On the Origin of Species* (1859) 'popular science', yet it brought the theory of evolution by natural selection to a wide audience. It was published by the literary giant John Murray and was written to capture the attention and imagination of the educated public as well as that of the scientific practitioners with whom Darwin corresponded. Its success in so doing helped to shape the scientific debates that followed, also often conducted among wide audiences and within larger public debates about the progress of civilization (Ruse). It is only the best known of many nineteenth-century philo-sophical and scientific works that encouraged general and specialist readers to reflect on the practices, implications and future possibilities of science. Secord has called these books 'reflective treatises', and his *Visions of Science* provides a compelling account of some important examples from the 1820s and 1830s (see also Secord J. 2009 460–62).

The reflective treatise spanned the putative divide between 'popular' and 'proper' science on several levels. Some were written by recognized scientific authorities, others were not. Some intervened directly in current debates within a scientific field: the human mind in George Combe's *Constitution of Man* (1828), the earth in Charles Lyell's *Principles of Geology* (1830–33), the stars in John Pringle Nichol's *Views of the Architecture of the Heavens* (1837). Here the boundary between a reflective treatise and a scientific treatise is impossible to draw with clarity. Others were framed as ambitious attempts to connect the sciences together: mathematics provided the unifying principle in Mary Somerville's *Connexion of the Physical Sciences* (1834), as did evolutionary development in *Vestiges*, the many works of Herbert Spencer, Edward Clodd's *Story of Creation* (1888) and Alice Bodington's *Studies in Evolution and Biology* (1890). Still others were framed as reflections on science's place in society, such as Charles Babbage's *Reflections on the Decline of Science in England* (1830) and John Herschel's *A Preliminary Discourse on the Study of Natural Philosophy* (1831).[10] Others considered various sciences' religious or spiritual implications (Topham 1998; Brooke and Cantor 176–200). The six Bridgewater Treatises of the 1830s discussed the evidence for God's power and benevolence in objects of scientific study, Balfour Stewart's and Peter Guthrie Tait's *The Unseen Universe* (1875) explored the connections between matter, electromagnetism and immortality, while Henry Drummond's *Ascent of Man* (1894) examined the moral and religious significance of evolution. Exegetically oriented or occult scientific treatises, such as Duncan's *Pre-Adamite Man* and biblical-literalist theories of the earth, also deserve considering as reflective treatises because of their shared ambition to connect disparate fields of evidence and reflect on the moral implications of scientific progress.

Secord considers these texts as a single genre because of their shared purposes. But, as he shows in *Visions of Science*, their generic affiliations and frameworks were as disparate as those within the field of 'popular science' described earlier. The reflective treatises he analyses include a set of dialogues opening with a fantastic dream-vision, a conduct-manual, a philosophical diatribe, a synthetic survey of physical science and a satirical mock-treatise authored by a fictitious philosopher. The range of narrative voices and modes employed is similarly varied and adventurous.

Many of these reflective treatises sold well, were reprinted cheaply and provided quotable passages for other writers. We might say that they 'popularized' science in the broad sense, because they made a wider public think about science in new ways. Yet, like Edward O. Wilson's *Sociobiology* or Richard Dawkins's *The Selfish Gene* from the 1970s, these works cannot be described as popularizations in the usual sense of the term. Rather than repackaging existing knowledge for the uninitiated, they invited readers to participate in a high-level philosophical discussion. Their effects were felt just as deeply within scientific communities as among the wider public. By negotiating science's cultural authority among a number of public constituencies, such works shaped the direction and communicative patterns of science – not least, in the 1820s and 1830s, by fostering the emergence of specialist forms of scientific publication (Secord J. 2014; Secord J. 2009; Yeo).

Some of the later works I have mentioned were aimed more at general readers than specialists, and some were marketed as 'popular', but this only underlines the vagueness of the label 'popular science' when we start to look at what texts were actually doing, even texts that I have described earlier in terms of popularization. Specialists were general readers, too, as soon as they stepped outside their specialisms (Whitley 6). They often found 'popular' accounts of their own field stimulating or instructive in their own scientific work (Topham 2009 17–18). For example, the Irish astronomer Agnes Clerke wrote several designedly 'popular' books explaining the methods and results of astronomy, but her eminent fellow-countryman Robert Ball felt that her work was so useful to other astronomers that it deserved better than the lowly label 'popular' (Lightman 2007 475–6). Furthermore, several works labelled then or now as 'popular science' offered new knowledge as well as repackaging existing knowledge, especially in the observational sciences and in the first half of the century. Miller's *Old Red Sandstone* is a good example, self-consciously aimed at both specialists and general readers. This book, too, shares with Secord's 'reflective treatises' a significant degree of moral and philosophical reflection on science's place in modern society.

Finally, some established literary genres shared the aim of awakening a thoughtful interest in science among general readers but did not need to be billed as 'popular science' because their own genre was already understood to combine instruction and reflection with an attractive narrative form. Journals by scientific travellers, for example, presented vivid personal narratives interlaced with scientific exposition and reflection, as in Charles Darwin's *Journal of Researches*

(1839) and Mary Kingsley's *Travels in West Africa* (1897). At the end of the century, more fantastic varieties of travel narrative became subsumed into scientific romance, which overlapped with the reflective treatise in its capacity to encourage a wide public to reflect philosophically on science, whether in a positive or negative vein.[11] Some of its most skilled practitioners, such as H. G. Wells, were also known for writing 'popular science' in the accepted sense. While it might seem easy at first to distinguish his 'popular science' non-fiction from his fictional scientific romances, this distinction blurs as soon as one reads the two sides of his *oeuvre* alongside each other. Fictional modes, often with a strong dose of fantasy, have always been an important aspect of science writing for the general reader, such as Miller's time-travel adventures in the Carboniferous forests in *The Old Red Sandstone*.

Engaging general readers in science, then, was something done not only through the popularization of existing knowledge but through many different kinds of literature and other media, overlapping significantly with several of the genres and formats examined in other chapters of this volume: scientific prose, periodicals, science for children, travel writing, life writing, poetry, the novel and the scientific romance. Perhaps a truly inclusive study of the process would include not only texts that aimed to do this but also the ways in which readers *derived* instruction and reflection on science from texts that originally had no such purpose.

Literary perspectives and 'literary' status

I have been using the term 'literature' to refer to all these texts, but we now need to consider whether it is appropriate. From one perspective, it is appropriate to study *any* text as literature in the sense of applying literary analysis to it, if that helps illuminate the text's meaning. This exercise is not dependent on a prior categorization of the text as literature. It requires only a text, and, as Charlotte Sleigh puts it, 'even the driest experimental account is susceptible to a literary analysis' (9). But, in practice, and in casual conversations that reflect disciplinary identities, most scholars use the word 'literature' to refer to an exclusive set of genres associated with so-called imaginative writing: poetry, fiction, drama, and a few favoured essays, memoirs, histories and travel writings that are felt to have a strong personal voice or narrative momentum. The fact that one can analyse an experimental account as discourse does not, for most of us, turn that experimental account into literature. Most non-fiction, especially science writing, tends to be excluded from this realm by definition. The strength of this assumption can be seen in the currency of the contrasting pair 'scientific and literary writing', so often invoked and so rarely explained. Even scholars who challenge this dichotomy in theory have often ended up supporting it in practice by focusing their critical attention almost entirely on novels, poems and dramas.[12] One or two of the works of Darwin are an occasional exception, and I suspect that this is more because of the critical mass of superb literary analysis of Darwin's writing, coupled with his retrospective giant stature, rather than because his writing is somehow more 'literary' than other science writing (for examples see Beer; Smith; Levine 2011).

The general exclusion of science writing from 'literature' matters, because most nineteenth-century texts outside the literary cordon are not studied or read at all.[13] On the surface, this exclusion seems justified by the word's semantic history. In the eighteenth century, when literary culture was the preserve of a small elite, the word 'literature' included all forms of educated writing: not just fiction, drama and poetry, but also history, biography, essays, philosophical papers and treatises. As reading publics broadened and diversified in the nineteenth century, the word 'literature' began to be used in a restricted sense, connoting writing that claimed attention on the grounds of aesthetic or emotional effect (this sounds old-fashioned, but it is not far from today's equivalent, 'imaginative writing'). The timing of this transition is unclear.

Jon Klancher has placed it around 1820 (524), and his views have become semi-canonical, but others hint at later dates, up to the 1860s, when the *Oxford English Dictionary* recorded this shift as very recent (Secord J. 2009 473–4, *OED* s.v. *literature* 3a). In practice, though, the word continued to be used by publishers and booksellers throughout the century in both its old, inclusive sense of 'all texts' and its new sense connoting aesthetic value, sometimes by the same people and heedless of the possible contradiction (O'Connor 2007A 14–15, 242–3, Dawson and Lightman I:vii–viii).[14]

But the new, restricted sense was gaining ground. A sense of 'distinct literary and scientific spheres', reflecting new power-bases for scientific authority and the rise of separate career paths and identities for the man (or woman) of science and letters respectively, can be seen in the proliferation of special-interest periodicals that treated literary or scientific pursuits in isolation (although many general-interest periodicals continued to cover both areas), and the special treatment of non-fiction in some review journals (Dawson, Noakes and Topham 11–12; Topham 2004 66). The mutual self-definition of science and literature suited both emerging professions, as Paul White has argued in his acute study of the many faces of Thomas Henry Huxley. Huxley systematically downplayed or denigrated the value of book-learning for scientific training as part of his crusade to professionalize the sciences, even as he recommended the study of literature (in the new sense) as part of a well-rounded education (White 2003 67–99).

But to state that science and literature became defined against each other as separate cultural *activities* is not the same as to say that science *writing* no longer qualified as 'literature' in the new, restricted sense of the word. Literature in this new sense no longer had room for technical papers on buzzards' gizzards or textbooks in advanced soil chemistry, but it manifestly did have room for many of the texts discussed in this chapter. Late Victorian literary historians agreed: George Saintsbury's *History of Nineteenth Century Literature* (1896) included at least some of the more prominent writers on physical science within its purview (Hutchison 167), as did Edmund Gosse's *Short History of Modern English Literature* (1897). Publishers and reviewers of science writing for general readers continued to comment on such works' stylistic qualities as well as on the arguments presented, whether the authors were eminent scientific figures (Somerville, Darwin, Huxley), non-elite contributors to scientific debates (Chambers, Duncan) or writers reinterpreting elite science in their own ways (Miller, Buckley, Clodd). Works claiming a popularizing function were especially amenable to treatment as literature, hence the high literary reputation achieved by Miller in the 1840s. Huxley himself was quick to highlight the literary and imaginative aspects of works he viewed as 'popular science', emphasizing their poetic and narrative appeal as 'wonderful stories' when reviewing them for the *Westminster Review*. As White points out (2003 71–2), Huxley did this partly to put 'popular science' in its place as mere dissemination, distinct from elite science, but also in order to recommend the best 'popular' writings to a wide readership who would benefit from a general understanding of science's importance, and who would recognize the authority of scientific elites as a result. In any case, Huxley prided himself on his own literary attainments as a communicator of science, as reviewers continued to notice; he was his own best popularizer.

Self-described 'popular science' had not always connoted imaginative writing. When the category was invented for the weekly miscellanies of the late 1820s, its contents soon ended up isolated from original fiction and were often hived off into separate publications. By the 1840s, however, as Fyfe has noted (2004 56), both the label 'popular science' and the verb 'to popularize' implied not merely replicating information in cheaper formats or forums (as they had in the 1820s) but conveying new knowledge afresh in attractive and easily understandable forms, composed especially for this purpose. This project required literary skill. For the many works aimed at dual readerships of specialists and everyone else, self-consciously 'literary' languages of poetic evocation, dramatization and personal response had to alternate with more technical discourses of description and reasoning, in what I have elsewhere called the 'swerve effect'. For example, Miller warned his two target readerships in *The Old Red Sandstone* to bear with his long passages of 'minute detail' aimed at geologists and his recourse to 'pictorial effect' and 'familiar phenomena' for the benefit of general readers (O'Connor 2007A 23–6, quoting Miller vii–viii). Ann B. Shteir has documented a particularly striking example in Phebe Lankester's additions to James Edward Smith's multi-volume manual of plant classification, *English Botany*, in the 1860s. To make the work more appealing for a wide audience, Lankester augmented the dry passages of technical description with reflections in an accessible and personal style; the two parts of each description were labelled the 'purely technical matter' and the 'popular part' (Shteir 2003 160–61, quoting Syme). The fact that the latter part was also referred to as the 'literary' part shows the close association between popular science and imaginative writing. It also exemplifies the divergence of scientific languages in this period between those deemed literary and non-literary, folding along lines of audience specialization.

It is thus possible to speak of science writing for the general reader as existing on a spectrum of 'literariness', depending on both the outlook of the reader and the nature of the text itself. More importantly, it is clear that literary frameworks, techniques and allusions were an essential component of both 'popular science' as the Victorians understood it, and of other science writings aimed at general readers. This means that literary analysis is an essential interpretative tool, not only in the sense of understanding meanings and discursive structures but also in the fuller sense in which scholars of English literature apply it to imaginative writing: attentive to playfulness and ambiguity as well as to intended and unintended meanings, and alive to a text's place in literary history, not just to its place in the history of whatever debate or field it happens to be 'about'. Seeing these writings as literature does not mean ignoring the increasing bifurcation of scientific and literary identities (professions, careers, elites, loci of authority, subcultures), nor does it mean making grand claims about 'culture' in general. It simply means recognizing the fact that literary non-fiction was one of the primary sites of public science in the period.

The various historical studies already cited have shed much light on science writings for the general reader in terms of their coverage, argument, ideological underpinnings, bibliographic nature, textual layout, readerships and circulation. Only a fraction of the corpus has been studied, but this includes some of the most important texts, providing an excellent basis for future research. The more specifically literary aspects of these texts – narrative voice and focalization, imagery, rhythm and prosody, narrative structure, genre, their place in literary history – have been much slower to come into focus. For any historically oriented discussion, these aspects are just as important as the 'hard facts' of print runs and typeface. If a text's meaning is constructed not only by *what* is said but *how* it is said, and especially if recognizable literary techniques have been used to write a text, then literary analysis is needed to understand what it means.

When the attempt has been made, the results have been extremely illuminating. For example, the explanatory or evocative potential of certain myths, metaphors and deep structures in science writing has received considerable attention, with foundational studies by Beer and Greg Myers ('Nineteenth-Century Popularizers') in the 1980s and by Misia Landau, Barbara Gates (34–65) and James Paradis in the 1990s. In recent years, several scholars have traced the way in which the 'fairyland' metaphor has been woven through science writing.[15]

Descriptive techniques and languages of visualization have come under sustained scrutiny.[16] The construction of narrative voices (familiar, authoritative, participatory, emotionally laden, fully fictive) is one aspect of science writing popularly thought to belong only to fiction and poetry, but it is becoming an increasing object of analysis in its own right.[17] The same is the case for science writers' experimentation with diverse narrative forms and structures, from dialogue,[18] travelogue, and personal narrative to epic, romance and fairytale.[19] In general, children's writing and women's writing on science has proved particularly amenable to literary analysis, benefiting from the more inclusive outlook towards what counts as literature in the fields of children's studies and women's studies (see Bark, in this volume).[20]

This varied body of scholarship has made a big difference to the historiography of nineteenth-century science, but it has had little impact on received understandings of English literary history. Even in much scholarship on science and literature, studies like those mentioned above are far more often cited than engaged with. This is partly because most literary scholars are more interested in novels, poetry and drama. But we are still only beginning to understand what the literary forms of science writing for the general reader are, considered as part of a generic system or a space on the literary-historical map. Preliminary attempts have been made but not yet in any comprehensive or systematic way.[21]

More problematically, some of the most useful generic categories employed by historians of science arguably do not denote genres at all but aspects of the texts' content, focus of attention or narratorial stance. This does not prevent them from being valuable tools for analysis but it is worth emphasizing that (for example) most so-called 'evolutionary epics' (Secord J. 2000 461, Lightman 2007 219–94) contain strikingly little past-tense narrative,[22] that Myers's 'narrative of nature' (1990) and Gates's 'narrative of natural theology' (39) are often conducted in non-narrative ways, and that both James Secord's 'reflective treatise' (2014) and Shteir's 'familiar format' (1996 81–3) contain several different genres within themselves, as both scholars implicitly acknowledge. Literary accounts of science writing still only rarely connect these texts up meaningfully with the larger framework of English literary history. The far-reaching insights to which such connections can lead are exemplified by recent discussions of how the conventions of historical fiction were manipulated in early nineteenth-century cosmological and geological writing, resulting in the rise of new ways of thinking about the history of life and the earth (Secord J. 2000 77–110; Buckland).

This question may nevertheless arise: why should literary scholars care about the literary aspects of science writing when there is so much other fascinating literature out there to study, much of it written in forms which are easier to enjoy, assimilate and relate to a critical tradition? One obvious reason is that, if these writings were part of the English literary tradition in our period, then general claims about that tradition (I use the old-fashioned word advisedly) need to take the full range of genres into account – especially where the relevant works were numerous, widely read and much talked about.

Another reason is that broadening the canon of literary works deemed worthy of serious critical attention to include previously neglected authors (or genres) sharpens not only our sense of literary history but also our understanding of canonical works. Besides showing them to be fascinating authors in their own right, the recent critical rehabilitation of Fanny Burney and Ann Radcliffe has greatly enriched our understanding of the novels of Jane Austen, who avidly read them both. Now, despite calls by some scholars to view the relationship between literature and science as working in both directions, most scholarship on nineteenth-century science and literature still proceeds in a fairly monodirectional manner, exploring how novelists, poets or dramatists have reimagined scientific ideas or perspectives (see Marsden). This situation is unlikely to change in a hurry, so long as fiction, poetry and drama are seen as artistically and semiotically 'richer' than non-fiction genres. The very label 'non-fiction', a twentieth-century coinage, implies deficit (Anderson ix). But the most compelling examples of this kind of research are those that do not treat the science in question solely as ideas in the air but examine widely available scientific *texts* in which it was discussed – either in order to compare fictional or poetic treatments with those in scientific texts, or in order to suggest possible sources for the fictional or poetic treatment.

The more attentive critics are to what is going on in the scientific literature, the more perceptive will be their account of its reimagining.

The classic proof of the benefits involved here is Gillian Beer's *Darwin's Plots* (1983). Her close attention to the literary texture of Darwin's prose not merely informs but is a precondition for the important insights into evolutionary aspects of fiction by Victorian novelists in the book's second half. Studies of science in Victorian fiction by Sally Shuttleworth (Shuttleworth 1984; Shuttleworth 1996), George Levine (Levine 1984), Paul White (White 2013) and Adelene Buckland all derive depth and precision from their insightful discussions of textual strategies in relevant evolutionary, geological and physiological writings. Alan Rauch's analysis of nineteenth-century novels about knowledge and education (notably *Frankenstein*) (Rauch 2001) is made compelling by being grounded in close readings of early nineteenth-century educational texts. And the value of Steven McLean's study of H. G. Wells's scientific romances lies in the fact that he sets them in the context of Wells's own 'popular science' articles of the same period, both fiction and non-fiction. By moving beyond a view of science writing as a disembodied intellectual background and attending to the descriptive techniques, literary allusions and narrative voices of specific texts that informed contemporary public debates and ideologies, these scholars have gained a clearer sense of what the real issues were. They have got under the skin of the novels they discuss. The literature of science for general readers is thus not the property of historians of science but is a rich and powerful resource for literary scholars too.

Finally, much of this literature is simply worth reading and enjoying for its own sake. Gilbert White's *Natural History of Selborne* is perennially popular (Secord A. 2013 xxvi–xxviii), but why not its nineteenth-century successors? Levine (2011 4–5) rightly emphasizes the pleasure to be had in reading Darwin's *Origin*, but if that may appear an acquired taste, more immediate enjoyment may be had from less hallowed works: the surreal natural-history anecdotes of Frank Buckland, the time-travelling adventures of Hugh Miller, the rapt visions of Humphry Davy and John Pringle Nichol, the minute but emotionally laden nature-contemplation of Henry Gosse and Richard Jefferies and the dazzling play of narrative perspectives in the writings of Charles Lyell and Arabella Buckley. Modern 'creative non-fiction' gurus sometimes seem to think that what is often called 'creative science writing' is a recent addition to the fold, but it has a long and distinguished history. These nineteenth-century writings all require some imaginative exertion on the part of the reader, but the same is true of most serious literature of the period. Like Romantic lyric poetry and Victorian realist fiction, science writing like this demonstrates the imaginative force and intensity of nineteenth-century authors' engagement with the 'real world' and allows its readers, even today, to experience it themselves.

Notes

1 A wide sampling of this literature from the 1830s to the 1900s, with helpful analysis, is provided in Dawson and Lightman. Future references will be to specific volumes in this set.
2 On the self-definition of literary and scientific elites as complementary aspects of high 'culture', see White 2003 67–99.
3 See Debbie Bark, 'Science for Children', in this volume.
4 Examples of overlap between the strategies of textbooks and 'popular science' works include Robert Hunt's physics textbook (Keene 2013) and William Carpenter's physiology textbooks (White 2013 23).
5 A representative sampling of Victorian-era titles can be found in the bibliography of primary sources in Lightman 2007.
6 See also Gowan Dawson, 'Science in the Periodical Press', in this volume.
7 The literary affiliations and techniques of most of the works discussed in this paragraph are discussed or mentioned in Merrill, Gates, Lightman 2007, O'Connor 2007A, Rauch 2011 (and other essays in Talairach-Vielmas), O'Connor 2011–12 and Keene 2015.

8   Michael A. Taylor and I are currently preparing a new annotated edition for publication. On Miller's career, see Taylor.
9   The best short surveys are Topham 2000 and Secord J. 2009.
10 For discussion of the works listed here, see Secord J. 2014 (for those published before 1835), Secord J. 2000 (for Nichol and Chambers) and Lightman 2007 (for those published after 1850).
11 See Adam Roberts, 'From Gothic to Science Fiction', in this volume.
12 Sleigh's book is an example, theorizing the possibility of a 'literary' status for science writing in the introduction, but focusing in the rest of the book mostly on highbrow novels, perhaps reflecting the main interests and expectations of a major segment of her target audience (English literature students).
13 See also Jonathan Smith, 'Writing Science: Scientific Prose', in this volume.
14 Dawson has discussed these matters more fully in his unpublished paper 'To the Writers on Science', and I am grateful to him for sending me a copy.
15 For example, Bown, Gooday 2007 255–62, Keene 2015, O'Connor 2011–12 and Talairach-Vielmas.
16 For example, Merrill, Lightman 2000, Armstrong, Smith, O'Connor 2007A, Fyfe 2007 and Levine 2011 37–73.
17 For example, Secord J. 2000 77–110, Secord J. 2014, Neeley, Brooke and Cantor 176–200, O'Connor 2007A, Levine 2011, Buckland, Rauch 2011, Somerset, and Fyfe 2004 110–19.
18 For example, Myers 1989, Shteir 1996 81–103 and Gates 37–44.
19 For example, Gates 44–65, Secord J. 2000, Shteir 2003, Lightman 2007 219–94, O'Connor 2009, O'Connor 2011–12, Somerset, Amigoni and Elwick, and Keene 2015.
20 For example, Shteir 1996, Shteir 1997, Shteir 2003, Gates and Shteir, Gates, Fyfe 2003 and Talairach-Vielmas.
21 For the second half of the century, see Lightman 2007. For women's writing on natural history, see Shteir 2003 and Gates 35–65. For geology in the first half of the century, see O'Connor 2007A.
22 This has been noted for one such writer by Somerset. These texts' demotion of narrative to only one among many discursive elements is mentioned by Amigoni and Elwick xiv.

Bibliography

Amigoni, David, and James Elwick, eds. *The Evolutionary Epic*. Vol. 4 of *Victorian Science and Literature*. Gen. eds Gowan Dawson and Bernard Lightman. 8 vols. London: Pickering & Chatto, 2011–12.

Anderson, Chris. 'Introduction: Literary Nonfiction and Composition'. *Literary Nonfiction: Theory, Criticism, Pedagogy*. Ed. Chris Anderson. Carbondale, IL: Southern Illinois University Press, 1989, ix–xxvi.

[Anon.]. 'The Old Red Sandstone'. *Presbyterian Review* 14 (1841–42): 208–17. Reprinted in *Science as Romance*. Ed. Ralph O'Connor. Vol. 7 of *Victorian Science and Literature*. Gen. eds Gowan Dawson and Bernard Lightman. 8 vols. London: Pickering & Chatto, 2011–12, 3–12.

Armstrong, Isobel. 'The Microscope: Mediations of the Sub-Visible World'. *Transactions and Encounters: Science and Culture in the Nineteenth Century*. Eds. Roger Luckhurst and Josephine McDonagh. Manchester: Manchester University Press, 2002, 30–54.

Beer, Gillian. *Darwin's Plots: Evolutionary Narrative in Darwin, George Eliot and Nineteenth-Century Fiction*. Cambridge: Cambridge University Press, 1983.

Bown, Nicola M. *Fairies in Nineteenth-Century Art and Literature*. Cambridge: Cambridge University Press, 2001.

Brock, Claire, ed. *New Audiences for Science*. Vol. 5 of *Victorian Science and Literature*. Gen. eds Gowan Dawson and Bernard Lightman. 8 vols. London: Pickering & Chatto, 2011–12.

Brooke, John, and Geoffrey Cantor. *Reconstructing Nature: The Engagement of Science and Religion*. New York, NY: Oxford University Press, 2000.

Buckland, Adelene. *Novel Science: Fiction and the Invention of Nineteenth-Century Geology*. Chicago, IL: University of Chicago Press, 2013.

Cooter, Roger, and Stephen Pumfrey. 'Separate Spheres and Public Places: Reflections on the History of Science Popularization and Science in Popular Culture'. *History of Science* 32.3 (1994): 237–67.

Dawson, Gowan, Richard Noakes and Jonathan R. Topham. 'Introduction'. *Science in the Nineteenth-Century Periodical: Reading the Magazine of Nature*. Eds. Geoffrey Cantor *et al*. Cambridge: Cambridge University Press, 2004, 1–34.

Dawson, Gowan, Richard Noakes and Jonathan R. Topham. '"To the Writers on Science": The Royal Literary Fund's Anniversary Dinners and the Use of Actors' Categories in Literature and Science Studies'. Unpublished lecture, King's College, Aberdeen, 8 July 2012.

Dawson, Gowan, Richard Noakes, Jonathan R. Topham and Bernard Lightman, gen. eds. *Victorian Science and Literature*. 8 vols. London: Pickering & Chatto, 2011–12.

Finnegan, Diarmid A. 'Exeter-Hall Science and Evangelical Rhetoric in Mid-Victorian Britain'. *Journal of Victorian Culture* 16 (2011): 46–64.

Fyfe, Aileen, ed. *Science for Children.* 7 vols. Bristol: Thoemmes, 2003.

Fyfe, Aileen. *Science and Salvation: Evangelical Popular Science Publishing in Victorian Britain.* Chicago, IL: University of Chicago Press, 2004.

Fyfe, Aileen. 'Reading Natural History at the British Museum and the *Pictorial Museum*'. *Science in the Marketplace: Nineteenth-Century Sites and Experiences.* Eds. Aileen Fyfe and Bernard Lightman. Chicago, IL: University of Chicago Press, 2007, 196–230.

Fyfe, Aileen. 'Natural History and the Victorian Tourist: From Landscapes to Rock-Pools'. *Geographies of Nineteenth-Century Science.* Eds. David N. Livingstone and Charles W. J. Withers. Chicago, IL: University of Chicago Press, 2011, 371–98.

Fyfe, Aileen. *Steam-Powered Knowledge: William Chambers and the Business of Publishing, 1820–1860.* Chicago, IL: University of Chicago Press, 2012.

Fyfe, Aileen, and Bernard Lightman, eds. *Science in the Marketplace: Nineteenth-Century Sites and Experiences.* Chicago, IL: University of Chicago Press, 2007.

Gates, Barbara T. *Kindred Nature: Victorian and Edwardian Women Embrace the Living World.* Chicago, IL: University of Chicago Press, 1997.

Gates, Barbara T., and Ann B. Shteir, eds. *Natural Eloquence: Women Reinscribe Science.* Madison, WI: University of Wisconsin Press, 1997.

Golinski, Jan. *Science as Public Culture: Chemistry and Enlightenment in Britain, 1760–1820.* Cambridge: Cambridge University Press, 1992.

Gooday, Graeme. 'Illustrating the Expert-Consumer Relationship in Domestic Electricity'. *Science in the Marketplace: Nineteenth-Century Sites and Experiences.* Eds. Aileen Fyfe and Bernard Lightman. Chicago, IL: University of Chicago Press, 2007.

Gooday, Graeme. *Domesticating Electricity: Technology, Uncertainty and Gender, 1880–1914.* London: Pickering & Chatto, 2008.

Heringman, Noah. *Sciences of Antiquity: Romantic Antiquarianism, Natural History, and Knowledge Work.* Oxford: Oxford University Press, 2013.

Hutchison, Hazel. '"The Telegraph has Other Work to do": Reading and Consciousness in Henry James's *In the Cage*'. *Uncommon Contexts: Encounters Between Science and Literature, 1800–1914.* Eds. Ben Marsden, Hazel Hutchison, and Ralph O'Connor. London: Pickering & Chatto, 2013, 167–86.

Keene, Melanie. 'An Active Nature: Robert Hunt and the Genres of Science Writing'. *Uncommon Contexts: Encounters Between Science and Literature, 1800–1914.* Eds. Ben Marsden, Hazel Hutchison, and Ralph O'Connor. London: Pickering & Chatto, 2013, 39–54.

Keene, Melanie. *Science in Wonderland: The Scientific Fairytales of Victorian Britain.* Oxford: Oxford University Press, 2015.

Klancher, Jon P. 'Romanticism and its Publics: A Forum: Introduction'. *Studies in Romanticism* 33 (1994): 523–5.

Landau, Misia. *Narratives of Human Evolution.* New Haven, CT: Yale University Press, 1991.

Levine, George. *Darwin and the Novelists: Patterns of Science in Victorian Fiction.* Cambridge, MA: Harvard University Press, 1984.

Levine, George. *Darwin the Writer.* Oxford: Oxford University Press, 2011.

Lightman, Bernard. '"The Voices of Nature": Popularizing Victorian Science'. *Victorian Science in Context.* Ed. Bernard Lightman. Chicago, IL: University of Chicago Press, 1997, 187–211.

Lightman, Bernard. 'The Visual Theology of Victorian Popularizers of Science: From Reverent Eye to Chemical Retina'. *Isis* 91 (2000): 651–80.

Lightman, Bernard. *Victorian Popularizers of Science: Designing Nature for New Audiences.* Chicago, IL: University of Chicago Press, 2007.

Luckhurst, Roger, and Justin Sausman, eds. *Marginal and Occult Sciences.* Vol. 8 of *Victorian Science and Literature.* Gen. eds Gowan Dawson and Bernard Lightman. 8 vols. London: Pickering & Chatto, 2011–12.

Marsden, Ben. 'Introduction'. *Uncommon Contexts: Encounters Between Science and Literature, 1800–1914.* Eds. Ben Marsden, Hazel Hutchison and Ralph O'Connor. London: Pickering & Chatto, 2013, 1–20.

McKitterick, David. 'Introduction'. *The Cambridge History of the Book in Britain: Volume VI, 1830–1914*. Ed. David McKitterick. Cambridge: Cambridge University Press, 2009, 1–74.

McLean, Steven. *The Early Fiction of H. G. Wells: Fantasies of Science*. Basingstoke: Palgrave Macmillan, 2009.

Merrill, Lynn R. *The Romance of Victorian Natural History*. New York, NY: Oxford University Press, 1989.

Miller, Hugh. *The Old Red Sandstone, or New Walks in an Old Field*. Edinburgh: Johnstone, 1841.

Morrell, Jack B., and Arnold Thackray. *Gentlemen of Science: The Early Years of the British Association for the Advancement of Science*. Oxford: Oxford University Press, 1981.

Myers, Greg. 'Nineteenth-Century Popularizers of Thermodynamics and the Rhetoric of Social Prophecy'. *Victorian Studies* 29 (1985): 35–66.

Myers, Greg. 'Science for Women and Children: The Dialogue of Popular Science in the Nineteenth Century'. *Nature Transfigured: Science and Literature, 1700–1900*. Eds. John Christie and Sally Shuttleworth. Manchester: Manchester University Press, 1989, 171–200.

Myers, Greg. *Writing Biology: Texts in the Social Construction of Scientific Knowledge*. Madison, WI: University of Wisconsin Press, 1990.

Myers, Greg. 'Discourse Studies of Scientific Popularization: Questioning the Boundaries'. *Discourse Studies* 5 (2003): 265–79.

Neeley, Katherine A. *Mary Somerville: Science, Illumination and the Female Mind*. Cambridge: Cambridge University Press, 2001.

O'Connor, Ralph. *The Earth on Show: Fossils and the Poetics of Popular Science, 1802–1856*. Chicago, IL: University of Chicago Press, 2007 [2007A].

O'Connor, Ralph. 'Young-Earth Creationists in Nineteenth-Century Britain? Towards a Reassessment of "Scriptural Geology"'. *History of Science* 45 (2007): 357–403 [2007B].

O'Connor, Ralph. 'From the Epic of Earth History to the Evolutionary Epic in Nineteenth-Century Britain'. *Journal of Victorian Culture* 14 (2009): 207–23.

O'Connor, Ralph, ed. *Science as Romance*. Vol. 7 of *Victorian Science and Literature*. Gen. eds. Gowan Dawson and Bernard Lightman. 8 vols. London: Pickering & Chatto, 2011–12.

Paradis, James. 'The Natural Historian as Antiquary of the World: Hugh Miller and the Rise of Literary Natural History'. *Hugh Miller and the Controversies of Victorian Science*. Ed. Michael Shortland. Oxford: Oxford University Press, 1996, 122–50.

Qureshi, Sadiah. *Peoples on Parade: Exhibitions, Empire, and Anthropology in Nineteenth-Century Britain*. Chicago, IL: University of Chicago Press, 2011.

Rauch, Alan. 'Parables and Parodies: Margaret Gatty's Audiences in the *Parables From Nature*'. *Children's Literature* 25 (1997): 137–52.

Rauch, Alan. *Useful Knowledge: The Victorians, Morality, and the March of Intellect*. Durham, NC: Duke University Press, 2001.

Rauch, Alan. 'The Pupil of Nature: Science and Natural Theology in Maria Hack's *Harry Beaufoy*'. *Science in the Nursery: The Popularisation of Science in Britain and France, 1761–1901*. Ed. Laurence Talairach-Vielmas. Newcastle: Cambridge Scholars, 2011, 69–90.

Rudwick, Martin J. S. *The Great Devonian Controversy: The Shaping of Scientific Knowledge Among Gentlemanly Specialists*. Chicago, IL: University of Chicago Press, 1985.

Ruse, Michael. *Monad to Man: The Concept of Progress in Evolutionary Biology*. Cambridge, MA: Harvard University Press, 1996.

Schaffer, Simon. 'Scientific Discoveries and the End of Natural Philosophy'. *Social Studies of Science* 16 (1986): 387–420.

Secord, Anne. 'Science in the Pub: Artisan Botanists in Early Nineteenth-Century Lancashire'. *History of Science* 32.3 (1994): 269–315.

Secord, Anne. 'Botany on a Plate: Pleasure and the Power of Pictures in Promoting Early Nineteenth-Century Scientific Knowledge'. *Isis* 93 (2002): 28–57.

Secord, Anne. 'Introduction'. Gilbert White, *The Natural History of Selborne*. Ed. Anne Secord. Oxford: Oxford University Press, 2013, ix–xxviii.

Secord, James A. *Victorian Sensation: The Extraordinary Publication, Reception and Secret Authorship of 'Vestiges of the Natural History of Creation'*. Chicago, IL: University of Chicago Press, 2000.

Secord, James A. 'Knowledge in Transit'. *Isis* 95 (2004): 654–72.

Secord, James A. 'Science, Technology and Mathematics'. *The Cambridge History of the Book in Britain: Volume VI, 1830–1914*. Ed. David McKitterick. Cambridge: Cambridge University Press, 2009, 443–74.

Secord, James A. *Visions of Science: Books and Readers at the Dawn of the Victorian Age*. Oxford: Oxford University Press, 2014.

Shteir, Ann B. *Cultivating Women, Cultivating Science: Flora's Daughters and Botany in England, 1760 to 1860*. Baltimore, MD: Johns Hopkins University Press, 1996.

Shteir, Ann B. 'Elegant Recreations? Configuring Science Writing for Women'. *Victorian Science in Context*. Ed. Bernard Lightman. Chicago, IL: University of Chicago Press, 1997, 236–55.

Shteir, Ann B. 'Finding Phebe: A Literary History of Women's Science Writing'. *Women and Literary History: 'For There She Was'*. Eds. Katherine Binhammer and Jeanne Wood. Newark, DE: University of Delaware Press, 2003, 152–66.

Shuttleworth, Sally. *George Eliot and Nineteenth-Century Science: The Make-Believe of a Beginning*. Cambridge: Cambridge University Press, 1984.

Shuttleworth, Sally. *Charlotte Brontë and Victorian Psychology*. Cambridge: Cambridge University Press, 1996.

Sleigh, Charlotte. *Literature and Science*. Basingstoke: Palgrave Macmillan, 2011.

Smith, Jonathan. *Charles Darwin and Victorian Visual Culture*. Cambridge: Cambridge University Press, 2006.

Snobelen, Stephen David. 'Of Stones, Men and Angels: The Competing Myth of Isabelle Duncan's *Pre-Adamite Man* (1860)'. *Studies in History and Philosophy of Biological and Biomedical Sciences* 32 (2001): 59–104.

Somerset, Richard. 'Bringing (Anti-)Evolutionism into the Nursery: Narrative Strategies in the Emergent "History of Life" Genre'. *Science in the Nursery: The Popularisation of Science in Britain and France, 1761–1901*. Ed. Laurence Talairach-Vielmas. Newcastle: Cambridge Scholars, 2011, 140–63.

Syme, J. T. Boswell, ed. *English Botany, or Coloured Figures of British Plants*. 3rd edn. 13 vols. London: Hardwicke, 1863–92.

Talairach-Vielmas, Laurence, ed. *Science in the Nursery: The Popularisation of Science in Britain and France, 1761–1901*. Newcastle: Cambridge Scholars, 2011.

Taylor, Michael A. *Hugh Miller: Stonemason, Geologist, Writer*. Edinburgh: NMS Enterprises, 2007.

Topham, Jonathan R. 'Beyond the "Common Context": The Production and Reading of the *Bridgewater Treatises*'. *Isis* 89 (1998): 233–62.

Topham, Jonathan R. 'Scientific Publishing and the Reading of Science in Early Nineteenth-Century Britain: An Historiographical Survey and Guide to Sources'. *Studies in History and Philosophy of Science* 31A (2000): 559–612.

Topham, Jonathan R. 'The *Mirror of Literature, Amusement and Instruction* and Cheap Miscellanies in Early Nineteenth-Century Britain'. *Science in the Nineteenth-Century Periodical: Reading the Magazine of Nature*. Eds. Geoffrey Cantor *et al*. Cambridge: Cambridge University Press, 2004, 37–66.

Topham, Jonathan R. 'Publishing "Popular Science" in Early Nineteenth-Century Britain'. *Science in the Marketplace: Nineteenth-Century Sites and Experiences*. Eds. Aileen Fyfe and Bernard Lightman. Chicago, IL: University of Chicago Press, 2007, 135–68.

Topham, Jonathan R. 'Rethinking the History of Science Popularization/Popular Science'. *Popularizing Science and Technology in the European Periphery, 1800–2000*. Eds. Faidra Papanelopoulou, Agustí Nieto-Galan and Enrique Perdiguero. Farnham: Ashgate, 2009, 1–20.

White, Paul. *Thomas Huxley: Making the 'Man of Science'*. Cambridge: Cambridge University Press, 2003.

White, Paul. 'The Experimental Novel and the Literature of Physiology'. *Uncommon Contexts: Encounters Between Science and Literature, 1800–1914*. Eds. Ben Marsden, Hazel Hutchison and Ralph O'Connor. London: Pickering & Chatto, 2013, 21–38.

Whitley, Richard. 'Knowledge Producers and Knowledge Acquirers: Popularisation as a Relation Between Scientific Fields and Their Publics'. *Expository Science: Forms and Functions of Popularisation*. Eds. Terry Shinn and Richard Whitley. Dordrecht: Reidel, 1985, 3–28.

Winter, Alison. 'Mesmerism and Popular Culture in Early Victorian England'. *History of Science* 32.3 (1994): 317–43.

Winter, Alison. *Mesmerized! Powers of Mind in Victorian Britain*. Chicago, IL: University of Chicago Press, 1998.

Yeo, Richard. *Defining Science: William Whewell, Natural Knowledge, and Public Debate in Early Victorian Britain*. Cambridge: Cambridge University Press, 1993.

12

SCIENCE IN THE PERIODICAL PRESS

Gowan Dawson

University of Leicester

Writing to Joseph Dalton Hooker in November 1869, Charles Darwin declared:

> I like all scientific periodicals, including poor 'Scientific Opinion', & I think higher
> than you do of 'Nature'. Lord what a rhapsody that was of Goethe, but how well
> translated—it seemed to me, as I told Huxley, as if written by the maddest English
> scholar. It is poetry, & can I say anything more severe?
>
> *(1985 17:488)*

The two naturalists had been discussing the sudden emergence of a host of new specialist journals, of which *Scientific Opinion*, founded in 1868, and *Nature*, begun by Norman Lockyer a year later, were only the most notable.[1] The loss-making *Scientific Opinion* would soon be taken over by the *English Mechanic*, while *Nature*, of course, would go on to become the international benchmark for modern science publishing. The advent of such specialist periodicals in the late nineteenth century – and the founding of *Nature* in particular – has often been assumed to signal the commencement of what Robert M. Young has called the 'fragmentation of the common intellectual context' (155). In this influential but now much-criticized thesis, general periodicals like the *Edinburgh Review* and the *Quarterly Review*, founded in 1802 and 1809 respectively, helped maintain a 'rich interdisciplinary culture' (Young 127) in which the sciences were fully integrated with other forms of knowledge including literature, before the 'popularity of *Nature* among increasingly professional scientists' (Young 156), along with other similar developments, shattered this unified intellectual culture and instead inaugurated a situation more akin to C. P. Snow's 'The Two Cultures'.[2] Indeed, with their authoritative processes of peer review, scientific periodicals such as *Nature* soon supplanted learned societies as the central institutions where the new forms of specialist expertise were adjudicated and guaranteed.[3]

Yet what is striking about Darwin's response to the opening number of *Nature* is not just the new journal's lack of popularity with Hooker, a leading specialist in botany, but also the confident designation of its very first article as verse, even if Darwin, who by this stage of his life was lamenting 'I cannot endure to read a line of poetry' (2002 84), remained teasingly equivocal as to whether he approved. The article that prompted Darwin's sardonic severity was a prose translation of Goethe's 'Aphorisms on Nature' by Thomas Henry Huxley, which closed with the sombre reflection that 'when another half-century has passed, curious readers of the back numbers of NATURE' will recognize that, 'long after the theories of the philosophers whose achievements are recorded in these pages, are obsolete, the vision of the poet' remains as a 'truthful and efficient symbol of the wonder and the mystery of Nature' (1869 11). In the opening pages of the specialist journal that, putatively at least, extinguished the common context of nineteenth-century culture, poetry was attributed a more enduring intellectual authority than science.

Needing to turn a profit in a crowded commercial marketplace, *Nature* was initially intended to appeal to both scientific practitioners and the general public, although it was soon apparent that the increasing specialization of science entailed that only the former could be accommodated by its emphasis on news of the latest research.[4] It was nevertheless not alone among the new breed of scientific periodicals in including literature within its otherwise specialist contents, with professional psychological journals such as *Mind*, founded in 1876, often carrying reviews of novels that illustrated particular mental states or could be regarded as case studies of psychiatric disorders.[5] Even while the new specialist journals continued to engage with the wider culture of late Victorian Britain, thereby stymying the main tenet of Young's fragmentation thesis, many prominent men of science shared Hooker's seeming distaste for the new mode of scientific communication and, as Melinda Baldwin has recently shown, preferred to continue publishing their research in general periodicals such as the *Fortnightly Review* and the *Nineteenth Century*, whose lengthy articles and monthly publication schedules contrasted with the weekly snippets permitted by *Nature*.[6]

But, if the changes in the late Victorian journalistic marketplace did not induce the decisive fragmentation envisioned by Young, his original conception of a 'common intellectual context', facilitated, at least in part, by the cultural centrality of periodicals, remains hugely significant for understanding the varied interconnections of science and literature across the nineteenth century. First adumbrated in the late 1960s (although not in published form), it affords an early precursor to the 'One Culture' model, based on the supposition that 'literature and science are mutually shaped by their participation in the culture at large' (Levine 5–6), that proved so influential from the 1980s onwards.[7] Unlike many uses of this ubiquitous model, however, the 'common intellectual context', divested of its problematic fragmentation coda, has the advantage of ascribing material reasons, rather than quasi-Foucauldian abstractions, for why this particular mutual shaping occurred so prominently in the nineteenth century. This was, after all, 'uniquely the age of the periodical', when, according to J. Donn Vann and Rosemary T. VanArsdel, the 'circulation of periodicals and newspapers was larger and more influential . . . than printed books, and served a more varied constituency in all walks of life' (3, 7). Both science and literature were vital components of the nineteenth-century periodical press, and even when they can be considered distinct entities rather than thinking of 'science *as* literature' (O'Connor 14) or 'literature *as* science' (Buckland 15), they were brought into conjunction by their proximity in periodicals.

The miscellaneous nature of most nineteenth-century periodicals, whether politically influential journals of the Romantic period such as *Blackwood's Edinburgh Magazine* or Victorian family-oriented shilling monthlies such as the *Cornhill Magazine*, provided a textual space in which scientific articles appeared alongside poetry, short stories and instalments of serial fiction. In *Blackwood's* during the 1810s, as William Christie notes, 'there are two contributions on science for every one on imaginative literature' (126–7), and, already notorious for its inconsistent but nonetheless fervently expressed opinions on Romantic poetry, *Blackwood's* 'can be as capricious on scientific topics and allegiances as it can on Coleridge and Wordsworth' (127). Half a century later, the *Cornhill* carried a regular feature entitled 'Our Survey of Literature, Science, and Art', which combined reviews of recent works of fiction with news of the latest scientific developments.[8] Notably, when published as part of a periodical rather than as a discrete book, scientific and literary writing had each to conform to the collective 'codes of discourse' (Brake 2001 18), including the format, politics and implied readership, of the particular journal in which they appeared. This was especially the case under the conditions of anonymous publication, when, as happened at both *Blackwood's* and the *Cornhill* as well as almost every periodical produced before the 1860s, the collective editorial 'we' subsumed the identities of individual contributors.[9]

From this perspective, the 'common language' that, as Gillian Beer proposed in *Darwin's Plots* (1983), during the 'mid-nineteenth century, scientists still shared . . . with other educated readers and writers of their time' (6) can be seen to derive largely from the conditions of periodical publication in which different kinds of articles had to accord with a particular editorial agenda and be accessible to the same general readership. Beer has subsequently acknowledged that the 'wonderful inclusiveness of generalist journals . . . meant that philosophers, lawyers, evolutionary theorists . . . astronomers, physicists, novelists, theologians, poets . . . all appeared alongside each other' (1996 202–3), although she suggests that these juxtapositions were more likely to produce an 'effect of bricolage than synthesis'. Such an eclectic bricolage, however, implies only a loose, and rather random, assemblage of identifiably distinct contents – as is found in the form of modern art made out of ready-made materials that uses the same designation – that overlooks the more corporate nature of periodicals in the nineteenth century. In fact, William Makepeace Thackeray, when taking on the editorship of the *Cornhill* in 1860, proposed that he would be the 'Conductor of a Concert, in which . . . many skilful performers will take part' (x), while John Morley, when editing the *Fortnightly* two decades later, similarly compared his role to that of a 'conductor' with 'his *bâton*' (1882 513) and vowed to treat the contributors as musicians in an orchestra who take 'their several parts in his performance'. Rather than an artistic bricolage, the nineteenth-century periodical more closely resembled a musical symphony.[10]

The periodical press was fundamental to both how literature and science were produced in the nineteenth century and how they are understood by modern readers. When what are traditionally conceived as disciplinarily distinct individual texts are placed back in their original periodical context, they become integrated into a multidisciplinary range of overlapping discourses that constitute what Laurel Brake has termed the 'larger magazine text' (1997 55). This hybrid 'magazine text' blurs the clear generic distinction between forms such as poetry or serial fiction and scientific essays, whether reviews of science books or published lectures, that have been made retroactively by the institutionalization of knowledge along disciplinary lines. For example, Alfred Tennyson's dramatic monologue 'Lucretius', which appeared as the lead item in the May 1868 number of *Macmillan's Magazine* and deals with the despairing suicide of the Roman poet and proponent of Epicurean atomism, engages with and implicitly contests the materialistic implications of articles published in the same periodical by men of science such as the physicist John Tyndall (who was a close friend of Tennyson's as well as the leader of a Victorian revival of atomism) and the physiological psychologist Alexander Bain. Once 'Lucretius' was published in book form as part of *The Holy Grail and Other Poems* (1870), though, this close connection with science became less evident and the poem can seem precisely the kind of self-contained, autonomous literary artefact valorized by a traditional Romantic hermeneutics.[11] The 'study of the periodical press', as Lyn Pykett has observed, 'is inevitably interdisciplinary. It not only challenges the boundaries between hitherto separately constituted fields of knowledge, but also challenges the internal hierarchies and sub-divisions within discrete academic disciplines' (4). In fact, the disciplinary organization of knowledge, which insists on the distinction between literature and science, is predicated upon the marginalization of the periodical press as a subject of academic study. By eschewing periodicals, English literature could be constructed as an autonomous and ostensibly apolitical subject untarnished by connections with less venerable discourses such as science.[12]

A greater emphasis on the nineteenth-century periodical press in scholarship over the last two decades, especially on science in periodicals, now makes it possible to recover the continuities and overlaps between literature and science that, as in the case of a text like Tennyson's 'Lucretius', would have been self-evident to most of their original readers.[13] Taking its lead from such scholarship, this chapter will offer a further, more detailed case study of how, when literary writing is restored to the periodical context in which it initially appeared, it can be seen to both invoke and actively participate in the scientific discussions taking place within the pages of the same journal. The particular example is an especially hard case for tracing literary and scientific interconnections: the studiedly aesthetic prose of Walter Pater, whose impressionistic accounts of the cultural rebirth of fifteenth-century Italy appeared in the *Fortnightly*, under Morley's orchestral editorship, before being collected in book form as *Studies in the History of the Renaissance* (1873). Pater was a diffident Oxford don and his writing, which championed subjective aesthetic experience, is often located outside of contemporary nineteenth-century culture in a transcendental realm of pure art untarnished by quotidian concerns like science. Denis Donoghue, for instance, has suggested that 'Pater's England is not the country as given; on that he has no purchase' (219), and instead Donoghue avers that Pater 'invents a place responsive to his desires, [and] calls it England'. Relocating Pater's essays as part of Morley's self-consciously symphonic editorial performance, alongside contributions from the likes of Huxley and Tyndall, reveals the limitations, and even distortions, of such a formalist approach. A further case study will take an opposing perspective and examine how, when scientific writing is read in its original periodical context, its literary dimensions, or at least its rhetorical and figurative features, are foregrounded. Indeed, the example of the geologist David Thomas Ansted's contributions to the highbrow *Dublin University Magazine* and the more populist *Temple Bar* shows how far what men of science could say, and how they said it, was determined by both the commercial imperatives and generic conventions of periodical publishing. Sometimes this even involved saying the exact opposite of what one had said elsewhere.

These two case studies come from the mid-nineteenth-century boom in periodical publishing that occurred when more effective methods of printing and papermaking, as well as speedier distribution networks, considerably lowered production costs, and the so-called 'taxes on knowledge' imposed earlier in the century, especially the stamp duty that, since 1815, was used to make newspapers all but unaffordable to readers who might be tempted into sedition, were finally repealed.[14] While Young finds the exemplars of his 'common intellectual context' in expensive early-century periodicals such as the *Edinburgh* and the *Quarterly*, which incorporated science into an elite highbrow culture, my case studies suggest that this shared context becomes much more evident – and 'common' in all senses of the word – in the veritable explosion of new reviews and magazines that, from the 1850s onwards, brought both science and literature to much broader audiences. At the same time, while the system of anonymous authorship that had been customary in journalism since the eighteenth century came under increasing strain after the 1850s, it was still strictly enforced even in new and otherwise innovative titles such as the *Cornhill* or *Temple Bar*, and thus the vital significance anonymity had for both scientific and literary writing that becomes manifest in each of my case studies has particular relevance for the earlier part of the century as well. Above all, both case studies demonstrate that, when considered in relation to the nineteenth-century periodical press, literature becomes tangibly more scientific and science more literary.

Pater and the *Fortnightly Review*

The *Fortnightly Review* was, paradoxically, neither fortnightly nor a review. Founded in 1865, it shifted to a less-demanding monthly schedule in the following year, and, from the very beginning, published freestanding articles that made no pretence about being reviews of recent books. Significantly, however, the *Fortnightly* was the first major periodical to disavow the previously 'sacred principle of the Anonymous' (Morley 1882 513) and enforce authorial responsibility by a strict policy of signature. Although the platform of signature ostensibly afforded complete freedom to contributors, the *Fortnightly* in fact pursued a distinctively liberal agenda that was evident in almost everything it published. Morley, whose editorship lasted from 1867 to 1882, recalled that 'our miscellany of writers and subjects was soon taken by prejudiced observers to disclose an almost sinister unity in spirit and complexion. This unity was in fact the spirit of Liberalism in its most many-sided sense' (1917 1:86). Support for modern science was one of the defining characteristics of the *Fortnightly*'s liberalism, and its tolerance of freethinking and secularism attracted an array of notable men of science who were willing to advance their often contentious findings and theories without the protection previously afforded by anonymity. At the same time, Morley contested the necessary connection of literature and ethics in his editorial 'Causeries' column, and solicited contributions from aesthetic writers such as Algernon Charles Swinburne, as well as poetry from Dante Gabriel Rossetti. Such rarefied – and morally suspect – aesthetes had little in common with practical and scrupulously respectable men of science, but the *Fortnightly*, under Morley's editorship, evinced 'a certain pervading atmosphere . . . a certain undefinable concurrence among writers coming from different schools and handling very different subjects' (1882 519). Nowhere was this remarkable concurrence more evident than in the series of essays on Renaissance art that Pater contributed to the *Fortnightly* between 1869 and 1871.

In Victorian Britain the most influential view of the Renaissance was the extravagant antipathy that John Ruskin had expressed in *The Stones of Venice* (1851–53). According to Ruskin, the Renaissance was a kind of biblical fall from the grace of the Gothic Middle Ages, which had unleashed the evil and corruption that now characterized contemporary Western society. The very different conception of fifteenth-century Italy advanced in Pater's *Fortnightly* articles was instead informed by the work of the French historians Edgar Quinet and Jules Michelet, who apotheosized the Renaissance as a period of discovery and heroic individualism.[15] Like them, Pater depicts the Renaissance as a time of rebellion against the conventional authority and superstition of the medieval Church. Significantly, this defiant assertion of human reason over received opinion was, for Pater, intimately connected with the rise of empirical science. In 'Notes on Leonardo Da Vinci', which appeared in the November 1869 number of the *Fortnightly*, Pater declares that this disavowal of orthodox theology had heralded 'the coming of what is called the modern spirit, with its realism, its appeal to experience' (499). In its insistence on the necessary limitation of knowledge to the data of experience, this new direction in human thought presaged the epistemology of nineteenth-century science that was championed so vociferously elsewhere in the *Fortnightly*'s pages.

Leonardo's remarkable prescience of modern science is emphasized throughout Pater's article. He asserts of the artist:

> we find him often in intimate relations with men of science . . . His observations and experiments fill thirteen volumes of manuscript; and those who can judge describe him as anticipating long before, by rapid intuition, the later ideas of science.
>
> *(Pater 1869 499)*

Notably, Pater meekly defers to the scientific authority of those he considers properly qualified to judge, and one of the most prominent scientific writers at the *Fortnightly* to whose opinion Pater might be expected to accede was Morley's predecessor as editor, George Henry Lewes. In an earlier article for the *Westminster Review*, a liberal quarterly that was founded in the 1820s, Lewes had asserted: 'Leonardo da Vinci, that great artist . . . anticipated discoveries which made Galileo and Kepler . . . and even some modern geologists, famous' (1852 479). While Pater's comments seem to defer to the scientific authority of this distinguished *Fortnightly* contributor, he nevertheless goes on implicitly to contest Lewes's literary judgement. In the same *Westminster* piece, Lewes contended: 'Poets and Men of Science, in all times, formed two distinct classes, and never, save in one illustrious example, exhibited the twofold manifestation of Poetry and Science working in harmonious unity: that single exception is Goethe' (1852 479). In 'Notes on Leonardo Da Vinci', Pater disputes this sanguine view of Goethe's writing (which would later be reiterated by Huxley in the first number of *Nature*), instead averring: 'the name of Goethe . . . reminds one how great for the artist may be the danger of over-much science . . . in the second part of Faust, [he] presents us with a mass of science which has no artistic character at all' (1869 501). One consequence of ending the convention of anonymity, as Morley later reflected, was that 'if an article is to be signed, the editor will naturally seek the name of an expert of special weight and competence on the matter in hand' (1882 514). While Pater was willing to submit to Lewes's scientific expertise in his own signed article, he clearly considered the interconnection of science and literature to be a subject on which he could legitimately question Lewes's judgement.

Pater's impressionistic reverie on 'Leonardo's masterpiece' (1869 506) *La Gioconda* (commonly known as the *Mona Lisa*) was famously excerpted and recontextualized as the opening poem of W. B. Yeats's *The Oxford Book of Modern Verse* (1936). When read in its original periodical context, however, it becomes evident that Pater's celebrated description is informed by recent scientific suppositions published in the *Fortnightly*. Leonardo's portrait, Pater contends, evinces 'a beauty wrought out from within upon the flesh, the deposit, little cell by cell, of strange thoughts and fantastic reveries and exquisite passions' (1869 506). The modern theory of cells advanced by Max Schultze and other German cytologists was discussed by Lewes in 'Mr. Darwin's Hypotheses', which appeared in the *Fortnightly* in 1868. Lewes affirmed that 'according to the most recent investigations . . . a Cell is "a nucleus with surrounding protoplasm"' (1868 61). The cell, Lewes went on,

> may lead an isolated life as plant or animal, or it may be united with others and lead a more or less corporate existence . . . we see animal forms of which the web is woven out of myriads upon myriads of cells, with various cell-products, processes, fibres, tubes.
>
> *(1868 62)*

Akin to Lewes, Pater suggests that the weary flesh of *La Gioconda*, which embodies 'the animalism of Greece, the lust of Rome . . . [and] the sins of the Borgias' (1869 506–7), is a web woven out of countless tiny organic cells. Significantly, the evocative suggestion that she has 'trafficked for strange webs with Eastern merchants' (Pater 1869 507) echoes the language of Lewes's article. Pater's contention that 'strange thoughts . . . and exquisite passions' (1869 506) are 'wrought out from within upon the flesh', moreover, conforms to Lewes's organicist conception, expressed in another of his contributions to the *Fortnightly* entitled 'The Heart and the Brain', of the indissoluble interweaving of the mind and the physical tissue of the body within the continuous web of the living organism.[16] The 'two systems', according to Lewes, 'interlace, interpenetrate each other, so that the slightest modification of the one is followed by a corresponding change in the other' (1865 67). Even Pater's most famously purple passage of aesthetic prose has important overlaps, conceptual as well as linguistic, with the scientific writing that was published alongside it in the *Fortnightly*.

By October 1871, Pater's essay 'Pico Della Mirandola' was appearing as the *Fortnightly*'s lead item, and, as with his earlier article on Leonardo, its account of the fifteenth-century translator of Plato again reflected the broader agenda of Morley's liberal review. Having escaped the harsh asceticism of medieval religion, Pico, according to Pater, does not doubt the 'dignity of human nature, the greatness of man' (1871 381). At the same time,

> Pico's theory of that dignity is founded on a misconception of the place in nature both of the earth and of man. For Pico the earth is the centre of the universe; and around it, as a fixed and motionless point, the sun and moon and stars revolve like diligent servants or ministers. And in the midst of all is placed man, *nodus et vinculum mundi*, the bond or copula of the world.
>
> *(Pater 1871 381–2)*

In describing Pico's idealist cosmography, Pater's language alludes, unmistakably, to what Huxley, in *Evidence as to Man's Place in Nature* (1863), had insisted was

> the question of questions for mankind—the problem which underlies all others, and is more deeply interesting than any other . . . the ascertainment of the place which Man occupies in nature and of his relations to the universe of things.
>
> *(57)*

Huxley's book, which displaced man from the centre of the universe and argued for his descent from apelike ancestors, had provoked considerable controversy, and he continued its incendiary arguments in his contributions to the *Fortnightly*. In 'On the Advisableness of Improving Natural Knowledge', published under Lewes's editorship in 1866, Huxley, like Pater with Pico, depicted anthropocentrism as the defining aspect of earlier understandings of the universe. He observed that 'uncultured man, no doubt, has always taken himself as the standard of comparison, as the centre and measure of the world' (Huxley 1866 633). In the more advanced nineteenth century, by contrast, 'the astronomers discover in the earth no centre of the universe, but an eccentric spark, [and] the naturalists find man to be no centre of the living world, but one amid endless modifications of life' (Huxley 1866 635). In his *Fortnightly* essays, Pater considers Renaissance figures such as Pico from the evolutionary and rigorously naturalistic perspective established by Huxley and, regardless of its continuing contentiousness in other quarters, adhered to by almost all of Morley's contributors.

With the same haughty disdain for earlier understandings of the universe exhibited by Huxley, Pater asserts:

> That whole conception of nature is so different from our own. For Pico the world is a limited place, bounded by actual crystal walls and a material firmament; it is like a painted toy, like that map or system of the world, held as a great target or shield in the hands of the grey-headed father of all things . . . How different from this childish dream is our conception of nature, with its unlimited space, its innumerable suns, and the earth but a mote in the beam.
>
> *(1871 382)*

The vast immensity of the universe was a frequent topic of discussion in the pages of the *Fortnightly* during the 1860s. Huxley observed that

> Astronomy,—which of all sciences . . . has more than any other rendered it impossible for [men] to accept the beliefs of their fathers . . . tells them that this so vast and seemingly solid earth is but an atom among atoms, whirling no man knows whither, through illimitable space.
>
> *(1866 634)*

Similarly, Tyndall began an article on 'The Constitution of the Universe' in the December 1865 number of the *Fortnightly* by declaring:

> We cannot think of space as finite, for wherever in imagination we erect a boundary we are compelled to think of space as existing beyond that boundary. Thus by the incessant dissolution of limits we arrive at a more or less adequate idea of the infinity of space.
>
> *(129)*

Akin to Tyndall, Pater also imaginatively represents the almost inconceivable magnitude of the universe. In fact, the imagery of his metaphor of the earth as a mote in a beam most likely derives from Tyndall's experimental work on the germ theory of disease, in which, as he described it in *Fraser's Magazine* in 1870, 'however well the tubes might be washed and polished . . . the electric beam infallibly revealed signs and tokens of dirt . . . The floating motes resembled minute particles of liquid' (310). This 'scientific use of the imagination', as Tyndall termed it in the title of a lecture from 1870, allowed Pater to continue to engage with the leading currents of modern science even in his ostensibly descriptive and impressionistic biographical sketches of Renaissance artists and writers.

Unlike contemporaries such as George Eliot or Thomas Hardy, Pater's concern with science is not particularly conspicuous, with neither his letters nor the records of his library borrowings in Oxford yielding much evidence of an engagement with modern scientific thought.[17] Formalist critics like Donoghue, moreover, suggest that his writing deliberately eschewed any involvement whatsoever with contemporary issues. Such a view is tenable when Pater's work is read in the discrete volumes in which it was subsequently collected and, as with Yeats's *Oxford Book of Modern Verse*, recontextualized. In such formats it can even seem, as Gene H. Bell-Villada has proposed with remarkable inaccuracy, that Pater's belletristic prose 'would have been completely out of sorts in the jungle of journalism, with its business pressures and demands for constant and varied output' (82). When Pater's essays are returned to the periodical context where, notwithstanding Bell-Villada's wilful obliviousness, they were originally published, their close involvement with science, and especially Lewes's organicism and Huxley's and Tyndall's evolutionary naturalism, becomes evident.[18] Such clear overlaps in a journal like the *Fortnightly* make sense as part of what Young was long ago describing as the 'common intellectual context'. And if such an approach can work with a hard case like Pater, it offers a particularly effective means of tracing the engagement with science of a number of other nineteenth-century essayists, novelists and poets.

Ansted and the *Dublin University Magazine* and *Temple Bar*

One of the reasons Pater began writing for the *Fortnightly* in 1869, having previously contributed to the *Westminster*, was that it offered better rates of pay, and, tellingly, his very first letter to

Morley began 'Many thanks for the cheque' (1970 6). Men of science, no less than literary writers, were similarly dependent on the financial rewards of journalism, especially as, for much of the nineteenth century, there were very few salaried posts in science. David Thomas Ansted had occupied one of these rare posts as professor of geology at King's College London, where he had succeeded Charles Lyell, but he stepped down in 1853 to work as a consulting mining engineer and by the end of the decade was increasingly reliant on his earnings from writing for a wide range of periodicals. In order to make freelance journalism a remunerative career, commercial writers had to regularly tailor their opinions to the demands of the literary marketplace, having to conform, for instance, to the varying editorial agendas and implied audiences of the disparate periodicals whose commissions were necessary to earn a living. There were few repercussions for a literary writer like Pater in echoing Huxley's evolutionary naturalism in the *Fortnightly* while also venerating the antiquity of the Church when writing for the Anglican newspaper, the *Guardian* ('a little to my own surprise', as he told a friend (1970 84)). Science, however, was predicated – or at least was purported to be – on a commitment to intellectual truth that could not accommodate such expedient equivocations. Some science writers did amend their views in different publications, with the still-anonymous author of *Vestiges of the Natural History of Creation* (1844), Robert Chambers, contradicting that book's arguments on the fall of sea levels in his *Ancient Sea Margins* (1848), which carried his name. This, as James Secord has argued, was necessary for Chambers to strategically distance himself from the fierce controversy provoked by *Vestiges*.[19] Ansted, on the other hand, exploited the same veil of anonymity for more pragmatically commercial considerations, simultaneously publishing contrasting opinions of the validity of one of the most significant geological axioms of the nineteenth century.

Georges Cuvier's law of correlation proposed that each element of an animal corresponds mutually with all the others, so that a carnivorous tooth must be accompanied by a particular kind of jawbone, neck, stomach and so on that facilitates the consumption of flesh. Since the 1790s, it had enabled geologists to reconstruct prehistoric animals, with a high degree of accuracy, from only fragmentary fossil remains (sometimes even just a single bone), demonstrating the vast differences between these strange creatures and present faunas and thereby providing compelling evidence for the previously disputed concept of extinction.[20] The efficacy and accuracy of Cuvier's law had begun to be questioned following his death in 1832, but throughout the 1840s and into the 1850s Ansted enthusiastically accepted that there 'exist[s] a principle . . . which, properly developed, is capable of banishing all doubt and uncertainty' in reconstructing prehistoric animals, and '*by means of which each species may be identified by even a fragment of any one of them*' (1844 1:74). Cuvier's conception of the harmonious correlation of the animal frame, in which no part could change without altering all the others, viewed each species as fixed, and much of its intellectual authority was inevitably undermined by the adaptational understanding of structure inaugurated by Darwin's *On the Origin of Species* (1859). Despite his earlier distaste for the transmutationist speculations of *Vestiges*, Ansted was an early convert to Darwin's evolutionism, and after 1859 his erstwhile enthusiasm for Cuvier's renowned law of correlation declined noticeably.

Explaining Darwin's new theory and 'observing the mode in which it affects the great Palæontological inquiry' for readers of the *Dublin University Magazine* in July 1860, Ansted still referred to the 'remarkable law . . . of correlation of form and structure' (30). But, whereas he had previously agreed with Cuvier that the same law enjoyed a geometrical level of clarity and certainty in which, 'to use a mathematical but very apt illustration—the equation of a curve involves all the properties of the curve; and . . . the curve may be drawn when we know one root of the equation' (Ansted 1844 1:76), he now proposed that it had for its 'basis an admission of perfect mutual adaptability of every part, combined with an obscure and mysterious reference to some general typical structure' (1860 30). The obscurity and convolution that had eclipsed mathematical precision were those of Richard Owen's homological conception of skeletal type. The much-vaunted ability, exhibited by both Cuvier and Owen, to identify a creature from just a fragment of its remains was now, in Ansted's revised estimate, predicated on a vague sense of correspondence between particular types of skeleton as much as an unerring correlation between anatomical parts. This purported morphological unity was, for Ansted, mired in such gloomy opaqueness that it blotted out even the lucidity of the celebrated law of correlation. In any case, neither Cuvier's nor Owen's methods were compatible with the 'adoption of Mr. Darwin's view of adaptation [of an organism's structure], by a natural method, to existing circumstances' (Ansted 1860 24), on which subject, Ansted noted, 'Mr. Darwin . . . is decidedly at issue with Professor Owen' (1860 33). Rather than directly repudiating the Cuvierian axiom of correlation, Darwin's natural selection had, in Ansted's opinion, simply superseded it, as well as Owen's enigmatic archetypal plan, as a means of understanding anatomical structure.

Significantly, however, while Ansted was willing to acknowledge such reservations regarding correlation in the pages of a highbrow periodical such as the *Dublin University Magazine*, in his more overtly popular writings he continued to acclaim Cuvier's law as if Darwin's *Origin* had never been published. In his regular articles for the newly founded shilling monthly *Temple Bar* he endeavoured to broach current scientific topics 'without going into the intricacies of the great question that now agitates the natural history world, and discussing "the Origin of Species"' (Ansted 1861A 541). This was especially evident when Ansted turned his attention to 'The Pre-Adamite World' only a year after his article on the same theme, rendered more prosaically as 'Palæontology', in the *Dublin University Magazine*. In the animal skeleton, he informed readers of *Temple Bar* in November 1861, a 'particular form and structure of tooth' is suited 'to a particular kind of food' (Ansted 1861B 364), while the 'stomach must also correspond to the teeth . . . so there is a certain mutual fitness of all parts, by means of which any one ought to be sufficient to enable us to describe and picture the whole animal'. Indeed, if even the most fragmentary

> bone . . . is shown to a competent naturalist,—a person whose study it is to find out
> and describe these mutual relations,—he is often able to tell at once to what animal
> it belonged . . . In the hands of the comparative anatomist, nature may be interrogated
> till she replies with complete accuracy.
>
> *(Ansted 1861B 365)*

This was, of course, the oft-repeated claim made by Cuvier in the early decades of the nineteenth century and subsequently by Owen, but now expunged of the doubts and complications that it had accrued steadily since the end of the previous decade and which Ansted himself had discussed at length in only the previous year. And he would recycle exactly the same words two years later in his popular geological treatise *The Great Stone Book of Nature* (1863).[21] Whatever Ansted might say about obscurity, mystery or Darwinian adaptation to the few thousand readers drawn mostly from the Protestant Anglo-Irish community who paid 2*s.* 6*d.* for the *Dublin University Magazine* he was not willing to repeat to the middle-class family audience regularly exceeding 30,000 who bought *Temple Bar*, in its eye-catching purple wrapper, for a single shilling.[22]

Ansted had cultivated a knack of treating scientific subjects as diverse as colour blindness and volcanoes in a lively and accessible way that, at the beginning of the 1860s, was perfectly suited to the new, and highly successful, genre of shilling monthly periodicals, which combined novels, stories and poems with non-fictional essays that steered clear of controversial issues.

His rhetorical ease and store of entertaining anecdotes brought Ansted numerous, remunerative commissions from editors eager to secure what the *Economist* called 'that limited share of public attention which is requisite to keep a monthly shilling magazine afloat' (quoted in Cantor *et al.* 127). In this crowded marketplace, the attention of the public was, inevitably, more likely to be attracted by accounts of astonishing feats of scientific induction that afforded an indubitably accurate picture of prehistoric life than arcane discussions of structural adaptations. The popular appeal of Cuvierian correlation, as opposed to the difficulties and dangers of Darwinism, was certainly appreciated by *Temple Bar*'s editor, the literary journalist, novelist and playwright George Augustus Sala. In one of his own contributions to the magazine, Sala drolly acknowledged that 'I am neither Professor Owen nor a medical student' (1863 56) and the 'study of comparative anatomy seldom leads a man so far as to induce him to convert his pockets into depositories for bones'. The discovery of some 'osseous fragments' in his waistcoat pocket while eating breakfast in bed had nevertheless led him to an exercise of 'domestic paleontology' (Sala 1863 57) through which the identity of the bones was gradually surmised (they turn out to be those of exotic frogs eaten at a dinner hosted by the Acclimatisation Society). As a journalist of long experience, Sala was well aware, as he observed in 1860 soon after assuming *Temple Bar*'s editorship, that the majority of readers wanted only to 'while away an idle hour' (v) and would tolerate 'neither a treatise on Metaphysics, [nor] a dissertation upon the Origin of Species'. The putative clarity of Owen's palaeontological accomplishments was in all ways preferable to the esoteric intricacies of Darwinism.

Ansted was a respected member of the scientific community, having served as vice secretary of the Geological Society of London and editor of its *Quarterly Journal*.[23] Without the inherited wealth enjoyed by the likes of Lyell and Darwin, however, he, and his growing family, remained reliant on the income from his popular journalism, and it made financial sense to continue to endorse the palaeontological procedures Sala evidently found so alluring and conveniently eschew the challenges recently posed to the correlative principle by Darwin's evolutionary approach to organic structure. At the *Cornhill Magazine*, the first and still the most successful of the shilling monthlies, Lewes, prior to taking on the editorship of the *Fortnightly*, had received a lavish twenty-five shillings a page while providing first-hand accounts of his 'very great . . . astonishment' at having seen 'Professor Owen' give a 'precise and confident description' of a 'curious fossil' as the 'third molar of the under-jaw of an extinct species of rhinoceros . . . before even it had quitted' the hands of the collector who had brought it to him (1860 438). But once Lewes began to espouse Darwinism, no matter how equivocally, his remunerative association with the *Cornhill* was terminated by its publisher, George Smith.[24] Like Lewes, Ansted was personally highly committed to the challenging new theories presented in the *Origin* and was quite literally indebted to the book's author, having persuaded Darwin, in the mid-1850s, to invest in a failed business venture.[25] With his contributions to *Temple Bar*, though, the commercial imperatives of the periodical marketplace, and the need to conform to the collective 'codes of discourse' of particular journals, outweighed even Ansted's intellectual allegiance – as well as his financial obligation – to the controversial evolutionist.

Conclusion

James Secord has recently lamented that while 'only naïve readers of *Jane Eyre* or *Pickwick Papers* assume that the author speaks directly through the voice of the narrator, we do just that in reading' nineteenth-century scientific texts, even where 'such works are engaging precisely because they do create a profound sense of a narrative presence' (2014 245–6). Attending to scientific writing in the periodical press, where, no less than literature, much of it was first published, helps enforce precisely the kind of literary-critical engagement with issues such as narrative voice that Secord calls for. It would, after all, be impossible – or at least extraordinarily naïve – to consider Ansted as speaking in a single, consistent authorial voice across his varied, and indeed contradictory, contributions to the *Dublin University Magazine* and *Temple Bar*. Instead, he can be seen to fashion distinct narrative personas attuned to the respective genres, editorial agendas, and implied audiences of the two journals, which enabled him to maintain entirely different outlooks. Such shifting narratorial personas were far from unusual in the nineteenth-century press, especially under the system of anonymous authorship that was inherited from the eighteenth century and lasted until the 1860s, and still survives even today in particular formats such as newspaper editorials. Previously, however, these fluctuating authorial personas have only been acknowledged with reference to literary writers such as Thackeray or, indeed, as I mentioned earlier, Pater.[26] Pater's own impressionistic essays on Renaissance aesthetics come to seem more closely related to science when read in the context of their original publication in the *Fortnightly*, and likewise Ansted's contributions to the *Dublin University Magazine* and *Temple Bar*, when considered alongside each other, assume an indeterminate narratorial presence and sensitivity to generic convention more usually associated with literature. Both, as this chapter has argued, can be brought together as part of a rehabilitated 'common intellectual context' based on the vibrancy and ubiquity of the periodical press in the nineteenth century.

Notes

1 On these new journals, see Barton.
2 For criticisms of Young's thesis, see Topham, Cantor *et al.* 3–5 and Dawson 2007 218–21.
3 On these developments, see Csiszar.
4 See Roos and Baldwin 2012.
5 See Shuttleworth and Taylor xv. Examples in *Mind* include Sully and Goodwin.
6 See Baldwin 2014.
7 On what Young has called the 'curious . . . underground life' of his thesis before its eventual publication in the 1980s, see 161–3.
8 See Cantor *et al.* 147.
9 On the implications of anonymity for how nineteenth-century periodicals conceived of themselves and actually operated, see Liddle and Buurma.
10 Of course, as Joanne Shattock has cautioned, 'the whole business of editing was a far more haphazard and chaotic affair . . . the existence of a periodical was often a precarious one, and the emergence of each new issue an event quite literally brought about more by good luck than by good management' (130), but both Thackeray and Morley seem confident that the aspiration for each number to resemble a concert or symphony could nonetheless be maintained.
11 On the particular significances of publishing 'Lucretius' in *Macmillan's*, see Ledbetter 80–83. There are nevertheless some critical readings of the poem that recognize its engagement with science without acknowledging its periodical publication; see O'Neill and Holmes 252–6.
12 See Brake 1994 xiv on this process.
13 On the scholarly focus on science in periodicals in the last twenty years, see White, Cantor *et al.*, Cantor and Shuttleworth, Henson *et al.*, Mussell, Christie, and Ruston 28–62.
14 See Cantor *et al.* 16–19.
15 See Bullen.
16 On Lewes's organicism, see Shuttleworth 1–23.
17 See Pater 1970 and Inman.
18 For a similar argument regarding Pater's 1885 novel *Marius the Epicurean*, see Dawson 2005.
19 See Secord 2000 394.
20 The history and broader significance of the law of correlation are explored in Dawson 2016.
21 See also Ralph O'Connor, 'Genres of Popular Science', in this volume.
22 On the respective characters of the two journals, see Brake and Demoor 183–4 and 618–19.
23 On Ansted's career, see Brake and Demoor 19.

24 See Cantor *et al.* 130–32.
25 On Ansted's debt to Darwin, see Dawson 2014 95.
26 On Thackeray, see Pearson.

Bibliography

Primary texts

Ansted, David Thomas. *Geology, Introductory, Descriptive and Practical*. 2 vols. London: John Van Voorst, 1844.
Ansted, David Thomas. 'Palæontology'. *Dublin University Magazine* 56 (1860): 20–34.
Ansted, David Thomas. 'Giants and Dwarves'. *Temple Bar* 1 (1861): 533–43 [1861A].
Ansted, David Thomas. 'The Pre-Adamite World'. *Temple Bar* 3 (1861): 363–75 [1861B].
Darwin, Charles. *The Correspondence of Charles Darwin*. Ed. Frederick Burkhardt *et al.* 22 vols. Cambridge: Cambridge University Press, 1985.
Darwin, Charles. *Autobiographies*. Eds Michael Neve and Sharon Messenger. London: Penguin, 2002.
Goodwin, Alfred. '*Marius the Epicurean: His Sensations and Ideas*, by Walter Pater'. *Mind* 10 (1885): 442–7.
Huxley, Thomas Henry. *Evidence as to Man's Place in Nature*. London: Williams and Norgate, 1863.
Huxley, Thomas Henry. 'On the Advisableness of Improving Natural Knowledge'. *Fortnightly Review* 3 (1866): 626–37.
Huxley, Thomas Henry. 'Nature: Aphorisms by Goethe'. *Nature* 1 (1869): 9–11.
Lewes, George Henry. 'Goethe as a Man of Science'. *Westminster Review* 2 n.s. (1852): 479–506.
Lewes, George Henry. 'Studies in Animal Life'. *Cornhill Magazine* 1 (1860): 438–47.
Lewes, George Henry. 'The Heart and the Brain'. *Fortnightly Review* 1 (1865): 66–74.
Lewes, George Henry. 'Mr. Darwin's Hypotheses'. *Fortnightly Review* 4 n.s. (1868): 61–80.
Morley, John. 'Valedictory'. *Fortnightly Review* 32 n.s. (1882): 511–21.
Morley, John. *Recollections*. 2 vols. London: Macmillan, 1917.
Pater, Walter. 'Notes on Leonardo Da Vinci'. *Fortnightly Review* 6 n.s. (1869): 494–508.
Pater, Walter. 'Pico Della Mirandola'. *Fortnightly Review* 10 n.s. (1871): 377–86.
Pater, Walter. *Letters of Walter Pater*. Ed. Lawrence Evans. Oxford: Clarendon, 1970.
Sala, George Augustus. *Make Your Game*. London: Ward and Lock, 1860.
Sala, George Augustus. 'Breakfast in Bed'. *Temple Bar* 9 (1863): 56–66.
Sully, James. 'George Eliot's Art'. *Mind* 6 (1881): 378–94.
Thackeray, William Makepeace. 'A Letter from the Editor to a Friend and Contributor'. *Cornhill Magazine* 1 (1860): x.
Tyndall, John. 'The Constitution of the Universe'. *Fortnightly Review* 3 (1865): 129–44.
Tyndall, John. 'Dust and Disease'. *Fraser's Magazine* 1 n.s. (1870): 302–10.
Tyndall, John. 'On the Scientific Use of the Imagination'. *Fragments of Science for Unscientific People*. 2nd edn. London: Longmans, Green, 1871. 127–67.

Secondary texts

Baldwin, Melinda. 'The Shifting Ground of *Nature*: Establishing an Organ of Scientific Communication in Britain, 1869–1900'. *History of Science* 50 (2012): 125–54.
Baldwin, Melinda. 'The Successors to the X Club? Late Victorian Naturalists and *Nature*, 1869–1900'. *Victorian Scientific Naturalism: Community, Identity, Continuity*. Eds. Gowan Dawson and Bernard Lightman. Chicago, IL: University of Chicago Press, 2014, 288–308.
Barton, Ruth. 'Just Before *Nature*: The Purposes of Science and the Purposes of Popularization in Some English Popular Science Journals of the 1860s'. *Annals of Science* 55 (1998): 1–33.
Beer, Gillian. *Darwin's Plots: Evolutionary Narrative in Darwin, George Eliot and Nineteenth-Century Fiction*. London: Routledge and Kegan Paul, 1983.
Beer, Gillian. 'Parable, Professionalization, and Literary Allusion in Victorian Scientific Writing'. *Open Fields: Science in Cultural Encounter*. Oxford: Clarendon, 1996, 196–215.
Bell-Villada, Gene H. *Art for Art's Sake and Literary Life: How Politics and Markets Helped Shape the Ideology and Culture of Aestheticism 1790–1990*. Lincoln, NE: University of Nebraska Press, 1996.

Brake, Laurel. *Subjugated Knowledges: Journalism, Gender and Literature in the Nineteenth Century*. London: Macmillan, 1994.

Brake, Laurel. 'Writing, Cultural Production, and the Periodical Press in the Nineteenth Century'. *Writing and Victorianism*. Ed. J. B. Bullen. London: Longman, 1997, 54–72.

Brake, Laurel. '"The Trepidation of the Spheres": The Serial and the Book in the Nineteenth Century'. *Print in Transition: Studies in Media and Book History*. Basingstoke: Palgrave Macmillan, 2001, 3–26.

Brake, Laurel, and Marysa Demoor, eds. *Dictionary of Nineteenth-Century Journalism*. London: British Library, 2009.

Buckland, Adelene. *Novel Science: Fiction and the Invention of Nineteenth-Century Geology*. Chicago, IL: University of Chicago Press, 2013.

Bullen, J. B. *The Myth of the Renaissance in Nineteenth-Century Writing*. Oxford: Clarendon, 1994.

Buurma, Rachel Sanger. 'The Anonymous System: Anonymity and Corporate Authority in Nineteenth-Century British Literary Culture'. PhD thesis, University of Pennsylvania, 2005.

Cantor, Geoffrey, Gowan Dawson, Graeme Gooday, Richard Noakes, Sally Shuttleworth and Jonathan R. Topham. *Science in the Nineteenth-Century Periodical: Reading the Magazine of Nature*. Cambridge: Cambridge University Press, 2004.

Cantor, Geoffrey, and Sally Shuttleworth, eds. *Science Serialized: Representations of the Sciences in Nineteenth-Century Periodicals*. Cambridge, MA: MIT Press, 2004.

Christie, William. '*Blackwood's Edinburgh Magazine* in the Scientific Culture of Early Nineteenth-Century Edinburgh'. *Romanticism and 'Blackwood's Magazine': 'An Unprecedented Phenomenon'*. Eds. Robert Morrison and Daniel S. Roberts. Basingstoke: Palgrave Macmillan, 2013, 125–36.

Csiszar, Alex Attila. 'Broken Pieces of Fact: The Scientific Periodical and the Politics of Search in Nineteenth-Century France and Britain'. PhD thesis, Harvard University, 2010.

Dawson, Gowan. 'Walter Pater's *Marius the Epicurean* and the Discourse of Science in *Macmillan's Magazine*: "A Creature of the Nineteenth Century"'. *English Literature in Transition 1880–1920* 48 (2005): 38–54.

Dawson, Gowan. *Darwin, Literature and Victorian Respectability*. Cambridge: Cambridge University Press, 2007.

Dawson, Gowan. 'Darwin Decentred'. *Studies in History and Philosophy of Biological and Biomedical Sciences* 46 (2014): 93–6.

Dawson, Gowan. *Show Me the Bone: Reconstructing Prehistoric Monsters in Nineteenth-Century Britain and America*. Chicago, IL: University of Chicago Press, 2016.

Donoghue, Denis. *Walter Pater: Lover of Strange Souls*. New York, NY: Alfred A. Knopf, 1995.

Donn Vann, J., and Rosemary T. VanArsdel, eds. *Victorian Periodicals and Victorian Society*. Aldershot: Scolar, 1994.

Henson, Louise, Geoffrey Cantor, Gowan Dawson, Richard Noakes, Sally Shuttleworth and Jonathan R. Topham, eds. *Culture and Science in the Nineteenth-Century Media*. Aldershot: Ashgate, 2004.

Holmes, John. *Darwin's Bards: British and American Poetry in the Age of Evolution*. Edinburgh: Edinburgh University Press, 2009.

Inman, Billie Andrew. *Walter Pater's Reading: A Bibliography of His Library Borrowings and Literary References, 1858–1873*. New York, NY: Garland, 1981.

Ledbetter, Kathryn. *Tennyson and Victorian Periodicals: Commodities in Context*. Aldershot: Ashgate, 2007.

Levine, George. 'One Culture: Science and Literature'. *One Culture: Essays in Science and Literature*. Ed. George Levine. Madison, WI: University of Wisconsin Press, 1987, 3–32.

Liddle, Dallas. 'Salesmen, Sportsmen, Mentors: Anonymity and Mid-Victorian Theories of Journalism'. *Victorian Studies* 41 (1997): 31–68.

Mussell, James. *Science, Time and Space in the Late Nineteenth-Century Periodical Press*. Aldershot: Ashgate, 2007.

O'Connor, Ralph. *The Earth on Show: Fossils and the Poetics of Popular Science, 1802–1856*. Chicago, IL: University of Chicago Press, 2007.

O'Neill, Patricia. 'Victorian Lucretius: Tennyson and the Problem of Scientific Romanticism'. *Writing and Victorianism*. Ed. J. B. Bullen. London: Longman, 1997, 104–19.

Pearson, Richard. *W. M. Thackeray and the Mediated Text: Writing for Periodicals in the Mid-Nineteenth Century*. Aldershot: Ashgate, 2000.

Pykett, Lyn. 'Reading the Periodical Press: Text and Context'. *Investigating Victorian Journalism*. Eds. Laurel Brake, Aled Jones, and Lionel Madden. London: Macmillan, 1990, 3–18.

Roos, David A. 'The "Aims and Intentions" of *Nature*'. *Victorian Science and Victorian Values: Literary Perspectives*. Eds. James Paradis and Thomas Postlewait. New Brunswick, NJ: Rutgers University Press, 1985, 159–80.

Ruston, Sharon. *Creating Romanticism: Case Studies in the Literature, Science and Medicine of the 1790s*. Basingstoke: Palgrave Macmillan, 2013.

Secord, James A. *Victorian Sensation: The Extraordinary Publication, Reception, and Secret Authorship of 'Vestiges of the Natural History of Creation'*. Chicago, IL: University of Chicago Press, 2000.

Secord, James A. *Visions of Science: Books and Readers at the Dawn of the Victorian Age*. Oxford: Oxford University Press, 2014.

Shattock, Joanne. 'Editorial Policy and the Quarterlies: The Case of *The North British Review*'. *Victorian Periodicals Newsletter* 10 (1977): 130–39.

Shuttleworth, Sally. *George Eliot and Nineteenth-Century Science: The Make-Believe of a Beginning*. Cambridge: Cambridge University Press, 1984.

Shuttleworth, Sally, and Jenny Bourne Taylor, eds. *Embodied Selves: An Anthology of Psychological Texts 1830–1890*. Oxford: Clarendon, 1998.

Topham, Jonathan R. 'Beyond the "Common Context": The Production and Reading of the Bridgewater Treatises'. *Isis* 89 (1998): 233–62.

White, Paul. 'Cross-Cultural Encounters: The Co-Production of Science and Literature in Mid-Victorian Periodicals'. *Transactions and Encounters: Science and Culture in the Nineteenth Century*. Eds. Roger Luckhurst and Josephine McDonagh. Manchester: Manchester University Press, 2002. 75–95.

Young, Robert M. 'Natural Theology, Victorian Periodicals, and the Fragmentation of a Common Context'. *Darwin's Metaphor: Nature's Place in Victorian Culture*. Cambridge: Cambridge University Press, 1985, 126–63.

13

SCIENCE FOR CHILDREN

Debbie Bark

University of Reading

The story of the history of children's literature is most often said to begin in London in the early 1740s with John Newbery's publication of *A Little Pretty Pocket-Book* (1744).[1] Although books ostensibly for a juvenile audience had been published before this date, Newbery is widely considered to be the first publisher to produce a book specifically to amuse and instruct its young readers. Small in size to rest snugly in the hands of a child, with a brightly coloured cover and pages of poems, proverbs and alphabet-related rhymes, *A Little Pretty Pocket-Book* was immensely popular and opened out the commercial market for books that were designed to be read by (and to) children. Newbery's pioneering work established what would become a flourishing trade in children's books, as writers looked to engage their child readers in a way that stimulated curiosity in the world around them and taught valuable lessons in morality and devotion.

What is particularly interesting about this narrative of the history of children's literature is that from its conception it is intrinsically connected to that of science for children. In 1761 Newbery published what is considered to be the first science book for children, 'Tom Telescope's' *The Newtonian System of Philosophy Adapted to the Capacities of Young Gentlemen and Ladies*, in which the young 'Tom Telescope' presents a series of six lectures on natural philosophy to his friends. The book's conversational format and the connection of scientific concepts to everyday situations proved popular, and the 'Tom Telescope' books were republished across several editions well into the mid-nineteenth century. Newbery's success as a publisher of children's books in no small part helped to establish science writing for children as a vibrant genre during the late eighteenth and early nineteenth century, with the popularity of science-themed books written for children reflecting the status of science as cultural currency at this time. With the dissemination of scientific discourse through public lectures and demonstrations, as well as through the publication of articles and books aimed at a general rather than specialist readership, science could be considered part of everyday experience rather than a distant intellectual activity. Writing for children was an opportunity to educate the next generation in matters scientific, and as science did not appear on school curricula until the end of the nineteenth century this education came from books that were read at home and that were often read for pleasure, necessitating a focus on both instruction and entertainment.

Science for children took a variety of forms and covered an array of scientific disciplines and themes. Natural history was particularly popular, with animals proving to be a useful tool for instruction. In an extension of animal-based narrative traditions such as Aesop's fables and the medieval bestiaries which linked animal characteristics to moral lessons, natural history provided the perfect platform for affirming humanity's place in a world created by a benevolent God. By extension, hierarchical relationships between animals, and between animals and humans, could be used to explain the social order in ways which upheld particular religious or political perspectives. Object narratives, such as those found in Annie Carey's *The Wonders of Common Things: Autobiographies of a Lump of Coal; a Grain of Salt; a Drop of Water; a Bit of Old Iron; a Piece of Flint* (1870), were another popular genre, bringing complex scientific concepts into the home and helping young readers to conceptualize science as something that formed and informed their everyday world. Other books were marketed on the back of successful public lectures for children, such as those given by John Henry Pepper at the Royal Polytechnic Institution,[2] during which dramatic demonstrations, often involving the loud explosions and vivid colours of chemical experimentation, made science both exciting and appealing. The accompanying books, such as *Scientific Amusements for Young People* (1861), provided instructions for experiments that could be conducted at home, making each reader a scientist in their own right. In pointing outwards to the world beyond the nursery, yet keeping its roots firmly in the familiar, science writing for children became an important tool for socialization; a conveyor of cultural and social values that showed a child how to respond to the world around them.

The recent broadening of research to include the large body of popular science literature published during the nineteenth century has brought new audiences into focus for critical discussion, with the value of writing for children being particularly recognized for what it reveals about the way science and scientists were perceived during the nineteenth century. Indeed, rather than being seen as peripheral, popularized forms of science writing are now being recognized for their importance as 'cultural signifiers and cultural forces' (Rauch 2001 6). As James A. Secord notes:

> Children's books deserve an important place in the history of science. Carefully interpreted, they provide invaluable indicators of the changing social, religious, and moral values carried by scientific knowledge in different circumstances. Just as historians now routinely use textbooks to trace the outlines of the contemporary consensus within the scientific community, so at a much more general level can children's books be used to portray the acceptable face of science as seen by much wider social groupings within the book buying public.
>
> *(128)*

In drawing attention to the market forces of the book-buying public, Secord raises an important methodological question, for, as with any retrospective attempt to group and classify texts in generic terms, 'science for children' does need some qualification. First, which 'children' do we mean? The cost of early books, such as those published by Newbery, would have meant that science for children was, at least initially, the domain of the middle and upper classes. Only once the demands of an increasingly literate population had led to the development of the popular printing press would books and periodical publications become in any way accessible to the majority of families, leading to a more inclusive audience. Second, there is the question of how far any book can be said to be *for* children at all when it is written by, published by, marketed by, purchased by and often read to the child by an adult. The inevitable presence of the adult in children's literature means that in studying science for children a far broader and more culturally significant strand of nineteenth-century scientific engagement can be explored. As books written for children were read aloud by adults, shared with children and, in many cases, consulted independently when a workable, accessible explanation of a scientific concept or subject was required, science for children draws in a far wider audience and enables the status of science in the broader population to be assessed. As Greg Myers has observed, popularizations of science 'do not simply transmit or water down the writing of professionals, they transform scientific knowledge as they put it in new textual forms and relate it to other elements of non-scientific culture' (171). Furthermore, recent research has begun to explore how science for children as a textual form was appropriated, often by women writers, to challenge the theories of respected scientists (see Rauch 1997, Murphy). Writing for children meant that young readers were likely to be open to ideas and not yet entrenched in a particular way of thinking; it also meant that a wide audience would be reached in terms of the adult purchasers and sharers of the texts and that a didactic style could be adopted that might have otherwise been resisted.

By drawing out key examples and considering scholarship to date, this chapter looks to trace the changing forms of popular science writing for children across the nineteenth century. Although organized thematically, there is inevitably a chronological impetus as methods of communicating science to children can be seen to change as ideas of childhood change and as frames of reference, particularly relating to religion and to humanity's place in the universe, are adjusted as a result of scientific enquiry over the century. From the earliest books that talk *at* the observing child about science, albeit through constructed dialogue that creates the impression of conversation rather than didacticism, to the child *as* science in later fictional texts that use the position of childhood to reflect on what it means to be human in the great chain of being, the history of science and the history of children's literature are inexorably entwined.

Dialogues and conversations

From the 'Tom Telescope' books onwards, dialogue and conversational narratives were popular methods of communicating scientific information to young readers in a way that entertained and engaged. Rather than a series of dry facts, science was a theatrical performance of knowledge, articulated by a narrator in a position of some authority (often a fictional parent or relative), whose theories and observations were extended and developed through the interjections and well-placed questions of a child. In her study of popular forms of science writing for children in the nineteenth century, Aileen Fyfe suggests that 'conversations were a useful form because they allowed authors to dramatise their works, with action either implied from the dialogue, or described in the accompanying narrative. Instruction was thus conveyed within a story, which helped to make learning fun' (2008 213). The popularity of the dialogue genre peaked in the 1820s. Titles reprinted several times include: Thomas Day's *The History of Sandford and Merton* (1783–89), Maria and Richard Lovell Edgeworth's *Harry and Lucy* (1801), Jeremiah Joyce's *Scientific Dialogues, Intended for the Instruction and Entertainment of Young People* (1805), Samuel Parkes's *The Chemical Catechism* (1808) and Jane Marcet's *Conversations on Chemistry* (1806), *Conversations on Natural Philosophy* (1819) and *Conversations on Vegetable Physiology* (1829).

Striking differences of style emerge when considering individual works. Books using the 'catechism' tag drew on religious conventions of learning by rote, as in George Roberts's *A Catechism of Electricity* (1822) from Pinnock's *Catechism* series, published as cheap, educational works during the 1820s. By assuming positions of ignorance and knowledge the text conveys the facts of science in a step-by-step format:

Q: What is Electricity?
A: Electricity treats of the *phenomena* exhibited by the operations of a very *subtile* [*sic*] fluid, which is one of the principal agents in nature.

Q: Whence is the term Electricity derived?
A: It is derived from the Greek word *electron*, or amber.

Q: Why is it so termed?
A: Because the attractive power which electric bodies acquire by *friction*, was first observed in amber.

(Roberts 3)

The demarcation between adult and child is more explicit in books such as Marcet's *Conversations on Chemistry*, as two children, Emily and Caroline, ask questions of their teacher, Mrs Bryant:

Mrs B: . . . I wish to direct your observations chiefly to the chemical operations of Nature; but those of Art are certainly of too high importance to pass unnoticed. We shall therefore allow them also some share of our attention.
Emily: Well then, let us now set to work regularly. I am very anxious to begin.
Mrs B: The object of chemistry is to obtain a knowledge of the intimate nature of bodies, and of their mutual action on each other. You find therefore, Caroline, that this is no narrow or confined science, which comprehends every thing material within our sphere.
Caroline: On the contrary, it must be inexhaustible; and I am a loss [*sic*] to conceive how any proficiency can be made in a science whose objects are so numerous.

(6–7)

As well as enlivening the narrative, this conversational format implies the active engagement of the child in science. It encourages the child reader to imagine herself asking questions and positions science as a topic to be shared and discussed. The use of two children is a familiar trope in these conversational narratives; one child is often less engaged in the topic, offering the opportunity for moral education and lessons in 'good' and 'appropriate' behaviour. In Marcet's *Conversations* series, Emily asks informal questions while Caroline offers a more considered and reflective perspective.

The role of dialogue narratives in the popularization of science writing for children is considered at length by Myers in 'Science for Women and Children: The Dialogue of Popular Science in the Nineteenth Century'. Using the Edgeworths' *Harry and Lucy* as an early case study, Myers explores theoretical questions raised by the dialogue form and considers how science and scientific authority are represented, particularly as, in its separation of teacher and learner, the dialogue embodies the 'separation of the production and consumption of knowledge' (176). Dialogues introduce a fiction to teach about facts, leading to questions about form, and by providing a narrative these dialogues 'ask what a scientist does, what behaviour is scientific' (Myers 176). Myers points to the constructedness of dialogues in which the 'boundless ignorance of the questioners and equally boundless knowledge of the teachers' (178) is relied upon to sustain the impetus of a developmental narrative that depends 'on the symmetry of questions and answers, on the asking of questions of the right kind in the right order' (179). Myers compares the fictional dialogue in *Harry and Lucy* with actual conversations between a father and son recorded by Maria Edgeworth for inclusion in the appendix to *Essays on Practical Education* (1798), noting that in the recorded conversations the child has 'an independent view of the world that the adults are trying to get him or her to discover' (180), whereas in dialogue narratives such as *Harry and Lucy* there can be no question for which Mother and Father do not have an answer. And there can be no answer, no lesson, that Mother and Father want to teach, that will not be raised by the children's direct experience of the world.

This separation of teacher and learner reveals a problematic contradiction in dialogue narratives: they appear 'to teach learning by experience' (Myers 181) but actually they 'teach learning from authority' as learning occurs in 'a closed world of knowable answers'. Myers concludes:

> If the purpose is to show how scientists work then the form of the dialogue contradicts this intention. But, if the function of popular science is to separate those who practice science from those who just know something about it, then the dialogue form effectively embodies this separation.
>
> *(181)*

Although the popularity of the dialogue form peaked in the 1820s and 1830s before falling out of favour, there is a notable exception: educational readers. In his article 'Reading Science: Images of Science in Some Nineteenth-Century Reading Lesson Books', David Layton surveys the representation of science in educational readers used in England, beginning with those published by the Commissioners of National Education for Ireland, which were widely used in English schools throughout the century. Layton notes that, while early readers were largely information-based, those published in the last decades of the nineteenth century often used dialogue, with fictional adults offering scientific information to children by way of conversation. Whereas other types of writing for children moved away from the dialogue format towards a more naturalistic representation of the child's experience of science, educational texts retained this method of instruction, suggesting a divide between what was considered to be for 'instruction' and for 'delight' as the century progressed. Where the more obviously didactic intentions of educational readers could accommodate the conversational format, the 'general trend in children's literature towards more realistic representations of children, and more complex modes of narration' (Fyfe 2008 213) meant that dialogue narratives fell out of favour. If dialogue was to be used, the questions needed to be less contrived and an attempt made to make the dialogue sound as if it could have come from a small child, as in Eliza W. Payne's *Peeps at Nature* (1850), which aimed, in Payne's words, to follow 'the actual workings of a child's mind' (Preface).

The science of common things: object narratives

Throughout the nineteenth century, and in all its forms, science writing for children was led by the assumption that children are innately curious about the world around them. By the 1840s 'a whole genre developed of the "science of common things" or "things familiar"' (Fyfe 2008 213), which placed this curiosity firmly in the domestic setting to explain scientific concepts through the medium of the everyday. Object narratives were particularly popular, with a range of familiar objects such as candles, feathers, teapots and stones acting as narrators to entertain and inform child readers in matters scientific. Virginia Zimmerman's work on the role of domestic spaces and images in nineteenth-century science writing for children considers the popularity of object narratives which had been used in the eighteenth century to present 'satirical social commentary to an adult audience' (416) but which became 'more earnest in tone' in nineteenth-century science writing for children. By using everyday items to present potentially overwhelming scientific processes, object narratives were a 'vehicle to alleviate anxiety' (Zimmerman 417) as 'those processes were diminished and domesticated'.

Zimmerman's reading of Annie Carey's 'Autobiography of a Lump of Coal', one of several object narratives from the collection *The Wonders of Common Things* (1870), shows how science positioned in the home becomes part of a child's immediate experience. As the story opens, the children are at home, sitting around the fire, calling to mind the settings of oral story-telling traditions. One of the children reaches for a poker to stoke the coals when, to their surprise, 'a deep voice from out of the middle of the coal' (Carey 10) addresses the gathered children: 'as you all seem to wish for a fairy tale, I will, if you please, tell you one which I think is as wonderful as any fairy tale can be, and that is — "my own history"'. Zimmerman suggests that the generic labels of 'history' and 'fairy tale' used by the coal 'articulates the dual purpose of the object narrative as both fantastical and educational' (419), fulfilling the requirement to both instruct and delight child readers: the children gathered by the fireside ask questions that the lump of coal answers, in an adaptation of the dialogue narratives made popular earlier in the century. Zimmerman suggests that although 'the coal's history spans the lifetime of the earth – an unimaginable space of time' (419), its presentation 'in the first person as an individual's life story' renders this history imaginable to the fictional children in the book, and to those reading it. Moreover, by using conversation and setting the narrative on the hearth, 'Carey makes her object narrative domestic in its form, as well as in its setting and imagery' (Zimmerman 420).

Similarly, by using an atom as his narrator in 'The Story of an Atom', from *The Fairy Tales of Science* (1859), John Cargill Brough 'confines a vast expanse of history into the frame of a single life' (Zimmerman 417), which is set 'within the familiar space of the home'. For, although the atom has passed through many forms, he narrates in the appealing and familiar guise of his present-day form: a sugar cube. In Zimmerman's reading, the atom's opening sentences – '"I am an atom of carbon. The members of my family are innumerable, and are disseminated throughout the universe"' (Brough 53) – 'establish the two poles of the narrative: the universe, and the family' (417–18) through which 'elusive and possibly threatening concepts like the universe and deep time' are mitigated by the familiar and reassuring context. In this way, the form of the text enables science to be received and processed from a basis of calm reassurance, rather than unleashing innumerable questions about the precarious nature of humanity's place in the universe. By tracing changes over millions of years, from rock to coal to sugar cube through the perspective of this friendly little atom, potentially overwhelming geological concepts are rendered knowable and unthreatening.

By bridging the gap between science as a professional and intellectual activity and science as part of everyday experience, these object narratives frame science in terms that are familiar and which use the domestic spaces of childhood to encourage curiosity and stimulate discussion. Moreover, they demonstrate the ways in which 'children are very much part of science. They may not contribute to scientific study, but they coexist with the stuff of science' (Zimmerman 421). Another popular children's book which explores this coexistence is *The House I Live In* (1837), edited by Thomas C. Girtin. The title was originally published in America under the name 'Dr Alcott', with Girtin reworking the vocabulary and idioms to suit the British reader, leading to impressive sales over at least ten editions up to 1872. Using the metaphor of the body as a house, the book encourages children to look after their bodies just as one might be concerned with the upkeep of a domestic space:

> 'The House I live in' is a curious building, one of the most curious in the world. Not that it is the largest, or the oldest, or the most beautiful, or the most costly; or that it has the greatest number of rooms, or that it is supplied with the most fashionable

furniture. But it is nevertheless one of the most wonderful buildings in the world, on account of the skill and wisdom of the great Master Workman who planned it.

(Alcott 17)

The individual chapters focus on the house frame (bones), its covering (skin and hair), hinges (joints), windows (eyes) and doors (ears, nose and mouth), and then the apartments and furniture (internal organs, blood, digestive, respiratory and circulatory systems) in order to 'open up the body to those outside the medical profession' (Brock 142). The domestic metaphors help readers connect with the workings of their own bodies, bringing science and the child together as both the observer and the observed. Behind all observations, all experiments and all scientific reasoning in *The House I Live In* stands 'the great Master Workman', for, as Fyfe notes, 'one thing that did not change during the first half of the nineteenth century was that it remained utterly standard for children's authors to present the sciences as the study of God's creation' (2008 214). In contextualizing the omnipresence of God as the agent of science in early nineteenth-century science writing for children, Fyfe suggests that as 'secular, let alone atheistic, science had revolutionary overtones' (214), children, 'supposedly the most delicate members of society', 'had to be carefully protected from dangerous ideas'. Studying the sciences revealed the complexity of the universe and in reinforcing the belief that every aspect had been carefully designed by an omnipotent Creator anxiety could be assuaged. By describing the beauty and abundance of nature laid out at the feet of God's supreme creation – human beings – the emphasis on natural theology in writing for children reassured while simultaneously reinforcing Christian ideals in the minds of the next generation.

Writing natural history

In *The Romance of Victorian Natural History*, Lynn Merrill suggests that natural history was considered to be a particularly appropriate topic for children's literature because 'it is visual, concrete, deals with things on a small scale, and enters other worlds' (39) – all of which were considered to be attractive to children. Natural history writing also 'feature[d] animals readily anthropomorphized or otherwise adaptable as characters', enabling scientific narratives to be constructed as entertaining stories or dialogues, which offered 'ample chances to teach lessons of piety, duty, and hard work'. While studies of plants and minerals were popular, books written about animals were in abundance. Up until the middle of the nineteenth century, natural history was almost exclusively used to convey an understanding of the order of creation, to encourage an appreciation of God's work and to teach children to treat animals kindly. Animals tended to be described one-by-one, and in relation to their usefulness to humankind, rather than organized according to type. 'For the most part,' Harriet Ritvo notes, 'animals were not even important enough to merit a moral judgement unless they somehow influenced human experience' (87). Titles such as *The Natural History of Domestic Animals: Containing an Account of Their Habits and Instincts and of the Services They Render to Man* (1821) make this connection explicit, but, as Ritvo observes in her discussion of the role of animals in science writing for children, the phraseology of virtually every natural history book written in the first half of the nineteenth century suggests that the entire purpose of an animal's existence is either to serve humans, to be eaten or worn by humans or to act as a moral exemplar to humans.

Ritvo goes on to suggest that 'the most important lesson taught by animal books was less directly acknowledged by their authors. This was the lesson about the proper structure of human society' (80), as 'the animal kingdom, with man in his divinely ordained position at its apex, offered a compelling metaphor for the hierarchical human social order' (81). In this way, representations of domestic animals in terms of servitude 'symbolized appropriate and inappropriate relations between human masters and servants' (Ritvo 85), while wild animals that seemed to have no immediate benefit to humans could be cast as alien groups that needed to be defended against. Ritvo cites the example of birds of prey, whose 'carnivorous way of life disposed them to challenge man rather than to serve or flee him: they were rebels who refused to accept his divinely ordained dominance' (87). In natural history books, therefore, birds of prey were often presented as 'both dangerous and depraved, like socially excluded or alien human groups which would not acknowledge the authority of their superiors' (Ritvo 87). Similarly, tigers and wolves were described as ferocious and insatiable, with a fondness for human flesh. 'Man-eating,' notes Ritvo,

> offered a serious lesson as well as an armchair thrill. It provided a graphic and extreme illustration of the consequences that might follow any weakening of the social hierarchy, any diminution of respect and obedience on one side and firmness and authority on the other.
>
> *(89)*

Discussions of natural history writing for children tend to pivot on the publication of Darwin's *On the Origin of Species* in 1859. Of course, debates surrounding evolutionary theory were rumbling long before this date, and the impact of Darwin's work had a gradually pervasive, rather than immediate, impact on writing for children. It is not possible to say that before 1859 all science writing for children operated in the traditions of natural theology, and that afterwards none did; Darwin's theories were far from universally accepted or welcomed and this is reflected in the kinds of science writing for children produced in the 1860s and beyond, as tensions between religious and scientific explanations of the world were debated across society more broadly. Critical analyses of children's books published after 1859 frequently draw on biographical material to interpret responses to evolutionary theory in a particular author's work, leading with an author's religious affiliation, or their responses to Darwin recorded in diaries or letters. The impact of Darwin's work on children's literature more broadly can be seen in the possibility that it opened up for nature and the natural world to be viewed and represented differently. In her discussions of evolutionary theory in 1860s children's fiction, Ruth Murphy suggests that post-Darwinian texts had 'a choice between two visions of nature – nature as a forum for physical, moral and religious progression towards perfection, or nature as a violent, chaotic struggle for life in the face of extinction' (14). These possibilities were variously played out in writing for children in the second half of the nineteenth century, including natural history narratives which started to position humans as part of nature rather than as the superior, divinely ordained beings around which the natural world was organized.

Arabella B. Buckley's *Winners in Life's Race, or, The Great Backboned Family* (1882) exemplifies the shift in emphasis post-Darwin. A sequel to Buckley's earlier work on invertebrates, *Life and Her Children* (1880), this survey of vertebrate life uses accurate zoological terms to mark a move away from descriptions of animals that focus on their relationship with humans in order to convey social and moral lessons. Buckley sets out to 'sketch in broad outline how structure and habit have gone hand in hand in filling every available space with living beings' (vi) and goes on to organize vertebrates according to function and development, rather than creature-by-creature according to their usefulness to man. Significantly, Buckley includes 'active, thinking, tool-making man' (343) in her survey of vertebrates. Although still elevated above the animals with 'powers which made him superior to all around him' (Buckley 344), this marks a departure from earlier narratives of natural history, a move that is inevitable, Ritvo suggests, if Darwinian evolution were to be acknowledged (90). Buckley's writing typifies the complex relationship between scientific and religious interpretations of the natural world at this time, as even this apparent alignment with Darwinian classification is set within a narrative framework that seeks to 'awaken in young minds a sense of the wonderful interweaving of life upon the earth, and a desire to trace out the ever-continuous action of the great Creator in the development of living beings' (viii). This coexistence of traditional narratives of natural theology alongside a tentative acceptance of evolutionary ideas became the pattern for instructional narratives, such as Buckley's, and also for the fantasy-based novels of Charles Kingsley and Lewis Carroll, which heralded a new way of presenting scientific concepts to children during the 1860s and beyond.

Fiction

During the 1860s imaginative fiction for children grew in popularity, replacing the more overtly didactic forms of writing for children which had come before. In critical terms, the decade marked the beginning of what has come to be known as the first 'Golden Age' of children's literature, with Charles Kingsley's *The Water-Babies* (1863) and Lewis Carroll's *Alice's Adventures in Wonderland* (1865) most often cited as initiators of a more child-focused, fantastical form of instruction through delight. Published in the years immediately following Darwin's *On the Origin of Species*, Kingsley's and Carroll's responses to scientific debates of the 1860s have been widely discussed in terms of the relationships between scientific and religious explanations of the natural world played out in their texts, most particularly the position of the human race in a Darwinian narrative of evolution. Recent critical evaluations of *The Water-Babies* and *Alice's Adventures in Wonderland* have begun to consider more closely how these texts function as books written for children, how constructions of the child in the text, and the assumed readership of the text, can reveal anxieties and attitudes towards science, specifically evolutionary theory, in the mid-nineteenth century. As Jessica Straley observes, in texts such as these the child no longer exists outside of nature as God's supreme being observing the world around him or her (as in the narrative of natural theology); instead, children are 'integral parts of an often base and bestial natural world' (586). In both *The Water-Babies* and *Alice's Adventures in Wonderland*, the child protagonist experiences 'an erasure of self that subjects them to a series of interrogations about who and what they are' (Straley 603), rehearsing questions about the nature of humanity through allegories of evolutionary processes. In *The Water-Babies*, Tom falls into the river and 'drowns', taking the form of a newt-like figure before working his way back into human form through a series of moral judgements that lead to progressive self-improvement; in *Alice's Adventures in Wonderland*, Alice tumbles down a rabbit hole and undergoes changes to her physical form as she learns to survive in a chaotic and often violent new world. In Straley's terms, these are texts that 'question the nature of humanity and invite their readers to do the same' (603).

Straley's stimulating analysis of *The Water-Babies* reads the child in the text against Herbert Spencer's use of recapitulation theory as a method of educational reform, showing how Kingsley's fantasy narrative situates the moral cultivation of the child in relation to contemporary evolutionary debates. By drawing attention to narrative constructions of scientific processes that parody empirical learning, Straley shows how Kingsley's text explores 'what it meant to *be* human and, more precisely, what every child needed to learn to *become* human' (585), a question that is 'as much about literature and pedagogy as about evolutionary science'. By 'fusing Tom's natural education with the reader's simultaneous literary one', Straley suggests that Kingsley created 'a place for literary experience as an educational process alongside, or in place of, empirical observation and scientific discovery' (597). In doing so, *The Water-Babies* 'not only made evolutionary theory an essential part of children's literature; it made literature itself the critical step in the child's evolution' (Straley 587). Listing *The Water-Babies* alongside *Alice's Adventures in Wonderland* and other fictional works published for children in the second half of the nineteenth century, Straley notes commonalities of form: 'a blend of natural history and fantasy, scientific accuracy and nonsense, pedagogy and play' (604), which suggest that 'children's fiction must become more fantastic, more parodic and more nonsensical – certainly not more realist – if it is to remain pedagogically effective for the animal child struggling to become human in a scientific age'. In this new way of writing for children, 'linguistic play transformed nature into a liberating allegorical space in which a human nature . . . could be explored and assured' (Straley 603) as children's literature 'took shape as a discourse in which the imaginative romps of fantasy concealed culturally embedded arguments about the location, and re-location, of the human' (603–4).

The relationship between evolutionary theory and literary form is further explored by Murphy, who considers the ways in which children's literature of the 1860s was appropriated to interrogate and discuss changing constructions of nature brought about by the burgeoning interest in evolutionary processes. By reading constructions of the child in *The Water-Babies*, *Alice's Adventures in Wonderland*, and Margaret Gatty's *Parables From Nature*,[3] Murphy suggests that these texts 'address fundamental issues raised by Darwin and evolutionary theory: what is nature and what is the child?' (6) and also 'how should nature be used to educate and understand the child in the post-Darwinian world?'. Indeed, these texts which are 'ostensibly *for* children, are in fact more *about* children' (Murphy 5), educating 'both the child and the adult reader about what childhood and children are in the wake of Darwinian challenges to popular understanding of nature, the child and the role of science-based literature'. Murphy's persuasive readings offer thoughtful reflections on the allegorical potential of the developing child in evolutionary narratives and on the theoretical possibilities opened out by considering constructions of the child as science in nineteenth-century works.

Although children's literature in the latter half of the nineteenth century moved towards fantasy in its engagements with science, it retained aspects of the didactic traditions of science writing that had been so popular early in the century, such as dialogue narratives. Now, though, rather than an authoritative adult responding to the staged, precise questions of the curious child, chaos reigns. Myers's research draws attention to the subversion of the dialogue form in texts such as *Alice's Adventures in Wonderland*, citing the Mad Hatter's riddles without an answer ('"Why is a raven like a writing-desk?"') and Alice's sleepy wonderings ('"Do cats eat bats? Do cats eat bats?" and sometimes, "Do bats eat cats?" for, you see, as she couldn't answer either question, it didn't really much matter which way she put it') as instances where dialogue leads to confusion rather than clarity (195). Myers suggests that like earlier dialogue narratives Carroll's text has a 'plot based on the exploration of the world, questions and answers, a concern with definition of terms, and awareness of roles and proper behaviour, even suggestions on scientific methods' (195), yet in *Alice's Adventures in Wonderland* each element 'is reversed, taken from the point of view of a child, in a world that is not designed for our convenience in reading its meanings'. Through these fictional narratives, which focus on the child in nature and the child *as* nature, the certainties embedded within an observational natural theology that proposed order, purpose and human centrality are called into question.

Directions for future research

Critical studies that consider the intersections of children's literature and science form a relatively new area of research that is bringing to the fore previously unknown or little-known authors and their works. Writers such as Margaret Gatty, whose *Parables From Nature* was published in five volumes between 1855 and 1871 and remained immensely popular throughout the latter half of the nineteenth century, are now being considered alongside more canonical works as critics consider the ways in which children's literature was used as a vehicle for scientific enquiry and debate. As Murphy notes, 'on first reading, the *Parables From Nature* appear to be exactly what the title suggests, that is, a series of short Christian allegories, using animals, plants and personified natural forces to teach moral and religious lessons to children' (8), which 'emphasise proper social behaviour and [to] endorse the power relations of Victorian society by establishing a natural and beneficial hierarchy'. Yet as Murphy, Rauch and others have begun to discuss, Gatty chose the medium of children's literature, rather than more factually based science writing, such as her *British Sea-Weeds* (1863), to present a sustained challenge to Darwin and theories of natural selection in order to 'cultivate a new generation of biologists who would find in her stories the strength to resist the growing materialist explanations' (Rauch 1997 140). Murphy suggests that children's fiction 'provided a space for [Gatty] to be more subversive, even as she authorized the hierarchy which excluded her voice from scientific debate' (9). Moreover, children's literature 'was an acceptable medium in which women could act as authority figures' (Murphy 9) and publicly challenge male authors on matters scientific. In writing 'for an ostensibly young audience that would not yet have taken sides' (Rauch 1997 140), Gatty 'found a way to resist evolutionary and materialist thought' without 'calling attention to either herself or her purpose' (141).

Rauch's research draws on reader response theory to highlight the inherent complexities of audience in writing 'for children' and suggests that *Parables From Nature* reveals 'a very sophisticated sense of "audience" at work' (1997 140). As Rauch observes, Gatty clearly understood 'that by writing for children she gained a set of audiences that would give her work lasting significance' and that by writing for a child the inevitable presence of the adult reader, purchaser or sharer of the text could be similarly reached:

> Thus although her 'parables' were amusing and pleasant to read, Victorian children were also learning from them that God and nature could never be treated separately. At the same time, their parents were being gently persuaded to take sides in a debate that, for many, was very much unresolved in their own minds. Although seldom referred to in 'serious' histories of evolutionary debate, Gatty undoubtedly helped shape popular attitudes about evolution in the latter part of the century.
>
> *(1997 140–1)*

The majority of the texts discussed in this chapter, with the exception of *The Water-Babies* and *Alice's Adventures in Wonderland*, have received little critical attention, either as children's literature or as nineteenth-century science writing, and there are countless others yet to be recovered. Recent projects such as Volume 5 of *Victorian Science and Literature: New Audiences for Science: Women, Children, and Labourers* and the *Science for Children* series are valuable resources which reproduce sections of primary works alongside critical introductions to place the texts in a cultural and literary context. Collections such as these, along with the increasing availability of primary material online, extend the possibilities for research into a far broader range of writers and texts, enabling a more nuanced interpretation of nineteenth-century science writing for children to be formed.

This chapter has considered research relating to dialogue narratives, object narratives, natural history writing and fiction, but there remain other popular genres of science for children to be considered, such as scientific biographies and public lectures, scientific games and toys, and scientific instruments produced specifically for children. Useful work is being done in the area of children's magazine publications, such as Richard Noakes's research tracing scientific content in the *Boy's Own Paper* and other late Victorian juvenile magazines, Claire Brock's discussions of the gendering of scientific discourse in her introduction to the *Girl's Own Paper* articles reproduced in *Victorian Science and Literature* and Jochen Petzold's work looking at the relationship between humans and primates as circulated in juvenile periodicals after Darwin. Research drawing on educational theory, as seen in the articles by Straley and Secord, is another potentially fruitful strand of investigation to explore how various concepts of the developing child are negotiated in science written for children. With the significance of popular forms of science writing now being recognized and new audiences considered, science for children is an expansive and culturally significant area of study. In its appeal to both the child and to the adult who purchases and often reads the text, children's literature offers a unique insight into the communication and reception of science. The development of innovative forms of scientific expression that incorporate both instruction and delight lends richness to discussions of literature and science and at the same time demonstrates the interrelationship between conceptions of science and conceptions of childhood throughout the nineteenth century.

Notes

1   For histories of children's literature see Townsend, Darton, Jackson, and Hunt.
2   See Iwan Rhys Morus, 'Staging Science', in this volume.
3   It is difficult to date precisely Gatty's *Parables From Nature*, although critics tend to agree on 1855 as the date of the first series, with various reprints and editions following. Murphy bases her discussions on the third series (1861–64).

Bibliography

Primary texts

Alcott, William A. *The House I Live In: Or, Popular Illustrations of the Structure and Function of the Human Body. For the Use of Families and Schools*. Ed. Thomas C. Girtin. London: John W. Parker, 1837.

Brough, John Cargill. *The Fairy Tales of Science: A Book for Youth*. London: Griffith and Farran, 1859.

Buckley, Arabella B. *Winners in Life's Race, or, The Great Backboned Family*. London: Edward Stanford, 1882.

Carey, Annie. *The Wonders of Common Things: Autobiographies of a Lump of Coal; a Grain of Salt; a Drop of Water; a Bit of Old Iron; a Piece of Flint*. London: Cassell, Petter, and Galpin, 1870.

Carroll, Lewis. *Alice's Adventures in Wonderland*. London: Macmillan, 1865.

Darwin, Charles. *On the Origin of Species by Means of Natural Selection, or the Preservation of Favoured Races in the Struggle for Life*. London: J. Murray, 1859.

Day, Thomas. *The History of Sandford and Merton*. London: Darton, 1783.

Edgeworth, Maria, and Richard Lovell Edgeworth. *Essays on Practical Education*. London: J. Johnson, 1798.

Edgeworth, Maria, and Richard Lovell Edgeworth. *Harry and Lucy: Part I. Being the First Part of Early Lessons*. London: J. Johnson, 1801.

Gatty, Margaret. *British Sea-Weeds* [1863]. London: Bell and Daldy, 1872.

Gatty, Margaret. *Parables From Nature*. 5 vols [1855–71]. London: J. M. Dent, 1907.

Joyce, Jeremiah. *Scientific Dialogues: Intended for the Instruction and Entertainment of Young People: In Which the First Principles of Natural and Experimental Philosophy are Fully Explained*. London: J. Johnson, 1805.

Kingsley, Charles. *The Water-Babies: A Fairy Tale for a Land Baby*. London: Macmillan, 1863.

Marcet, Jane. *Conversations on Chemistry: In Which the Elements of That Science are Familiarly Explained and Illustrated by Experiments*. London: Longman, Hurst, Rees, and Orme, 1806.

Marcet, Jane. *Conversations on Natural Philosophy, by the Author of Conversations on Chemistry*. London: [n. p.], 1819.

Marcet, Jane. *Conversations on Vegetable Physiology, by the Author of Conversations on Chemistry*. 2 vols. London: [n. p.], 1829.

Newbery, John. *A Little Pretty Pocket-Book.* London: J. Newbery, 1744.

Parkes, Samuel. *The Chemical Catechism: With Notes, Illustrations, and Experiments.* 3rd edn. London: [n. p.], 1808.

Payne, Eliza W. *Peeps at Nature: Or, God's Works and Man's Wants.* London: Religious Tract Society, 1850.

Pepper, John Henry. *Scientific Amusements for Young People.* London: Routledge, Warne, and Routledge, 1861.

Roberts, George. *A Catechism of Electricity: Being a Short Introduction to That Science, Written in Easy and Familiar Language. Intended for the Use of Young People.* London: G. and W. B. Whittaker, 1822.

'Telescope, Tom'. *The Newtonian System of Philosophy Adapted to the Capacities of Young Gentlemen and Ladies, and Familiarized and Made Entertaining by Objects with Which They are Intimately Acquainted: Being the Substance of Six Lectures Read to the Lilliputian Society.* London: J. Newbery, 1761.

Secondary texts

Brock, Claire, ed. *New Audiences for Science: Women, Children, and Labourers.* Vol. 5 of *Victorian Literature and Science.* Gen. eds. Gowan Dawson and Bernard Lightman. 8 vols. London: Pickering & Chatto, 2011–12.

Darton, F. J. Harvey. *Children's Books in England: Five Centuries of Social Life.* 3rd edn, rev. Brian Alderson. Cambridge: Cambridge University Press, 1982.

Fyfe, Aileen, ed. *Science for Children.* 7 vols. Bristol: Thoemmes, 2003.

Fyfe, Aileen. 'Tracts, Classics and Brands: Science for Children in the Nineteenth Century'. *Popular Children's Literature in Britain.* Eds. Julia Briggs, Dennis Butts, and M. O. Grenby. Aldershot: Ashgate, 2008, 209–28.

Hunt, Peter. *An Introduction to Children's Literature.* Oxford: Oxford University Press, 1994.

Jackson, Mary V. *Engines of Instruction, Mischief and Magic: Children's Literature in England From its Beginnings to 1839.* Aldershot: Scolar, 1989.

Layton, David. 'Reading Science: Images of Science in Some Nineteenth-Century Reading Lesson Books'. *Paradigm* 10 (1993). <http://faculty.education.illinois.edu/westbury/paradigm/Layton.html>. Accessed 21 March 2014.

Merrill, Lynn. *The Romance of Victorian Natural History.* Oxford: Oxford University Press, 1989.

Murphy, Ruth. 'Darwin and 1860s Children's Literature: Belief, Myth or Detritus'. *Journal of Literature and Science* 5.2 (2012): 5–21.

Myers, Greg. 'Science for Women and Children: The Dialogue of Popular Science in the Nineteenth Century'. *Nature Transfigured: Science and Literature, 1700–1900.* Eds. John Christie and Sally Shuttleworth. Manchester: Manchester University Press, 1989, 171–200.

Noakes, Richard. 'The *Boy's Own Paper* and Late-Victorian Juvenile Magazines'. *Science in the Nineteenth-Century Periodical: Reading the Magazine of Nature.* Eds. Geoffrey Cantor *et al.* Cambridge: Cambridge University Press, 2004, 151–71.

Petzold, Jochen. '"How Like us is That Ugly Brute, the Ape!": Darwin's "Ape Theory" and its Traces in Victorian Children's Magazines'. *Reflecting on Darwin.* Eds. Eckart Voigts, Barbara Schaff, and Monika Pietrzak-Franger. Farnham: Ashgate, 2014, 57–71.

Rauch, Alan. 'Parables and Parodies: Margaret Gatty's Audiences in the *Parables From Nature*'. *Children's Literature* 25 (1997): 137–52.

Rauch, Alan. *Useful Knowledge: Victorians, Morality, and the March of Intellect.* Durham, NC: Duke University Press, 2001.

Ritvo, Harriet. 'Learning From Animals: Natural History for Children in the Eighteenth and Nineteenth Centuries'. *Children's Literature* 13 (1985): 72–93.

Secord, James A. 'Newton in the Nursery: Tom Telescope and the Philosophy of Tops and Balls, 1761–1838'. *History of Science* 23 (1985): 127–51.

Straley, Jessica. 'Of Beasts and Boys: Kingsley, Spencer, and the Theory of Recapitulation'. *Victorian Studies* 49.4 (2007): 583–609.

Townsend, John Rowe. *Written for Children: An Outline of English-Language Children's Literature.* London: Garnet Miller, 1965.

Zimmerman, Virginia. 'Natural History on Blocks, in Bodies, and on the Hearth: Juvenile Science Literature and Games, 1850–1875'. *Configurations: A Journal of Literature, Science and Technology* 19.3 (2011): 401–30.

Further reading

Cosslett, Tess. '"Animals Under Man"? Margaret Gatty's *Parables from Nature*'. *Women's Writing* 10.1 (2003): 137–52.

Fyfe, Aileen. 'Young Readers and the Sciences'. *Books and the Sciences in History*. Eds. Marina Frasca-Spada and Nick Jardine. Cambridge: Cambridge University Press, 2000, 276–90.

Harper, Lilia Marz. 'Children's Literature, Science and Faith: *The Water-Babies*'. *Children's Literature: New Approaches*. Ed. Karin Lesnik-Oberstein. Basingstoke: Palgrave Macmillan, 2004, 118–43.

Keene, Melanie. 'Domestic Science: Making Chemistry Your Cup of Tea'. *Endeavor* 32.1 (2008): 16–19.

Layton, David. *Science for the People: The Origins of the School Science Curriculum in England*. London: George Allen and Unwin, 1973.

Lightman, Bernard. *Victorian Popularizers of Science: Designing Nature for New Audiences*. Chicago, IL: University of Chicago Press, 2007.

Lovell-Smith, Rose. 'Eggs and Serpents: Natural History Reference in Lewis Carroll's Scene of Alice and the Pigeon'. *Children's Literature* 53 (2007): 27–53.

MacLeod, Anne Scott. 'From Rational to Romantic: The Children of Children's Literature in the Nineteenth Century'. *Poetics Today* 13 (1992): 141–53.

Rauch, Alan. 'A World of Faith on a Foundation of Science: Science and Religion in British Children's Literature: 1761–1878'. *Children's Literature Quarterly* 14.1 (1989): 13–19.

Rauch, Alan. 'Mentoria: Women, Children, and the Structures of Science'. *Nineteenth-Century Contexts* 27 (2005): 335–51.

Shuttleworth, Sally. *The Mind of the Child: Child Development in Literature, Science, and Medicine, 1840–1900*. Oxford: Oxford University Press, 2010.

14

STAGING SCIENCE

Iwan Rhys Morus

Aberystwyth University

On the day before Christmas 1862, Professor John Henry Pepper delivered his 'Strange Lecture' at London's Royal Polytechnic Institution as part of the annual seasonal celebrations. The Polytechnic had a reputation for spectacular Christmas performances and this one was no exception. Pepper's lecture, according to the *Times*, consisted of 'a series of the most wonderful optical delusions ever placed before the public' and culminated with a performance 'intended to illustrate Charles Dickens's idea of the haunted man'. It was a spectacular and baffling performance, in which the

> spectres and illusions are thrown upon the stage in such a perfect embodiment of real substance that it is not till the haunted man walks through their apparently solid forms that the audience can believe in their being optical illusions at all.

Even as the audience knew they were being fooled,

> it is almost difficult to imagine that the whole is not a *wonderful* trick, for people cling hard to the old saying that 'seeing is believing', and if ever mere optical delusions assumed a perfect and tangible form, they do so in this 'strange lecture'
>
> *('Royal Polytechnic Institution')*

By the beginning of March, according to the *Daily News* at least, more than 100,000 people had passed through the Polytechnic's doors to marvel at this amazing display of spectres ('Drama').

Pepper's performances at the Polytechnic in December 1862 offer a useful way in to starting to understand some of the confluences of science and theatre during the nineteenth century – a set of relationships that is now coming under increasing scrutiny, as well as discussions of the deployment of scientific ideas on stage (Watt Smith; Lightman 2007; Goodall).[1] The performance that so impressed the *Times* correspondent was the first public appearance of the illusion that came to be known as Pepper's ghost, and versions of it soon became a standard component of Victorian stage illusion. The Polytechnic, where the performance took place, was by the 1860s London's main site of popular scientific performance (Brooker; Weeden; Altick). That raising ghosts might be regarded as a quintessentially scientific activity might seem peculiar to modern sensibilities, but the scientific status of Pepper's ghost was not in doubt for Victorian audiences. Neither was the assumption that such scientific performances might be listed along with comic plays, magic lantern shows, operas and pantomimes as popular Christmas entertainment.[2] The ghost shows how Victorian science and Victorian theatre intersected in important ways – not least in the spaces and technologies they shared. It shows too how, on scientific stages, like theatrical stages, a well-honed and polished performance was the key to generating authority. The spectre drama needed careful choreography to make the illusion convincing and make sure that the audience did not see through the illusion. Scientific performances, like any other piece of theatre, were the product of hard work and careful preparation.

Scrutinizing scientific performances has become an increasingly important focus of research in the history of science and Victorian performances have attracted particular attention in this respect (Morus 2010A; Morus 2010B). Historians are now interested in understanding science as a set of practices rather than simply as a body of knowledge. As a result they have started to pay more attention to the places where scientific activities are undertaken, to the skills and technologies of knowledge production and circulation and to the different embodiments of scientific knowledge (Golinski 1998). Looking at scientific performances offers new ways of understanding the routes through which practitioners of science make their work public, generate trust and represent authority. Looking at science as performance highlights the ambiguities of knowledge-creation as well. While putting on a good show has often been an essential part of the process of scientific persuasion, it is also open to charges of imposture. The accusation that all they are doing is pulling the wool over audiences' eyes through artful manipulation is one that has often been levelled at scientific performers, after all (Schaffer; Golinski 1992; Morus 2009). This offers another good reason why historians should pay close attention to these sorts of performances and the work that goes into them. They offer windows onto the strategies that performers and their audiences adopt to discriminate between fact and artefact, illusion and reality.

Scientific performers throughout the nineteenth century were themselves quite self-conscious about the ambiguities of performance – and unsurprisingly so given the highly contested nature of scientific authority throughout the nineteenth century. For much of the century, the cultural space that men of science tried to carve out for themselves was both porous and precarious (Lightman 1997). The performers inhabiting this ambiguous space knew that their performances were key to fashioning themselves as men of science. Inhabiting as they did a public culture where appearance and the proper management of the body and its gestures were taken to be important indicators of moral value, Victorian performers were keenly aware that what they did on stage was central to forging the right kind of relationship between themselves and their audiences (Lightman 2007). Such scientific performers were also well aware of the slippage between the appeal to reason and the appeal to sensation that was built into their performances. Victorian scientific performances across the spectrum were calculated to generate wonder. Playing with illusions and challenging audiences to discriminate between fact and fantasy were common elements in the repertoire. The point of such performance was to generate matters of fact. Ironically, of course – and as seasoned scientific performers understood very well – the business of revealing truth on stage often seemed indistinguishable from (and certainly shared the same techniques and technologies as) the business of illusion.

This chapter will take a closer look at Pepper's 'Strange Lecture', the work that went into its performance and the place that it occupied on the spectrum of mid-Victorian spectacles, scientific and otherwise. Looking at Pepper's performance, along with those of others such as Michael Faraday and his successor John Tyndall at the Royal Institution, should help underscore the artful nature of Victorian public science. They were choreographed – and deliberately so – to place science in particular ways before their publics. One important consequence of the turn towards investigating practice in the history of science has been a concern with knowledge's embodiments and the bodies of its performers (Lawrence and Shapin). Performers' bodies were integral to the production of convincing spectacle. Other bodies, dead and alive, often had parts to play in mounting scientific spectacle as well. Putting bodies on display offered a strategy for instructing audiences in the art of reading from the body as well as for managing and familiarizing difference.

The choreography of knowledge

It is worth looking at Pepper's 'Strange Lecture' in some detail to get a sense of the work that went into its performance and of what the audience saw. The lecture started with the announcement to the audience that it would 'instruct and entertain you this day with some curious experiments, which are intended to cheat and deceive your vision' (Pepper 1863 3). Pepper then guided his audience through a sequence of experiments designed to illustrate the phenomenon of persistence of vision, such as Geissler tubes arranged on a wheel and illuminated by a battery and induction coil. He demonstrated John Ayrton Paris's thaumatrope, Joseph Plateau's phenakistoscope, and the kalotrope, an apparatus designed 'for showing the illusions of the phenakistiscope and kindred devices to a numerous audience' (Pepper 1881 374). This section of the performance climaxed with a display of Thomas Rose's photodrome, similarly designed with a view to demonstrating persistence of vision effects (or spectres) to a large audience.[3] These were all devices that produced the illusion of movement by presenting the eye with a rapid succession of images. Pepper then showed off a variety of experiments for producing ghosts and spectres by means of reflection, culminating in the performance of Dickens's *The Haunted Man*,

> consisting of a large and small plate of glass, in which the illuminated images of all kinds of objects, dead or living are reflected, and which instantly become invisible by cutting off the light which is used to illuminate them.
>
> *(1863 5)*

What the audience saw on stage during the spectre drama was an apparent scene of masculine domesticity. A man sat in his study, deep in thought, only to be disturbed by the sudden appearance of a ghost to disturb his reveries. Victorian audiences were used to ghosts, of course, but the ghost on the Polytechnic's stage seemed capable of a whole new repertoire of behaviour (Luckhurst and Morin; McCorristine). It could appear and disappear at will. Apparently solid objects could pass through it. The effect had originally been developed by Henry Dircks and adapted by Pepper for use in the Royal Polytechnic Institution's lecture theatre. To make the illusion work, a large sheet of plate glass was suspended at an angle at the front of the stage between the audience and the performers. The actor actually playing the ghost was in the pit, out of sight of the audience. When he was illuminated by a bright light from a magic lantern his reflection appeared on the plate glass, looking from the audience's perspective as if it were on the stage itself. The ghost could be made to appear and disappear simply by uncovering and covering the light. The effect depended on a combination of a clever, if simple, piece of optical trickery, along with very careful choreography to make sure that the actors on stage and in the pit moved appropriately with regard to each other (Pepper 1890; Lamb; Steinmeyer).

The play was based, loosely, on Charles Dickens's Christmas story *The Haunted Man*, first published in 1848. It was, presumably, a familiar story to Pepper's audience at the Royal Polytechnic. Both the original story and Pepper's performance played with the Victorian public's fascination with ghosts. Both story and performance also played with the role that ghosts had in contemporary discussions about the nature of knowledge. At least some astute readers of Dickens's stories (and certainly their author) would have been aware of these sorts of epistemological ramifications (Morus 2006). Pepper's audience at his 'Strange Lecture' would also have been familiar with these concerns, not least because they formed the basis for the lecture's narrative. The spectre drama was a play within a performance that was aimed at inviting its audience to worry about the evidence of their eyes. Its audience was being offered a story about how to understand the relationship between sensation and authority. Seeing a ghost on stage was both a challenge to decipher the mechanism behind its appearance and an invitation to be seduced by the verisimilitude of the sensation. Victorian scientific performers quite deliberately and self-consciously played with this not-quite-paradox between contrived performance and the revelation of truth in order to bolster the appearance of authority.[4]

More broadly, scientific performances at places like the Royal Polytechnic Institution, where Pepper's ghost drama was staged, occupied a strategic place in a repertoire of display designed to elicit wonder. Dramatist and critic Edmund Yates listed both the Adelaide Gallery, featuring 'Perkins's steam-gun, which discharged a shower of bullets' (1:135), and the Royal Polytechnic, 'with its diving-bell, the descent in which was so pleasantly productive of imminent head-splitting' (1:136), in his list of youthful amusements, which ranged from Cremorne and Vauxhall Gardens through city theatres to a collection of dubious coffee-houses. His recollection shifted seamlessly from 'lectures, in which Professor Bachhoffner was always exhibiting chemistry to "the tyro"', to 'dissolving views of "A Ship", afterwards "on fi-er", and an illustration of—as explained by the unseen chorus—"The Hall of Waters—at Constant—nopull—where an unfort—nate Englishman—lost his life"'. To their audiences, there was a sense in which all these performances looked the same. The chemistry lectures, or the displays of electric shocks and sparks, required the same kind of careful practice and preparation – the same sort of behind the scenes work – as the magic lantern dissolving views or the theatrical productions. Whether what was on show was to be understood as nature or as illusion (or sometimes both), the performances belonged to similar networks of production and distribution.

Michael Faraday's early attempts to refashion himself as a lecturer fitted to perform for a cultured metropolitan elite at the Royal Institution offer an interesting example of the work required for successful performance, as well as underlining that a concern with appearance was not restricted to performers in the more commercially-minded scientific spaces (Gooding and James). Faraday took elocution lessons from Benjamin Smart, at the time one of London's most sought-after teachers. Smart's *Practice of Elocution* (1819) offered its readers a guide through the different degrees of effective public performance, emphasizing the gestural as much as the vocal in evoking the 'notion or feeling of *laying*, as it were, his facts or truths *before* his auditors'. Smart hammered home the repetitive work needed to turn artificial gesture into natural expression. Experimental work behind the scenes was as important as bodily preparation in this respect. Faraday and his assistants laboured hard in the Royal Institution's basement laboratory to make sure that experiments worked seamlessly before they were carried upstairs into the lecture theatre (Gooding). Faraday's rival performer William Sturgeon emphasized the importance of visibility and the care needed to make sure that the audience could see properly the sometimes fleeting and liminal phenomena that lecturers laboured to elicit from their apparatus. Lecturers competed to produce – and audiences expected to see – spectacle.

Faraday and John Tyndall, his successor on stage at the Royal Institution, both recognized this. Following his discovery of electromagnetic induction in the late summer of 1831, one of Faraday's main experimental preoccupations was to turn the phenomenon into a reliable on-stage performance. Writing to the *Philosophical Magazine* about his experiments – and at least in part in an effort to assert his priority in discovery over a pair of Italian rivals – Faraday emphasized the reliability of the experimental arrangement he had developed by describing how he had 'shown it brilliantly to two or three hundred persons at once, and over all parts of the theatre of the Royal Institution' (1832 413). The experimental work of discovery and of demonstration could sometimes appear interchangeable. In this respect, public performance was knowledge production. The same can be said of Tyndall's experiments with sensitive flames thirty-six years later. The discovery of flames' responsiveness to acoustic vibrations was the outcome of rehearsal for a key scientific performance. As William Barrett, then Tyndall's assistant, described it,

> while preparing the experiments for one of the Christmas lectures at the Royal Institution, I noticed that the higher harmonics of a brass plate (which I was sounding with a violin-bow in order to obtain Chladni's figures) had a remarkable effect on a tall and slender gas-flame that happened to be burning near. At the sound of any shrill note the flame shrank down several inches, at the same time spreading out sideways into a flat flame, which gave an increased amount of light from the more perfect combustion of the gas.
>
> *(216)*

The way in which that phenomenon was then transformed into a piece of scientific stage theatre emphasizes the self-consciousness of mid-Victorian natural philosophers' performances. Tyndall's favoured technique for putting the 'most marvellous flame hitherto discovered' through its paces in front of the Royal Institution's audience was to read a passage from Spenser's *The Faerie Queene*.[5] His audience could see and hear how the 'flame picks out certain sounds from my utterance; it notices some by the slightest nod, to others it bows more distinctly, to some its obeisance is very profound, while to many sounds it turns an entirely deaf ear' (Tyndall 1867 241). One newspaper account described the flames as 'the most extraordinary and apparently magical things', which responded 'in a striking way' to 'the creaking of boots, the rustle of a lady's dress, the patter of rain, and the tick of a watch' ('Curiosities of Sound' 48). Witnesses saw how, like

> a living being, the flame trembles and cowers down at a hiss – it crouches and shivers as if in agony at the crisping of this metal foil, though the sound is so faint as scarcely to be heard; it dances in tune to the waltz played by this musical box – and, finally, it beats time to the ticking of my watch.
>
> *('On Musical and Sensitive Flames' 8)*

Performances like this one illustrate the careful preparation and self-conscious choreography that made for a successful stage presence. Tyndall's own persona, as much as the phenomenon on show, was the product of careful experiment and meticulous preparation. His lectures were expected to be the product of his own research and to be accompanied by experiments, as well as accessible to the institution's well-heeled audiences. His efforts were not always successful, in his early years at least. The Royal Institution's secretary, John Barlow, admitted to Faraday that 'I dread the tendency of Tyndall's Lectures to become abstruse' (Faraday 1991–2012 5:543). Tyndall had Faraday's model before him – and it was clearly a model that he took seriously and aimed to emulate. He noted how 'the habit of self-control became a second nature to him' (Tyndall 1879 2:452). One of the most striking features of contemporary accounts of Tyndall's sensitive flames was the pervasive anthropomorphism. The flame's response to sound was 'like that of a sensitive, nervous person uneasily starting and twitching at every little noise' (Barrett 219). Like the ghost, the flame was another actor on stage with the lecturer, and its presence and performance there was designed to draw attention to the presence and performance of the lecturer on the stage.

Bodies on display

In 1818 the chemist (and later author of the *Philosophy of Manufactures*) Andrew Ure published an account of electrical experiments he had carried out in Glasgow on the body of Matthew Clydesdale following his execution (Farrar). He was following the example of the Italian experimenter Giovanni Aldini who had carried out experiments on the corpse of George Forster in London about fifteen years earlier. Ure's experiments were clearly conducted with sensation in mind. Ure's account of them dwelt on the dead body's responses to electrical stimulation in ways that seemed calculated to appeal to his readers' taste for the fantastic. Every 'muscle in the body was immediately agitated with convulsive movements, resembling a violent shuddering from cold' (Ure 1818 289). Later, 'the leg was thrown out with such violence, as nearly to overturn one of the assistants, who in vain attempted to prevent its extension'. With appropriate modifications to the apparatus, full, 'nay laborious breathing, instantly commenced. The chest heaved, and fell; the belly was protruded, and again collapsed, with the relaxing and retiring diaphragm' (Ure 1818 290). Clydesdale's corpse was being used as a prop in a macabre performance designed to persuade the audience of the reality of Ure's claim that electricity could revive the dead. It was startling enough a performance that 'several of the spectators were forced to leave the apartment, and one gentleman fainted'. Embellished accounts (some even claiming that Clydesdale had indeed been revivified) were still being circulated a generation later.

This was a phantasmagoria with a real body – and phantasmagoria performances themselves were shows that quite deliberately and knowingly toyed with the confluence of science and titillation. The language Ure used to convey the atmosphere of his experimental play with Clydesdale is in this respect quite revealing. When 'every muscle in his countenance was simultaneously thrown into fearful action; rage, horror, despair, anguish, and ghastly smiles, united their hideous expression in the murderer's face', the effect produced outdid 'the wildest representations of a Fuseli or a Kean' (Ure 1818 290). It is difficult to read this as anything other than an account of a piece of theatre, and one deliberately designed to appeal to contemporary readers' own senses of what theatre was. Edmund Kean was celebrated as one of the leading actors of the period, particularly known for portrayals of Shakespearean villains. Henry Fuseli was famous for his grotesque and sexually charged paintings of contorted bodies as well as being a contributor to John Boydell's Shakespeare Gallery in London. Their performances on stage and canvas arranged bodies for spectacle, and Ure was trying to deploy Clydesdale in the same sort of way. His experiments provided him, as an ambitious philosophical performer trying to forge a career for himself (he would soon abandon Glasgow for London), with flesh-and-blood props to bolster his own stage presence.

Ure's example shows how bodies on stage could be used to frame authority. Clydesdale in this performance was an automaton, just as Ure would portray factory workers conditioned to work to the rhythms of machinery in his paean to the factory system fifteen years later (Ure 1835). His body (like their bodies) was meant to perform in such a way as to make authority tangible. Performers' own bodies could be deployed like this too. Faraday, as we have seen, worked hard at turning his body into a philosophical instrument on stage and John Tyndall noted discipline as a key feature of his mentor's performances (Tyndall 1868). Faraday's success as a purveyor of polite knowledge to the metropolitan upper classes was a product of his ability to use his own body to direct the audience's engagement. As one auditor put it,

> It was an irresistible eloquence, which compelled attention and insisted upon sympathy . . . A gleaming in his eyes which no painter could copy, and which no poet could describe . . . His body then took motion from his mind; his hair streamed out from his head; his hands were full of nervous action; his light, lithe body seemed to quiver with its eager life. His audience took fire with him, and every face was flushed.
>
> *(Pollock 302)*

What Lady Pollock described was an act calculated to embody the appeal to sensation.

Tyndall also clearly worked hard at the business of performance. The kind of dextrous interplay between the stage philosopher and his props described in accounts of his lectures on sensitive flames needed careful preparation. Careful choreography was essential whenever performers' bodies were integral to the production of scientific spectacle. The original spectre drama and the illusions that succeeded it at the Royal Polytechnic during the 1860s and 1870s needed to be seamless if they were to fool their audiences. They featured bodies appearing from nowhere or transforming from one form to another and depended on careful stage arrangement. Performances such as the physioscope in which Thomas Tobin, Pepper's assistant and fellow inventor of optical illusions, had his head magnified on screen to drink a glass of wine and wink at the audience, needed substantial physical stamina as well (Pepper 1881). Victorian middle-class audiences – in and out of theatres and lecture halls – were very good at reading bodies and recognizing the significance of how bodies were arranged. Spectacles in which bodies appeared mutable – a young woman turning into an old man in Pepper's metempsychosis illusion, for example – toyed with such audiences' expectations that bodies be conformable. By making their own bodies part of the illusion, their audiences were being challenged to decipher, performers such as Pepper displayed their own capacity for manipulation at its most challenging.

One of the reasons that these kinds of bodily spectacles are so revealing is that they make explicit the extent to which stage theatre and stage science shared in the same material and gestural culture of performance. They drew on the conventions of science and of showmanship. It seems clear that contemporary audiences understood this too. Just as with other theatrical performances, it is worth considering the identity of stage scientific performers. Historians of science are used to talking about the invisible technicians whose hidden labour make experiments work (Shapin). Stage performances depend on similarly hidden performers too. Without Sergeant Anderson, William Barrett or Thomas Tobin, Faraday's, Tyndall's or Pepper's stage identities would not have been able to carry out their performances. But at the same time, their success as performers depended on keeping their assistants invisible, even when they were physically present on the stage (as illustrations show Anderson to have been at the Royal Institution, for example).[6] Rendering others' bodies docile was often a prerequisite of performances involving others' on-stage presence. Clydesdale's body was a good stage prop for Ure's experiments precisely since it could only act as Ure directed it. When supposedly docile bodies took control of scientific performances, the result could often be the undermining of the experimenter's agency and their authority as arbiter of the demonstration's outcome.

John Elliotson's mesmeric performances with Elizabeth O'Key provide a nice example in this respect (Winter). As a radical supporter of the new science of mesmerism during the late 1830s, Elliotson took advantage of his position as professor at the London University and physician to the University College Hospital to use the hospital's patients in public demonstrations of mesmerism. One of these patients, Elizabeth O'Key, proved to be a remarkably adept and versatile mesmeric subject. She carried out diagnoses of other patients' bodily states, identified concealed objects and carried out sometimes flirtatious conversations with Elliotson while under the mesmeric influence. Her mesmeric act with Elliotson brought the question of agency into sharp focus, as Elliotson's many critics made clear. It looked to many as if O'Key, rather than Elliotson, was the main performer. Similar questions about the identity of the performer and the distribution of performative authority could be raised in other instances where living bodies were constitutive elements of scientific theatre. These were problems of agency. Bodies such as Joseph Merrick, the Elephant Man, for example, were not docile but often rather active participants in the creation of scientific spectacle around themselves. Performances with bodies depended for their success on the maintenance and display of power between performers (Durbach; Qureshi).

Performers' bodies and the relationships between them were there to be read by their audiences. Strategies for performing might shift throughout the nineteenth century as conventions and expectations changed but audiences still understood that the ways bodies moved and gestured were meant to tell them something about the person of the performer. Victorian theatre increasingly leaned towards realism, with actors trying to mimic nature rather than depend on stylized gestures. Performers were expected to act naturally. Stage actors studied physiognomy in their efforts to develop techniques of self-expression that conveyed the reality of their performances to their audiences. Physiognomists such as Lavater offered rules for reading facial expressions that performers could adopt as guides to how to arrange their bodies to convey specific types of character (Pearl). By the second half of the century, anthropometrists were developing detailed lexicons that related outward expression to inner emotion. They developed these sorts of technologies as tools for classifying the dangerous and the different (Sera-Shriar). These tools for disciplining errant bodies also provided tools that stage performers could use to school themselves and their audiences in the art of reading bodies properly. They offered a gestural language that performers and audiences alike understood. Interestingly in this context, Charles Darwin used photographs of an actor to illustrate *The Expression of the Emotions in Man and Animals*.

Philosophers of science have tended to portray science as a quintessentially disembodied way of going on. Practitioners' bodies cannot matter to the production of knowledge precisely because scientific knowledge is bodiless – it exists everywhere and nowhere (Lawrence and Shapin). Historians and sociologists have challenged this notion by drawing attention to the importance of embodied practice in making knowledge. Victorian men of science would have agreed with the historians and sociologists. They were clearly keenly aware that their performances – and therefore their bodies – were important tools for making and distributing knowledge. Faraday, Tyndall and Pepper understood very well that how they acted had a great deal to do with how the knowledge they wanted to impart to their audiences would be understood. Just as dramatic actors knew that they needed to arrange their bodies in very specific ways if their performances were to carry the appropriate air of authenticity, scientific actors knew that their act constituted their authority too. Audiences knew how to identify the marks of authenticity that performers' bodies carried with them whether they were watching a scientific lecture at the Royal Polytechnic Institution or the latest play at the Adelphi Theatre. The same sorts of rules of bodily deportment applied. Just as there was an aesthetic to theatre performance, there was an aesthetic to public experiment as well.

Scientific theatre

In June 1898 the scientific periodical *Nature* published an article on 'Science in the Theatre'. 'The assimilation of nature on the stage! To what extent is assimilation possible, and what are the necessary methods and appliances for obtaining a satisfactory assimilation?' it asked ('Science in the Theatre' 164). Science in the theatre, from the author's perspective, was meant to bring nature onto (or, rather, into) the stage. Drama was meant to take place in realistic surroundings and failures in realism could only lead to failures in the integrity of the performance. Stage scenery depended on illusion but it needed to fool the eye decisively. If what the audience saw was a 'sky that looks like so much blue calico hanging on a wash-line, a horizon with angles, a tree that looks like a piece of cardboard, or a moon which suddenly rushes into the sky and then remains stationary' ('Science in the Theatre'), they would not be deceived into thinking they were seeing nature. Theatre managers needed to take realism literally and employ the latest scientific techniques to increase the verisimilitude of their stage productions. It was not just a matter of generating realistic renditions of nature on canvas – as panorama and diorama producers realized too – the proper arrangement of light and shadow, or the movement of scenery, had to be arranged to aid the illusion of reality as well (Oetterman).

Scientific spectacle could be right at the heart of theatrical performance. In 1849, for example, Her Majesty's Theatre mounted a production of the ballet *Electra; or, the Lost Pleiade*, with the newly invented electric arc light as the centrepiece.[7] The result, according to the *Examiner*, was 'some beautiful and most extraordinary effects of light and shade' ('The Theatrical Examiner'). The *Standard* described how 'a capital close is accomplished by the uprising of a large star, generated by the electric process, which pales every other light in the house, and is too intense to be looked upon' ('Her Majesty's Theatre'). The electric process in question was the arc light recently invented and patented by William Edwards Staite. Staite had put his arc light on show in Trafalgar Square the previous year as well. His arc light illuminated 'the whole of Trafalgar-square, the rays reaching as far as Northumberland-house . . . The Nelson column, which was selected as the principal point, being frequently as conspicuous as at noonday' ('Electric Light'). Staite, like many electrical inventor-entrepreneurs, was, and needed to be, a consummate showman. The episode offers a nice example of the place theatrical exhibition could play in such strategies for self-promotion, while offering theatre managers and their publics a new spectacle that displayed new levels of stage naturalism for their productions (Beauchamp).[8]

Just as technologies such as the electric light could move easily between being demonstrations in a scientific lecture and stage props in a theatrical production, new effects such as Pepper's ghost could move between different kinds of stages to support different narratives. Scientific demonstrations of the power of illusion at the Royal Polytechnic Institution used Pepper's ghost to make a point about epistemology, but like theatrical productions in other spaces they were also designed to generate an alternative reality for their audiences. Even as they were being told that what they saw was an illusion they were being primed to accept what they saw as real. They were being invited to be seduced by the illusion. When Pepper's ghost appeared as part of other theatrical productions outside the Polytechnic it carried some of the connotations of scientific performance with it. When the ghost appeared at the Adelphi Theatre in June 1863, critics recognized that it brought something of the Polytechnic with it. The *Times* thought that 'the maintenance of the illusion on so large an arena as the stage of the New Adelphi Theatre may be regarded as a great triumph on the part of Mr. Pepper's shadowy protégé' ('Adelphi Theatre'). Others complained that the stage managers had not done enough to hide the illusion's mechanism. *Lloyd's Weekly Newspaper* sneered that 'sensible people always doubted the propriety of acting pantomimes at the Polytechnic, and are now doubting the wisdom of acting Polytechnics at the playhouses' ('Public Amusements').

The *Lloyd's Weekly* complaint rested on competing views about where the realism of theatrical performances should reside – with the acting or with the stage machinery. Pepper's ghost (outside the Polytechnic in any case) was 'a startling addition to stage machinery, and a terrible extension of a new style of attraction which has sprung up during the last ten years'. Its critic's argument was simply that the 'attractions, as well as the merits, of a performance should depend as much as possible upon the actors – as little as possible upon accessories which properly belong to other branches of art' ('Public Amusements'). These kinds of tensions between conflicting accounts of how illusion should be generated in the theatre that are evident here went to the heart of Victorian performative culture. In January 1905, the scientific conjuror John Nevil Maskelyne (in many ways Pepper's successor as impresario of scientific spectacle) mounted a theatrical performance of Edward Bulwer-Lytton's *The Coming Race* at St George's Hall. It was designed to be a showpiece for Maskelyne's particular brand of stage science, and which in other settings had clearly drawn in the audiences (Steinmeyer). This production flopped, though, with the *Times* critic complaining about the lacklustre stage effects, featuring 'two mine explosions, both of them disappointing; one airship, an inferior and undeceptive vessel; one flying boy and one flying Princess', along with vril-staves that looked like 'bicycle pumps of nickel loaded with squibs' ('St. George's Hall').

Maskelyne's production of *The Coming Race* seems to have failed (it closed after only eleven weeks) because of a failure in realism. The scientific spectacle around which it was organized simply was not enough to carry the necessary weight. What worked in the context of a stage performance of scientific conjuring, designed to challenge the audience's capacity to distinguish between reality and fantasy, failed to work as theatre. The drama itself, as the *Times* put it, was a 'poor and sometimes ridiculous romantic play' ('St. George's Hall') and Maskelyne's illusions failed to bridge the gap between what audiences to his conjuring shows expected and the expectations of dramatic theatre. This was the same kind of problem as the one that some critics had identified in efforts to translate Pepper's ghost from one performative space to another. What seemed spectacular in one setting failed to elicit a response in another. Getting the spectacle right on stage mattered. Critics of the Adelphi's efforts to turn the ghost to their own theatrical purposes complained 'that sufficient care was not shown to conceal the means by which the illusion is produced . . . [I]t tended to "disillusionise" those less informed on the subject, and to deprive the spectacle of some portion of its mystery' ('Adelphi Theatre', *Morning Post*).

Scientific topics themselves do not appear to have had a very significant role to play in nineteenth-century theatre. There were no Victorian equivalents of Christopher Marlowe's *Doctor Faustus* or Ben Jonson's *The Alchemist*, in terms of popularity at least. Relatively few plays were produced that explored or dealt directly with scientific concepts, or featured experimenters and natural philosophers as protagonists (Shepherd-Barr 2006). Nevertheless, when plays such as Henry Milner's adaptation of Mary Shelley's *Frankenstein*, subtitled *The Man and the Monster*, were performed, they clearly did raise questions about philosophical identity and responsibility. Like the novel on which it was based, Milner's play asked its audience to think about how the natural philosopher ought to act. Farces such as J. Sterling Coyne's *The Phrenologist*, staged at Dublin's Theatre Royal in 1835, or James Planché's *Lavater – The Physiognomist; or, Not a Bad Judge*, at London's Royal Lyceum Theatre in 1849, poked fun at scientific practices that were already on the fringes of orthodoxy. Even so, they still toyed with notions of who the man of science was and the place of science in culture.

Plays like this, though, shared in the culture of scientific spectacle as much as they engaged with scientific ideas (Milner). Later in the century, ideas about evolution, for example, contributed to the intellectual framework of dramatic performances (Shepherd-Barr 2015). When popular authors such as Wilkie Collins, who played with scientific ideas in their novels, dabbled with theatre, they played with science there too. In 1857, for example, Collins co-authored a play with Charles Dickens on the fate of the Franklin expedition to discover the North-West

Passage. Collins also wrote plays based on *The Woman in White* (staged in 1871) and *The Moonstone* (staged in 1877), both of which, like the novels on which they were based, played with contemporary scientific ideas about mind and matter (Lycett). Right at the end of the century, Arthur Conan Doyle collaborated with the American actor and playwright William Gillette to put the ultimate scientific detective Sherlock Holmes on stage (interestingly, Planché's play had portrayed the physiognomist Lavater as a detective too). By and large, though, science entered the theatre through a shared culture of spectacle and performance, and that has been my main focus here. Theatrical science and scientific theatre shared similar challenges in engaging with their audiences – and their audiences in turn moved quite easily from one arena to the other.

Conclusion

In 1895, shortly after its publication in book form, the electrician and pioneer of animated photography Robert Paul contacted H. G. Wells with a proposal to turn his novel *The Time Machine* into a spectacular sensory entertainment that would combine elements of diorama, kinetoscope, magic lantern and stage technologies to offer audiences an entirely novel kind of immersive experience (Sherborne 112). Though the proposal came to nothing in the end it offers an instructive example of the culture and technology of display that proliferated during the nineteenth century. It demonstrates the extent to which particular practices and technologies underpinned a common performative culture. Scientific lectures and theatrical productions, clearly, were not the same kinds of performances, but they belonged nevertheless on the same spectrum (Watt Smith). Actors, effects and technologies moved around this shared performative culture – not without effort and not always successfully. Places such as the Adelphi Theatre, the Royal Polytechnic Institution or even the Royal Institution were nodes in the same networks. They shared technologies of display and they shared audiences. To make successful performances at these places required labour, material resources and skills. Whether the actor was John Tyndall playing with sensitive flames at the Royal Institution, Mr Phillips playing the ghost at the Adelphi or Charles Kean striving for Shakespearean authenticity at the Princess Theatre, everything they did required careful management and preparation.

It is clear that some Victorian scientific performers, at least, understood the contingencies of performance (Morus 2010B). Faraday and Tyndall rehearsed carefully, after all. The lesson of contingency is an important one for historians of science too. We are familiar with the ways in which the work of experimentation in the laboratory depends on a range of contingencies. We need to understand that science disseminated remains knowledge in the making too (Secord). Putting science on stage was as much a matter of making knowledge as was the work of laboratory experiment. How that knowledge was understood and how the authority of its makers was recognized was contingent upon the spaces and contexts in which it was performed. Looking at Victorian science and its performances in particular helps underline the extent to which much Victorian scientific practice was informed by a broader culture of spectacle. It was visual and geared to the generation of effects that were meant to appeal to the sensations as much as to the intellect. Audiences went to be seduced by sensational performances as much as to be persuaded by sober argument. What they understood science to be was generated by their experiences of these carefully prepared and choreographed performances. If we want to correctly identify the ways in which Victorian scientific culture proliferated and came to play a constitutive role in the origins of modernity, we need to be sensitive to the place occupied by scientific appeals to sensation in defining science to Victorian publics.

Notes

1 See also Debbie Bark, 'Science for Children', in this volume.
2 A glance at the listings of Christmas entertainments to be found in any of the major mid-century London papers makes the point.
3 The photodrome was one of a number of mid-nineteenth-century devices designed to project persistence of vision effects so that they could be witnessed by larger numbers. It is described in Pepper (1881 375–6). Like many such devices it was worked in conjunction with a magic lantern.
4 The showman P. T. Barnum's career offers a nice example in this respect (Harris; Cook).
5 'Her yuorie forhead, full of bountie braue, / Like a broad table did it selfe despred, / For Loue his loftie triumphes to engraue, / And write the battailes of his great godhed: / All good and honour might therein be red: / For there their dwelling was. And when she spake, / Sweete wordes, like dropping honny she did shed, / And twixt the perles and rubins softly brake / A siluer sound, that heauenly musicke seemd to make' (Book II, Canto iii, Stanza 24).
6 Anderson can be seen at the back of the stage in the well-known painting by Alexander Blaikley of Faraday's Christmas lecture before the Prince Consort and the future King Edward VII in 1856.
7 See Stella Pratt-Smith, 'Electricity', in this volume.
8 The classic Schivelbusch (1988) and Otter (2008) offer contrasting views of the place of light and illumination in Victorian culture.

Bibliography

Primary texts

'Adelphi Theatre'. Times, 22 June 1863.
'Adelphi Theatre'. Morning Post, 22 June 1863.
Barrett, William F. 'Note on Sensitive Flames'. Philosophical Magazine 33 (1867): 216–22.
'Curiosities of Sound'. The Intellectual Observer 12 (1868): 42–8.
'Drama'. Daily News, 3 March 1863.
'Electric Light'. Patent Journal 6 (1849): 80.
Faraday, Michael. 'On the Electro-motive Force of Magnetism by Signori Nobili and Antinori, With Notes by Mr. Faraday'. Philosophical Magazine 11 (1832): 401–14.
Faraday, Michael. The Correspondence of Michael Faraday. Ed. Frank A. J. L. James. 6 vols. London: Institution of Electrical Engineers, 1991–2012.
'Her Majesty's Theatre'. Standard, 19 April 1849.
Milner, Henry. Frankenstein, or, The Man and the Monster. London: Thomas Hayles Lacy, 1867.
'On Musical and Sensitive Flames'. Chemical News 3 (1868): 6–8.
Pepper, John Henry. A Strange Lecture, in Illustration of the Haunted Man. London: McGowan and Danks, 1863.
Pepper, John Henry. The Boy's Playbook of Science. Rev. edn. London: Routledge, 1881.
Pepper, John Henry. The True History of the Ghost; and All About Metempsychosis. London: Cassell, 1890.
Pollock, Juliet. 'Michael Faraday'. St. Paul's Magazine 6 (1870): 293–303.
'Public Amusements'. Lloyd's Weekly Newspaper, 28 June 1863.
'Royal Polytechnic Institution'. Times, 27 December 1862.
Spenser, Edmund. The Faerie Queene. Ed. A. C. Hamilton. 2nd edn, rev. Hiroshi Yamashita and Toshiyuki Suzuki. Harlow: Pearson Longman, 2007.
'Science in the Theatre'. Nature 58 (1898): 164–5.
'St. George's Hall'. Times, 3 January 1905.
'The Theatrical Examiner'. Examiner, 21 April 1849.
Tyndall, John. A Course of Eight Lectures on Sound. London: Longmans, 1867.
Tyndall, John. Faraday as a Discoverer. London: Longmans, 1868.
Tyndall, John. Fragments of Science. 6th edn. 2 vols. London: Longmans, 1879.
Ure, Andrew. 'An Account of Some Experiments Made on the Body of a Criminal Immediately After Execution, With Physiological and Practical Observations'. Quarterly Journal of Science 6 (1819): 283–94.
Ure, Andrew. The Philosophy of Manufactures. London: Charles Knight, 1835.
Yates, Edmund. His Recollections and Experiences. 2 vols. London: Richard Bentley, 1884.

Secondary texts

Altick, Richard. *The Shows of London*. Cambridge: Belknap, 1978.

Beauchamp, K. G. *Exhibiting Electricity*. London: Institution of Electrical Engineers, 1997.

Brooker, Jeremy. *The Temple of Minerva: Magic and the Magic Lantern at the Royal Polytechnic Institution, London, 1837–1901*. London: Magic Lantern Society, 2013.

Cook, James W. *The Arts of Deception: Playing with Fraud in the Age of Barnum*. Cambridge, MA: Harvard University Press, 2001.

Durbach, Nadja. *The Spectacle of Deformity: Freak Shows and Modern British Culture*. Berkeley, CA: University of California Press, 2009.

Farrar, W. V. 'Andrew Ure, F. R. S., and the Philosophy of Manufactures'. *Notes & Records of the Royal Society* 27 (1973): 299–324.

Golinski, Jan. *Science as Public Culture: Science and Enlightenment in Britain, 1760–1820*. Cambridge: Cambridge University Press, 1992.

Golinski, Jan. *Making Natural Knowledge: Constructivism and the History of Science*. Cambridge: Cambridge University Press, 1998.

Goodall, Jane. *Performance and Evolution in the Age of Darwin: Out of the Natural Order*. London: Routledge, 2002.

Gooding, David. 'In Nature's School: Faraday as an Experimentalist'. *Faraday Rediscovered: Essays on the Life and Work of Michael Faraday, 1791–1867*. Eds. David Gooding and Frank A. J. L. James. London: Macmillan, 1985, 105–35.

Gooding, David, and Frank A. J. L. James, eds. *Faraday Rediscovered: Essays on the Life and Work of Michael Faraday, 1791–1867*. London: Macmillan, 1985.

Harris, Neil. *Humbug: The Art of P. T. Barnum*. Boston, MA: Little & Brown, 1973.

Lamb, Geoffrey. *Victorian Magic*. London: Routledge & Kegan Paul, 1976.

Lawrence, Christopher, and Steven Shapin, eds. *Science Incarnate: Historical Embodiments of Natural Knowledge*. Chicago, IL: University of Chicago Press, 1998.

Lightman, Bernard, ed. *Victorian Science in Context*. Chicago, IL: University of Chicago Press, 1997.

Lightman, Bernard. *Victorian Popularizers of Science: Designing Nature for New Audiences*. Chicago, IL: University of Chicago Press, 2007.

Luckhurst, Mary, and Emilie Morin, eds. *Theatre and Ghosts: Materiality, Performance and Modernity*. London: Palgrave Macmillan, 2014.

Lycett, Andrew. *Wilkie Collins: A Life of Sensation*. London: Windmill, 2014.

McCorristine, Shane. *Spectres of the Self: Thinking About Ghosts and Ghost-seeing in England, 1750–1920*. Cambridge: Cambridge University Press, 2010.

Morus, Iwan Rhys. 'Seeing and Believing Science'. *Isis* 97 (2006): 101–10.

Morus, Iwan Rhys. 'Radicals, Romantics and Electrical Showmen: Placing Galvanism at the End of the English Enlightenment'. *Notes & Records of the Royal Society* 63 (2009): 263–75.

Morus, Iwan Rhys. 'Placing Performance'. *Isis* 101 (2010): 775–8 [2010A].

Morus, Iwan Rhys. 'Worlds of Wonder: Sensation and the Victorian Scientific Performance'. *Isis* 101 (2010): 806–16 [2010B].

Oetterman, Stephan. *The Panorama: History of a Mass Medium*. New York, NY: Zone, 1997.

Otter, Chris. *The Victorian Eye: A Political History of Light and Vision in Britain, 1800–1910*. Chicago, IL: University of Chicago Press, 2008.

Pearl, Sharrona. *About Faces: Physiognomy in Nineteenth-century Britain*. Chicago, IL: University of Chicago Press, 2010.

Qureshi, Sadiah. *Peoples on Parade: Exhibitions, Empire and Anthropology in Nineteenth-century Britain*. Chicago, IL: University of Chicago Press, 2011.

Schaffer, Simon. 'Natural Philosophy and Public Spectacle in the Eighteenth Century'. *History of Science* 21 (1983): 1–43.

Schivelbusch, Wolfgang. *Disenchanted Night: The Industrialization of Light in the Nineteenth Century*. Berkeley, CA: University of California Press, 1988.

Secord, James. 'Knowledge in Transit'. *Isis* 95 (2004): 654–72.

Sera-Shriar, Efram. *The Making of British Anthropology, 1813–1871*. London: Pickering & Chatto, 2013.

Shapin, Steven. 'The Invisible Technician'. *American Scientist* 77 (1989): 554–63.

Shepherd-Barr, Kirsten. *Science on Stage: From Doctor Faustus to Copenhagen*. Princeton, NJ: Princeton University Press, 2006.

Shepherd-Barr, Kirsten. *Theatre and Evolution from Ibsen to Beckett.* New York, NY: Columbia University Press, 2015.

Sherborne, Michael. *H. G. Wells: Another Kind of Life.* London: Peter Owen, 2010.

Steinmeyer, Jim. *Hiding the Elephant: How Magicians Invented the Impossible.* London: Arrow, 2005.

Watt Smith, Tiffany. *On Flinching: Theatricality and Scientific Looking From Darwin to Shellshock.* Oxford: Oxford University Press, 2014.

Weeden, Brenda. *The Education of the Eye: History of the Royal Polytechnic Institution, 1838–1881.* London: Granta, 2008.

Winter, Alison. *Mesmerized: Powers of Mind in Victorian Britain.* Chicago, IL: University of Chicago Press, 1998.

Further reading

Kember, Joe, John Plunkett, and Jill A. Sullivan, eds. *Popular Exhibitions, Science and Showmanship, 1840–1910.* London: Pickering and Chatto, 2012.

Lachapelle, Sofie. *Conjuring Science: A History of Scientific Entertainment and Stage Magic in Modern France.* London: Palgrave, 2015.

Willis, Martin, ed. *Staging Science: Scientific Performance on Stage, Street and Screen.* London: Palgrave, 2016.

III

Mathematical and physical sciences

15

MATHEMATICS

Alice Jenkins

University of Glasgow

Mary Poovey has observed, strikingly, that '[f]or a late twentieth-century literary critic, numbers constitute something like the last frontier of representation' (1998 xi). Certainly the study of relationships between nineteenth-century literature and mathematics is something of a Cinderella field compared to the much larger and more established study of literature and sciences such as biology, geology or psychology. This contrast is especially strong in historicist scholarship. The models that underlie much historicist work in literature and science studies have proven better adapted to support research into the less mathematical sciences and do not translate easily to equivalent work in literature and mathematics.

For scholars of nineteenth-century literature in particular, awareness of the diffusion of mathematical knowledge, terms and ideas through culture is especially helpful. This is not because British mathematics was especially vibrant or innovative, in the first half of the century at least; on the contrary, it is usually seen as rather stagnant compared with the advances being made in Continental Europe. But, in the age of emerging mass education, mathematics was one of the areas of knowledge to which almost all educated people were exposed in some degree. Even more than Latin, mathematics (albeit often only in the very basic form of arithmetic) was a shared experience and a 'common knowledge' for nineteenth-century readers and writers, and its impact on society and culture was immense.

Much less work has been done on the distribution of mathematical knowledge than on patterns of verbal literacy, but it is clear that formal mathematical education in England was extremely patchy through most of the century. As with almost all aspects of school education in the nineteenth century, mathematical provision and standards varied enormously not only from one historical moment to another but also from one institution to another. Dame schools for the poor generally taught nothing but reading and sewing; only a small number taught arithmetic (see Gardner 20–22). Middle-class boys in commercial schools learnt basic arithmetic; most national schools offered little more than the rudiments of arithmetic, though elementary geometry was taught in a few (Allsobrook 15, 161). Differences between Scottish and English school provision were frequently commented on, with Scottish mathematical education generally far better regarded. 'The teaching of mathematics in English schools is rarely satisfactory,' noted the report of the 1868 Schools Inquiry Commission (1:30). A few schools specialized in mathematics or had particularly good reputations in that subject, such as the Royal Mathematical School at Christ's Hospital, founded in 1673, and Sir Joseph Williamson's Mathematical School in Rochester, Kent, founded in 1701. As examinations began to be instituted for entrance to the professions and the training schools, above all the Civil Service and the Royal Military Academies, reforming mathematical teaching in the schools feeding those professions became a matter of urgency (Howson 128). This impetus did not apply, of course, to mathematical education for girls, who would not be taking the relevant examinations; Teri Perl has noted that female mathematical literacy fell sharply behind as mathematical knowledge became increasingly necessary for trades and professions (48).

The kinds of mathematics taught in the English and Scottish universities, and the level to be achieved, varied considerably from institution to institution, and according to the type of degree being studied (Rice; Mann and Craik; Crilly; Hannabus 2011). Access to advanced mathematics for women and girls was more difficult. The story of Philippa Fawcett, who beat all the male students in the Cambridge mathematics tripos exams in 1890 but was not permitted to qualify with a degree (women were not awarded degrees at Cambridge until 1948), is well known (see, for example, Gould 155–7, Warwick 281–2). But the two most famous female mathematicians of the nineteenth century, Mary Somerville and Ada Lovelace, were both tutored at home (Neeley; Toole).

In their most elementary forms, mathematical concepts were more widely spread through all classes of nineteenth-century society than any other kind of knowledge. The effects on literary culture of this diffusion of mathematical terms, ideas and experiences have not yet been systematically explored. However, though there are as yet few monograph-length historicist studies of nineteenth-century literature and mathematics, interest in the subject is growing in romanticist and Victorianist circles. The mathematician who is best known to literary studies, of course, is Charles Dodgson: for introductory accounts of Dodgson in the context of the Oxford mathematics of his period, see Hannabus 2000 and Hannabus 2011; for views of Carroll/Dodgson from history of mathematics, see Pycior 1984A, Wilson, and Richards M.; and, rather more technically, Abeles 2005 and Abeles 2008. Elizabeth Throesch's essay 'Nonsense in the Fourth Dimension of Literature' is an interesting recent example of literary-critical engagement with Dodgson's mathematics.

Beyond Dodgson, a number of notable essays and articles on other topics in nineteenth-century literature and mathematics have appeared, and many briefer discussions of the subject occur in work from a wide range of disciplines. Interestingly, and again, perhaps, rather in contrast with literature and science studies, there is also a body of work exploring literature and mathematics from outside the conventional bounds of academic literary criticism, including by mathematicians. For example, Springer has produced a number of series dealing with very wide-ranging aspects of mathematics and culture (including visual arts, literature, film, theatre and other forms). These collections print very brief papers by international groups of researchers from varying disciplinary backgrounds (Emmer and Manaresi; Manaresi; a series of annual volumes from the same publisher uses a similar format (Emmer 2004–9); see also Emmer 2012–15). As well as this, a flourishing field of criticism of mathematical literature (broadly understood) has been written by and aimed at teachers and educationalists (for example, Hunsader; Nesmith and Cooper). Rather than attempt to catalogue all these, this chapter first outlines some of the methodological considerations that have affected the appearance of the historicist study of literature and mathematics, then offers very brief sketches of some of the work done on three particular nineteenth-century conjunctions of the two disciplines. These three conjunctions are poetry and mathematics, long fiction and probability and statistics, and short fiction and non-Euclidean geometry. While these three topics are not, of course, the only meeting-points of the period's mathematics and literature, they are among those that have received the most critical attention, and they give a sense of the variety of ways in which scholars have dealt with the methodological problems that I discuss below.

Methodology in literature and mathematics studies

Rachel Feder has rightly pointed out that historicist work on literature and mathematics helps to 'recover a conceptual genealogy that has been effaced by disciplinary divisions' (169). This is not to say, however, that we should expect to find a mathematical equivalent of the 'One Culture', which, even if it was never a historical reality, has been a powerfully enabling model in literature and science studies. The familiar 'One Culture' argument that Victorian readers were able to move almost seamlessly between literary and scientific content in periodicals does not apply in the same way to mathematical material; though there was some mathematical content in most of the major reviews, it was much scarcer and often of a different nature from the scientific articles and reports. In turn, this means that the assumptions about Victorian general readers and their reading practices, which are familiar in literature and science studies, do not transfer easily to literature and mathematics. By consequence, historicist studies in the tradition of Beer 1983, Levine 1988, Shuttleworth, Dawson 2007, and O'Connor have given comparatively little attention to the mathematical sciences, and especially to mathematics itself. Mathematics, indeed, has sometimes been seen as an obstacle to the flow of ideas, terms and information between literature and science. Gillian Beer argued in the 1990s, for example, that 'the mathematicization of scientific knowledge' (1996 321) over the past 200 years had expedited communication between scientists to such a degree as to exclude non-specialists from the conversation.

If the assumption of a 'common context' is not available to the study of Victorian literature and mathematics, neither is an equivalent of the strong support that literature and science studies has drawn from history of science. The changes that in recent decades have foregrounded externalist and contextualist approaches to the history of science are not so apparent in history of mathematics, where a 'cultural turn' has been resisted repeatedly. Philip Kitcher and William Aspray describe the history of mathematics as characterized by 'the firmly seated belief that mathematicians but not their ideas may be affected by external factors' (24). Thomas Hankins, similarly, noted in his biography of the nineteenth-century Irish mathematician William Rowan Hamilton that 'mathematics is the most "internalist" of all the sciences. It has a logical coherence independent of physical events or their chronology' (xviii). Nevertheless, there are, and have been for many decades, influential calls for history of mathematics to have a closer relationship with other forms of cultural history. The pioneering historian of mathematics George Sarton, for example, argued in the 1930s that 'the history of mathematics should really be the kernel of the history of culture' (4). A generation or so later, Ivor Grattan-Guinness decried the 'vicious self-generating circle' of disciplinary isolation in which

> generations and communities of mathematicians and educators [practised] their subject as if they had created it all themselves, and remaining completely ignorant of its historical and cultural background – or even that it has such a background in the first place.
>
> *(155–6)*

An influential essay of 2001 on mathematics education and democracy calls for research in this area to 'deconstruct internalism and . . . provide alternative conceptualizations of the discipline' (Skovsmose and Valero 42).

The general tendency in the field now is indeed in this direction. In a focus section in *Isis* in 2011, Joan L. Richards, whose work in history of nineteenth-century mathematics has been and remains a key source for literary scholars interested in these topics, stressed that mathematics is highly culturally contingent: its 'purview and powers are constantly being adjusted and defined by the cultures in which it is being pursued' (510). Amir Alexander, writing in the same section on the relationship between history of mathematics and history of science, argued that 'after decades of separation' (480), the two fields 'are now drawing closer together', though 'the terms of their new union are still being negotiated'. An example of this methodological shift is the use of actors' categories in the history of mathematics. As early as 1984, Helena M. Pycior urged that 'while immersing themselves in particular scientific subcultures, historians [should] abide by the differentiation of scientific and non-scientific factors *made by the subculture's members*' (1984B 427; my italics). Drawing on emerging work in the sociology of scientific knowledge, Pycior proposed stepping aside from the internal/external debate to work instead with the categories used and understood by writers and practitioners of particular sciences in particular historical periods. A similar shift towards emphasizing actors' categories was already being made in the history of science, where it is now mainstream; and the move towards actors' categories and the decisions about scope and approach that it tends to exemplify have also become more common in history of mathematics. In the introduction to her 2001 book *Ancient Mathematics*, for example, Serafina Cuomo wrote 'I trust that I have used actors' categories throughout: if a land-surveyor says that mathematics is part of his job, that is enough for me' (2). Cuomo's approach steps decisively away from the traditional history of ancient mathematics she criticizes for focusing on internalist topics and avoiding 'trying to relate mathematical practices to their historical contexts' (1). Instead she seeks to provide 'insight into everyday, everyperson, "popular" views of mathematics'.

This kind of approach produces a history of mathematics that makes room for other kinds of cultural history, and that is thus likely to support the development of historicist literature and mathematics studies. The editors of a recent, wide-ranging and exceptionally useful collection on *Mathematics in Victorian Britain*, for example, structure the volume's essays partly based on geography, with chapters on Cambridge, Oxford, London, Scotland, Ireland and the Empire (Flood, Rice and Wilson). Together with a chapter on mathematical journals and societies, these surveys reflect the importance of understanding the networks, institutions and groups involved in nineteenth-century mathematics; tracing these networks necessarily connects mathematicians and their work with wider educational, scientific and cultural activities in the period. Longer histories of particular mathematical networks based in geographical and institutional contexts also illustrate this ability to connect with wider contexts. Andrew Warwick's history of mathematical physics at Cambridge University, for instance, includes an influential chapter on aspects of masculinity and the body in student life during the nineteenth century, while A. D. D. Craik's study of one of the mathematics tutors in Victorian Cambridge gives an illuminating picture of the ways in which Cambridge's peculiar curriculum produced generations of a highly mathematically-educated social, political, religious and professional elite. As these and other studies indicate, the history of mathematics in nineteenth-century education is profoundly affected by mathematics's interactions with religious faith, not only because of the close connections between the English universities and the Church of England but also because of the importance of religious faiths of various kinds in the biographies of many of the period's mathematicians. William Kingdon Clifford, an extraordinarily interesting and controversial figure, is a rare example of an outspokenly atheist nineteenth-century mathematician (see Small 2004; Dawson 2004). For others, including James Clerk Maxwell, James Joseph Sylvester and Charles Dodgson, religious faith and affiliation were highly important. Daniel J. Cohen argues in *Equations*

From God (2007) that the mathematical history of the nineteenth century is closely interwoven with the period's religious history, suggesting, for example, that 'It is no coincidence that symbolic logic arose in the wake of Catholic Emancipation, the beginning of Jewish emancipation, and the Oxford Movement' (11). A very recent collection of essays on *Mathematicians and Their Gods* includes four chapters on interactions between religion, theology and nineteenth-century mathematics (Lawrence; Richards M.; Lewis; Bayley 2015). Work of this kind integrates the history of mathematics into a much broader pattern of intellectual, cultural, and material history, and thus provides some of the underpinnings for the development of historicist studies of mathematics and literature.

While these underpinnings are as yet not fully developed, the very long-established and productive field of the philosophy of mathematics has provided support for postmodernist and theoretical approaches to this field. A recent essay collection gives a helpful sample of some of the current philosophically and theoretically based approaches to literature and mathematics. *Circles Disturbed: The Interplay of Mathematics and Narrative* (Doxiadis and Mazur) includes chapters by mathematicians and historians and philosophers working on mathematics as well as by literary scholars. Among the essays likely to be of interest to nineteenth-century scholars is Uri Margolin's on narratology, which sets out 'six areas of significant contact or meaningful comparison' (481) between narrative in fiction and mathematics. These include 'the use of numerical or geometrical formulas, procedures, or patterns to determine the composition or architecture of a narrative, and their occasional predominance over mimetic or thematic factors' (483) and 'mathematical concepts, models and methods in theories of narrative' (504).

Work in this theoretical strand sometimes draws its examples from past texts, including nineteenth-century ones, but often seeks to explore conceptual relationships between number and words rather than particular historical instantiations of such relationships. Barbara Herrnstein Smith and Arkady Plotnitsky note in the introduction to their influential 1997 edited collection on this topic that mathematics and cultural theory are sometimes imagined to be polar opposites: theory is thought to have no purchase on 'the manifest objectivity of mathematical entities and the manifest transcultural, transhistorical validity of mathematical knowledge' (2). Smith and Plotnitsky aim to show, however, that 'the relations – both historical and conceptual – between mathematics and post-classical theory [including postmodernism and deconstruction] are on the whole quite cordial' (3). Indeed, literary theory has a long-standing interest in the relationship between mathematics and language, particularly whether poststructuralist ways of understanding language also apply to mathematical expressions (essential texts include Derrida; Badiou). This interest opens up a fundamental question in the philosophy of mathematics: putting it in a very basic form, is mathematics best understood as a set of formal expressions that refer only to one another, or – as one critic has put it – as a 'noise-free description of a mind-independent reality' (Brubaker 869)? Vicki Kirby, in an illuminating article, explains the problem for literary theorists who do not accept the latter claim:

> The assumption that mathematics is the language of Nature, or that its representational veracity evidences the guidance of divine authorship, appears throughout history, and is even today held by many practitioners; it is the basis of the perception that mathematics is discovered rather than invented, and its endurance suggests that it is not easily countered.
>
> *(419)*

The problem of countering deep assumptions about the irrelevance of theoretical or cultural ideas to mathematics is not limited to literary theorists or to scholars of literature and mathematics;

rather, it is a hard or extreme case of a question that is raised by any historicist literature and science study, or, by analogy, by any study that reads literary texts as something other than closed, inward-facing systems.

Brian Rotman is one of the most innovative and influential writers in this theoretical strand of research. In a series of books published from the late 1980s, he explores mathematical concepts in relation to narratology and semiotics.[1] Rotman argues against the 'referentialist' account of mathematics, which sees it as 'about "things" – numbers, points, lines, spaces, functions and so on – that are somehow considered to be external and prior to mathematical activity' (1987 2). Instead, he regards mathematics as fundamentally a human construct, 'a symbolic activity conducted solely through thought experiments on ideal, invisible objects' (Rotman 2008 58). Rotman's work is philosophical and theoretical, and at the same time historicist. Thus his most-cited book considers the 'introduction of the mathematical zero sign into Western consciousness in the thirteenth century' and asks

> what is the impact on a written code when a sign for Nothing, or more precisely when a sign for the absence of other signs, enters its lexicon? What can be said through the agency of such a sign that could not be said, was unsayable, without it?
>
> (Rotman 1987 1–2)

This central gesture of tracing the impact of a new scientific idea or entity on wider cultural formations is very familiar to literature and science research. But, in focusing on responses to innovative ideas, it is easy to miss the importance of old, outdated or simply well-established science for contemporary readers and writers. This is perhaps especially true of mathematics, because of the relatively low level of general mathematical knowledge. Many nineteenth-century readers and writers who engaged directly with new scientific ideas, such as – for example – the theory of evolution by natural selection, engaged only with old mathematical ideas such as the rules of arithmetic and the basics of plane geometry. Tracing cultural responses to knowledge that is widespread and taken for granted is crucial but very challenging for both literature and science studies and literature and mathematics studies. Matthew Wickman's Literature *After Euclid* (2016) is a very interesting example of a project that grapples with the uneven chronologies of literary responses to mathematics, in this case geometry. In Wickman's account, geometry comes to mean not only a preoccupation with forms but also a kind of politics of knowledge that is often defiantly oppositional, backward-looking and yet artistically enabling. This key insight launches Wickman on a series of investigations of how this 'geometrical' imagination shaped – and Wickman's book is highly alert to the profound epistemological strategies underlying such geometrical puns as 'shaping' – Scottish narrative, poetry and criticism in the late eighteenth and nineteenth centuries. The geometry is often left far behind, but the geometrical imagination is not; one of the contributions of Wickman's book, indeed, is its proposal that the geometrical imagination is far more capacious and more radical than one might have imagined from, for example, George Elder Davie's work on Scottish geometry in *The Democratic Intellect* (1961).

Steven Connor's work on literature and number is another bridge between theoretical and historicist approaches. Connor criticizes the academic humanities for having failed to come to grips with mathematical and numerical ideas, techniques and materials. He argues that this unhelpful belief in 'the encroaching horror of the domain of number' (Connor 2014B 1) has blinded literary criticism to the fact that although 'modern literature and culture may try to get themselves on the other side of number' (Connor 2013 7), still their 'obsession with this anumerical project' reveals them as deeply concerned with mathematics. Connor proposes to remedy this failure in literary and cultural criticism through attention to what he terms the 'quantical' (2013 2), that is, 'the tendency or aspiration to render things in terms of quantity'. Attention to the quantical, for example, leads Connor to write that 'Dickens's comedy, like his writing practice in general' is 'impregnated with number from top to bottom' (2014A 8). A quantical attitude in literary criticism does not necessarily require advanced mathematical ability, only a willingness 'to make out quantitative relations' (Connor 2013 2); indeed, Connor's own work deals mainly with basic numerical ideas.

Mary Poovey's wide-ranging and groundbreaking work offers a series of highly influential historicist approaches to mathematics (broadly understood) and culture, including literary culture. In *A History of the Modern Fact*, Poovey explores 'the story of how one kind of representation – numbers – came to seem immune from theory or interpretation' (xii). She argues that 'In contrast to ancient facts, which referred to metaphysical essences, modern facts are assumed to reflect things that actually exist, and they are recorded in a language that seems transparent' (Poovey 1998 29). Poovey reads the history of the modern fact (a term in which every word, she acknowledges, is debatable (1998 xii–xiii)) as profoundly shaped by the 'problem of induction', a term that covers various kinds of scepticism about the validity of inductive reasoning. Her central concern is with the question of whether it is possible to separate description from interpretation, a question that looks towards artistic representations as well as towards scientific method. Thus the book's final chapter, for example, which deals with the emergence of statistics in the early nineteenth century as part of the development of inductive scientific reasoning, also includes a section on Romantic poetry and its claims to produce systematic knowledge in ways that marginalize or exclude the amassing of 'facts'. One of Poovey's potentially fertile contributions in this book is to extend the boundaries of what can be classed as mathematical for the purposes of a cultural study. Her phrase 'gestural mathematics' (Poovey 1998 172) refers to 'an invocation of mathematical processes and language' in texts that may have nothing directly to do with mathematics and are employing these invocations for a metaphorical or rhetorical purpose. Poovey's application of the phrase to certain passages, and thus her identification of mathematical qualities in them, has been criticized as too broad (Gallagher 445). But too broad a concept of the 'mathematical' is, arguably, a lesser obstacle to the development of our understanding of nineteenth-century literature than too narrow a concept. The comparative lack of attention paid to mathematics in historicist literary scholarship means that we risk being unresponsive to a wide range of mathematical metaphors, terms and structures that shape an extraordinary number of the century's literary, political and polemical writings. Acknowledging 'gestural mathematics' may be one way of recouping these losses.

Poetry and mathematics

The relationship between mathematics and poetry is one of the most venerable topics in the study of mathematics and literature. Perhaps the *locus classicus* of nineteenth-century poetry's engagement with mathematics is in the sections (in Books 5 and 6) of *The Prelude* that discuss Euclid's *Elements of Geometry*. Wordsworth's deep interest in mathematics and especially geometry has been explored in several studies (Baum; Gaull; Simpson; Owens; Ottinger; Tomalin; see also Brown, especially pp. 1–9). In particular, there is a long critical tradition dealing with the 'Arab dream', in which a stone stands for the *Elements* and a shell for a poetic ode, perhaps poetry in general. The conjunction of the two symbols has prompted critical discussion of the relationship between poetry and geometry (often extended to logical or scientific reason). In the 1950s, Newton P. Stallknecht's *Kenyon Review* essay 'On Poetry and Geometric Truth' argued that, although these two categories 'represent the two poles of our awareness' (19), still it is 'the goal, perhaps the unattainable goal, of the philosophical poet' (20) to bring together 'logical rigour and metaphorical subtlety'. Writing from a quite different theoretical perspective, J. Hillis Miller also found a tension between the stone and shell that holds them together as well as apart. For Miller, they form 'a unit in which each member is neither simply itself nor its opposite, and yet is both at once' (94). The two disciplines are not opposed, but related: 'Poetry is a transformation of the kind of reason that produces geometry' (Miller 94). A recent essay by Imogen Forbes-MacPhail sees early nineteenth-century poetry and mathematics as different but related approaches to the same problem of finding 'a semiotic system which would not only describe the world, but allow [poets and mathematicians] to forge a link between abstract symbolic systems and the human body or the workings of machines' (153).

Daniel Brown's *The Poetry of Victorian Scientists* (2013) shows that, despite the methodological difficulties I have outlined above, a historicist and close reading approach can be applied to mathematical material with great success. Brown's book deals with poetry and writing about poetry by a group of nineteenth-century figures who were primarily, or at least partly, mathematicians, including Hamilton, Maxwell and Sylvester. By examining both literature and mathematics by the same writer, Brown neatly sidesteps some of the problems of traditional historicist work in literature and science studies, such as the difficulties of tracing the exact routes by which scientific ideas enter literary texts and vice versa. He explores the ways in which these writers' interests in form, space and dynamics take shape in their poetic and critical writing as well as in their mathematical and scientific writing. Of Sylvester, for example, Brown writes 'He not only makes [his mathematical and poetical work] companionate by citing poetry in his lectures on mathematics, but intimate through a shared terminology' (207). Unlike the historian of mathematics J. D. North, who wrote sarcastically 'No doubt it will be of help to those who deal in catalecticants [a polynomial used in invariant theory] to remember that the name was inspired by the iambicus trimeter catalecticus [a poetic meter]' (90), Brown takes the transfer of terms and ideas from poetry to Sylvester's mathematics seriously, including by exploring the ways in which poetic metre especially drew the attention of nineteenth-century writers with mathematical interests and expertise. John Herschel and William Whewell, for example, also published poetry and were active participants in a Victorian debate over the use of hexameters in English (Hall 10–11, Brooke-Smith). The towering figure of James Clerk Maxwell is also relevant to this discussion; his work in mathematical physics was accompanied by a lifelong interest in poetry, which is now beginning to be explored by both literary critics and historians (Brown; Pratt-Smith; McCartney).

Brown's work on Sylvester illustrates and contributes to a very interesting moment of change in the relationship between literary history and history of mathematics. Sylvester is one of a very small number of major nineteenth-century mathematicians whose work has been explored by historians of mathematics with a strong interest in cultural history. In the past, historians of mathematics tended to see Sylvester's poetic interests as peripheral and even faintly risible, and not to credit that there could be much real traffic of ideas between the two areas. In his *History of Mathematics*, for example, Florian Cajori dismissively noted that 'Sylvester sometimes amused himself writing poetry' (343), describing his treatise on prosody *The Laws of Verse* as 'a curious booklet'. But Karen Hunger Parshall's 2006 biography of Sylvester has shown that his fascination with poetry should be understood along with, and not as a diversion from, his mathematics. Though she treats his poetry and criticism as biographically separate from his serious mathematical work, Parshall, like Brown, sees a deep connection between his mathematics, poetry and philosophy (218, 7, 214).

Comparisons of poetry and mathematics often mention two features both kinds of writing are thought to share: compression and concision of expression and a fundamental basis in counting

– in poetry's case, via metre. Partly because of these features, discussions of the relationship between poetry and mathematics tend towards formalism, sometimes reaching into philosophical topics. In the case of historicist studies, the primary sources themselves often deal with formalism, as in Brown's study of Sylvester's prosodic writings.

Long fiction: statistics and probability

Over the last two decades or so, interest in the ways in which statistical thinking reacted with realist fiction has generated some highly ambitious and innovative work on Victorian novels. Though most of the critical interest in statistics and literature has focused on fiction, some attention has been given to poetry: Fowler, for example, makes a case for Browning as a poet 'of quasi-mathematical self-consciousness' (11), a category in which she also includes Hopkins and Keats.

Statistics has its roots in the mid-eighteenth century, but it was not until the nineteenth that methods for collecting and analysing data began to draw on emerging probability theory, producing powerful new methods for undertaking planning in government and laying the foundations for various academic fields including sociology (Halsey). Key texts on the history of statistics include Theodore M. Porter's *The Rise of Statistical Thinking* (1986) and, more recently, Ian Hacking's *The Emergence of Probability* (2006). Particularly useful histories for Victorianists are Joan L. Richards's article 'The Probable and the Possible in Victorian England', Porter's biographical study *Karl Pearson: The Scientific Life in a Statistical Age*, Mary Poovey's 'Figures of Arithmetic' and *A History of the Modern Fact* (308–16), and M. Eileen Magnello's essays 'Vital Statistics' and 'Darwinian Variation'.

Some Victorian writers thought of statistical averages as expressing cosmic laws so powerful that they might be thought of as a kind of design principle: Dickens, for example, is said to have worried during a year of comparatively few traffic deaths in London: 'is it not dreadful to think that before the last day of the year some forty or fifty persons *must* be killed – and killed they will be' (quoted in Fitzgerald 1:207; on Dickens and statistics, see Kent; Bayley 2007; Adams). This is a well-known instance of 'statistical determinism': the belief that statistics reveals laws of human behaviour and that individuals are somehow compelled to act in accordance with these laws. Probably the most important figure in discussions of Victorian statistical determinism is the historian Henry Thomas Buckle, whose uncompleted but highly influential positivist *History of Civilisation in England* appeared in two volumes in the middle of the century. (For a fertile discussion of statistical determinism, including in Buckle and Dickens, see Kern 292–5.) Buckle's *History* was controversial and widely discussed and had a considerable impact on some contemporary literary writers as well as historians. Helen Small's essay on Buckle and Thomas Hardy gives an excellent account of Victorian responses to his belligerent and radical attempt to remake history as a scientific discipline, and Odin Dekkers's study of the late nineteenth-century critic John Mackinnon Robertson includes a substantial discussion of Buckle's contribution to the emergence of a scientific literary criticism (see also Porter 1994).

Most of the criticism investigating probability and statistics in the Victorian novel deals with ways of understanding and constructing relationships between individuals and larger groups. In a recent essay, Michael Klotz asks 'what similarities there might be between mentally considering fictional characters . . . and contemplating the predicted individuals derived from a prescribed average, ratio, or frequency' (214). Klotz points out that 'Victorian fiction routinely exploits this type of double vision in representing individuals as unique, separable members of a community and at the same time instantiations of one or more classes or categories of persons' (216). Audrey Jaffe's full-length study of statistics, the stock market and Victorian fiction explores relationships between individuals and the ideal of the statistically average in major canonical texts including *Middlemarch*, *David Copperfield* and John Stuart Mill's *Autobiography*, arguing that the internalization of ideas about averageness had a profound effect on Victorian ideas of selfhood, with 'any notion of a single or stable interiority . . . replaced by an image of relationship: the self as an effect of a comparison with others' (85).

In a fascinating recent essay on *Daniel Deronda* and statistics, Jesse Rosenthal shows that one of the strengths of investigating fiction via an interest in statistics is that it allows one to make an argument that is both formal and historicist. The argument of Rosenthal's essay is formal in that it deals with 'the unity of the interrelations in the book' (804) and the possibilities and requirements of the 'large novel', and historicist in that it traces *Deronda*'s themes and techniques to 'a conceptual shift in Eliot's thinking about the relation of the one and the many' (777), influenced, among other ways, by her reading on mathematical and statistical topics (see, for example, 799, 810). Rosenthal points out that the scale of a novel such as *Deronda* allows a writer to fulfil a reader's expectation of a balanced, just or fitting fictional universe *overall* while still denying individual characters' hopes for balance, justice or fit in their *particular* case (804). In the case of *Deronda*, Rosenthal reads this contrast between the outcome of one story and the overall effect of many stories as part of 'a representational strategy . . . that is closely tied to the novel's imagining of large and small numbers' (782). Studies such as those by Rosenthal, Klotz and Jaffe indicate how historicist scholarship of literature and mathematics can open up fresh perspectives on key topics in representation and subjectivity.

Short fiction: non-Euclidean geometry and the fourth dimension

Other than arithmetic, geometry was perhaps the mathematical subject to which the largest number of nineteenth-century readers and writers were exposed, and its cultural prestige was far greater than any other branch of mathematics. Almost all men in the nineteenth century whose education reached university level were familiar, to some extent, with Euclidean geometry, which was often considered a *sine qua non* of a liberal education and an indispensable training in reasoning. Outside the university-educated elite, knowledge of Euclidean geometry became increasingly widespread as textbooks and courses were provided for adult self-educators, attenders at Mechanics' Institutes, and the children of the middle classes (Jenkins 2007B; for a history of mathematics in nineteenth-century education, see Howson).

Joan L. Richards's *Mathematical Visions* (1988) has made nineteenth-century geometry and its extraordinary cultural capital accessible to readers outside the history of mathematics, and is the essential reading on the subject. *Mathematical Visions* is deeply interested in geometry's relationship with broader intellectual culture and has been a key source for literary historians interested in mathematics in British culture. (Richards's 'The Geometrical Tradition' is an extremely helpful shorter account; on Victorian Euclideanism more generally, see Jenkins 2007A 141–75 and Jenkins 2014.) The breadth of exposure to Euclidean geometry, and the range of methods used for gaining access to it, are illustrated by a quick survey of some nineteenth-century writers. Wordsworth studied geometry in the most canonical English way, at Cambridge University, and revered it; Thackeray also met it at Cambridge, but loathed it; Harriet Martineau studied it at home, between sessions of gardening and drawing; Carlyle learnt it at Edinburgh University and taught and wrote on mathematics for several years afterwards; George Eliot went to geometry classes for women given by John Henry Newman's brother Francis; Ruskin learnt it with a tutor and was so inspired that he spent a great deal of leisure time trying to trisect an angle, a task that is impossible using Euclidean methods (Jenkins 2008; Moore; Ruskin 55–6).

Despite the immense cultural prestige of Euclidean geometry in the nineteenth century, modern literary criticism and history have as yet paid relatively little attention to the impact of this mathematical *lingua franca* on cultural production. Critics have, however, become interested in the imaginative implications of the revolutionary mathematical work pioneered by Gauss, Bolyai and Lobachevsky on non-Euclidean geometry and ideas of a fourth dimension, sometimes referred to as higher dimensional or hyperspace thinking (for the history of non-Euclidean geometry, see Gray 2004; Gray 2010). The best-known of the Victorian fourth dimension texts is Edwin Abbott's *Flatland* (1884). Several new editions containing very helpful notes have appeared in the last few years. Essays published in the 1980s and 1990s read *Flatland* primarily as an intervention in theological debates and the relationship of science and religion (Jann; Smith, Berkove and Baker); Mark McGurl's wide-ranging essay on *Flatland* and Henry James, on the other hand, discusses it in the context of late nineteenth- and early twentieth-century theorizations of realist fiction. A more recent essay by the mathematician K. G. Valente gives a helpful account of the mathematical and scientific context (Valente, 'Transgression and Transcendence'; see also Valente 2008; Bayley 2015). Andrea Henderson makes an ambitious case for *Flatland* as a document of aestheticism and a critique of realism, arguing that it illustrates the way in which 'developments in Victorian geometry prompted a rethinking of how representation worked' (469).

Steven Connor's brief 'Afterword' to a collection of essays on the Victorian supernatural is an excellent starting point for an exploration of fourth dimension writing beyond *Flatland*. Connor gives a clear outline of the introduction of non-Euclidean geometry into British culture, and of the controversies that surrounded it. He uses a discussion of one of the most dedicated British hyperspace theorists, Charles Howard Hinton, to introduce a number of late nineteenth-century thinkers who found the idea of a fourth dimension productive in occultist and other speculative contexts. Hinton, who was educated and worked in Britain until he was convicted of bigamy, emigrated to the United States to become a teacher of mathematics at Princeton and later at other institutions. Hinton has recently become a subject of critical interest in literary studies because of an intriguing group of fiction and non-fiction texts he wrote mainly during the 1880s, in which he proposed a kind of mind-expanding practice of imaginative engagement with the fourth dimension; he hoped that such engagement would lead to a freer moral and intellectual being. Elizabeth Throesch has explicated Hinton and highlighted his influence on late nineteenth- and early twentieth-century literature ('Charles Howard Hinton's Fourth Dimension' and 'Nonsense in the Fourth Dimension of Literature'; for other recent treatments of hyperspace thinkers and their influence on late Victorian writers including H. G. Wells, see Kreisel; Blacklock; Smajic 157–80; White). Linda Dalrymple Henderson is a key source on higher-dimension thinking in visual culture and has also been very influential in literary scholarship.

Probably few of the late nineteenth-century literary writers who commented on or drew imaginatively on non-Euclidean geometry and associated concepts thoroughly understood the mathematics. Many of their borrowings and transformations of mathematical ideas would not have met with acceptance by mainstream mathematicians, but they tie themselves too closely to this mainstream to fit into Mary Poovey's category of 'gestural mathematics', and because of their tendency towards esotericism and occultism they cannot be thought of as the 'everyday, everyperson, "popular" views' of mathematics that Serafina Cuomo emphasizes. Perhaps instead we can think of them as analogous to the fringe or pseudo-sciences of the nineteenth century such as mesmerism or phrenology. Scholarship exploring this topic extends the range of literature and mathematics studies and asks interesting questions about what 'counts' as mathematics for the purposes of the field.

Afterword: counting in literary criticism

This chapter has been chiefly concerned with historicist studies of literature and mathematics employing what one might think of as conventional methodologies. A small but significant challenge to the dominance of these methodologies, and one that encourages us to return to counting, is presented by the experimental quantitative techniques made possible by digital technologies. James F. English argued in 2010 that 'literary studies is not . . . a "counting" discipline' (xii), especially at the present time: as universities become 'ever more committed to numerical data, imposing on us ever more stringent quantificational regimes of value and assessment' (xiii), 'literary studies has shouldered much of the burden of critique and resistance to this encroachment, defending qualitative models and strategies'. Whether or not it is really part of 'an urgent struggle for survival' (English xiii), there is evidence to suggest that literary studies has become less numerically minded: in a study of a very large corpus of critical articles, Andrew Goldstone and Ted Underwood found that 'literary scholars now mention numbers only about 60 percent as often as we did in the early twentieth century' (359).

Franco Moretti's work on 'distant reading' is perhaps the best-known example of number-based techniques in nineteenth-century scholarship, but there are numerous others. For example, the online 'pamphlets' produced by the Literary Lab at Stanford University, co-directed by Moretti, include work dealing with large corpuses of nineteenth-century texts (particularly Allison *et al.* and Heuser and Le-Khac). *Victorian Studies* recently published a 'Forum on Evidence and Interpretation in the Digital Age' (Stauffer), which included articles by Moretti and other members of the Literary Lab group (see also Rogers). Also in this forum, Frederick W. Gibbs and Daniel J. Cohen describe a number of less complicated research projects using the Google Books database of more than a million Victorian books. They conclude in conciliatory fashion that 'the best humanities work will come from synthesizing "data" from different domains; creative scholars will find ways to use text mining in concert with other cultural analytics' (Gibbs and Cohen 76). Despite the novelty of the methods and the extraordinary scale of the resources now in use, there may be something vestigially familiar to more conventional nineteenth-century scholars in some of these practices: if scholars of the present have shied away from counting, many Victorian critics did not. As Jonathan Farina has pointed out in a stimulating essay on Victorian critical uses of quantitative methods, 'Counting was one of the generic practices of Victorian literary criticism'. We do not, of course, have to adopt a mathematical method in order to investigate nineteenth-century literature, but, for a fuller picture of the period's ideas about rationality, structure, certainty and the nature of knowledge, we need to understand the interactions of that literature with mathematics.

Note

1   Conversely, the narrative theorist David Herman offers a mathematical approach to narratology.

Bibliography

Primary texts

Abbott, Edwin A. *The Annotated Flatland: A Romance of Many Dimensions.* Intro. and notes Ian Stewart. Oxford: Perseus, 2002.
Abbott, Edwin A. *Flatland: A Romance of Many Dimensions.* Ed. Rosemary Jann. Oxford: Oxford University Press, 2006.

Abbott, Edwin A. *Flatland: An Edition with Notes and Commentary*. Eds William F. Lindgren and Thomas F. Banchoff. Cambridge: Cambridge University Press; Washington, DC: Mathematical Association of America, 2010.

Buckle, Henry Thomas. *History of Civilisation in England*. 2 vols. Vol. 1: London: Parker, 1857; vol. 2: London: Parker, Son, and Bourn, 1861.

Fitzgerald, Percy Hetherington. *The Life of Dickens as Revealed in His Writings*. 2 vols. London: Chatto and Windus, 1905.

Ruskin, John. *Praeterita*. Ed. Francis O'Gorman. Oxford: Oxford University Press, 2012.

Schools Inquiry Commission: Report of the Commissioners. 21 vols. London: Eyre and Spottiswoode, 1868.

Wordsworth, William. *The Thirteen-Book Prelude*. Ed. Mark L. Reed. Ithaca, NY: Cornell University Press, 1991.

Secondary texts

Abeles, Francine F. 'Lewis Carroll's Ciphers: The Literary Connections'. *Advances in Applied Mathematics* 34 (2005): 697–708.

Abeles, Francine F. 'Lewis Carroll's Visual Logic'. *History and Philosophy of Logic* 28 (2007): 1–17.

Adams, Maeve E. 'Numbers and Narratives: Epistemologies of Aggregation in British Statistics and Social Realism, c. 1790–1880'. *Statistics and the Public Sphere: Numbers and the People in Modern Britain, c. 1800–2000*. Eds. Tom Crook and Glen O'Hara. Abingdon: Routledge, 2011, 103–20.

Alexander, Amir. 'The Skeleton in the Closet: Should Historians of Science Care About the History of Mathematics?'. *Isis* 102 (2011): 475–80.

Allison, Sarah, Ryan Heuser, Matthew Jockers, Franco Moretti and Michael Witmore. 'Quantitative Formalism: An Experiment'. *Pamphlets of the Stanford Literary Lab* 1. 2011. <http://litlab.stanford.edu/LiteraryLabPamphlet1.pdf>. Accessed 24 July 2015.

Allsobrook, David Ian. *Schools for the Shires: The Reform of Middle-Class Education in Mid-Victorian England*. Manchester: Manchester University Press, 1986.

Badiou, Alain. *Number and Numbers*. Cambridge: Polity, 2008.

Baum, Joan. 'On the Importance of Mathematics to Wordsworth'. *Modern Language Quarterly* 46 (1985): 390–406.

Bayley, Melanie. '*Hard Times* and Statistics'. *British Society for the History of Mathematics Bulletin* 22 (2007): 92–103.

Bayley, Melanie. 'Faith and Flatland'. *Mathematicians and Their Gods: Interactions Between Mathematics and Religious Beliefs*. Eds. Snezana Lawrence and Mark McCartney. Oxford: Oxford University Press, 2015, 249–78.

Beer, Gillian. *Darwin's Plots: Evolutionary Narrative in Darwin, George Eliot, and Nineteenth-Century Fiction*. London: Routledge and Kegan Paul, 1983; Cambridge: Cambridge University Press, 2000, 2009.

Beer, Gillian. 'Square Rounds and Other Awkward Fits: Chemistry as Theatre'. *Open Fields: Science in Cultural Encounter*. Oxford: Oxford University Press, 1996, 321–32.

Blacklock, Mark. 'The Higher Spaces of the Late Nineteenth-Century Novel'. *19: Interdisciplinary Studies in the Long Nineteenth Century* 17 (2013). <www.19.bbk.ac.uk/articles/10.16995/ntn.668>. Accessed 24 February 2015.

Brooke-Smith, James. '"A Great Empire Falling to Pieces": Coleridge, Herschel, and Whewell on the Poetics of Unitary Knowledge'. *Configurations* 20 (2012): 299–325.

Brown, Daniel. *The Poetry of Victorian Scientists: Style, Science and Nonsense*. Cambridge: Cambridge University Press, 2013.

Brubaker, Anne. 'Between Metaphysics and Method: Mathematics and the Two Canons of Theory'. *New Literary History* 39 (2008): 869–90.

Cajori, Florian. *A History of Mathematics*. 5th edn. New York, NY: Chelsea, 1991.

Cohen, Daniel J. *Equations from God: Pure Mathematics and Victorian Faith*. Baltimore, MD: Johns Hopkins University Press, 2007.

Connor, Steven. 'Afterword'. *The Victorian Supernatural*. Eds. Nicola Bown, Carolyn Burdett, and Pamela Thurschwell. Cambridge: Cambridge University Press, 2004, 258–77.

Connor, Steven. 'Quantum Writing: Literature and the World of Numbers'. 2013. <http://stevenconnor.com/quantumwriting/quantumwriting.pdf>. Accessed 15 January 2015.

Connor, Steven. 'Hilarious Arithmetic'. 2014. <http://stevenconnor.com/wp-content/uploads/2014/09/hilarious.pdf>. Accessed 15 January 2015 [2014A].

Connor, Steven. 'The Horror of Number: Can Humans Learn to Count?'. 2014. <http://stevenconnor.com/wp-content/uploads/2014/10/Horror-of-Number.pdf>. Accessed 15 January 2015 [2014B].

Craik, Alex D. D. *Mr Hopkins' Men: Cambridge Reform and British Mathematics in the Nineteenth Century*. London: Springer, 2008.

Crilly, Tony. 'Cambridge'. *Mathematics in Victorian Britain*. Eds. Raymond Flood, Adrian Rice and Robin Wilson. Oxford: Oxford University Press, 2011, 17–34.

Cuomo, Serafina. *Ancient Mathematics*. London: Routledge, 2001.

Davie, George Elder. *The Democratic Intellect: Scotland and Her Universities in the Nineteenth Century*. Edinburgh: Edinburgh University Press, 1961.

Dawson, Gowan. 'Victorian Periodicals and the Making of William Kingdon Clifford's Posthumous Reputation'. *Science Serialized: Representations of the Sciences in Nineteenth-Century Periodicals*. Eds. Geoffrey Cantor and Sally Shuttleworth. Cambridge, MA: MIT Press, 2004, 259–84.

Dawson, Gowan. *Darwin, Literature and Victorian Respectability*. Cambridge: Cambridge University Press, 2007.

Dekkers, Odin. *J. M. Robertson: Rationalist and Literary Critic*. Aldershot: Ashgate, 1998.

Derrida, Jacques. *Edmund Husserl's Origin of Geometry: An Introduction*. Transl. John P. Leavey, Jr. Hassocks: Harvester, 1978; Lincoln, NE: University of Nebraska Press, 1989.

Doxiadis, Apostolos, and Barry Mazur, eds. *Circles Disturbed: The Interplay of Mathematics and Narrative*. Princeton, NJ: Princeton University Press, 2012.

Emmer, Michele, ed. *Mathematics and Culture I to VI*. Berlin: Springer, 2004–09.

Emmer, Michele, ed. *Imagine Math to Imagine Math 3*. Berlin: Springer, 2012–15.

Emmer, Michele, and Mirella Manaresi, eds. *Mathematics, Art, Technology, and Cinema*. Berlin: Springer, 2003.

English, James F. 'Everywhere and Nowhere: The Sociology of Literature After "The Sociology of Literature"'. *New Literary History* 41 (2010): v–xxiii.

Farina, Jonathan. 'The New Science of Literary Mensuration: Accounting for Reading, Then and Now'. *Victorians Institute Journal* Digital Annexe 38 (2010). <www.nines.org/exhibits/Literary_Mensuration>. Accessed 24 July 2015.

Feder, Rachel. 'The Poetic Limit: Mathematics, Aesthetics, and the Crisis of Infinity'. *English Literary History* 81 (2014): 167–95.

Flood, Raymond, Adrian Rice, and Robin Wilson, eds. *Mathematics in Victorian Britain*. Oxford: Oxford University Press, 2011.

Forbes-MacPhail, Imogen. 'The Enchantress of Numbers and the Magic Noose of Poetry: Literature, Mathematics, and Mysticism in the Nineteenth Century'. *Journal of Language, Literature and Culture* 60 (2013): 138–56.

Fowler, Rowena. 'Blougram's Wager, Guido's Odds: Browning, Chance, and Probability'. *Victorian Poetry* 41 (2003): 11–28.

Gallagher, Catherine. 'Matters of Fact'. *Yale Journal of Law and the Humanities* 14 (2002): 441–7.

Gardner, Phil. *The Lost Elementary Schools of Victorian England*. Beckenham: Croom Helm, 1984.

Gaull, Marilyn. 'Romantic Numeracy'. *The Wordsworth Circle* 22 (1991): 124–30.

Gibbs, Frederick W., and Daniel J. Cohen. 'A Conversation with Data: Prospecting Victorian Words and Ideas'. *Victorian Studies* 54 (2011): 69–77.

Goldstone, Andrew, and Ted Underwood. 'The Quiet Transformations of Literary Studies: What Thirteen Thousand Scholars Could Tell Us'. *New Literary History* 45 (2014): 359–84.

Gould, Paula. 'Models of Learning? The "Logical, Philosophical, and Scientific Woman" in Late Nineteenth-Century Cambridge'. *Teaching and Learning in Nineteenth-Century Cambridge*. Eds. Jonathan Smith and Christopher Stray. Woodbridge: Boydell, 2001, 150–64.

Grattan-Guinness, I. 'Does History of Science Treat of the History of Science? The Case of Mathematics'. *History of Science* 28 (1990): 149–73.

Gray, Jeremy. *Janos Bólyai, Non-Euclidean Geometry, and the Nature of Space*. Cambridge, MA: MIT Press, 2004.

Gray, Jeremy. *Worlds Out of Nothing: A Course on the History of Geometry in the Nineteenth Century*. London: Springer, 2010.

Hacking, Ian. *The Emergence of Probability: A Philosophical Study of Early Ideas About Probability, Induction and Statistical Inference*. Cambridge: Cambridge University Press, 2006.

Hall, Jason David. 'Introduction: A Great Multiplication of Meters'. *Meter Matters: Verse Cultures of the Long Nineteenth Century*. Ed. Jason David Hall. Athens, OH: Ohio University Press, 2011, 1–25.

Halsey, A. H. *A History of Sociology in Britain*. Oxford: Oxford University Press, 2004.

Hankins, Thomas. *Sir William Rowan Hamilton*. Baltimore, MD: Johns Hopkins University Press, 1980.

Hannabus, Keith. 'The Mid-Nineteenth Century'. *Oxford Figures: 800 Years of the Mathematical Sciences*. Eds. John Fauvel, Raymond Flood and Robin Wilson. Oxford: Oxford University Press, 2000, 187–202.

Hannabus, Keith. 'Mathematics in Victorian Oxford: A Tale of Three Professors'. *Mathematics in Victorian Britain*. Eds. Raymond Flood, Adrian Rice and Robin Wilson. Oxford: Oxford University Press, 2011, 35–52.

Henderson, Andrea. 'Math for Math's Sake: Non-Euclidean Geometry, Aestheticism, and "Flatland"'. *PMLA* 124 (2009): 455–71.

Henderson, Linda Dalrymple. *The Fourth Dimension and Non-Euclidean Geometry in Modern Art*. Princeton, NJ: Princeton University Press, 1983; 2nd edn. Cambridge, MA: MIT Press, 2013.

Herman, David. 'Formal Models in Narrative Analysis'. *Circles Disturbed: The Interplay of Mathematics and Narrative*. Eds. Apostolos Doxiadis and Barry Mazur. Princeton, NJ: Princeton University Press, 2012, 447–80.

Heuser, Ryan, and Long Le-Khac. 'A Quantitative Literary History of 2,958 Nineteenth-Century British Novels: The Semantic Cohort Method'. *Pamphlets of the Stanford Literary Lab* 4. 2012. <http://litlab.stanford.edu/LiteraryLabPamphlet4.pdf>. Accessed 24 July 2015.

Howson, Geoffrey. *A History of Mathematics Education in England*. Cambridge: Cambridge University Press, 1982.

Hunsader, Patricia D. 'Mathematics Trade Books: Establishing Their Value and Assessing Their Quality'. *The Reading Teacher* 57 (2004): 618–29.

Jaffe, Audrey. *The Affective Life of the Average Man: The Victorian Novel and the Stock-Market Graph*. Columbus, OH: Ohio University Press, 2010.

Jann, Rosemary. 'Abbott's "Flatland": Scientific Imagination and "Natural Christianity"'. *Victorian Studies* 28 (1985): 471–90.

Jenkins, Alice. *Space and the March of Mind: Literature and the Physical Sciences in Britain, 1815–1850*. Oxford: Oxford University Press, 2007 [2007A].

Jenkins, Alice. 'What the Victorians Learned: Geometry'. *Journal of Victorian Culture* 12 (2007): 267–73 [2007B].

Jenkins, Alice. 'George Eliot, Geometry and Gender'. *Literature and Science*. Ed. Sharon Ruston. Woodbridge: Brewer, 2008, 72–90.

Jenkins, Alice. 'Genre and Geometry: Victorian Mathematics and the Study of Literature and Science'. *Uncommon Contexts: Encounters Between Science and Literature, 1800–1914*. Eds. Ben Marsden, Hazel Hutchison and Ralph O'Connor. London: Pickering and Chatto, 2014, 111–23.

Kent, Christopher. 'The Average Victorian: Constructing and Contesting Reality'. *Browning Institute Studies* 17 (1989): 41–52.

Kern, Stephen. *A Cultural History of Causality: Science, Murder Novels, and Systems of Thought*. Princeton, NJ: Princeton University Press, 2004.

Kirby, Vicki. 'Enumerating Language: "The Unreasonable Effectiveness of Mathematics"'. *Configurations* 11 (2003): 417–39.

Kitcher, Philip, and William Aspray. 'An Opinionated Introduction'. *History and Philosophy of Modern Mathematics*. Eds. William Aspray and Philip Kitcher. Minneapolis, MN: University of Minnesota Press, 1988, 3–60.

Klotz, Michael. 'Manufacturing Fictional Individuals: Victorian Social Statistics, the Novel, and *Great Expectations*'. *Novel* 46 (2013): 214–33.

Kreisel, Deanna K. 'The Discreet Charm of Abstraction: Hyperspace Worlds and Victorian Geometry'. *Victorian Studies* 56 (2014): 398–410.

Lawrence, Snezana. 'Capital G for Geometry: Masonic Lore and the History of Geometry'. *Mathematicians and Their Gods: Interactions Between Mathematics and Religious Beliefs*. Eds. Snezana Lawrence and Mark McCartney. Oxford: Oxford University Press, 2015, 167–90.

Levine, George. *Darwin and the Novelists: Patterns of Science in Victorian Fiction*. Cambridge, MA: Harvard University Press, 1988.

Levine, George. *Dying to Know: Scientific Epistemology and Narrative in Victorian England*. Chicago, IL: University of Chicago Press, 2002.

Lewis, Elizabeth F. 'P. G. Tait, Balfour Stewart, and *The Unseen Universe*'. *Mathematicians and Their Gods: Interactions Between Mathematics and Religious Beliefs*. Eds. Snezana Lawrence and Mark McCartney. Oxford: Oxford University Press, 2015, 213–48.

Magnello, M. Eileen. 'Darwinian Variation and the Creation of Mathematical Statistics'. *Mathematics in Victorian Britain*. Eds. Raymond Flood, Adrian Rice and Robin Wilson. Oxford: Oxford University Press, 2011, 283–302 [2012A].

Magnello, M. Eileen. 'Vital Statistics: The Measurement of Public Health'. *Mathematics in Victorian Britain*. Eds. Raymond Flood, Adrian Rice and Robin Wilson. Oxford: Oxford University Press, 2011, 261–82 [2012B].

Manaresi, Mirella, ed. *Mathematics and Culture in Europe: Mathematics in Art, Technology, Cinema, and Theatre*. Berlin: Springer, 2007.

Mann, A. J. S., and A. D. D. Craik. 'Scotland'. *Mathematics in Victorian Britain*. Eds. Raymond Flood, Adrian Rice, and Robin Wilson. Oxford: Oxford University Press, 2011, 77–101.

Margolin, Uri. 'Mathematics and Narrative: A Narratological Perspective'. *Circles Disturbed: The Interplay of Mathematics and Narrative*. Eds. Apostolos Doxiadis and Barry Mazur. Princeton, NJ: Princeton University Press, 2012, 481–507.

McCartney, Mark. 'The Poetic Life of James Clerk Maxwell'. *British Society for the History of Mathematics Bulletin* 26 (2011): 29–43.

McGurl, Mark. 'Social Geometries: Taking Place in Henry James'. *Representations* 68 (1999): 59–83.

Miller, J. Hillis. *The Linguistic Moment: From Wordsworth to Stevens*. Princeton, NJ: Princeton University Press, 1985.

Moore, Carlisle. 'Carlyle, Mathematics and "Mathesis"'. *Carlyle Past and Present: A Collection of New Essays*. Eds. K. J. Fielding and Rodger L. Tarr. London: Vision, 1976, 61–95.

Neeley, Kathryn A. *Mary Somerville: Science, Illumination, and the Female Mind*. Cambridge: Cambridge University Press, 2001.

Nesmith, Suzanne, and Sandy Cooper. 'Trade Books in the Mathematics Classroom: The Impact of Many, Varied Perspectives on Determinations of Quality'. *Journal of Research in Childhood Education* 24 (2010): 279–97.

North, J. D. Review of Karen Hunger Parshall, *James Joseph Sylvester: Life and Work in Letters*. *Historia Mathematica* 30 (2003): 88–90.

O'Connor, Ralph. *The Earth on Show: Fossils and the Poetics of Popular Science, 1802–1856*. Chicago, IL: University of Chicago Press, 2007.

Ottinger, Aaron. 'Geometry, the Body, and Affect in Wordsworth's *The Ruined Cottage*'. *Essays in Romanticism* 21 (2014): 159–78.

Owens, Thomas. 'Wordsworth, William Rowan Hamilton, and Science in *The Prelude*'. *The Wordsworth Circle* 42 (2011): 166–9.

Parshall, Karen Hunger. *James Joseph Sylvester: Jewish Mathematician in a Victorian World*. Baltimore, MD: Johns Hopkins University Press, 2006.

Perl, Teri. 'The Ladies' Diary or Woman's Almanack, 1704–1841'. *Historia Mathematica* 6 (1979): 36–53.

Poovey, Mary. 'Figures of Arithmetic, Figures of Speech: Discourses of Statistics in the 1830s'. *Critical Inquiry* 19 (1993): 256–76.

Poovey, Mary. *A History of the Modern Fact: Problems of Knowledge in the Sciences of Wealth and Society*. Chicago, IL: University of Chicago Press, 1998.

Porter, Theodore M. *The Rise of Statistical Thinking: 1820–1900*. Princeton, NJ: Princeton University Press, 1986.

Porter, Theodore M. 'From Quetelet to Maxwell: Social Statistics and the Origins of Statistical Physics'. *The Natural Sciences and the Social Sciences: Some Critical and Historical Perspectives*. Ed. I. Bernard Cohen. Dordrecht: Springer, 1994, 345–62.

Porter, Theodore M. *Karl Pearson: The Scientific Life in a Statistical Age*. Princeton, NJ: Princeton University Press, 2004.

Pratt-Smith, Stella. 'Boundaries of Perception: James Clerk Maxwell's Poetry of Self, Senses and Science'. *James Clerk Maxwell: Perspectives on His Life and Work*. Eds. Raymond Flood, Mark McCartney, and Andrew Whitaker. Oxford: Oxford University Press, 2014, 233–57.

Pycior, Helena M. 'At the Intersection of Mathematics and Humor: Lewis Carroll's "Alices" and Symbolical Algebra'. *Victorian Studies* 28 (1984): 149–70 [1984A].

Pycior, Helena M. 'Internalism, Externalism, and Beyond'. *Historia Mathematica* 11 (1984): 424–41 [1984B].

Rice, Adrian. 'Mathematics in the Metropolis: A Survey of Victorian London'. *Historia Mathematica* 23 (1996): 376–417.

Richards, Joan L. *Mathematical Visions: The Pursuit of Geometry in Victorian England.* San Diego, CA: Academic Press, 1988.

Richards, Joan L. 'The Probable and the Possible in Victorian England'. *Victorian Science in Context.* Ed. Bernard Lightman. Chicago, IL: University of Chicago Press, 1997, 51–71.

Richards, Joan L. 'The Geometrical Tradition: Mathematics, Space, and Reason in the Nineteenth Century'. *The Cambridge History of Science.* Vol 5: The Modern Physical and Mathematical Sciences. Ed. Mary Jo Nye. Cambridge: Cambridge University Press, 2003, 449–67.

Richards, Joan L. '"This Compendious Language": Mathematics in the World of Augustus De Morgan'. *Isis* 102 (2011): 506–10.

Richards, Mark. 'Charles Dodgson's Work for God'. *Mathematicians and Their Gods: Interactions Between Mathematics and Religious Beliefs.* Eds. Snezana Lawrence and Mark McCartney. Oxford: Oxford University Press, 2015, 191–212.

Rogers, Helen, ed. Special Issue. *Journal of Victorian Culture* 13 (2008): 56–107.

Rosenthal, Jesse. 'The Large Novel and the Law of Large Numbers; or, Why George Eliot Hates Gambling'. *English Literary History* 77 (2010): 777–811.

Rotman, Brian. *Signifying Nothing: The Semiotics of Zero.* Basingstoke: Macmillan, 1987.

Rotman, Brian. *Becoming Beside Ourselves: The Alphabet, Ghosts, and Distributed Human Being.* Durham, NC: Duke University Press, 2008.

Sarton, George. *The Study of the History of Mathematics.* Cambridge, MA: Harvard University Press, 1936.

Shuttleworth, Sally. *George Eliot and Nineteenth-Century Science.* Cambridge: Cambridge University Press, 1984.

Simpson, Michael. 'Strange Fits of Parallax: Wordsworth's Geometric Excursions'. *The Wordsworth Circle* 34 (2003): 19–24.

Skovsmose, Ole, and Paola Valero. 'Breaking Political Neutrality: The Critical Engagement of Mathematics Education with Democracy'. *Sociocultural Research in Mathematics Education: An International Perspective.* Eds. Bill Atweh, Helen Forgasz and Ben Nebres. Mahwah: Lawrence Erlbaum, 2001, 37–55.

Smajic, Srdjan. *Ghost-Seers, Detectives, and Spiritualists: Theories of Vision in Victorian Literature and Science.* Cambridge: Cambridge University Press, 2010.

Small, Helen. 'Buckle, Hardy, and the Individual at Risk'. *Literature, Science, Psychoanalysis, 1830–1970: Essays in Honour of Gillian Beer.* Eds. Helen Small and Trudi Tate. Oxford: Oxford University Press, 2003. 64–85.

Small, Helen. 'Science, Liberalism, and the Ethics of Belief: *The Contemporary Review* in 1877'. *Science Serialized: Representations of the Sciences in Nineteenth-Century Periodicals.* Eds. Geoffrey Cantor and Sally Shuttleworth. Cambridge, MA: MIT Press, 2004, 239–58.

Smith, Barbara Herrnstein, and Arkady Plotnitsky. 'Introduction: Networks and Symmetries, Decidable and Undecidable'. *Mathematics, Science, and Post-Classical Theory.* Eds. Barbara Herrnstein Smith and Arkady Plotnitsky. Durham, NC: Duke University Press, 1997, 1–16.

Smith, Jonathan, Lawrence I. Berkove, and Gerald A. Baker. 'A Grammar of Dissent: "Flatland", Newman, and the Theology of Probability'. *Victorian Studies* 39 (1996): 129–50.

Stallknecht, Newton P. 'On Poetry and Geometric Truth'. *Kenyon Review* 18 (1956): 1–20.

Stauffer, Andrew, ed. 'Forum on Evidence and Interpretation in the Digital Age'. *Victorian Studies* 54 (2011): 63–94.

Throesch, Elizabeth. 'Charles Howard Hinton's Fourth Dimension and the Phenomenology of the Scientific Romances (1884–1886)'. *Foundation: The International Review of Science Fiction* 36 (2007): 29–48.

Throesch, Elizabeth. 'Nonsense in the Fourth Dimension of Literature: Hyperspace Philosophy, the "New" Mathematics, and the *Alice* Books'. *Alice Beyond Wonderland.* Ed. Cristopher Hollingsworth. Iowa City, IA: University of Iowa Press, 2009, 37–52.

Tomalin, Marcus. 'William Rowan Hamilton and the Poetry of Science'. *Romanticism and Victorianism on the Net* 54 (2009). <www.erudit.org/revue/ravon/2009/v/n54/038763ar.html>. Accessed 3 June 2015.

Toole, Betty Alexandra. *Ada, The Enchantress of Numbers: A Selection from the Letters of Lord Byron's Daughter and Her Description of the First Computer.* Mill Valley, CA: Strawberry, 1992.

Valente, K. G. 'Transgression and Transcendence: *Flatland* as a Response to "A New Philosophy"'. *Nineteenth-Century Contexts* 26 (2004): 61–77.

Valente, K. G. '"Who Will Explain the Explanation?": The Ambivalent Reception of Higher Dimensional Space in the British Spiritualist Press, 1875–1900'. *Victorian Periodicals Review* 41 (2008): 124–49.

Warwick, Andrew. *Masters of Theory: Cambridge and the Rise of Mathematical Physics*. Chicago, IL: University of Chicago Press, 2003.

White, Christopher. 'Seeing Things: Science, the Fourth Dimension, and Modern Enchantment'. *American Historical Review* 119 (2014): 1466–91.

Wickman, Matthew. *Literature After Euclid, Before Scott*. Philadelphia, PA: University of Pennsylvania Press, 2016.

Wilson, Robin J. *Lewis Carroll in Numberland: His Fantastical Mathematical Logical Life; An Agony in Eight Fits*. London: Allen Lane, 2008.

Further reading

Poovey, Mary. *Genres of the Credit Economy: Mediating Value in Eighteenth and Nineteenth-Century Britain*. Chicago, IL: University of Chicago Press, 2008.

Rotman, Brian. *Ad Infinitum: The Ghost in Turing's Machine: Taking God Out of Mathematics and Putting the Body Back In*. Stanford, CA: Stanford University Press, 1993.

Rotman, Brian. *Mathematics as Sign: Writing, Imagining, Counting*. Stanford, CA: Stanford University Press, 2000.

16

ASTRONOMY

Pamela Gossin

University of Texas at Dallas

From the moment that the *Principia* appeared in 1687, Isaac Newton's achievements in astronomy and physics demanded the muse and the muse responded. Natural philosophers who were as well-versed in mathematics and science as they were in literary tradition (Edmund Halley first among them) praised his insights – seemingly divine – into the deeper order and hidden design of nature so long unseen by lesser humans' eyes and minds. Poets of all ages and skill levels, whether word-masters such as Alexander Pope, youthful apprentices such as Richard Glover or worthy journeymen such as James Thomson, were inspired to try their hand at crafting poetical descriptions of his new vision of natural phenomena. In doing so, they joined a long line of poets who from ancient times had sung or penned songs that recorded observations about the physical world, including what they could see, ponder and appreciate of the heavens from the earth.

Writers of Newtonian literature were important participants in humankind's extensive tradition of inventing and adapting elevated forms of linguistic and artistic expression to convey and preserve for posterity hard-won treasures of natural knowledge and cosmic understanding. This 'literature of astronomy' would take many forms, from formal and informal verse, prose fiction, drama, satire, essays, popular and technical texts, and biographical and historical narratives to works of literary-critical scholarship. The unprecedented story of Newton's life and mind generated equally unprecedented cultural reception and response. Like the long tail of one of the 'great' comets of his own era, the glowing narrative of his ideas and achievements grew and grew until it dominated the intellectual horizon, arching across the next two centuries. (For an extended discussion of this theme, see Hankins; Dobbs and Jacob; Gossin 1989; Gossin 2007 57–102; Gossin 2015A).

For creative writers attracted to post-Newtonian astronomy and cosmology, classical texts by writers such as Hesiod, Aratus and Lucretius and the authoritative epic poems of Dante and Milton provided still-powerful literary models for expressing and harmonizing humanity's aesthetic, scientific and religious efforts to understand the cosmos. Newton's resolution of one scientific revolution and his ushering-in of another inspired a wide diversity of such works. Contemporary writers praised Newton's genius and satirized it; astronomical poems, essays and sermons with primarily physico-theological purposes appeared in numbers matched, if not surpassed, by texts that emphasized social, political, philosophical and Newtonian critique (e.g. Glover; Halley; Desaguliers; Thomson; Blackmore; Derham; Pope; Swift). Such materials quickly attracted the attention of biographers, historians and literary scholars, with their initial and subsequent research contributing to the development of both the history of science (Higgitt) and 'literature and science' as new academic domains.

Within narratives of the history of science, Newton's contribution was described as at once representing the resolution of ancient problems in astronomy and cosmology and as setting the agenda for future investigations. His approach to 'natural philosophy' transformed the scope and practice of physics, literally redefining our understanding of φύσις 'the nature' of nature. His development of mathematical analysis and gravitational theory transformed astronomy, effectively treating Earth as one planetary body (and mathematical variable) among many and shifting astronomical focus from direct, observational practices that emphasized description, positional mapping and measurement to new techniques of an emergent astrophysics that increasingly depended upon mathematical calculations and modelling, theoretical and experimental methods. Before Newton, stargazing shepherds, observational astronomers and star-crossed lovers and their poets had all found inspiration in the simple, universal human act of looking up. They had contemplated together the same set of physically visible celestial objects: the sun, moon, stars, the five 'naked-eye' planets and the occasional comet. By the dawn of the nineteenth century, the upward gaze of observational and mathematical astronomers and perceptive British Romantic writers and thinkers would be drawn into spaces far beyond such patently observable phenomena to consider the meaning of the new, invisible realities of the post-Newtonian universe.

Literature and astronomy in the first post-Newtonian century

The historical, cultural and literary impact of Newtonian astronomy and science was extensive, complex and continuous. The essential difficulty in assessing and discussing just *how* extensive is the still-troublesome fact that with the publication of his masterwork Newton issued a substantial discursive and representational challenge. With the carefully chosen words of the title, *Principia Mathematica,* **Mathematical** *Principles of Natural Philosophy* (my emphasis), he alerted all potential readers that the natural philosophy contained therein required new ways of seeing and employed new ways of describing the long-accepted, but now demonstrably superficial, appearances of things. Newton's insights demanded the mathematical muse, even if he had to coinvent her himself. In future, where even the most sophisticated poetical descriptions and prose popularizations might falter, the elegance of mathematical expression would not.

When Newton looked up, he saw what other human eyes could visually observe and measure (with or without the telescope); he then enhanced his vision with what he could also mathematically discern and describe. In that abstract space, he saw that the interactions of invisible forces between physical bodies (both seen and unseen), whether down here upon Earth or scattered throughout our solar system, could be represented and analysed by mathematical symbols and equations. His new theory of nature united astronomy and physics, observational data with mathematical analysis, to provide a powerful synthesis of celestial dynamics with terrestrial mechanics. According to Newton's truly *uni*-versal story, all moved as one with the lawful force of gravitational attraction judiciously governing a now unified solar *system* comprised of our life-sustaining star, its planets, their moons and the formerly wayward and menacing but now (thanks to his inverse-square law) no longer unpredictable comets.

Within the discursive culture of his most mathematical readership, the momentum of Newton's ideas grew rapidly as astronomers and natural philosophers picked up threads of his mathematics and followed them up in their own theoretical and experimental programmes, testing and applying the power of his descriptive laws to ever-increasing classes of natural phenomena. Just past the middle of the eighteenth century, new calculations of the 'three-body problem' and the date of the return of Halley's Comet by astronomer-mathematicians such as Alexis Clairaut, Joseph Jérôme de Lalande and others confirmed and reaffirmed the unprecedented accuracy and predictive power of Newton's laws. Now his mathematical prophecies extended to the outer reaches of the solar system and, seemingly, beyond the grave.

Within another few years, however, those who looked up through the long lens of his 'mathematical principles of natural philosophy' would see a far different cosmic story unfolding in an astronomical and physical space that Newton himself would scarcely have recognized. The natural philosophical speculations of Immanuel Kant and the mathematical analyses of planetary dynamics by Pierre-Simon Laplace and others began to express alternative – and cosmology-shifting – explanations to Newton's concerning the probable cause of order and predictability in the universe. Their revised calculations had cosmic consequences. When Newton did the mathematics, his interpretation of the equations enabled him to envision the continuous action of divine will maintaining cosmic stability, occasionally sending comets to fine-tune planetary orbits and prevent their decay and collapse into the sun. Variations of the nebular hypothesis told, instead, of a solar system that had formed through the operation of still-unidentified but likely indifferent natural processes (the gradual evolution of celestial gas clouds over long aeons of time) and that perturbations detected within individual planetary orbits cancelled each other out. Such ordered regularity resulted without need for direct action by the invisible hand of Newton's God.

Near the close of the eighteenth century, the improved optics of Newtonian reflecting telescopes greatly extended the limits of human vision and allowed observers to fathom previously unimagined depths of space and time. Such instruments enabled astronomers to discover one new planet visually (Uranus), confirm the mathematical calculations that predicted Neptune's existence and apply Newton's gravitational law to their orbits and those of numerous (and even greatly more distant) binary and multiple-star systems. This revision of cosmic scale and the expansive effective descriptive range of Newton's mathematics held equal potential to amaze or demoralize. At the same time, early indications of evolutionary development and change in nebulae, 'island universes' and star clusters cast doubt upon previous narratives of cosmogony and cosmology and fuelled speculation about the 'eternal' heavens' beginning *and* end. By the opening decades of the nineteenth century, the optics of Newton's rainbow aided astrophysical investigations of the chemical composition of our Sun, the stars, planets and nebulae through spectroscopy, which revealed them to be celestial engines burning with darkness visible and *finite* energy, as liable as mere earthly things to physical decay and demise.

Near the three-hundredth anniversary of Newton's great work, theoretical physicist and Nobel laureate Steven Weinberg voiced his domain's view that 'all that has happened [in physics] since 1687 is a gloss on the *Principia*' (quoted in Hawking and Israel 16). Yet the fact remains that in all eras since its publication the great majority of general ('literary') readers have not been able to follow either Newton's original mathematical demonstrations or the post-Newtonian applications inspired by his work. Among his contemporaries or near-contemporaries, even the most determined and mathematically minded creative writers and philosophers struggled to comprehend it (e.g. Arbuthnot, Swift, Locke and Voltaire). Such difficulties motivated myriad attempts to translate and popularize Newton's ideas, preferably in narratives that relied upon little maths and less geometry, spurring a virtual cottage industry in the emerging genre of 'popular science writing' that continues to this day (e.g., most immediately, Pemberton; Gregory; Algarotti; Émilie du Châtelet; Voltaire; Martin; and, much more recently, Guicciardini; Chandrasekhar; Donahue; Pask).

Some of Newton's aspiring readers consulted more highly skilled friends to check their mathematics (such as Locke with Huygens). Others puzzled-out the details of Newton's thought for themselves, supplemented by the content of ongoing physico-theological debates, learned correspondence and the latest coffee-house exchanges. Among these was Samuel Johnson, whose essays and letters show that he achieved a sophisticated enough understanding of Newton's physics to question assertively the first generation of Newtonians' moral and spiritual reliance upon the powers of human reason over divine law, disciplined conscience and right action (Schwartz). In the aftermaths of the American and French Revolutions, William Blake generated poetic, visionary opposition to what he saw as the negative energy of Newtonian reason that had hardened a universe of infinitely imaginative and spiritual possibilities into a merely material and mathematically delimited world (as in 'Mock on, Mock on, Voltaire, Rousseau', *An Island in the Moon*, *Vala*, *Jerusalem*, *Book of Urizen* and *Milton* – see discussions by Ault; Kirk; Worrall; Miner; Bronowski; Peterfreund; Greenberg; Valiunas). Blake's reaction was no doubt conditioned by nearly a century of Enlightenment Newtonianism and its various forms of Newton-worship. Had he been aware of the still-secret details of Newton's life and personality, especially his obsessive, physically and emotionally tortured investigations into occult mysteries of alchemy and unorthodox theology, his poetic and artistic depictions of Newton might have represented a much more kindred spirit.

The gap that emerged between technical astronomy and cosmology and the general public's construction of a mathematically and scientifically informed world view during the first century after Newton remains unresolved today. Despite the wide range of historical and critical engagement with many aspects of post-Newtonian astronomy and physics, scholars of literature and science have still not adequately addressed the meaning and consequences of Newton's literary and linguistic choice to develop a new mathematical language in which to express his new world view. This situation has especially strong implications for the study of nineteenth-century interrelations of literature and astronomy because the continuing and interconnected story of the physical sciences in this period (and beyond) turns precisely on how the poetics of equations enabled investigations of Newton's view of nature ever further out into space and across all known and then-yet-undiscovered scales of material reality. Unfortunately, the linguistic territories of geometry, algebra, trigonometry and calculus remain places where even the most intrepidly interdisciplinary of literature and science scholars fear to tread (although see Flood, Rice and Wilson; Woolley; Toole; Warrick; Simon for recent scholarship in this area).[1] As things currently stand, lack of mathematical and technical expertise combined with inadequate understanding of the history of physical science (its content and contexts) not only distorts but prevents otherwise interested and insightful scholars from seeing and identifying 'what's out there' in astronomical literature and producing astronomically and historically informed analyses and interpretations.

Like all post-Newtonian audiences before them, most twenty-first-century readers can more readily 'see' and read astronomical references and tropes involving physically visible phenomena (stars, sun, moon, planets, Milky Way etc.) as they appear in various kinds of texts; although, even then, they may not always accurately identify and interpret them. So rarely have allusions to and literary treatments of gravitational, theoretical and mathematical astronomy been identified and discussed by scholars that some have considered them ostensibly non-existent (Jones; Nicolson). Given the ongoing nature of such constraints, the full story of the literature of astronomy in nineteenth-century Britain cannot be told here, but the omission of references to post-Newtonian gravitational and theoretical astronomy going forward should not be mistaken as evidence for its absence, rather as an indication of all the dark matter/dark energy yet to be detected amid the celestial array of light-emitting objects that have more readily captured scholarly attention.

Literature and astronomy in nineteenth-century Britain

For rising intellects at the dawn of the nineteenth century, Newton was still a force to be reckoned with. William Wordsworth saw beyond previous images of Newton as a materialistic, hard-as-rock law-giver to recognize something deeper, even Romantic, in his gaze:

> [L]ooking forth by light
> Of moon or favouring stars, I could behold
> The antechapel where the statue stood
> Of Newton with his prism and silent face,
> The marble index of a mind for ever
> Voyaging through strange seas of Thought, alone.
>
> *(The Prelude (1850)) (Book 3, ll. 58–63)*

Significantly, he would abstain from the mockingly 'merry' anti-Newtonian comments famously made by Charles Lamb and John Keats at Benjamin Haydon's studio in 1817 (see Fara 124). While viewing their host's large-scale painting-in-progress, *Christ's Entry Into Jerusalem*, the inebriated Lamb objected to Newton's depiction among the crowd of believers, remarking that the philosopher had more faith in mathematics than the divinity and, further, had 'destroyed all the Poetry of the rainbow'. Keats then proposed a toast to Newton's health and 'the confusion of mathematics'. While fun as an anecdote, the incident serves to poorly represent the already intricately braided interrelations of and attitudes towards art, religion, literature, mathematics and science present in the scene and in the complex social world around it.

The works of the British Romantics, such as Keats ('Lamia'), Wordsworth ('Star-Gazers'), Coleridge ('Religious Musings'), Byron (*Don Juan, Cain*) and both Shelleys (*Prometheus Unbound, Queen Mab; Frankenstein, The Last Man*) are replete with their understanding of the natural world and the emergent natural sciences, including astronomy (Almeida; Owens; Levere; Mitchell R.; Ruston; Chow; Grabo; Morton; Underwood; Knellwolf and Goodall; Holmes R.). Although Keats misrepresented the earthly history in 'On First Looking Into Chapman's Homer' (it was Balboa rather than Cortez), he got the history and poetics of the astronomy right. William Herschel often adapted nautical language to his telescopic observations, describing them as attempts to 'fathom' the profundity of space (1817). Keats's 1816 allusion to his discovery of Uranus – a full year before his infamous toast – was subtly in sync with the astronomer's own view of himself as a sailor peering into the depths of the sea, as a 'new planet swims into his ken' (l. 10), referring here to the field of vision seen through a telescope's eyepiece.

Yet for all of his almost preternatural powers of prediction, neither Newton (nor any of the Newtonians) saw William Herschel coming. A former military band musician and mostly self-educated German émigré, it would be hard to imagine a figure less likely to usher in the greatest age of observational astronomy since Galileo. Herschel's highly visual and publicly acclaimed work in telescope-building, astronomical observation and cosmological theory had been in progress for several decades when Keats wrote his sonnet. Although his discovery of a new planet merely identified him as the only human since ancient times to do so, brought him broad international public and scientific attention, and attracted the notice of King George III (for whom he wanted to name the new orbiting body), along with royal funding for a bigger, better telescope, Herschel himself thought of that achievement as quite trifling compared to his others.

Since he primarily studied far more exotic and less understood celestial objects far beyond our solar system, it is hard to disagree with him.

On William Herschel's watch the population of night sky objects familiar since ancient times grew to include one new planet, numerous new comets, moons and asteroids, hundreds of 'double' stars and thousands of star clusters, nebulae and island universes. His careful tracking of stellar pairs over several decades revealed many to be physical binaries, a fact that confirmed Newton's gravitational law as a truly 'universal' one and extended its applicability to the motions of even very distant objects. The keenness of Herschel's observational powers was matched by that of his theorizing. Through his telescopes he *saw* the relationship of deep space to deep time: the more distant the objects we observe, the longer their light has taken to reach our eyes and therefore the further back into time we are seeing. He boldly attempted to map the distribution of stars in our Milky Way and gauge stellar and cosmic distances. His work more than doubled the size of our solar system and increased the estimated bounds of the universe exponentially. Even more profoundly, he saw evidence for evolutionary development, growth and change, and life and death cycles in his newly resolved nebulae and, thus, in the heavens themselves (see Gossin 2007 57–102; Gossin 2010).

After his death in 1822, William's work was continued by his sister, Caroline, and extended by his son, John. In addition to her devoted assistance to her brother's observations, Caroline independently discovered eight comets, numerous nebulae and star clusters and performed tens of thousands of calculations that organized and corrected many decades of observational data into the best astronomical catalogues since those of Astronomer Royal, John Flamsteed (whose work had informed Newton's theory-building). She kept detailed observatory notes and diaries that provide unprecedented insights into the family's daily and nightly working lives. Recognized as an 'exceptional' woman (with all of that designation's attendant connotations, both positive and negative), she was awarded national medals of honour and became a scientific celebrity, attracting notable visitors from around the world. Members of the royal family, curious social luminaries, working astronomers and other learned ladies, such as the novelist Fanny Burney, flocked to her home to see her latest telescopic discoveries first-hand. Her nephew, John, grew up to confirm his father's mapping of the northern sky and then observed, catalogued and mapped the stars, nebulae and apparent void spaces of the southern hemisphere.

The elder Herschel experienced periodic fluctuations in scientific and social favour, with both his technical achievements and theoretical ideas marginalized at times. He speculated openly, for example, about the plurality of worlds and their inhabitants as well as possible correlations between sunspots and wheat prices (both open questions in current astrobiology, space weather and climate science). Hailing from Hanover, William always remained a 'resident' of England and not native, as the King upon occasion reminded him. His son John's acclaimed astronomical expedition to the Cape of Good Hope, however, would make him an undisputed *national* 'hero of science', second only to Newton in glory and fame. Throughout his long life, John established a strong network of scientifically minded friends and held numerous offices within scientific societies and organizations. A gifted scholar of the Classics and poetry as well as Continental calculus, he drew on both sets of expressive skills to reach out to the general public with elegantly crafted – at times, sublime – popularizations of his own astronomical work, mathematics and other sciences.

The collective achievements of the Herschels in observational astronomy – William's readable and often eye-catchingly illustrated papers and essays, Caroline's professional astronomical catalogues, journals and memoirs, John's expeditions, essays, popularizations and poetry – along with their highly visible public personae all effectively served as vehicles of astronomical dissemination and public education. The Herschels made no special demands on the muse. They themselves stood as their own best texts: living exempla of curious minds reading and writing the universe. Their appearance on the nineteenth-century astronomical horizon provided some relief for popular understanding of the science even as research in their field grew increasingly mathematical, statistical and experimental. Some of their findings would appear quite naturally in the literary works of writers with first-hand experience of astronomy. Tennyson drew upon nebular imagery in *In Memoriam* and Hardy's account of the workaday details of his fictional astronomer in *Two on a Tower* owes much to John Herschel's real life. Other writers responded poetically, instead, to the difficulties of crafting literary representations of abstract technical astronomy, as in American poet Walt Whitman's account of 'proofs, figures . . . charts and diagrams' (ll. 2–3) in 'When I Heard the Learn'd Astronomer'.

Popular understanding during the long Herschel era was also aided by the fact that the nineteenth-century day and night skies were unusually busy astronomically, with the apparitions of numerous 'great' (large, naked-eye) comets, solar and lunar eclipses, Transits of Venus and related astronomical expeditions almost continuously providing newspapers and popular periodicals with reasons to post celestial headlines. Such news may have shaped Dickens's notion of 'telescopic charity' and the 'cosmic' complexity and perspective of the narration of *Bleak House*, as well as George Eliot's references to personal experiences with astronomical study and telescopic observation in *Mill on the Floss* and *Daniel Deronda*. Other major developments in astronomy remained less amenable to literary adaptation or effective popularizations: the confirmation of stellar parallax in the 1830s, early calculations of stellar distances and proper motion, the invention and application of spectroscopy, the *mathematical* 'discovery' of Neptune (although the subsequent international priority dispute was plenty popular, if not approaching 'vulgar'), measurements of radial velocity and the detection of red shift; calculations of stellar brightness and magnitude, and the rise of solar chemistry and solar physics. Near the *fin de siècle*, Norman Lockyer's cosmological concepts of the 'heat-death of the universe' and his 'principle of degradation' figured strongly in H. G. Wells's *The Time Machine* (1895), as did the cosmological implications of entropy and thermodynamics in numerous other late works (see Myers 1985; Myers 1989; Brantlinger; Alexander; MacDuffie).

Nineteenth-century British literature and astronomy in other forms

Partly eclipsed in the minds of twentieth-century scholars by the more down-to-earth narratives of Darwinism and the Industrial Revolution, the powerful new visions of the universe constructed by nineteenth-century astronomers have recently begun to attain the level of attention they deserve within historical, literary and cultural treatments of science. Early biographies memorialized the Herschels' achievements and stand as still-useful attempts to contextualize the intellectual, social, national and even moral significance of their lives and works (see Herschel Mrs J.; Holden; Proctor R. 1872; Lubbock). Of more immediate interest for current literature and science studies may be eminent historian of astronomy Michael Hoskin's comprehensive and compelling accounts of the Herschel family business. Read together, Hoskin's numerous publications bring into sharp focus the collective nature of the astronomical achievements of William, Caroline and John. Feminist historians of science have successfully advocated for richer recognition of Caroline's individual contributions (Ogilvie; Brock; Schiebinger). Even more recently, best-selling 'popular' or 'hybrid' histories of science have offered novelistic accounts of the diverse investigations of the inquisitive personalities at work within this unusual scientific household and the social, political and cultural spheres around them (Holmes R.; Lemonick). Through such scholarship and narratives, we can gain a vibrant sense of the importance of studying literature and astronomy as and through the artefacts of lived experience. Such remains of living personalities matter; they 'tell' and communicate in ways that the formal publication of words on paper never can. How did it feel to spend time with the Herschels' contagious astronomical enthusiasm (if not mania)? To be whisked off from piano lessons with your master to look through his twenty-foot reflector? To meet William while travelling and have him set up his trusty seven-footer at the next convenient stop? Such creative humans, fully living their science, embodied the best potential of the 'literature of astronomy' as they put words and images into action by offering first-hand experiences that transformed one star-struck heart and mind at a time (see early memoirs and biographies as well as eyewitness accounts by Fanny Burney as in Lubbock; Watson; Mitchell M. 1889; Mitchell M. 1896).

Not everyone, of course, could meet the Herschels personally to experience astronomical poetry in motion for themselves. Fortunately, we can also recover some sense of their lives and work and those of their contemporaries through a whole constellation of popular accounts. These texts – of special interest to scholars of the rhetoric of science – range from essay-length to book-length biographies to general texts written for the reading public, some by astronomical practitioners or peer-practitioners, others by variously skilled 'popularizers' or 'familiarizers'. Among these materials, there was something for almost everyone, from gentleman-scholars and serious amateurs to rising middle-class readers, women and children. Some related the past achievements of astronomy within the context of emerging nineteenth-century research and discoveries, such as Robert Small's *Astronomical Discoveries of Kepler* (1804), Scottish astronomer James Ferguson's *Astronomy Explained Upon Sir Isaac Newton's Principles* (1809) and John Bonnycastle's *An Introduction to Astronomy* (which was published in numerous editions between 1786–1816 and was known to Keats). Some translated and transmitted significant texts from other languages into English, such as Mary Somerville's translation and popularization of Laplace's *Mécanique céleste*, *The Mechanism of the Heavens* (1831), which provided cogent syntheses and explications of technical science, as did her *On the Connexion of the Physical Sciences* (1834) (Patterson; Neeley; Arianrhod; Chapman 2004).

Others attempted to harmonize the results of new astronomy and the nebular hypothesis with established religious interpretations of cosmogony, cosmology and belief. William Whewell's 1833 Bridgewater Treatise, *Astronomy and General Physics With Reference to Natural Theology*, had direct impact (in both spoken and written forms) on the development of important intellects of the day, mostly notably Tennyson and Darwin. Many practitioner accounts conscientiously addressed natural theological concerns within their prose tours of the universe, including John Herschel's *A Treatise on Astronomy* (1833) and *Outlines of Astronomy* (1849) (see Myers 1985; Myers 1989; Ruskin). Scottish educator and astronomer John Pringle Nichol, an effective public lecturer and writer, did much to explain and popularize the nebular hypothesis. His *Views of the Architecture of the Heavens* (1837) and *Phenomena and Order of the Solar System* (1838) were intently read by De Quincey and George Eliot (Smith J.). His works often reported the latest in astronomical evidence or theory, as in *Thoughts on Some Important Points Relating to the System of the World* (1846), *The Stellar Universe* and *The Planet Neptune: An Exposition and History* (both 1848).

Practitioner-popularizer Richard Anthony Proctor was practically a one-person factory of popular astronomy, producing over a dozen books and other publications, many of which were well-known to Hardy and other writers, including *Other Worlds Than Ours* (1870), *Essays on Astronomy* (1872) and *The Poetry of Astronomy* (1880) (Gossin 2007). He also contributed astronomical articles to popular periodicals and founded his own science magazine, *Knowledge*, in which he brought recent astronomy to an even broader audience (Lightman 1996; Henson *et al.*). His daughter, Mary Proctor, followed in both his astronomical and his literary footsteps, publishing numerous works of popular astronomy across several decades. The honour of having the 'last word' for the century's astronomical achievements, however, goes to the highly regarded work of Irish-born Agnes Clerke (Brück; Lightman 1997A). In her *A Popular History of Astronomy During the Nineteenth Century*, first published in 1885, she poignantly summed up the life of William Herschel as a model of learning who opened up astronomical inquiry to all, not just those who through chance circumstances had been born or deemed 'great'. While she may have specifically targeted then-current (and still post-Newtonian) social and national bias within the increasingly 'professional' astronomical community, her insight has had lasting prescience, as, even in the age of 'big' science, amateur astronomers of modest financial and social means still make significant contributions.

Interdisciplinary studies of nineteenth-century British literature and astronomy

Neither twentieth-century 'influence' studies of the impact of science upon literature nor history of science models of 'trickle-down' popularization enable us to describe and analyse adequately the sheer volume and variety of the sites, levels, means, modes and methods active in promoting exchange between so-called 'literary' and 'scientific' cultures among which astronomy was but a part. New interdisciplinary cultural studies approaches are turning valuable scholarly attention to the materials of popular culture and science in the press, public spaces, institutions and 'marketplaces'. Such volumes often include at least one or two essays that make specific reference to the intermixing cultures and discourses of literature and astronomy (Fyfe and Lightman; Lightman 2007; Lightman 1997B; Henson *et al.*; Bowler). This scholarship reveals that a fascinating range of popular texts and products were consumed by an equally diverse range of readers. Figures we know as 'literary' writers did not then always loftily occupy separate spheres but often lived and moved and were social beings in a bustling agora of ideas, actively consuming and participating in a wide variety of scientific and intellectual texts and cultural activities. Yet, even through this multiplicity of processes and products, most members of the generally educated public (especially the marginally literate) never gained more than a passing, elementary textbook-level understanding of astronomical phenomena and their cosmological consequences. That was far from true, however, for a more highly sensitive and educated reader and writer such as Tennyson or a more highly motivated, self-educated one such as Hardy. Tennyson's 'gradualist' resolution of grief in *In Memoriam* was equally dependent upon his knowledge of the 'deep time' of both geology and astronomy, while in *The Princess* his vision of a feminist utopian educational system incorporates first-hand astronomical observation into the curriculum (Korg; Gossin 1996A). Hardy drew upon his popular reading of astronomy and evolutionary theories in his complex fictional world-building, epistemology and social philosophy, equally apparent in his descriptions of the practical 'shepherd's' knowledge of the night sky exhibited by Gabriel Oak in *Far From the Madding Crowd* and of Swithin St Cleeve's sophisticated understanding of deep space/time and cosmic evolution in *Two on a Tower* (see Gossin 2007). A similar integration of astronomical allusions, from direct naked-eye observations to cosmological speculations, also appears in his verse (Gossin 2010).

There is no full survey of nineteenth-century British literature per se – referring here now to novels, poetry and other forms of creative writing – that engages astronomy and cosmology. A. J. Meadows's *The High Firmament* (1969), however, offers a broad sweep of astronomy and British literature, medieval through Victorian. Modelling his study closely upon Francis Johnson's historical account of *Astronomical Thought in Renaissance England* (1937), Meadows expanded its time frame while providing the literary side of the story. The penultimate chapter offers a useful overview of British literature and Newton. In the final one, he examines the view that astronomy (often described as the most central of sciences within most of the history of science) later found itself (in his view) diminished in status with the rise of the Industrial and Darwinian revolutions. Throughout this discussion, he refers to the astronomical writing of Kingsley, Blake, Hazlitt, Burney, Keats, Disraeli, Coleridge, P. B. Shelley, Tennyson, Hardy, Swinburne, Yeats and Wells. Of special interest may be his account of William Wordsworth's particular brand of Newtonianism and both his and Dorothy's regular habit of stargazing and planet-hunting; Shelley's astronomical references in the text and notes to *Queen Mab*; and Tennyson's lifelong interest in astronomical observation (despite his own poor eyesight), personal acquaintances with astronomers such as Lockyer and Le Verrier and deep understanding of the interconnectedness of the 'evolution' of celestial and terrestrial realms. Meadows also discusses poets who successfully reconciled nineteenth-century discoveries of cosmic vastness with their religious beliefs, such as Meredith, Hopkins and Francis Thompson.

In *The 'Scientific Movement' and Victorian Literature* (1982), Tess Cosslett describes the incorporation of Darwin's, Huxley's and Tyndall's science within Tennyson's *In Memoriam*, Eliot's *Middlemarch*, Meredith's 'Melampus' and 'Meditation Under Stars', and Hardy's *A Pair of Blue Eyes*. All of these works, however, also have deep astronomical and cosmological content that could provide significant insight into how Victorian writers synthesized natural history with astronomy and cosmology in their personal world views. This perspective informs Jacob Korg's masterful 'Astronomical Imagery in Victorian Poetry' (1985), in which he details the important cultural work such texts performed in 'reconciling the scientific and humanistic traditions' (137). Drawing a useful comparison with the metaphysical poets, Korg describes how many Victorian writers worked towards new philosophical syntheses of natural and theological explanations, as evidenced by such works as Clough's 'The New Sinai' and 'Easter Day', Robert Browning's 'Abt Vogler' and *The Ring and the Book* (among others), Tennyson's long and shorter verse, Thompson's 'Carmen Genesis', 'The Hound of Heaven' and 'Orient Ode' and Meredith's 'Lucifer in Starlight'.

Given the obvious challenges of achieving proficiency in the technical details of nineteenth-century astronomy and astrophysics as mathematical and experimental sciences, and developing appreciation for their complex historical contexts before attempting to identify and explicate allusions to their methods, theories and discoveries within the novels or poetry of the period, it is understandable why most literature and astronomy scholars limit the scope of their studies to the works of a single author or a selection of closely related texts. Scholars must often painstakingly sift through dozens of contemporary primary astronomical texts, textbooks and dictionaries, and specialized studies by historians of nineteenth-century astronomy, to learn adequately about the usage of even one seemingly transparent allusion or term. Astronomical terminology was under rapid, constant – and contradictory – construction. 'Variable stars', for instance, could refer to stellar or planetary phenomena and be used quite differently by amateur practitioners and emerging professionals, and those usages, in turn, varied sharply from even early twentieth-century technical ones, let alone current ones. Attention to how linguistic changes occur under historical pressures can prevent anachronistic missteps and make or break textual analysis in literature and astronomy.

With hopes of resolving (or, at least, minimizing) such issues, a small cluster of recent scholars have attempted to produce historically and technically informed studies of the literature of astronomy either through their own extensive multidisciplinary training or through careful consultation with historians of astronomy or working astronomers. In *Virginia Woolf and the Discourse of Science: The Aesthetics of Astronomy* (2003), Holly Henry traces Woolf's early interest in astronomy through that of her father, Leslie Stephen, and her own reading of Hardy's *Two on a Tower*. Later, popular astronomy texts, lectures and the BBC radio broadcasts of James Jeans made even stronger impressions, enabling Woolf to develop a 'global aesthetic' (Henry 2) within her literary narratives that took on the problem of 'modernist human decentering and re-scaling' (Henry 3). Henry provides an overview of contemporary astronomical discoveries and cosmology and gives effective readings of 'The Searchlight', 'Solid Objects' and three major works, *The Waves*, *The Years* and *Three Guineas*. Through these analyses she suggests that critical study of the Bloomsbury circle should also consider the interdisciplinary cultural and celestial 'globe' around it (Gossin 2006). In *Thomas Hardy's Novel Universe: Astronomy, Cosmology and Gender in the Post-Darwinian World* (2007), I propose a hybrid 'literary history of science' methodology and apply it to the study of Hardy's knowledge and use of astronomical discoveries, theories, and cosmological ideas in his long prose fiction. Attending to the cosmological settings, narrative structure, astronomical tropes, characterization and representations of gender in seven of his novels, I argue that Hardy's personal synthesis of ancient and contemporary astronomy with mythopoetic and scientific cosmologies enabled him to write as a 'literary cosmologist' for his age, bringing the findings of both astronomy and geology fully to bear upon his vision of nineteenth-century life on this planet. The profound new myths that comprise Hardy's novel universe can be read as a sustained set of literary thought experiments within which he tests out, observes and describes complex interrelations of character and environment, local and universal forces, regional and cosmic scales of being, psychological and social realities, the natural and supernatural, the masculine and the feminine.

In *Science, Time and Space in the Late Nineteenth-Century Periodical Press: Movable Types* (2007), James Mussell expands our view of 'literature and astronomy' by focusing our critical gaze away from writers' consideration of celestial objects and their cosmological meanings and towards representations of time and space in the materials of print culture. His analysis effectively incorporates the historical, philosophical, social and literary-critical insights of such scholars as Stephen Kern, Henri Lefevre, Bruno Latour, Ron Schleiffer and Mark Turner. Further extending our notion of 'astronomy' to include problems of astrophysics and cosmological theory, Barri J. Gold demonstrates in *ThermoPoetics: Energy in Victorian Literature and Science* (2010) how writers such as Tennyson, Dickens, Spencer, Stoker and Wilde engaged the ideas and language of nineteenth-century physics, energy, force and entropy in relation to concerns of faith, evolutionary theory, empire, race, class, gender and sexuality. Such studies explicate the cultural transformation of 'literature and astronomy' during the final years of Victoria's reign, when the results of the Michelson–Morley experiment of 1887 (which heralded in the age of special relativity) gave subjects across her waning empire early warning that the centre of their seemingly stable Newtonian world view might not hold. Late Victorian and emergent modern writers such as Yeats, Joyce and Beckett would soon rise to the challenges of producing literary representations of Einstein's relativistic cosmology instead (Price; Ebury; Friedman and Donley).

Most book-length studies of literature and astronomy conscientiously and effectively combine historically and technically accurate information with textual analysis and literary interpretation. Given the complications cited above, most future scholarship in literature and astronomy will likely continue to appear as articles or book chapters, such as Katharine Anderson's work on almanacs (as in Henson *et al.*) and Elizabeth Patterson's on translation, or individual case studies such as that of John Cartwright on Housman, Anne DeWitt on Hardy and Arnold, Linda Marshall on Christina Rossetti or Nicholas Russell on emergent science professions in Victorian and Edwardian novels. Other current scholars who employ nuanced interdisciplinary approaches include David Schroeder, Dometa Wiegand Brothers, Gillian Daw, Meegan Hasted, Mona Narain and Karen Gevirtz, Michelle Boswell and Garrett Peck.

One perhaps unexpected subcategory of nineteenth-century literature and astronomy studies has been inspired by recent developments in planetary science and space exploration projects.

In the wake of SETI (the Search for ExtraTerrestrial Intelligence) and the Voyager missions, historian of science Michael J. Crowe published two extensive works on the extraterrestrial life debate, which include chapters on British views of the plurality of worlds before and after 1750 (Crowe 1986; Crowe 2008). Aaron Parrett's *The Translunar Narrative in the Western Tradition* (2004) and Bernd Brunner's *Moon: A Brief History* (2010) revisit lunar voyages, literature and lore, with plentiful reference to nineteenth-century writers and artists. Of more immediate social and political concern, as discussions of climate change attempt to gauge the still-unfolding consequences of the Industrial Revolution, Robert Markley draws together planetary astronomy with literary, cultural and ecological perspectives in *Dying Planet: Mars in Science and the Imagination* (2005). He discusses Percival Lowell's canal controversy and science-fictional representations of Mars from 1880–1913, including H. G. Wells's most famous ones. In *Imagining Mars* (2011), Robert Crossley attempts a 'comprehensive literary history of Mars in the English-speaking world' (4), from about 1600 forward, including Swedenborg, the Romantics, Flammarion and Verne, William Herschel, Proctor, Tennyson, Lowell, Wells and Stapledon. Martin Willis explicitly connects the 'visual epistemologies' of the Victorians to our own in Part 2 of *Vision, Science and Literature, 1870–1920* (2011), where he also considers astronomical debates about Mars and Lowell's role within them. In *Geographies of Mars: Seeing and Knowing the Red Planet* (2010), Maria Lane provides planetary perspectives from both popular culture and science, with special attention to geography, cartography, visual and literary representations and their rhetoric, as well as the geopolitics of the nineteenth into early twentieth centuries. Collectively, such interdisciplinary studies may yield real-world benefits by helping us to better understand how we created our planetary dilemma and by suggesting ways to imagine our way out of them (or off it).

Today, as in nineteenth-century Britain, working astronomers actively popularize their science and actively contribute to the discourse of 'literature and astronomy'. Co-discoverer of Comet Shoemaker–Levy David Levy offered his practitioner perspective on astronomy-inspired literary works in *Starry Night: Astronomers and Poets Read the Sky* (2001), including texts by many Romantic and Victorian writers. Texas State University astronomer Donald Olson recently published his findings in literary, artistic and historical 'forensic astronomy' in *Celestial Sleuth* (2013) and he frequently publishes in popular astronomy magazines as well. Professional and amateur astronomers and historians of astronomy also regularly make themselves and their online resources available to the public (see especially Andrew Fraknoi). The HASTRO-L (History of Astronomy Listserv) is invaluable for collegial (usually!) interdisciplinary conversations, insights into internalist debates, 'fun facts' and the decoding of nebulous terminology, as well as advice on astronomical software and teaching technology. Increasingly, digital archives and online museums will aid research in the history and literature of astronomy (see the Newton Project). Long featured within planetarium shows and astronomical poetry readings, our future work may inform virtual reality simulation games and 3D educational immersion environments (aka 'holodecks'), and we will be able to take whole classes of our students back to specific astronomical times and spaces. Want to recreate Tess's experience at Stonehenge or see the spiral galaxies with Lord Rosse at Birr Castle? Soon we will be able to experience the cosmic views of such moments – fictional or historical – in interactive high-definition recreations of the night sky as our favourite poets and novelists envisioned or experienced it.

Sadly, however, such technological visualizations may prove increasingly necessary. As Blake saw the dark satanic mills of the Industrial Revolution closing in his English sky, so we too cannot help but see how the smog and light pollution produced by another two hundred years of global development now threatens the very possibility of directly observing the heavens. Many fear that in coming years our children may look up to see only a blankly dark night sky, apparently devoid of stars and any sense of the celestial deep. They may not ponder their place in the universe – wishing and dreaming by starlight – as has every human generation before them. IDA, the International Dark-Sky Association (although not, to my knowledge, named for Tennyson's astronomical 'Princess Ida'), actively advocates for the preservation of the 'poetics of astronomy', as she did. For the sake of future human understanding of cosmic profundity, so should we.

Note

1   See also Alice Jenkins, 'Mathematics', in this volume.

Bibliography

Primary texts

Algarotti, Francesco. *Sir Isaac Newton's Philosophy Explain'd for the Use of the Ladies: In Six Dialogues on Light and Colours* [*Il Newtonianismo per le dame*, 1737]. Transl. Elizabeth Carter. London: St John's Gate, 1739.

Blackmore, Richard. *Creation: A Philosophical Poem in Seven Books*. London: S. Buckley, J. Tonson, 1712.

Blake, William. *The Complete Poetry and Prose of William Blake*. Ed. David Erdmann. Berkeley, CA: University of California Press, 2008.

Bonnycastle, John. *An Introduction to Astronomy*. London: J. Johnson, 1786.

Byron, George Gordon. *Lord Byron: The Major Works*. Ed. Jerome McGann. Oxford: Oxford University Press, 2008 [See especially *Don Juan* and *Cain*].

Clerke, Agnes. *A Popular History of Astronomy During the Nineteenth Century*. Edinburgh: A. and C. Black, 1885.

Coleridge, Samuel Taylor. *The Collected Works of Samuel Taylor Coleridge*. Vol. 16: *Poetical Works (Reading Text)*. Ed. J. C. C. Mays. Princeton, NJ: Princeton University Press, 2001.

De Quincey, Thomas. *The Works of Thomas De Quincey*. Gen. ed. Grevel Lindop. 21 vols. London: Pickering and Chatto, 1999–2003.

Derham, William. *Physico-Theology: or, a Demonstration of the Being and Attributes of God, From the Works of His Creation*. London: [n. p.], 1713.

Derham, William. *Astro-Theology: or, a Demonstration of the Being and Attributes of God, From a Survey of the Heavens*. London: [n. p.], 1714.

Desaguliers, John. *The Newtonian System of the World, the Best Model of the World: An Allegorical Poem*. Westminster: [n. p.], 1728.

Dickens, Charles. *Bleak House*. Ed. Nicola Bradbury. London: Penguin, 2003.

Du Châtelet, Émilie, transl. *Principes mathématiques de la philosophie naturelle* [*Mathematical Principles of Natural Philosophy*]. London: [n. p.], 1759.

Eliot, George. *Mill on the Floss*. Ed. Gordon S. Haight. Oxford: Oxford University Press, 2008.

Eliot, George. *Daniel Deronda*. Ed. Graham Handley. Oxford: Oxford University Press, 2009.

Ferguson, James. *Astronomy Explained Upon Sir Isaac Newton's Principles, and Made Easy to Those who Have Not Studied Mathematics*. London: [n. p.], 1756.

Glover, Richard. 'A Poem on Sir Isaac Newton'. Following Preface to Henry Pemberton, *A View of Sir Isaac Newton's Philosophy*. London: S. Palmer, 1728.

Gregory, David. *Astronomiae Physicae and Geometricae Elementa*. Oxford: E. Theatro Sheldoniano, 1702.

Halley, Edmund. 'Ode to Isaac Newton'. Orig. Latin. Isaac Newton, *Philosophiae Naturalis Principia Mathematica*. London: [n. p.], 1687.

Halley, Edmund. 'Ode on This Splendid Ornament of Our Time and Nation, The Mathematico-Physical Treatise by the Eminent Newton'. Isaac Newton, *The Principia: Mathematical Principles of Natural Philosophy. A New Translation*. Transl. I. Bernard Cohen and Anne Whitman, with Julia Bundenz. Berkeley, CA: University of California Press, 1999.

Hardy, Thomas. *Far from the Madding Crowd*. Ed. Robert C. Schweik. New York, NY: W. W. Norton, 1986.

Hardy, Thomas. *A Pair of Blue Eyes*. New York, NY: Penguin, 1998.

Hardy, Thomas. *Two on a Tower*. Ed. Sally Shuttleworth. New York, NY: Penguin, 1999.

Herschel, Caroline. 'An Account of a New Comet'. Read at the Royal Society, 9 November 1786. London: J. Nichols, 1787.

Herschel, Caroline. 'Account of the Discovery of a New Comet'. In a Letter to Sir Joseph Banks. *Philosophical Transactions of the Royal Society of London*. Reprint. London: [n. p.], 1796, 5–6.

Herschel, Caroline. *Catalogue of Stars Taken from Mr. Flamsteed's Observations*. London: Royal Astronomical Society, 1798.

Herschel, John. *A Treatise on Astronomy*. London: Longman, Rees, Orme, Brown, Green, and Longman, 1833.

Herschel, John. *Outlines of Astronomy*. London: Longman, Brown, Green, and Longmans, and John Taylor, 1849.

Herschel, Mrs John, ed. *Memoir and Correspondence of Caroline Herschel*. London: John Murray, 1876.

Herschel, William. 'Account of a Comet [Planet, Uranus]'. *Philosophical Transactions of the Royal Society of London* 71 (1781). Reprinted in William Herschel, *Scientific Papers* 1:30–8.

Herschel, William. 'Account of Some Observations Tending to Investigate the Construction of the Heavens'. Read 17 June 1784. *Philosophical Transactions of the Royal Society of London* 74 (1784): 437–51. Reprinted in William Herschel, *Scientific Papers* 1:157–66.

Herschel, William. 'Catalogue of Double Stars' [Second]. Read 9 December 1784. *Philosophical Transactions of the Royal Society of London* 75 (1785). Reprinted in William Herschel, *Scientific Papers* 1:167–223.

Herschel, William. 'On the Construction of the Heavens'. Read 3 February 1785. *Philosophical Transactions of the Royal Society of London* 75 (1785): 213–66. Reprinted in William Herschel, *Scientific Papers* 1:223–59.

Herschel, William. 'Catalogue of One Thousand New Nebulae and Clusters of Stars'. Read 27 April 1786. *Philosophical Transactions of the Royal Society of London* 76 (1786): 457–99. Reprinted in William Herschel, *Scientific Papers* 1:260–94.

Herschel, William. 'Catalogue of a Second Thousand of New Nebulae and Clusters of Stars; With a Few Introductory Remarks on the Construction of the Heavens'. Read 11 June 1789. *Philosophical Transactions of the Royal Society of London* 79 (1789): 212–55. Reprinted in William Herschel, *Scientific Papers* 1:329–65.

Herschel, William. 'On Nebulous Stars, Properly So Called'. Read 10 February 1791. *Philosophical Transactions of the Royal Society of London* 81 (1791): 71–88. Reprinted in William Herschel, *Scientific Papers* 1:415–24.

Herschel, William. 'Description of a Forty-Foot Reflecting Telescope'. Read 11 June 1795. *Philosophical Transactions of the Royal Society* 85 (1795). Reprinted in William Herschel, *Scientific Papers* 1:485–527.

Herschel, William. 'On the Penetrating into Space by Telescopes; With a Comparative Determination of the Extent of That Power in Natural Vision, and in Telescopes of Various Sizes and Construction'. Read 21 November 1799. *Philosophical Transactions of the Royal Society of London* 90 (1800): 49–85. Reprinted in William Herschel, *Scientific Papers* 2:31–52.

Herschel, William. 'Experiments on the Solar, and on the Terrestrial Rays That Occasion Heat'. *Philosophical Transactions of the Royal Society of London* 90 (1800): 293–326.

Herschel, William. 'Catalogue of 500 New Nebulae, Nebulous Stars, Planetary Nebulae, and Clusters of Stars; With Remarks on the Construction of the Heavens'. Read 1 July 1802. *Philosophical Transactions of the Royal Society of London* 92 (1802): 477–528. Reprinted in William Herschel, *Scientific Papers* 2:199–234 [First suggestion of gravitational binaries].

Herschel, William. 'Astronomical Observations Relating to the Construction of the Heavens, Arranged for the Purpose of a Critical Examination, the Result of Which Appears to Throw Some New Light Upon the Organization of the Celestial Bodies'. Read 20 June 1811. *Philosophical Transactions of the Royal Society of London* 101 (1811): 269–336. Reprinted in William Herschel, *Scientific Papers* 2:459–97.

Herschel, William. 'Astronomical Observations Relating to the Siderial Part of the Heavens, and its Connection with the Nebulous Part; Arranged for the Purpose of a Critical Examination'. Read 24 February 1814. *Philosophical Transactions of the Royal Society of London* 104 (1814): 248–84. Reprinted in William Herschel, *Scientific Papers* 2:520–41.

Herschel, William. 'Astronomical Observations and Experiments Tending to Investigate the Local Arrangement of the Celestial Bodies in Space, and to Determine the Extent and Condition of the Milky Way'. Read 19 June 1817. *Philosophical Transactions of the Royal Society of London* 107 (1817): 302–31. Reprinted in William Herschel, *Scientific Papers* 2:575–91.

Herschel, William. 'Astronomical Observations and Experiments, Selected for the Purpose of Ascertaining the Relative Distances of Clusters of Stars, and of Investigating How Far the Power of our Telescopes may be Expected to Reach Into Space, When Directed to Ambiguous Celestial Objects'. Read 11 June 1818. *Philosophical Transactions of the Royal Society of London* 108 (1818): 429–70. Reprinted in William Herschel, *Scientific Papers* 2:592–613.

Herschel, William. *The Scientific Papers of Sir William Herschel*. Ed. J. L. E. Dreyer. Vols 1 and 2. London: [n. p.], 1912.

Herschel, William, and John F. W. Herschel, James South. 'On the Systems of Double Stars Which Have Been Demonstrated to be Binary Ones by the Observations of Sir W. Herschel, and Messrs. Herschel, and South'. Edinburgh: William Blackwood; London: T. Cadell, 1827.

Herschel, William, and J. L. E. Dreyer. 'Unpublished Observations of Messier's Nebulae and Clusters. With References to the Observations Published in the *Philosophical Transactions* 1800, 1814 and 1818'. Compiled by J. L. E. Dreyer from William Herschel's notes. Reprinted in William Herschel, *Scientific Papers* 2:651–60.

Johnson, Samuel. *The Yale Digital Edition of the Works of Samuel Johnson*. Gen. ed. Robert DeMaria, Jr. 2014. <www.yalejohnson.com/frontend/collections>. Accessed 19 January 2016.

Kant, Immanuel. *Allgemeine Naturgeschichte und Theorie des Himmels* [*Universal Natural History and Theory of the Heavens*]. Königsberg and Leipzig: Johann Friedrich Petersen, 1755.

Keats, John. *The Poems of John Keats*. Ed. Jack Stillinger. Boston, MA: Harvard University Press, 1978.

Keats, John. *The Letters of John Keats, 1814–1821, Volumes 1 and 2*. Ed. Hyder Edward Rollins. Boston, MA: Harvard University Press, 2002.

Laplace, Pierre-Simon. *Exposition du système du monde* [*Exposition of the System of the World*]. Paris: De l'Impr. du Cercle-Social, An IV. 1796.

Lockyer, [Joseph] Norman. *Elementary Lessons in Astronomy*. London: Macmillan, 1868.

Lockyer, [Joseph] Norman. *Elements of Astronomy: Accompanied With Numerous Illustrations, a Colored Representation of the Solar, Stellar, and Nebular Spectra, and Celestial Charts of the Northern and Southern Hemisphere*. New York, NY: D. Appleton, 1870.

Lockyer, [Joseph] Norman. *The Spectroscope and Its Applications*. London: Macmillan, 1873.

Lockyer, [Joseph] Norman. *Contributions to Solar Physics*. London: Macmillan, 1874.

Lockyer, [Joseph] Norman. *Studies in Spectrum Analysis*. New York, NY: D. Appleton, 1878.

Lockyer, [Joseph] Norman. *Astronomy*. New York, NY: D. Appleton, 1879.

Lockyer, [Joseph] Norman. *The Chemistry of the Sun*. London: Macmillan, 1887.

Lockyer, [Joseph] Norman. *The Dawn of Astronomy*. London: Cassell, 1894.

Lockyer, [Joseph] Norman. *Recent and Coming Eclipses: Being Notes on the Total Solar Eclipses of 1893, 1896 and 1898*. London: Macmillan, 1897.

Lockyer, [Joseph] Norman. *The Sun's Place in Nature*. London: Macmillan, 1897.

Lockyer, [Joseph] Norman. *Inorganic Evolution as Studied by Spectrum Analysis*. London: Macmillan, 1900.

Lockyer, [Joseph] Norman. *Stonehenge and Other British Stone Monuments Astronomically Considered*. London: Macmillan, 1906.

Lockyer, [Joseph] Norman, and G. M. Seabroke. *Stargazing*. London: Macmillan, 1878.

Lockyer, [Joseph] Norman, and Winifred Lucas Lockyer. *Tennyson, As a Student and Poet of Nature*. London: Macmillan, 1910.

Martin, Benjamin. *A Plain and Familiar Introduction to the Newtonian Philosophy*. London: W. Owen, 1751.

Martin, Benjamin. *The Young Gentleman and Lady's Philosophy*. 3 vols. London: W. Owen, 1759–82.

Mitchell, Maria. 'Reminiscences of the Herschels'. *Century Magazine* (October 1889): 903–9.

Mitchell, Maria. *Maria Mitchell: Life, Letters, and Journals*. Ed. Phebe Mitchell Kendall. Boston, MA: Lee and Shepard, 1896.

Newton, Isaac. *Philosophiae Naturalis Principia Mathematica*. London: Joseph Streater, 1687.

Newton, Isaac. *Opticks, or, A Treatise of the Reflexions, Refractions, Inflexions and Colours of Light: Also Two Treatises of the Species and Magnitude of Curvilinear Figures*. London: Samuel Smith and Benjamin Walford, 1704 [see also 2nd, 3rd and 4th edns].

Newton, Isaac. *The Chronology of Ancient Kingdoms Amended, to Which is Prefix'd a Short Chronicle From the First Memory of Things in Europe to the Conquest of Persia by Alexander the Great*. London: J. Tonson, 1728.

Newton, Isaac. *The Treatise of the System of the World*. London: F. Fayram, 1728.

Newton, Isaac. *The Mathematical Principles of Natural Philosophy*. Translated into English by Andrew Motte. To Which are Added, The Laws of the Moon's Motion, According to Gravity. By John Machin. 2 vols. London: Benjamin Motte, 1729.

Newton, Isaac. *The Correspondence of Isaac Newton*. Ed. H. W. Turnbull *et al.* 7 vols. Cambridge: Cambridge University Press, 1959–77.

Newton, Isaac. *The Mathematical Papers of Isaac Newton*. Ed. D. T. Whiteside. 8 vols. Cambridge: Cambridge University Press, 1967–81.

Newton, Isaac. *The Optical Papers of Isaac Newton: Vol. 1: The Optical Lectures, 1670–1672*. Ed. Alan Shapiro. Cambridge: Cambridge University Press, 1984.

Newton, Isaac. *The Principia: Mathematical Principles of Natural Philosophy*. Transl. I. Bernhard Cohen. Berkeley, CA: University of California Press, 2008.

Nichol, John Pringle. *Views of the Architecture of the Heavens*. Edinburgh: William Tait, 1837.

Nichol, John Pringle. *Phenomena and Order of the Solar System*. Edinburgh: William Tait, 1838.

Nichol, John Pringle. *Thoughts on Some Important Points Relating to the System of the World*. Edinburgh: William Tait, 1846.

Nichol, John Pringle. *The Planet Neptune: An Exposition and History*. Edinburgh: J. Johnstone, 1848.

Nichol, John Pringle. *The Stellar Universe*. Edinburgh: J. Johnstone, 1848.

Pemberton, Henry. *A View of Sir Isaac Newton's Philosophy*. London: S. Palmer, 1728.

Pope, Alexander. *The Poems of Alexander Pope: A One-Volume Edition of the Twickenham Text With Selected Annotations*. Ed. John Butt. London: Routledge, 2003 [see especially *The Dunciad*, 'Epitaph on Sir Isaac Newton'].

Proctor, Mary. *Stories of Starland*. New York, NY: Potter and Putnam, 1898.

Proctor, Mary. *Giant Sun and His Family* [1896]. New York, NY: Silver, Burdett, 1906.

Proctor, Richard. *Other Worlds Than Ours*. London: Longmans, Green, 1870.

Proctor, Richard. *Essays on Astronomy: A Series of Papers on Planets and Meteors, the Sun and Sun-Surrounding Space, Stars and Star Cloudlets; and a Dissertation on the Approaching Transits of Venus. Preceded by a Sketch of the Life and Work of Sir John Herschel*. London: Longmans, Green, 1872.

Proctor, Richard. *The Poetry of Astronomy*. London: Longmans, Green, 1880.

Shelley, Mary. *Frankenstein: The 1818 Text, Contexts, Nineteenth-Century Responses, Modern Criticism*. Ed. J. Paul Hunter. New York, NY: W. W. Norton, 1996.

Shelley, Mary. *The Last Man*. 3 vols. London: Henry Colburn, 1826.

Shelley, Percy Bysshe. *The Complete Poetry of Percy Bysshe Shelley*. Eds. Donald H. Reiman, Neil Fraistat and Nora Crook. 3 vols to date. Baltimore, MD: Johns Hopkins University Press, 1999 [see especially vol. 2: *Queen Mab; A Philosophical Poem, With Notes*].

Shelley, Percy Bysshe. *Shelley's Poetry and Prose: Authoritative Texts and Criticism*. Eds. Donald H. Reiman and Neil Fraistat. 2nd edn. New York, NY: W. W. Norton, 2002 [see especially *Prometheus Unbound*].

Small, Robert. *Astronomical Discoveries of Kepler*. London: J. Mawman, 1804.

Somerville, Mary, transl. *The Mechanism of the Heavens/Laplace's Mécanique céleste*. London: J. Murray, 1831.

Somerville, Mary. *On the Connexion of the Physical Sciences*. London: J. Murray, 1834.

Swift, Jonathan. *Jonathan Swift: The Complete Poems*. Ed. Pat Rogers. New Haven, CT: Yale University Press, 1983.

Swift, Jonathan. *Jonathan Swift*. Eds. Angus Ross and David Woolley. Oxford: Oxford University Press, 1984 [see especially *A Tale of a Tub*, *The Battel of the Books*, *A Discourse Concerning the Mechanical Operation of the Spirit*, *The Bickerstaff Papers*, and 'The Progress of Beauty'].

Swift, Jonathan. *Gulliver's Travels*. Ed. Albert J. Rivero. New York, NY: W. W. Norton, 1996 [see especially Book 3].

Swift, Jonathan. *The Essential Writings of Jonathan Swift*. Eds. Claude Rawson and Ian Higgins. New York, NY: W. W. Norton, 2009.

Tennyson, Alfred. *The Poems of Tennyson*. Ed. Christopher Ricks. Berkeley, CA: University of California Press, 1987 [see especially *In Memoriam* and *The Princess*].

Tennyson, Alfred. *In Memoriam*. Ed. Erik Gray. New York, NY: W. W. Norton, 2004.

Thomson, James. *'The Castle of Indolence' and Other Poems*. Ed. Alan Dugald McKillop. Lawrence, KS: University of Kansas Press, 1961 [see especially 'A Poem Sacred to the Memory of Sir Isaac Newton', 1727].

Thomson, James. *The Seasons* [1730]. Ed. James Sambrook. Oxford: Clarendon, 1981.

Voltaire [François-Marie Arouet]. *Oeuvres complètes de Voltaire* [*Complete Works of Voltaire*]. Multiple vols to date. Oxford: Voltaire Foundation, 1968 [see 'Essai sur la poésie épique'/'Essay on Epic Poetry' for the story of Newton's 'apple'].

Voltaire [François-Marie Arouet]. *Letters Concerning the English Nation*. Ed. Nicholas Cronk. New York, NY: Oxford University Press, 1994 [see especially Letter XIV, On Descartes and Newton; XVI, On Newton's *Opticks*; XVII, On Newton's infinities in geometry and his chronology; XXII; XXIV].

Voltaire [François-Marie Arouet] [and Émilie du Châtelet]. *The Elements of Sir Isaac Newton's Philosophy* [*Elémens de la philosophie de Neuton*, 1738]. Transl. John Hanna. London: Stephen Austen, 1738.

Watson, William. Letters, as in Herschel, William. *The Scientific Papers of Sir William Herschel.* Ed. J. L. E. Dreyer. Vols 1 and 2. London: [n. p.], 1912 [see correspondence from and other numerous references to Watson throughout].

Wells, H. G. *The Time Machine: An Invention.* London: William Heinemann, 1895.

Wells, H. G. *First Men in the Moon.* London: George Newnes, 1901.

Whewell, William. *The Bridgewater Treatises on the Power, Wisdom and Goodness of God as Manifested in the Creation. Treat. III: Astronomy and General Physics Considered with Reference to Natural Theology.* London: Pickering, 1833.

Whitman, Walt. *Complete Poetry and Collected Prose.* Ed. Justin Kaplan. New York, NY: Library of America, 1982 [see especially 'When I Heard the Learn'd Astronomer'].

Wordsworth, William. 'Star-Gazers'. *Poems, in Two Volumes, and Other Poems, 1800–1807.* Ed. Jared Curtis. Ithaca, NY: Cornell University Press, 1993, 234–5.

Wordsworth, William. *Wordsworth's Poetry and Prose.* Ed. Nicholas Halmi. New York, NY: W. W. Norton, 2013 [see *Prelude*, all versions, especially 1850].

Secondary texts

Alexander, Sarah C. *Victorian Literature and the Physics of the Imponderable.* London: Pickering and Chatto, 2015.

Almeida, Hermione de. *Romantic Medicine and John Keats.* Oxford: Oxford University Press, 1991.

Anderson, Katharine. 'Almanacs and the Profits of Natural Knowledge'. *Culture and Science in the Nineteenth-Century Media.* Eds. Louise Henson *et al.* Aldershot: Ashgate, 2004, 97–112.

Arianrhod, Robyn. *Seduced by Logic: Émilie du Châtelet, Mary Somerville and the Newtonian Revolution.* Oxford: Oxford University Press, 2012.

Ault, Donald. *Visionary Physics: Blake's Response to Newton.* Chicago, IL: University of Chicago Press, 1974.

Boswell, Michelle Suzanne Lang. *Beautiful Science: Victorian Women's Scientific Poetry and Prose.* PhD thesis, University of Maryland, College Park, 2014, #3627619.

Bowler, Peter J. *Science for All: The Popularization of Science in Early Twentieth-Century Britain.* Chicago, IL: University of Chicago Press, 2009.

Brantlinger, Patrick, ed. *Energy and Entropy: Science and Culture in Victorian Britain.* Bloomington, IN: Indiana University Press, 1989.

Brock, Claire. *The Comet Sweeper: Caroline Herschel's Astronomical Ambition.* Thriplow: Icon, 2004.

Bronowski, J. *William Blake and the Age of Revolution.* New York, NY: Harper and Row, 1965.

Brothers, Dometa Wiegand. *The Romantic Imagination and Astronomy: On All Sides Infinity.* New York, NY: Palgrave Macmillan, 2015 [see also Wiegand].

Brück, Mary. *Agnes Mary Clerke and the Rise of Astrophysics.* Cambridge: Cambridge University Press, 2008.

Brunner, Bernd. *Moon: A Brief History.* New Brunswick, NJ: Yale University Press, 2010.

Cartwright, John. '"Star-Defeated Sighs": Classical Cosmology and Astronomy in the Poetry of A. E. Housman'. *Journal of Literature and Science* 3:1 (2010): 71–96.

Chandrasekhar, S. *Newton's 'Principia': For the Common Reader.* Oxford: Oxford University Press, 2003.

Chapman, Allan. *Mary Somerville and the World of Science.* Bristol: Canopus, 2004.

Chow, Denise. 'Under a "Frankenstein" Moon: Astronomer Sleuths Solve Mary Shelley Mystery'. *SPACE.com.* 28 September 2011. <www.space.com/13112-frankenstein-moon-mary-shelley-mystery.html>. Accessed 19 January 2016.

Cosslett, Tess. *The 'Scientific Movement' and Victorian Literature.* Brighton: Harvester, 1982.

Crossley, Robert. *Imagining Mars: A Literary History.* Middletown, CT: Wesleyan University Press, 2011.

Crowe, Michael. *The Extraterrestrial Life Debate 1750–1900: The Idea of a Plurality of Worlds from Kant to Lowell.* Cambridge: Cambridge University Press, 1986.

Crowe, Michael. *The Extraterrestrial Life Debate, Antiquity to 1915: A Source Book.* South Bend, IN: Notre Dame University Press, 2008.

Daw, Gillian Jane. *The Victorian Poetic Imagination and Astronomy: Tennyson, De Quincey, Hopkins and Hardy.* PhD thesis, University of Sussex, 2012.

Daw, Gillian Jane. '"On the Wings of Imagination": Agnes Giberne and Women as the Storytellers of Victorian Astronomical Science'. *The Victorian* 2.1 (2014).

DeWitt, Anne. '"The Actual Sky is a Horror": Thomas Hardy and the Arnoldian Conception of Science'. *Nineteenth-Century Literature* 61.4 (2007): 479–506.

DeWitt, Anne. *Moral Authority, Men of Science, and the Victorian Novel*. Cambridge: Cambridge University Press, 2013 [see especially Chapter 3].

Dobbs, Betty Jo Teeter, and Margaret C. Jacob. *Newton and the Culture of Newtonianism*. Atlantic Highlands, NJ: Humanities Press International, 1995.

Donahue, William H., and Dana Densmore. *Newton's 'Principia': The Central Argument*. Santa Fe, NM: Green Lion, 2003. 3rd edn 2010.

Ebury, Katherine. *Modernism and Cosmology: Absurd Lights*. New York, NY: Palgrave Macmillan, 2014.

Fara, Patricia. *Newton: The Making of Genius*. New York, NY: Columbia University Press, 2002.

Flood, Raymond, Adrian Rice, and Robin Wilson, eds. *Mathematics in Victorian Britain*. Oxford: Oxford University Press, 2011.

Fraknoi, Andrew. 'About Andrew Fraknoi'. *Foothill College.* <www.foothill.edu/ast/fraknoi.php>. Accessed 19 January 2016.

Fraknoi, Andrew. *Music Inspired by Astronomy: A Selected Listing for the International Year of Astronomy*. 2008. PDF available at <http://en.wikipedia.org/wiki/Andrew_Fraknoi>. Accessed 19 January 2016.

Fraknoi, Andrew. *Science Fiction Stories with Good Astronomy & Physics: A Topical Index*. Version 6.1. 2014. <www.astrosociety.org/education/astronomy-resource-guides/science-fiction-stories-with-good-astronomy-physics-a-topical-index>. Accessed 19 January 2016.

Friedman, Alan J., and Carol C. Donley. *Einstein as Myth and Muse*. Cambridge: Cambridge University Press, 1989.

Fyfe, Aileen, and Bernard Lightman, eds. *Science in the Marketplace: Nineteenth-Century Sites and Experiences*. Chicago, IL: University of Chicago Press, 2007.

Gold, Barri J. *ThermoPoetics: Energy in Victorian Literature and Science*. Cambridge, MA: MIT Press, 2010.

Gossin, Pamela. *Poetic Resolutions of Scientific Revolutions: Astronomy and the Literary Imaginations of Donne, Swift and Hardy*. PhD thesis, University of Wisconsin, 1989, #9010301.

Gossin, Pamela. '"All Danaë to the Stars": Nineteenth-Century Representations of Women in the Cosmos'. *Victorian Studies* 40.1 (1996): 65–96 [1996A].

Gossin, Pamela. 'An Interdisciplinary Community of One?': Review of Holly Henry, *Virginia Woolf and the Discourse of Science: The Aesthetics of Astronomy*. *Arts and Humanities in Higher Education* 5 (February 2006): 107–12.

Gossin, Pamela. *Thomas Hardy's Novel Universe: Astronomy, Cosmology, and Gender in the Post-Darwinian World*. Aldershot: Ashgate, 2007.

Gossin, Pamela. 'Hardy's Poetic Cosmology and the "New Astronomy"'. *The Ashgate Research Companion to Thomas Hardy*. Ed. Rosemarie Morgan. Aldershot: Ashgate, 2010, 235–51.

Gossin, Pamela. 'The Challenges and Legacy of Newtonian Literature and Astronomy'. 2015. <https://utdallas.academia.edu/PamelaGossin/Papers>. Accessed 19 January 2016 [2015A].

Grabo, Carl. *Newton Among Poets: Shelley's Use of Science in 'Prometheus Unbound'*. Chapel Hill, NC: University of North Carolina Press, 1930.

Greenberg, Mark. 'Blake's "Vortex"'. *Colby Library Quarterly* 14 (1978): 198–212.

Greenberg, Mark. 'Blake's "Science"'. *Studies in Eighteenth-Century Culture* 12 (1983): 115–30.

Guicciardini, Niccolò. *Reading the 'Principia': The Debate on Newton's Mathematical Methods for Natural Philosophy From 1687 to 1736*. Cambridge: Cambridge University Press, 1999.

Hankins, Thomas L. *Science and the Enlightenment* [1985]. Cambridge: Cambridge University Press, 2007.

Hasted, Meegan. *Bright Star: John Keats and Romantic Astronomy*. PhD thesis, University of Sydney, 2014.

HASTRO-L. Listserv for the History of Astronomy Discussion Group. <https://listserv.wvu.edu/cgi-bin/wa?A0=HASTRO-L>. Accessed 19 January 2016.

Hawking, Stephen W., and Werner Israel, eds. *Three Hundred Years of Gravitation*. Cambridge: Cambridge University Press, 1989.

Henry, Holly. *Virginia Woolf and the Discourse of Science: The Aesthetics of Astronomy*. Cambridge: Cambridge University Press, 2003.

Henson, Louise, Geoffrey Cantor, Gowan Dawson, Richard Noakes, Sally Shuttleworth and Jonathan R. Topham, eds. *Culture and Science in the Nineteenth-Century Media*. Aldershot: Ashgate, 2004. See especially Chapter 8: Katharine Anderson, 'Almanacs and the Profits of Natural Knowledge', 97–112, and Chapter 16: Bernard Lightman, '*Knowledge* Confronts *Nature:* Richard Proctor and Popular Science Periodicals', 199–210.

Higgitt, Rebekah. *Recreating Newton: Newtonian Biography and the Making of Nineteenth-Century History of Science*. London: Pickering and Chatto, 2007.

Holden, Edward S. *Sir William Herschel: His Life and Works*. New York, NY: Charles Scribner's Sons, 1881.

Holmes, Richard. *The Age of Wonder: How the Romantic Generation Discovered the Beauty and Terror of Science*. New York, NY: Pantheon, 2008.

Hoskin, Michael, ed. *The Cambridge Concise History of Astronomy*. Cambridge: Cambridge University Press, 1999.

Hoskin, Michael. *Caroline Herschel's Autobiographies*. Cambridge: Science History Publications, 2003.

Hoskin, Michael. *Discoverers of the Universe: William and Caroline Herschel*. Princeton, NJ: Princeton University Press, 2011.

Hoskin, Michael. *The Construction of the Heavens: William Herschel's Cosmology*. Cambridge: Cambridge University Press, 2012.

Hoskin, Michael. *Caroline Herschel: Priestess of the New Heavens*. Sagamore Beach, MA: Watson, 2013.

IDA: The International Dark-Sky Association. <http://darksky.org>. Accessed 19 January 2016.

Johnson, Francis. *Astronomical Thought in Renaissance England: A Study of the English Scientific Writings From 1500 to 1645*. Baltimore, MD: Johns Hopkins University Press, 1937.

Jones, William Powell. *The Rhetoric of Science: A Study of Scientific Ideas and Imagery in Eighteenth-Century English Poetry*. Berkeley, CA: University of California Press, 1966.

Kirk, Eugene. 'Blake's Menippean Island'. *Philological Quarterly* 59 (1980): 194–215.

Knellwolf, Christa, and Jane Goodall, eds. *Frankenstein's Science: Experimentation and Discovery in Romantic Culture, 1780–1830*. Aldershot: Ashgate, 2008.

Korg, Jacob. 'Astronomical Imagery in Victorian Poetry'. *Victorian Science and Victorian Values: Literary Perspectives*. Eds. James Paradis and Thomas Postlewait. New York, NY: New York Academy of Sciences, 1981. 137–58.

Lane, K. Maria D. *Geographies of Mars: Seeing and Knowing the Red Planet*. Chicago, IL: University of Chicago Press, 2010.

Lemonick, Michael D. *The Georgian Star: How William and Caroline Herschel Revolutionized Our Understanding of the Cosmos*. New York, NY: W. W. Norton, 2009.

Levere, Trevor H. *Poetry Realized in Nature: Samuel Taylor Coleridge and Early Nineteenth-Century Science*. Cambridge: Cambridge University Press, 2002.

Levy, David. *Starry Night: Astronomers and Poets Read the Sky*. Amherst, NT: Prometheus, 2001.

Lightman, Bernard. 'Astronomy for the People: R. A. Proctor and the Popularization of Victorian Astronomy'. *Facets of Science*. Ed. Jitse van der Meer. Lanham: Pascal Centre for Advanced Studies in Faith and Science/University Press of America, 1996. 3, 13–45.

Lightman, Bernard. 'Constructing Victorian Heavens: Agnes Clerke and Gendered Astronomy'. *Natural Eloquence: Women Reinscribe Science*. Eds. Barbara T. Gates and Ann B. Shteir. Madison, WI: University of Wisconsin Press, 1997. 61–75 [1997A].

Lightman, Bernard, ed. *Victorian Science in Context*. Chicago, IL: University of Chicago Press, 1997 [1997B] [see especially pp. 187–211].

Lightman, Bernard. '*Knowledge* Confronts *Nature*: Richard Proctor and Popular Science Periodicals'. *Culture and Science in the Nineteenth-Century Media*. Eds. Louise Henson *et al*. Aldershot: Ashgate, 2004, 199–210.

Lightman, Bernard. *Victorian Popularizers of Science: Designing Nature for New Audiences*. Chicago, IL: University of Chicago Press, 2007.

Lubbock, Constance A. *The Herschel Chronicle: The Life-Story of William Herschel and His Sister, Caroline Herschel*. Cambridge: Cambridge University Press, 1933.

MacDuffie, Allen. *Victorian Literature, Energy, and the Ecological Imagination*. Cambridge: Cambridge University Press, 2014.

Markley, Robert. *Dying Planet: Mars in Science and the Imagination*. Durham, NC: Duke University Press, 2005.

Marshall, Linda E. 'Astronomy of the Invisible: Contexts for Christina Rossetti's Heavenly Parables'. *Women's Writing: The Elizabethan to Victorian Period* 2 (1995): 167–81.

Meadows, A. J. *The High Firmament: A Survey of Astronomy in English Literature*. Leicester: Leicester University Press, 1969.

Meadows, A. J. 'Astronomy and Geology, Terrible Muses! Tennyson and 19th-Century Science'. *Notes and Records of the Royal Society* 46 (1992): 111–18.

Miner, Paul. 'Visionary Astronomy'. *Bulletin of Research in the Humanities* 84 (1981): 305–36.

Mitchell, Robert. *Experimental Life: Vitalism in Romantic Science and Literature*. Baltimore, MD: Johns Hopkins University Press, 2013.

Morton, Timothy. *Shelley and the Revolution in Taste: The Body and the Natural World*. Cambridge: Cambridge University Press, 1995.

Mussell, James. *Science, Time and Space in the Late Nineteenth-Century Periodical Press: Movable Types*. Aldershot: Ashgate, 2007.

Myers, Greg. 'Nineteenth-Century Popularizations of Thermodynamics and the Rhetoric of Social Prophecy'. *Victorian Studies* 29.1 (1985) 35–66.

Myers, Greg. 'Science for Women and Children: The Dialogue of Popular Science in the Nineteenth Century'. *Nature Transfigured: Science and Literature, 1700–1900*. Eds. John Christie and Sally Shuttleworth. Manchester: Manchester University Press, 1989, 171–200.

Narain, Mona, and Karen Gevirtz, eds. *Gender and Space in British Literature, 1660–1820*. Aldershot: Ashgate, 2014 [see especially Kristine Larsen, 'Margaret Bryant and Jane Marcet: Making Space for "Space" in British Women's Science Writing', 67–84].

Neeley, Kathryn A. *Mary Somerville: Science, Illumination, and the Female Mind*. Cambridge: Cambridge University Press, 2001.

The Newton Project. Director Rob Iliffe, AHRC Newton Papers Project. <www.newtonproject. sussex.ac.uk/prism.php?id=1>. Accessed 19 January 2016.

Nicolson, Marjorie Hope. *Newton Demands the Muse: Newton's Opticks and the Eighteenth Century Poets*. Princeton, NJ: Princeton University Press, 1946.

Ogilvie, Marilyn B. *Searching the Stars: The Story of Caroline Herschel*. Stroud: History, 2008.

Olson, Donald. *Celestial Sleuth: Using Astronomy to Solve Mysteries in Art, History and Literature*. New York, NY: Springer/Praxis, 2014.

Owens, Thomas. 'Astronomy at Stowey: The Wordsworths and Coleridge'. *Wordsworth Circle* 43.1 (2012): 25–9.

Owens, Thomas. '"The Telescope of the Church": Coleridge, Astronomy, and Religion'. *The Coleridge Bulletin* 39.2 (2012): 37–47.

Owens, Thomas. 'Coleridge's Lost Marginalia to *Philosophical Transactions of the Royal Society* (November 1803-May 1809)'. *The Book Collector* 62.2 (2013): 241–53.

Owens, Thomas. 'Coleridge's Reading of Two Translations of *Galileo's Dialogue Concerning The Two Chief World Systems* (1632)'. *Notes and Queries* 258.2 (2013): 229–32.

Owens, Thomas. 'Did the Wordsworths Own a Telescope?'. *Notes and Queries* 258.2 (2013): 232–5.

Owens, Thomas. '"Form One Consciousness": Coleridge, Wordsworth, Milton and the Origins of the "Optic Tube" in Poetry'. *Notes and Queries* 258.2 (2013): 227–8.

Parrett, Aaron. *The Translunar Narrative in the Western Tradition*. Aldershot: Ashgate, 2004.

Pask, Colin. *Magnificent 'Principia': Exploring Isaac Newton's Masterpiece*. New York, NY: Prometheus, 2013.

Patterson, Elizabeth. 'The Case of Mary Somerville: An Aspect of Nineteenth-Century Science'. *Proceedings of the American Philosophical Society* 118.3 (1974): 269–75.

Patterson, Elizabeth. *Mary Somerville and the Cultivation of Science, 1815–1840*. Boston, MA: Martinus Nijhoff, 1983.

Peck, Garrett. 'Realism and Victorian Astronomy: The Character and Limits of Critique in Thomas Hardy's *Two on a Tower*'. *Pacific Coast Philology* 46 (2011): 28–45.

Peterfreund, Stuart. 'Blake on Space, Time and the Role of the Artist'. *Science/Technology and the Role of the Humanities* 2.3 (1979): 246–63.

Peterfreund, Stuart. 'Blake and Newton: Argument as Art; Argument as Science'. *Studies in Eighteenth-Century Culture* 10 (1981): 205–26.

Peterfreund, Stuart. *William Blake in a Newtonian World: Essays on Literature as Art and Science*. Norman, OK: University of Oklahoma Press, 1998.

Price, Katy. *Loving Faster Than Light: Romance and Readers in Einstein's Universe*. Chicago, IL: University of Chicago Press, 2012.

Ruskin, Steven. *John Herschel's Cape Voyage: Private Science, Public Imagination and the Ambitions of Empire*. Aldershot: Ashgate, 2004.

Russell, Nicholas. 'Science and Scientists in Victorian and Edwardian Literary Novels: Insights into the Emergence of a New Profession'. *Public Understanding of Science* 16 (2007): 205–22.

Ruston, Sharon. *Creating Romanticism: Case Studies in the Literature, Science and Medicine of the 1790s*. New York, NY: Palgrave Macmillan, 2013.

Schiebinger, Londa. *The Mind Has No Sex?: Women in the Origins of Modern Science*. Cambridge, MA: Harvard University Press, 1991 [see especially Chapters 3 and 9].

Schroeder, David Alan. *A Message from Mars: Astronomy and Late-Victorian Culture*. PhD thesis, Indiana University, 2002, #3054420.

Schwartz, Richard B. *Samuel Johnson and the New Science*. Madison, WI: University of Wisconsin Press, 1971.

Simon, Leslie Sandra. *Novel Relations: Dickens, Narrative Realism, and Nineteenth-Century Mathematics*. PhD thesis, Boston University, 2012, #3483508.

Smith, Jonathan. 'De Quincey's Revisions to "The System of the Heavens"'. *Victorian Periodicals Review* 26.4 (1993): 203–12.

Toole, Betty A. *Ada, The Enchantress of Numbers: Prophet of the Computer Age*. Mill Valley, CA: Strawberry, 1998.

Underwood, Ted. 'The Science in Shelley's Theory of Poetry'. *Modern Language Quarterly* 58.3 (1997): 299–321.

Valiunas, Algis. 'Scientists Fallen Among Poets'. *The New Atlantis: A Journal of Technology and Society* (Spring 2010). <www.thenewatlantis.com/publications/scientists-fallen-among-poets>. Accessed 19 January 2016.

Warrick, Patricia. *Charles Babbage and The Countess*. Bloomington, IN: AuthorHouse, 2007.

Wiegand, Dometa. 'Coleridge's "Web of Time": The Herschels, the Darwins, and "Psalm 19"'. *The Coleridge Bulletin* n.s. 28 (2006): 91–100.

Wiegand, Dometa. *On All Sides Infinity*. PhD thesis, Washington State University, 2006, #3206171 [see also Brothers].

Willis, Martin. *Vision, Science and Literature, 1870–1920: Ocular Horizons*. London: Pickering and Chatto, 2011.

Woolley, Benjamin. *The Bride of Science: Romance, Reason, and Byron's Daughter*. New York, NY: McGraw-Hill, 2002.

Worrall, David. 'The "Immortal Tent"'. *Bulletin of Research in the Humanities* 84 (1981): 273–95.

Further reading

Anderson, Katharine. *Predicting the Weather: Victorians and the Science of Meteorology*. Chicago, IL: University of Chicago Press, 2005.

Armstrong, Isobel. *Victorian Glassworlds: Glass Culture and the Imagination 1830–1880*. Oxford: Oxford University Press, 2008 [see especially chapters on lenses and the telescope].

Aubin, David, Charlotte Bigg, and H. Otto Sibum, eds. *The Heavens on Earth: Observatories and Astronomy in Nineteenth-Century Science and Culture*. Durham, NC: Duke University Press, 2010.

Barrows, Adam. *The Cosmic Time of Empire: Modern Britain and World Literature*. Berkeley, CA: University of California Press, 2011.

Becker, Barbara J. *Unravelling Starlight: William and Margaret Huggins and the Rise of the New Astronomy*. Cambridge: Cambridge University Press, 2011.

Berry, Arthur. *A Short History of Astronomy: From Earliest Times Through the Nineteenth Century* [1898]. New York, NY: Dover, 1961.

Chapman, Allan. *The Victorian Amateur Astronomer: Independent Astronomical Research in Britain 1820–1920*. Chichester: Wiley/Praxis, 1998.

Clark, Stuart. *The Sun Kings: The Unexpected Tragedy of Richard Carrington and the Tale of How Modern Astronomy Began*. Princeton, NJ: Princeton University Press, 2007.

Crowe, Michael J. *Modern Theories of the Universe from Herschel to Hubble*. New York, NY: Dover, 1994.

Danielson, Dennis Richard, ed. *The Book of the Cosmos: Imagining the Universe from Heraclitus to Hawking*. Cambridge: Perseus, 2000 [see especially Part 4].

Danielson, Dennis Richard. *'Paradise Lost' and the Cosmological Revolution*. Cambridge: Cambridge University Press, 2014.

Glass, I. S. *Victorian Telescope Makers: The Lives and Letters of Thomas and Howard Grubb*. Bristol: Institute of Physics, 1997.

Gossin, Pamela. 'Literature [and Astronomy]'. *History of Astronomy: An Encyclopedia*. Ed. John Lankford. New York, NY: Garland, 1996, 307–14 [1996B].

Gossin, Pamela, ed. *An Encyclopedia of Literature and Science*. Westport, CT: Greenwood, 2002 [see especially 'Introduction', 'Astronomy', 'Literature and Science – Chronological Periods', 'Two Cultures Debate'].

Gossin, Pamela. 'Literature and the Modern Physical Sciences'. *Cambridge History of Science, Volume 5: Modern Physical and Mathematical Sciences*. Ed. Mary Jo Nye. Cambridge: Cambridge University Press, 2003, 91–109.

Gossin, Pamela. Review of Stuart Clark, *The Sun Kings: The Unexpected Tragedy of Richard Carrington and the Tale of How Modern Astronomy Began*. Princeton, NJ: Princeton University Press, 2007. *Isis* 99.4 (2008): 848–9.

Gossin, Pamela. Review of David Aubin, Charlotte Bigg, and H. Otto Sibum, eds. *The Heavens on Earth: Observatories and Astronomy in Nineteenth-Century Science and Culture*. Durham, NC: Duke University Press, 2010. *Victorian Studies* 53.4 (2011): 770–2.

Gossin, Pamela. Review of Anna Henchman, *The Starry Sky Within: Astronomy and the Reach of the Mind in Victorian Literature*. Oxford: Oxford University Press, 2014. *Review of English Studies* 66 (2014): 189–91.

Gossin, Pamela. 'The Influence of Early Studies of Literature and Astronomy from the Scientific Revolution to the Nineteenth Century'. 2015. <https://utdallas.academia.edu/PamelaGossin/Papers>. Accessed 19 January 2016 [2015B].

Hoskin, Michael, ed. *The General History of Astronomy*. 4 vols. Cambridge: Cambridge University Press, 1984–95.

Iliffe, Rob, ed. *Literature and Science, 1660–1884: Vol. 6: Astronomy*. London: Pickering and Chatto, 2004.

Jenkins, Alice. 'Genre and Geometry: Victorian Mathematics and the Study of Literature and Science'. *Uncommon Contexts: Encounters Between Science and Literature, 1800–1914*. Eds Ben Marsden, Hazel Hutchison and Ralph O'Connor. London: Pickering and Chatto, 2013, 111–24.

Kern, Stephen. *The Culture of Time and Space, 1880–1918*. Cambridge: Harvard University Press, 1983. 2nd edn 2003.

Lankford, John, ed. *History of Astronomy: An Encyclopedia*. New York, NY: Garland, 1996.

Lindberg, David C., and Ronald L. Numbers, eds. *The Cambridge History of Science*. 8 vols. Cambridge, MA: Cambridge University Press, 2003–13 [see especially vol. 4: *Eighteenth-Century Science* and vol. 5: *The Modern Physical and Mathematical Sciences*].

North, John D. *Cosmos: An Illustrated History of Astronomy and Cosmology*. Chicago, IL: University of Chicago Press, 2008.

North, John D., and Roy Porter, eds. *The Norton History of Astronomy and Cosmology*. New York, NY: W. W. Norton, 1994.

Olby, Robert C., Geoffrey N. Cantor, John R. R. Christie and Michael Jonathan Session Hodge, eds. *Companion to the History of Modern Science*. London: Routledge, 1990.

Olson, Roberta J. M., and Jay M. Pasachoff. *Fire in the Sky: Comets and Meteors, the Decisive Centuries, in British Art and Science*. Cambridge: Cambridge University Press, 1998.

Otis, Laura, ed. *Literature and Science in the Nineteenth Century: An Anthology*. Oxford: Oxford University Press, 2002.

Pang, Alex. *Empire and the Sun: Victorian Solar Eclipse Expeditions*. Stanford, CA: University of Stanford Press, 2002.

Pycior, Helena M., Nancy G. Slack, and Pnina G. Abir-Am, eds. *Creative Couples in the Sciences*. New Brunswick, NJ: Rutgers University Press, 1996 [see especially Chapter 5: Barbara Becker on Margaret and William Huggins, and Chapter 16: Marilyn Bailey Ogilvie on the Campbells and Maunders].

Ratcliff, Jessica. *The Transit of Venus Enterprise in Victorian Britain*. London: Pickering and Chatto, 2008.

Schatzberg, Walter, Ronald A. White, and Jonathan K. Johnson, eds. *The Relations of Literature and Science: An Annotated Bibliography of Scholarship, 1880–1980*. New York, NY: MLA, 1987.

Smith, Crosbie. *The Science of Energy: A Cultural History of Energy Physics in Victorian Britain*. Chicago, IL: University of Chicago Press, 1998.

Stableford, Brian M. *Science Fact and Science Fiction: An Encyclopedia*. New York, NY: Taylor and Francis, 2006.

Thomas, Walter Keith, and Warren U. Ober. *A Mind for Ever Voyaging: Wordsworth at Work Portraying Newton and Science*. Edmonton: University of Alberta Press, 1989.

Tucker, Jennifer. *Nature Exposed: Photography as Eyewitness in Victorian Science*. Baltimore, MD: Johns Hopkins University Press, 2005.

Upgren, Arthur. *Night Has a Thousand Eyes: A Naked-Eye Guide to the Sky, Its Science and Lore*. New York, NY: Basic, 2000.

Westfall, R. S. *Never at Rest: A Biography of Isaac Newton*. Cambridge: Cambridge University Press, 1980.

17

GEOLOGY

Adelene Buckland
King's College London

In 1830, five chapters into his epoch-making *Principles of Geology* (1830–33), Charles Lyell invited his readers to imagine a descent into the underworld conducted by a gnome. 'A being, entirely confined to the nether world' (1:82), Lyell's gnome is a '"dusky melancholy sprite", like Umbriel', the gnome who travels to the depths of the Cave of Spleen (located in the heroine's abdomen) in Alexander Pope's mock-epic poem *The Rape of the Lock* (1714). Not Umbriel, but like him, Lyell's gnome is 'never permitted to "sully the fair face of light", and emerge into the regions of water and of air', trapped beneath the rocks and unable to see all the many kinds of geological change operating on the earth's surface or beneath the rivers and sea. The gnome cannot see erosion and weathering, for instance, which mainly affect the rocks on the earth's surface, nor can he see the deposition of the strata beneath the sea. Such a gnome, Lyell tells his readers, might 'busy himself in investigating the structure of the globe' and attempt to turn geologist. But if he did, he would probably 'frame theories the exact converse of those usually adopted by human philosophers'.

Examining the subterranean world from within, then, the geologist-gnome would only be able to see layers of rock descending from the surface of the earth ever-deeper towards its centre. Near the surface, those rocks would be neatly layered and full of shells and fossils – evidence that plant and animal life had been abundant when those rocks were formed. As the rocks got deeper, they would contain fewer and fewer fossils and the layers would become less obvious. At the bottom of the pile, the rocks would not be layered at all but would exist as masses of granite and basalt shot through with metallic veins. Were the gnome to write a history of the earth based on his observations, he would most likely conclude that 'Every year the strata' found at the top of the rock pile 'get broken and shattered by earthquakes, or melted by volcanic fire', 'fused and crystallized' until they turn into the granites and basalts of the deeps (1:82–3). At the dawn of the world, he would assume, all the rocks must have been layered in 'curiously-bedded formations' (the strata) (1:82), full of fossils. This primeval world must therefore have been more stable than the convulsive, volcano-and-earthquake-ridden present. For the gnome, trapped in the underground, the earth would seem to be getting hotter and more violent. From his perspective, life would seem to be getting progressively harder to sustain on the earth: one day, he might surmise, 'the whole globe shall be in a state of fluidity and incandescence' (1:83).

Lyell's gnome is designed to raise a smile, perhaps, and he occupies only a fleeting moment in the text. But he belongs to a rich literary and philosophical tradition of which Pope's

The Rape of the Lock is merely the most well-known example, and in which gnomes fulfilled a range of both serious and satiric purposes. In his 'dedicatory epistle', Pope cited *Le Comte de Gabalis* as his source, a book first written in 1670 by Abbé N. de Montfaucon de Villars, translated into English in 1680 and which satirized what Pope calls a poetic machinery 'raised on a very new and odd foundation, the Rosicrucian doctrine of spirits' (15). This 'doctrine', associated with a secret society called Rosicrucianism, was derived from the Renaissance philosopher and alchemist Paracelsus, who argued that Aristotle's four 'sublunary' elements comprising the world beneath the heavens – fire, air, water and earth – were tenanted by elemental beings: salamanders, sylphs, nymphs and gnomes. Challenging long-held belief that all sublunary spirits were fallen angels, Paracelsus advocated the first-hand, empirical study of nature and suggested that elementals were embodiments of the invisible processes by which inanimate matter was made to move, think, breathe and live. Though they were invisible, and therefore inaccessible to human observation, they were part of Paracelsus's argument for the detailed study of nature over adherence to received doctrine. This paradox was exploited by Villars in *Le Comte* to satiric ends, but many of its readers – including its English translators – missed the satire and took it for a serious exposition of a philosophical idea. Pope's Umbriel draws on this ambiguity, pretending to take *Le Comte* seriously in the 'dedicatory epistle' to the poem by poking fun at female readers who had thought it merely a novel (15), but having his gnome explain nothing about the natural world and journey to the centre of his victims' spleens instead of to the underworld of classical epic, where in comic mode he makes those victims pimply, red, suspicious and sick (Canto 4, ll. 67–78). As such, Pope's 'Rosicrucian' machinery both deploys and pokes fun at supernatural explanations of physical processes (see Veenstra, and also Latimer).

It is tempting to say that this is the sense in which Lyell's gnome operates in *Principles of Geology*. Lyell regularly debunks superstitious explanations of natural events as 'delusion[s]' common to human beings who were 'in an early stage of advancement' (1:76), mocking those who wittingly – or even unwittingly – resorted to supernatural explanations of the earth's phenomena. And Lyell's gnome inhabits a nonsense world, a world turned upside down, framing 'theories the exact converse of those usually adopted by human philosophers'. But it is quickly revealed that the gnome's nonsense is not the opposite of human 'sense'. Instead, it is a mirror for the equal kinds of nonsense produced by the 'human philosophers' confined to the earth's surface.

To characterize this nonsense, Lyell had a specific 'theory' in mind: an account of earth history that had just been given elegant exposition in *Consolations in Travel; or, The Last Days of a Philosopher* (1830), written by the famous chemist and philosopher Humphry Davy. In this text, an even more mysterious figure than Lyell's gnome, a 'stranger' called 'The Unknown', describes 'the early changes and physical history of the globe' (Davy 1830 132) in what he presents as chronological order:

> [T]he globe, in the first state in which the imagination can venture to consider it, was a fluid mass with an immense atmosphere revolving in space round the sun, and . . . by its cooling, a portion of its atmosphere was condensed in water which occupied a part of the surface. In this state, no forms of life, such as now belong to our system, could have inhabited it; and, I suppose the crystalline rocks, or as they are called by geologists, the primary rocks, which contain no vestiges of a former order of things, were the results of the first consolidation on its surface.
>
> *(134)*

Here, the 'primary' rocks – those masses of basalts and granite found at the bottom of the rock sequence – were considered the *first* rocks formed in the history of the earth. They were not fossiliferous, because the earth had been too hot to support life in this period. As it cooled, the Unknown continues, 'depositions took place, shell fish and coral insects of the first creation began their labours; and islands appeared in the midst of the ocean raised from the deep by the productive energies of millions of zoophytes' (Davy 1830 134). Palms and tropical plants 'similar to those which now exist in the hottest parts of the world' covered the globe, and fish and shellfish swam in the tropical seas. Sands were agglutinated from the cooling of the early globe's molten fluids, and the first 'secondary' rocks – rocks we now find layered in strata and filled with fossils – were laid down (Davy 1830 134–5). '[O]viparous reptiles' succeeded fish and corals, 'and the turtle, crocodile and various gigantic animals of the sauri kind seem to have haunted the bays and waters of the primitive lands' (Davy 1830 135), lands violent and volcanic, in which 'there was no order of events similar to the present': mountains were thrown up and lands thrown down from the ocean with frequency and speed since the earth's surface was so thin. Slowly, the earth became gentler and capable of supporting more complex forms of life, each leaving behind fossils in freshly laid layers of strata: 'the mammoth, megalonix, megatherium and gigantic hyena', all 'now extinct' (Davy 1830 136), superseded the reptiles as the eruptions and earthquakes grew fewer. There had been only one event in the earth's recent past to recall the horrors of the primeval world, dragging 'immense quantities of water, worn stones, gravel and sand' (Davy 1830 136) across the earth. But generally the earth's rate of change had slowed. When the ancient battle between fire and water was 'no longer to be dreaded, the creation of man took place; and since that period there has been little alteration in the physical circumstances of the globe' (Davy 1830 137).

Davy's story, of an earth getting progressively cooler, able to support increasingly more complex forms of life until stable geological conditions heralded the coming of man, had (unlike the gnome's 'converse' account), the advantage of being based upon three decades of European research into the history of the earth. In Britain this research was associated, at least in part, with the Geological Society of London, formed in 1807. The Society's definition of 'geology' was as an empirical science based on fieldwork and devoted to the mapping of the strata, the collection of rocks and fossils, and the production of 'columns' and 'sections' – vertical images of the layers of sequences of rock in particular regions. This research programme was powerful and successful, spawning the so-called 'heroic age of geology', in which all the major subdivisions of the stratigraphic column were thrashed out (see Rudwick 1985; Secord 1986; Oldroyd; Rudwick 2007 35–47).

In its early decades, this research had to be cautiously articulated. In the late eighteenth century, two rival theories of the earth had ruined reputations and caused fights and factions in the clubs of Edinburgh, one group arguing that the earth's geological processes were primarily powered by fire, the other that its strata had been deposed from particles floating in a primordial ocean that had once covered the entire globe (see Buckland 2013 33–40). By the early nineteenth century, these high-level theories and the arguments they had generated were threatening to bring science into disrepute: further research had revealed contradictions in both arguments, suggesting that geological processes were more complex than a single law (based on 'fire' or 'water') could explain. And controversy did little to inspire confidence in the authority of science. Furthermore, geology was contentious for other reasons, too. By 1800, many geologists believed that the earth was millions of years old, but other groups – including those who interpreted the Bible literally (by no means everybody who believed in the Gospels) – considered it merely a few thousand years old, suggesting that the new 'geology' flatly contradicted the Bible. Others argued that the Hebrew word for 'day' in the account of God's six-day Creation was metaphorical and could be stretched to include millions of years; still others that a 'gap' filled with geological aeons sat between verse 1 of Genesis – 'In the beginning God created the heavens and the earth' – and verse 2 – 'the earth was without form and void' (King James Bible). Many elite men of science at the Geological Society interpreted the puzzling boulders that had been dragged for miles across the earth away from the strata to which they belonged by reference to Noah's Flood, which offered the closest match in the historical records to the event that might have produced such phenomena (see O'Connor 2007B). But, by the 1830s, other explanations were becoming more convincing – Davy's Unknown suggests, for instance, that the 'immense quantities of water, worn stones, gravel and sand' (137) strewn across the earth's surface were attributable not to Noah's Flood but to the creation of a new continent south of the equator. In the years following the publication of Lyell's *Principles of Geology*, several leading geologists would publicly 'recant' their earlier belief in Noah's Flood as a geological event recorded in the earth's strata, but none of them, including Lyell, lost their faith in a world created by God (see Rudwick 2009 73–88). Nonetheless, geology did suggest a wide variety of troubling views of Creation, and of the fit between the scriptural and natural accounts of the world's construction and purpose.

Furthermore, in nineteenth-century Britain the study of rocks and fossils was widely associated with the scientific culture of pre-revolutionary France, with speculations that life was purely material. Another gnome-filled scientific poem employing Pope's 'Rosicrucian doctrine', written just as the French Revolution began, was Erasmus Darwin's *The Botanic Garden* (1791). This poem detailed the sex lives of the plants (in conventional, courtly but erotic fashion), espoused anti-slavery and libertarian politics and speculated on evolution. It was spectacularly successful: *The Botanic Garden* was perhaps the most widely read poem of the 1790s. And, unlike Pope's elemental beings, Darwin's had the unequivocally serious purpose of updating Paracelsus's elementals, which personified the invisible processes animating matter. For Darwin, the elementals were ancient 'hieroglyphs', allegorical representations of knowledge possessed by pre-literate societies, still useful to a modern poet seeking to 'visualize' invisible scientific processes such as those happening underground, 'and to attach them to concrete agencies as yet inaccessible to "rigorous" science', as Noah Heringman has put it (2004 223). Pope's gnomes allowed his readers – and men of science – to imagine processes that science had not yet observed. But in 1798 a parody of the poem, entitled *The Loves of the Triangles*, was published in a magazine called *The Anti-Jacobin Review*, making explicit the supposed connections between the poem's materialist explanations of physical processes, its sympathy with revolutionary politics and its evolutionary musings. Darwin's reputation suffered a blow from which it has still not quite recovered. By the time Lyell – whose father was a botanist and literary critic – wrote *Principles*, Darwin's poetry and evolutionary speculations both lay outside the bounds of respectability. (On Darwin's geology, see Heringman 2004 191–227; for detailed accounts of the *Anti-Jacobin Review* episode, see Priestman 193–216 and Fara 30–42).

Members of the Geological Society of London responded by focusing on empirical fieldwork, putting geology back into descriptive rather than speculative mode. Only once all the descriptions, maps and specimens were collected and amassed could any generalized geological law or system emerge, they argued. Nonetheless, by around 1830, geologists began to relax their strict adherence to enumerative induction, to the simple accumulation of 'facts' of nature, and to attempt to add up all their observations of the strata in the hope of actually turning to reality the long-held ideal of *universal* stratigraphic column, an idealized column which would represent all the strata of the globe in a single image (see O'Connor 2009; Laudan 537). The question was how to move between local observations to grander visions of the workings of nature without falling into the old traps: glossing over exceptions to the rule, oversimplifying the data or implying an evolutionary narrative of life on earth. Tellingly, then, right before he tells the progressive story that was emerging from all this research, the Unknown in Davy's

Consolations advocates caution. 'It is the general vice of philosophical systems,' he states, with clear reference to the eighteenth-century theories geology had worked hard to leave behind,

> that they are usually founded upon a few facts, which they well explain, and are extended by the human fancy to all the phenomena of nature, to many of which they must be contradictory. The human intellectual powers are so feeble that they can with difficulty embrace a single series of phenomena, and they consequently must fail when extended to the whole of nature.
>
> *(Davy 1830 131)*

This might seem a strange caveat given that we have just ranged with a mysterious figure whose identity is never disclosed to us over the entire history of the globe. As Jan Golinski writes, 'the Unknown was conceived as an unworldly, even otherworldly, being, not so much a flesh-and-blood person as a spirit or angel temporarily assuming human form' (2013 11), allowing him to authoritatively claim the reality 'of intellectual or spiritual life continuing beyond the death of the material body'. He also bears an uncanny resemblance to the 'Genius' of the first dialogue of *Consolations in Travel*, who appears to the protagonist in a vision and guides him through the solar system to the rings of Saturn, revealing myriads of celestial beings inhabiting the universe beyond earth – each increasingly intelligent, less dependent on the clumsy physical senses on which human beings depend in order to acquire knowledge, and closer to the angels and to God (Davy 1830 1–60). Together, the Unknown and the Genius offer radical shifts in perspective that argue for man's spiritual existence, and also enable temporary moments of transcendence over his inherently 'feeble' 'intellectual powers' in order to articulate wider visions of the nature of creation and its history, to consider the span of geological time beyond human history and to range into unseen depths of the universe. It is not quite that they, like Darwin's gnomes, help us visualize the unseen. It is rather that they help replace limited powers of human *sight*, and the inductive scientific method that relied upon it, with *vision*: the Unknown can attain a visionary perspective on creation that lies beyond the reach of those relying only on the evidence of their imperfect senses. In journeying with him, the human being can begin to imagine the mysteries of a universe the vast majority of which he cannot see or yet comprehend. (See Secord 2014 25–51 for an excellent account of Davy's *Consolations* and its visionary and imaginative appeal; for more on Davy's *Consolations*, see Golinski 2013; Knight 172–84; Lawrence 225–7).

Davy's *Consolations* was far from being the first Romantic literary text to imagine a voyage into space conducted by a mysterious spirit guide – Percy Bysshe Shelley's *Queen Mab* (1813) had contained similar passages, for instance. Most importantly for Davy, just eight years before the publication of Davy's *Consolations*, Lord Byron had written the sensational closet drama *Cain* (1821), which recasts the story of Cain and Abel as a tragedy of human *un*knowing. Cain is unable to reconcile the world in which, as a fallen human being, he 'seem[s] nothing' (Act 2, Scene 2, l. 420), with the 'Thoughts that arise within me' of the immensity of Creation and of the omnipotence and eternity of God (Act 1, Scene 1, l. 177). Asking to 'Let me but / Be taught the mystery of my being' (Act 1, Scene 1, l. 320), Cain is taken on an interplanetary journey by Lucifer to understand the insignificance of the earth when seen from space, before travelling into the depths of Hades to see 'enormous creatures . . . / Resembling somewhat the wild inhabitants / Of the deep woods of the Earth . . . / . . . but ten-fold / In magnitude and terror' (Act 2, Scene 2, ll. 132–8), the spirits of creatures who now 'lie / By myriads' (Act 2, Scene 2, ll. 143–4) in fossils beneath the earth's surface drawn from Byron's reading of the French comparative anatomist Georges Baron Cuvier. But Cain runs into problems. First, his human mind cannot comprehend the immensity of the universe with which he is confronted, too dependent as it is on the limited evidence of his senses: 'thou canst not / Speak aught of knowledge which I would not know, / And do not thirst to know, and *bear a mind / To know*' (Act 1, Scene 1, ll. 246–9; my italics), he tells Lucifer. As Lucifer puts it, in terms that would later be echoed by Davy's 'Unknown', 'matter cannot / Comprehend spirit wholly – but 'tis something to know / There are such realms' (Act 2, Scene 2, ll. 169–71). But, whereas for Davy's hero it truly is enough to know there are such realms, to prove that there are spirit worlds beyond these in which greater knowledge of the universe is made possible, this is both frightening and depressing for Cain. Lucifer cannot show him the secret of death or the nature of eternity – just myriad worlds piled up through space and time, each created only to be destroyed. In 'rage and fury against the inadequacy of his state to his conceptions' (1922 5:470), as Byron later wrote, Cain kills his brother.

Davy's journeys through geological time and across astronomical space, then, rewrite Cain's tragedy as triumph. Scientific discovery precipitates Cain's decline, reminding him both of his human insignificance in a universe of infinite age and size, and also of his human inability to see the eternal realms of God. But science only works in this way if we do not have faith that there are conditions of being that make knowledge of the eternal possible. For Davy, grasping the immensity of geological time is a *precursor* to comprehending worlds beyond human reach, science offering a new imaginative vista to the human imagination that marks a step forward in his intellectual and spiritual evolution.

The disturbed imagination

Lyell too was a fan of Byron's poetry. He wrote Byronic verse as an undergraduate and wrote to his father that he had spotted Byron rowing in his gondola while on holiday in Venice. He also quoted from Byron's *Childe Harold's Pilgrimage* twice in *Principles of Geology* (see Buckland 2013 94–6, 127–30). And he shares Byron's more pessimistic view of ruin, both in the natural world and of the human capacity to comprehend it. And so, despite the emerging scientific consensus on which Davy's progressionist vision was based, and despite the Unknown's careful presentation of the relationship between a global geology and detailed local research, in Chapter 9 of the first volume of *Principles of Geology* Lyell took progressionist geology to task, through a special attack on Davy, whom he quotes at length. This chapter, entitled the 'Theory of the Progressive Development of Organic Life', begins by noting that there are a 'very few exceptions' (Lyell 1:147) to the general rule that simpler species were found in the earliest strata and more complex ones in the most recently deposited rocks. While the carboniferous era contained hundreds of species of monocotyledonous plants (plants with only one cotyledon – a leaf-like part of the embryo of a plant), a handful of fossilized dicotyledonous (and therefore more complex) plants had been found there, Lyell informed his readers. Also recently discovered was 'a saurian' (a giant reptilian form) 'in the mountain limestone of Northumberland' (Lyell 1:129), appearing much earlier in the fossil record than it should according to the law of progress. 'These exceptions,' Lyell claims, 'are as fatal to the doctrine of successive development as if they were a thousand' (1:147).

It did not matter that these exceptions to progress were few and far between, Lyell argued, for two reasons. The first was obvious: one species out of sequence ruined the whole pattern, making the appearance of progress an illusion – or, as the very nature of Davy's visionary prose revealed, a delusion. Second, Lyell claimed that it was almost incredible that *any* reptiles or mammals had been fossilized at all. Fossils had to withstand millions upon millions of years of pressure, heat, chemistry and erosion, meaning that very few of them actually survived. So few fossils existed that it was madness to extrapolate a complete story of life on earth from their scanty remains.

Most importantly, the majority of the earth's strata had been formed under water and only later risen from beneath the lakes, rivers and seas. Lyell asked his readers to indulge their imaginations once again in order to grasp the implications of this point for any consideration of earth history. 'Suppose our mariners were to report,' he wrote,

> that on sounding in the Indian ocean near some coral reefs, and at some distance from the land, they drew up on hooks attached to their line portions of a leopard, elephant, or tapir; should we not be sceptical as to the accuracy of their statements; and if we had no doubt of their veracity, might we not suspect them to be unskilful naturalists?
>
> *(Lyell 1:149)*

Lyell continued,

> Can we expect for a moment that when we have only succeeded amidst several thousand fragments of corals and shells, in finding a few bones of *aquatic* or *amphibious* animals, that we should meet with a single skeleton of an inhabitant of the land?
>
> *(1:149)*

Hardly any aquatic animals survived in fossils, and they lived and died in the water where the rocks were formed, multiplying their chances of preservation. If we would laugh at the naiveté (or suspect the honesty) of a fisherman who claimed to catch leopards in his nets, then surely we would laugh at the geologist who found fossilized mammals in the mostly aqueous strata?

Lyell went on in this passage to compare this kind of geologist unfavourably to Clarence from Shakespeare's *Richard III*, whose prophetic dream of his own death revealed to him 'in the slimy bottom of the deep, / A thousand fearful wrecks; / A thousand men, that fishes gnaw'd upon; / Wedges of gold, great anchors, heaps of pearl' (quoted in Lyell 1:149). 'Had he also beheld,' Lyell continued,

> amid 'the dead bones that lay scatter'd by', the carcasses of lions, deer, and the other wild tenants of the forest and the plain, the fiction would have been deemed unworthy of the genius of Shakespeare. So daring a disregard of probability, so avowed a violation of analogy, would have been condemned as unpardonable even where the poet was painting those incongruous images which present themselves to a disturbed imagination during the visions of the night. But the cosmogonist is not amenable, even in his waking hours, to these laws of criticism; for he assumes either that the order of nature was formerly distinct, or that the globe was in a condition to which it can never again be reduced by changes which the existing laws of nature can bring about.
>
> *(1:149–50)*

Now, the geologist who has seen in the rock and fossil record evidence of progression is suffering visions even wilder than those of Clarence – imprisoned, terrified and dreaming. Even he would not have imagined lions and deer at the bottom of the sea. Lyell does not make the parallel explicit, but the first conversation of Davy's *Consolations*, featuring the spectral 'Genius' and his journey into space, also appears in a dream to the protagonist, who, like Clarence, knows himself to be dying. The connection between Davy's visionary elaboration of the widely held view of progress in the fossil record and Clarence's nightmare is easy to spot.

At the end of this passage Lyell slips from the specific claim that few species are preserved in the strata to a more general argument that 'the cosmogonist' (a heading under which Davy, and the entire category of 'human philosophers', are now subsumed) 'assumes . . . that the order of nature was formerly distinct' (1:150). The link between these two things might not be immediately obvious, but it is crucial. Looking for leopards at the bottom of the sea is too preposterous even for a dream. But looking for leopard-like creatures in the *fossil record* is only preposterous if you assume, as Lyell does, that the way the earth works now – the kinds of things that happen on earth and the power of individual geological events to create change – is the way it has *always* worked. As such, Lyell's central strategy in *Principles of Geology* was to suggest a new *methodology* for studying the past. Since human beings had not witnessed most of earth history, and the evidence they had with which to understand it was so incomplete, they would need to analyse 'causes *now* in operation' (my italics) – 'now' meaning 'within human history': geologists would analyse only those phenomena they could see happening around them every day, or that had at least occurred within human history. Events in the earth's past could only be understood by analogy with the present – and, for the analogy to hold, the geologist would need to assume that the past and present were similar enough to be compared, that geological events had always been of the same type, and of the same intensity, as those currently operating. By contrast, the idea of geological progress assumed that conditions in the past had been radically different from those in the present (the earth had once been much hotter and more volatile; disasters operated on a global scale). In Davy's account, it has only been since 'the creation of man took place' that 'there has been little alteration in the physical circumstances of the globe'. For Lyell, this was pure guesswork, an incredible assumption based on unreliable evidence about the earliest history of the world.

Even more radically than Davy, Lyell emphasizes the 'feeble' nature of human intellectual powers. And it is in this sense that Lyell invokes Pope's gnome in Chapter 5 of the first volume of *Principles*. The gnome is not the first fantastical creature Lyell asks his readers to imagine, but the last in a long passage in which Lyell has also unfavourably compared the human geologist to 'an amphibious being, who should possess our faculties' (1:82). This amphibian 'would more easily arrive at sound theoretical opinions in geology' than a human being, 'since he might behold, on the one hand, the decomposition of rocks in the atmosphere and the transportation of matter by running water; and, on the other, examine the deposition of sediment in the sea'. The amphibious being endowed with human intelligence sees *more* than the human being, because he can inhabit both air and water. But he cannot inhabit the depths of the earth, the 'rocks of subterranean origin', which are the sole province of the gnome. Sylph, nymph, gnome, salamander: each restricted to a single element, each is also restricted to a partial view of the earth and its workings. But the human, too, is trapped more hopelessly than the amphibian, just as hopelessly as the gnome.

The lesson of these imaginings for Lyell was clear: human beings were inadequate observers of the natural world, bound by their bodies to a small fraction of its surface, a tinier fraction of its depths and an even tinier portion of its history. They could not rely on their senses alone for the conduct of good scientific work, since they simply could not see into the past or into the depths of the earth or the sea. Even worse, humans could barely even rely on themselves to remember their own inadequacy. Many of Lyell's contemporaries had long held that the earth was millions of years old, but Lyell believed they had not fully comprehended the implications of the earth's antiquity. They had not *imaginatively* grasped the ways in which such age meant that the kinds of causes they saw operating all around them were enough to explain the wholesale geological changes that had taken place on the earth during its past. While Davy might have used supernatural beings to acknowledge and yet overcome the limitations of human perspective, to transcend our fixed positions on earth and show us things we could not otherwise see, Lyell's gnome was an attempt to make his readers confront the reality of just how limited that position really is: an objective correlative not, as we might expect, for the unseen processes of the underground but for the tragicomedy of human perception.

Imagining geology anew

Geology was a fashionable, controversial and exciting science in the nineteenth century, and its images and stories were a rich source of material for novelists and poets throughout that period. Literary critics have shown that Lyell's dramatic evocation of the power of minute causes to effect great change over a long enough span of time gave Alfred Tennyson a framework for exploring patterns of grief in *In Memoriam* (1850).[1] As navvies and colliers dug deep into the earth to build canals, railways and sewers, they revealed immense fossilized 'lizards' or prehistoric beasts, like the unwieldy megatherium who appears with regularity in William Makepeace Thackeray's fiction, or the megalosaurus who famously waddles up Holborn Hill in the first paragraph of Charles Dickens's 1853 novel *Bleak House*.[2] As other critics have revealed, geologists also described stranger, smaller primeval creatures, like the trilobite with whom Thomas Hardy's protagonist Henry Knight comes face to face when he falls off a cliff in the 1874 novel *Two on a Tower*, or found in May Kendall's poem 'The Lay of the Trilobite' (1885).[3] One of Lyell's most speculative arguments against progress in the fossil record – that if hotter conditions reappeared on the earth then it was not inconceivable that seemingly 'ancient' species like the iguanodon and the ichthyosaur might also live again – inspired a plot sequence in another novel playing fast and loose with the idea of geological progress, Arthur Conan Doyle's *The Lost World* (1912) (see Buckland 2013 (Chapter 5)). That novel ends with a pterodactyl flying around the streets of London. Periodical articles appealing to all kinds of 'ordinary' readers would exhort them to take hammer and sack and go collecting specimens and chiselling at rocks and cliff faces (or would gently satirize the people who actually did). And Charles Kingsley's novels *Yeast* (1848), *Alton Locke* (1850) and *Two Years Ago* (1853) have geologists for protagonists – an aristocrat, a working-class autodidact and a middle-class scapegrace returned from the Australian gold diggings – representing the wide social range of participants in nineteenth-century geological fieldwork, collecting and writing. Kingsley plots the last of these novels according to geological maps and columns: the novel's action takes place in the 'Devonian' strata of Devon, Wales and the Eifel region in Germany; it may also be the case that the developing sense of British regionality and of regional writing in the nineteenth century was shaped by the three-dimensional form of geological maps, each region lying on a rock that represented a block or blocks of time in the stratigraphic record. These examples (given more thorough treatment in Buckland 2013 131–220) form only a handful of the most famous of many literary engagements with the forms, ideas, practices, objects and images of geological science in this period.

Literary critics, too, have been repeatedly drawn to the geological passages in nineteenth-century poetry and fiction. In this critical tradition, geology has been associated with two basic plot patterns in nineteenth-century fiction: Lyell's 'uniformitarianism' (taken roughly to mean gradual change over a long span of time, and linked to slow political reform and the gradually developing plots of some forms of realism); and 'catastrophism', typically linked with Lyell's geological colleagues or with natural theology, and with revolutionary change and the broad strokes of literary 'romance' of various kinds (see Shuttleworth 52, 81; Smith 257, 120–21; Beer 181; Cosslett 4–5; Meckier 243–76 for examples). In *Novel Science*, I argue against this move (see especially the introduction and Chapters 3 and 6), for two reasons. First, there were not two 'schools' of geologists. Most of Lyell's contemporaries at the Geological Society of London agreed that the earth was millions of years old but disagreed that all geological events throughout earth history had operated at the same intensity as those now operating. Second, the tendency to view science as the purveyor of big 'ideas' or plot patterns tends to abstract it from the everyday practices, methods, instruments and debates that made that science possible. Some recent literary accounts of geology have thought much harder about these practices and literary engagements with them in more detailed and compelling ways. To take a key example, Noah Heringman has demonstrated the ways in which geologists used the traditional literary genre of antiquarian local history to describe the historicity of rocks and landforms; he has also detailed interrelationships between early geology and mineralogical tourism, travel literature and Romantic conceptions of the sublime and the picturesque, revealing that literature shaped the descriptive and historical practices of geology as much as it was shaped by them. Ralph O'Connor, too, studies 'science *as* literature, rather than science *and* literature', emphasizing that in the nineteenth century '[m]uch poetry, and still more prose fiction, continued to be valued for its factual content; equally, scientific writing continued to have an aesthetic dimension, both self-consciously and by default' (2007A 15). Most importantly, O'Connor reminds us that earth history was presented to its readers in a wide variety of textual forms, 'not only in scientific treatises, but also in pamphlets, magazine articles, epic poems, autobiographies, children's stories, travelogues, and sermons' (2007A 15–16).

This variety of textual forms and of practitioners of geology means that there is no clear-cut way to distinguish between 'scientific writing', poetry, prose fiction and non-fiction in geological writing. Nor is there any clear-cut sense in which writing can be distinguished from geo-logical practice: ordering and naming the rocks, taking notes in the field, corresponding with other geologists and writing and publishing geological books and papers were fundamental activities for many geological practitioners and were often shaped by the stories, narrative strategies and attitudes of contemporary writers such as Walter Scott and Lord Byron, as has been hinted at here. The sheer excitement surrounding geological discovery and the practice of fieldwork in the nineteenth century means that it is rich in evocative, beautiful and challenging texts, many of which are still ripe for exploration by literary critics.

While this chapter has concentrated on just two elite men of science, Humphry Davy and Charles Lyell, then, it has also attempted to demonstrate an important point about the nature of this kind of geological writing. As James A. Secord has recently shown, the 1820s and 1830s were crucial decades in the history of publishing, as the invention of the steam press and the production of cheap paper transformed the availability of printed matter (2014 1–23). Scientific writers sought to take advantage of the new plethora of print in a variety of ways, and Davy and Lyell clearly sought to define geological writing as a prestigious mode of authorship as philosophical and imaginative as poetry. In this, the two men were far more alike than they first appear. Just as Lyell attributed superstitious belief to man 'in an early stage of advance-ment', for instance, Davy too wrote that mankind in 'a state of nature' was 'a creature of almost pure sensation', 'harassed by superstitious dreams' or left 'to the mercy of nature and the elements' (1839–40 2:318–19). And why was early man so superstitious? Because, unlike modern, enlightened man, he had to rely only upon his fallible senses, the immediate impressions the world made upon him. 'How different is man,' Davy continued, 'informed by the beneficence of the Deity, by science and the arts!' (1839–40 2:319). Without a powerful, imaginative, literary culture, of which science was clearly a part, man could not grasp the deep structures and pro-cesses that made the world tick. His mind was only capable of what Davy's friend Samuel Taylor Coleridge called 'Fancy' (313): passively receiving information imprinted on and recorded by his mind, mechanically recalled, and giving him little insight into the workings of nature.[4]

For both Davy and Lyell, reinventing science as a literary mode of authorship, the answer to all this was the sophisticated human imagination. For Coleridge, imagination in its 'primary' form was 'the living Power and prime Agent of all human Perception . . . a repetition in the finite mind of the eternal act of creation in the infinite I AM' (313), a unifying and idealizing 'mysterious power' creatively conceiving of the world in repetition of God's creative acts, making sense of the world's deepest mysteries. As Davy put it, the gift of science offered precisely this kind of imaginative act – it had 'bestowed upon' mankind 'powers which may be called "creative"', so that he could 'modify and change the beings surrounding him, and by his experiments to interrogate nature with power' (1839–40 2:319). And Lyell, despite the fact that he was concluding his chapter on the critique of Davy and of geological progress, also claimed that science enabled an imaginative and creative engagement with the natural world in similarly Romantic terms. 'Although we are mere sojourners on the surface of the planet,' he consoled his readers, quoting John Dryden's translation of Virgil's *Georgics*,

> chained to a mere point in space, enduring but for a moment of time, the human mind is not only enabled to number worlds beyond the unassisted ken of mortal eye, but to trace the events of indefinite ages before the creation of our race, and is not even withheld from penetrating into the dark secrets of the ocean, or the interior of the solid globe; free, like the spirit which the poet described as animating the universe, 'Thro' Heav'n, and Earth and Oceans depth he throws / His Influence round, and kindles as he goes'.
>
> *(Lyell 1:166)*

Man could not observe these 'worlds' beyond 'mortal eye', these 'dark secrets' and inner spaces, but he could, if he worked hard enough and read his Lyell carefully, dare to imagine them. Astronomers, spiritualists, evolutionary theorists and poets the century long would describe worlds beyond the ken of human observation, citing Lyell's method of analogy between the world you *can* see and the worlds you *can't* – those worlds so far in the past, so deep into space, so big, so small or so inexplicable that human beings could not actually see them.[5] In order for Lyell to get his readers to reimagine the power minute causes might hold over millennia, then, he sought first and foremost to remind them of the profound imaginative challenges they had in apprehending anything about their place in the world, in history or in the universe, with any kind of certainty – without ever losing commitment to the notion that human beings were special possessors of a powerful and God-given imagination that, rightly trained, could transcend the limited evidence of the senses to echo (though never to fully attain) something like an omniscient view. As this example demonstrates, nineteenth-century geology participated in a broad literary and philosophical discussion about what the imagination was and what it was capable of. In doing so, geological writing depended upon imaginative literature for its inspiration and its claims to power and prestige and existed as a beautiful, evocative and challenging form of literature in its own right.

Notes

1   A wealth of scholarship exists on Tennyson and geology: see Dean; Armstrong; Tomko; Zimmerman; Snyder.
2   On Thackeray, see Dawson 2011; Dawson 2013. On Dickens, see Buckland 2013 247–75; Zimmerman (Chapter 5); Dawson 2015.
3   See Ingham's seminal essay, Radford; Buckland 2008.

4 For Davy as a poet, and for the relationships between Davy's writing and poetry, see Fullmer; Sharrock; Levere; Lawrence; Ruston.
5 See Buckland 2018.

Bibliography

Primary texts

[Anon.]. 'The Loves of the Triangles'. *Anti-Jacobin Review; Or, Weekly Examiner*. 16 April 1798: 179–82; 23 April 1798: 188–9; 7 May 1798: 204–6.

Byron, George Gordon. *Sardanapalus, A Tragedy; The Two Foscari, A Tragedy; Cain, A Mystery*. London: Murray, 1821.

Byron, George Gordon. *The Works of Lord Byron: Letters and Journals*. Ed. Rowland E. Prothero. Rev. edn. 6 vols. London: Murray, 1922.

Byron, George Gordon. *Lord Byron's Cain: Twelve Essays and a Text with Variants and Annotations*. Ed. Truman Guy Steffan. Austin, TX: University of Texas Press, 1968.

Coleridge, Samuel Taylor. *The Major Works*. Ed. H. J. Jackson. Oxford: Oxford University Press, 1985.

Darwin, Erasmus. *The Botanic Garden; A Poem, in Two Parts. Part I. Containing the Economy of Vegetation. Part II. The Loves of the Plants. With Philosophical Notes*. London: Johnson, 1791.

Davy, Humphry. *Consolations in Travel; or, The Last Days of a Philosopher*. London: Murray, 1830.

Davy, Humphry. *The Collected Works of Sir Humphry Davy*. Ed. John Davy. 9 vols. London: Smith, Elder & Co. Cornhill, 1839–40.

Lyell, Charles. *Principles of Geology: Being an Attempt to Explain the Former Changes of the Earth's Surface, by Reference to Causes Now in Operation*. 3 vols. London: Murray, 1830–33.

Pope, Alexander. *The Rape of the Lock: An Heroi-Comical Poem. In Five Cantos*. 2nd edn. London: Lintott, 1714.

Secondary texts

Armstrong, Isobel. 'Tennyson in the 1850s: From Geology to Pathology'. *Tennyson: Seven Essays*. Ed. Phillip Collins. London: Palgrave, 1992. 102–40.

Beer, Gillian. *Darwin's Plots: Evolutionary Narrative in Darwin, George Eliot and Nineteenth-Century Fiction*. London: Routledge, 1983.

Buckland, Adelene. '"A Product of Dorsetshire": Geology and the Material Imagination of Thomas Hardy'. *19: Interdisciplinary Studies in the Long Nineteenth Century* 6 (2008). <www.19.bbk.ac.uk/articles/abstract/469>. Accessed 10 January 2016.

Buckland, Adelene. *Novel Science: Fiction and the Invention of Nineteenth-Century Geology*. Chicago, IL: University of Chicago Press, 2013.

Buckland, Adelene. 'The World Beneath Our Feet: Insight and Imagination in Romantic Geological Writing', *Philological Quarterly*, (forthcoming 2018).

Cosslett, Tess. *The 'Scientific Movement' and Victorian Literature*. New York, NY: St Martin's, 1982.

Dawson, Gowan. 'Literary Megatheriums and Loose Baggy Monsters: Palaeontology and the Victorian Novel'. *Victorian Studies* 53 (2011): 203–20.

Dawson, Gowan. 'Like a Megatherium Smoking a Cigar: Darwin's *Beagle* Fossils in Nineteenth-Century Popular Culture'. *Darwin, Tennyson and Their Readers*. Ed. Valerie Purton. London: Anthem, 2013. 81–96.

Dean, Dennis R. *Tennyson and Geology*. Lincoln: Tennyson Society, 1985.

Fara, Patricia. *Erasmus Darwin: Sex, Science, & Serendipity*. Oxford: Oxford University Press, 2012.

Fullmer, June Z. 'The Poetry of Sir Humphry Davy'. *Chymia* 6 (1960): 102–26.

Fullmer, June Z. 'Romantic Science: Humphry Davy's *Consolations in Travel*'. Lindberg Lecture, UNH. 11 April 2013. <http://cola.unh.edu/sites/cola.unh.edu/files/media/thecollegeletter/2013/may/LindbergLectureTEXT.pdf>. Accessed 10 January 2016.

Heringman, Noah. *Romantic Rocks, Aesthetic Geology*. Ithaca, NY: Cornell University Press, 2004.

Ingham, Patricia. 'Hardy and *The Wonders of Geology*'. *Review of English Studies* 31 (1980): 59–64.

Knight, David. *Humphry Davy: Science and Power*. Cambridge: Cambridge University Press, 1998.

Latimer, Bonnie. 'Alchemies of Satire: A History of the Sylphs in *The Rape of the Lock*'. *Review of English Studies* 57 (2006): 684–700.

Laudan, Rachel. *From Mineralogy to Geology: The Foundations of Science, 1650–1830*. Chicago, IL: University of Chicago Press, 1985.

Lawrence, Christopher. 'The Power and the Glory: Humphry Davy and Romanticism'. *Romanticism and the Sciences*. Eds. Andrew Cunningham and Nicholas Jardine. Cambridge: Cambridge University Press, 1990.

Levere, Trevor H. '"The Lovely Shapes and Sounds Intelligible": Samuel Taylor Coleridge, Humphry Davy, Science and Poetry'. *Nature Transfigured: Science and Literature 1700–1900*. Eds. John Christie and Sally Shuttleworth. Manchester: University of Manchester Press, 1989.

Meckier, Jerome. *Hidden Rivalries in Victorian Fiction: Dickens, Realism, and Revaluation*. Lexington, KY: University Press of Kentucky, 1987.

O'Connor, Ralph. *The Earth on Show: Fossils and the Poetics of Popular Science, 1802–1856*. Chicago, IL: University of Chicago Press, 2007 [2007A].

O'Connor, Ralph. 'Young Earth Creationists in Early Nineteenth-Century Britain? Towards a Reassessment of "Scriptural Geology"'. *History of Science* 45 (2007): 357–403 [2007B].

O'Connor, Ralph. 'Facts and Fancies: The Geological Society of London and the Wider Public, 1807–1837'. *The Making of the Geological Society of London*. Eds. C. L. E. Lewis and S. J. Knell. London: Geological Society of London, 2009, 344–53.

Oldroyd, David. *The Highlands Controversy: Constructing Geological Knowledge Through Fieldwork in Nineteenth-Century Britain*. Chicago, IL: University of Chicago Press, 1990.

Priestman, Martin. *The Poetry of Erasmus Darwin: Enlightened Spaces, Romantic Times*. Ashgate, 2013.

Radford, Andrew. *Thomas Hardy and the Survivals of Time*. Aldershot: Ashgate, 2003.

Rudwick, Martin. *The Great Devonian Controversy: The Shaping of Scientific Knowledge Among Gentlemanly Specialists*. Chicago, IL: University of Chicago Press, 1985.

Rudwick, Martin. *Worlds Before Adam: The Reconstruction of Geohistory in the Age of Reform*. Chicago, IL: University of Chicago Press, 2007.

Rudwick, Martin. 'Uniformity and Progression: Reflections on the Structure of Geological Theory in the Age of Lyell'. *Perspectives in the History of Science and Technology*. Ed. Duane Roller. Norman, OK: University of Oklahoma Press, 2009. 209–27.

Ruston, Sharon. *Creating Romanticism: Case Studies in the Literature, Science and Medicine of the 1790s*. Basingstoke: Palgrave Macmillan, 2013.

Secord, James A. *Controversy in Victorian Geology: The Cambrian-Silurian Dispute*. Princeton, NJ: Princeton University Press, 1986.

Secord, James A. *Visions of Science: Books and Readers at the Dawn of the Victorian Age*. Chicago, IL: University of Chicago Press, 2014.

Sharrock, Roger. 'The Chemist and the Poet: Sir Humphry Davy and the Preface to *Lyrical Ballads*'. *Notes and Records of the Royal Society of London* 17 (1962): 57–76.

Shuttleworth, Sally. *George Eliot and Nineteenth-Century Science: The Make-Believe of a Beginning*. Cambridge: Cambridge University Press, 1984.

Smith, Jonathan. *Fact and Feeling: Baconian Science and the Nineteenth-Century Literary Imagination*. Madison, WI: University of Wisconsin Press, 1994.

Snyder, E. E. 'Tennyson's Progressive Geology'. *Victorian Network* 2 (2010): 27–48.

Tomko, Michael. 'Varieties of Geological Experience: Religion, Body, and Spirit in Tennyson's *In Memoriam* and Lyell's *Principles of Geology*'. *Victorian Poetry* 42 (2004): 113–33.

Veenstra, Jan R. 'Paracelsian Spirits in Pope's *Rape of the Lock*'. *Airy Nothings: Imagining the Otherworld of Faerie from the Middle Ages to the Age of Reason*. Eds. Karin E. Olsen and Jan R. Veenstra. Leiden: Brill, 2014. 213–40.

Zimmerman, Virginia. *Excavating Victorians*. Albany, NY: State University of New York Press, 2008.

Further reading

Bowler, Peter J. *Fossils and Progress: Palaeontology and the Idea of Progressive Evolution in the Nineteenth Century*. New York, NY: Science History Publications, 1976.

Golinski, Jan. 'Humphry Davy: The Experimental Self'. *Eighteenth-Century Studies* 45 (2011): 15–28.

Heringman, Noah, ed. *Romantic Science: The Literary Forms of Natural History*. Albany, NY: State University of New York Press, 2003.

Heringman, Noah. 'Picturesque Ruin and Geological Antiquity: Thomas Webster and Sir Henry Englefield on the Isle of Wight'. *The Making of the Geological Society of London*. Eds C. L. E. Lewis and S. J. Knell. London: Geological Society of London, 2009, 299–318.

Holmes, John. '"The Lay of the Trilobite": Rereading May Kendall'. *19: Interdisciplinary Studies in the Long Nineteenth Century* 11 (2010). <www.19.bbk.ac.uk/articles/575>. Accessed 10 January 2016.

Gossin, Pamela. *Thomas Hardy's Novel Universe: Astronomy, Cosmology and Gender in the Post-Darwinian World*. Burlington, VT: Ashgate, 2007.

King-Hele, Desmond. *Erasmus Darwin and the Romantic Poets*. Basingstoke: Macmillan, 1986.

King-Hele, Desmond. *Erasmus Darwin: A Life of Unequalled Achievement*. London: DLM, 1999.

O'Connor, Ralph. 'From the Epic of Earth History to the Evolutionary Epic in Nineteenth-Century Britain'. *Journal of Victorian Culture* 14 (2009): 207–23.

Porter, Roy. 'Gentlemen and Geology: The Making of a Scientific Career 1660–1920'. *The Historical Journal* 21 (1978): 809–36.

Rudwick, Martin. *Bursting the Limits of Time: The Reconstruction of Geohistory in the Age of Revolution*. Chicago, IL: University of Chicago Press, 2005.

Rupke, Nicholas A. 'Caves, Fossils and the History of the Earth'. *Romanticism and the Sciences*. Eds. Andrew Cunningham and Nicholas Jardine. Cambridge: Cambridge University Press, 1990, 241–61.

Secord, James A. 'King of Siluria: Roderick Murchison and the Imperial Theme in Geology'. *Victorian Studies* 25 (1982): 413–42.

Secord, James A. Introduction. Charles Lyell. *Principles of Geology*. Ed. James A. Secord. London: Penguin, 1997, ix–xliii.

Secord, James A. *Victorian Sensation: The Extraordinary Publication, Reception, and Secret Authorship of 'Vestiges of the Natural History of Creation'*. Chicago, IL: University of Chicago Press, 2000.

Trower, Shelley. 'Primitive Rocks: Humphry Davy, Mining and the Sublime Landscapes of Cornwall'. *Journal of Literature and Science* 7 (2014): 20–40.

Veneer, Leucha. 'Practical Geology in the Geological Society in its Early Years'. *The Making of the Geological Society of London*. Eds C. L. E. Lewis and S. J. Knell. London: Geological Society of London, 2009. 243–53.

Weindling, Paul Julian. 'Geological Controversy and its Historiography: The Prehistory of the Geological Society of London'. *Images of the Earth: Essays in the History of the Environmental Sciences*. Eds. Ludmilla Jordanova and Roy Porter. Oxford: Alden, 1997, 248–71.

18

CHEMISTRY

Sharon Ruston

Lancaster University

In his 1802 *Discourse, Introductory to a Course of Lectures*, Humphry Davy ascribes powers that had been formerly ascribed to alchemists to the modern chemist. Davy's chemist has:

> [P]owers which may almost be called creative; which have enabled him to modify and change the beings surrounding him, and by his experiments to interrogate nature with power, not simply as a scholar, passive and seeking only to understand her operations, but rather as a master, active with his own instruments.
>
> *(2:319)[1]*

Davy's modern chemist, like the ancient alchemist, sees the world as constantly transforming matter, shifting and changing from one form to another. Not content merely to witness these changes, the modern chemist can make them occur. The creative powers that Davy's chemist possesses can be likened to those of the philosopher's stone, which promised to change lead to gold, or of the elixir that promised eternal life, but for Davy this is a 'rational Alchemy' (letter to Jöns Jacob Berzelius, 10 July 1808; Berzelius 2:7–8). This chapter argues that chemical philosophy offers a foundation upon which much important Romantic poetic theory was based. While Davy's influence on Wordsworth's revised preface to *Lyrical Ballads* is well known (see Sharrock), I show that Davy's chemical philosophy helped other Romantic writers to conceptualize the act of creativity itself.

According to John Herschel in 1831, chemistry was the 'most popular' and 'one of the most extensively useful' (299) of the sciences. It was specifically the 'wonderful and sudden transformations with which chemistry is conversant' and 'above all, the insight it gives into the nature of innumerable operations which we see daily carried on around us' that accounted for its popularity. Herschel dates the emergence of 'modern chemistry' (301) from Antoine Laurent Lavoisier's experiments. Alchemy's promise to turn base metal into gold had given way to more moderate – but nonetheless sublime – transformations, such as Lavoisier's decomposition of water into its component parts and its recomposition back into water in 1785 (Golinski 1992 133).[2] At the start of the nineteenth century, chemistry was difficult to define. Indeed, Davy 'thought of himself as a natural philosopher or a chemical philosopher expounding a world view, and not as a specialist' (Knight 5). As the century progressed, chemistry became more clearly demarcated into different branches and subdivisions, such as organic and industrial chemistry. In the early part of the century it was more aligned to natural history, using a biological language of 'form and function' to describe molecules rather than the later 'mechanical language of matter, motion and force' that was more indebted to physics (Nye 121). It was not until the development of the periodic law in the 1870s that chemists finally devised 'a comprehensive classificatory system of elements' (Brock xvii). Throughout the century, literary texts engaged with contemporary developments in chemistry, such as recognition of the importance of carbon, yet they continued to emphasize chemistry's potential for transformation. Even into the twentieth century, atomic physics could be described as 'modern alchemy', with radium identified as the 'philosopher's stone' (Morrisson 2011 2).

While Davy's legacy may have waned among men of science after his death in 1829, his popular appeal remained. Twenty-five years after Davy's death, Henry Mayhew wrote a part-fictionalized account of his youth 'for boys', *The Wonders of Science; Or, Young Humphry Davy* (1854), to encourage them to take up scientific pursuits. Davy is mentioned in Charles Dickens's *Nicholas Nickleby* (1838–9), Anne Brontë's *The Tenant of Wildfell Hall* (1848) and George Eliot's *Middlemarch* (1871–2) (see Flaherty). On 3 July 1897 Thomas Hardy visited his grave in Cimetière de Plainpalais in Geneva with his first wife, Emma (Hardy 208; see Brown 169). After my discussion of early nineteenth-century poetic theory, I move on to consider four nineteenth-century novels that reveal the persistence of Davy's ideal of modern chemistry as realizing alchemy's aspirations to transform and to create. Mary Shelley's *Frankenstein* (1818), Edward Bulwer-Lytton's *Zanoni* (1847), Honoré de Balzac's *The Alkahest* (1834) and Robert Louis Stevenson's *Strange Case of Dr Jekyll and Mr Hyde* (1886) represent nineteenth-century chemistry as a science poised to make great discoveries about the nature of life itself. In all four texts, the new chemistry, like alchemy before it, has the power to reveal hidden unities in nature and thereby to transform matter from one form into another.

Chemistry and the Romantic poet

Davy was the pre-eminent English chemist of the early nineteenth century and he continued to be cited as an authority in literary texts published throughout the nineteenth century. He refused to accept some of Lavoisier's claims and continued to question the newly proposed nomenclature. Davy was motivated partly by nationalism to prove the French chemist wrong when he showed that soda and potash were not elements – as Lavoisier had thought – but could be further decomposed using the electric battery to isolate the real elements, sodium and potassium (Levere 2001 89). Despite isolating them, Davy did not believe that these were definitely elements or that these so-called elements had unique properties or characteristics. As Trevor Levere writes, 'Instead, Davy speculated that there might be one kind of ultimate atom . . . This kind of atom would make transmutation a theoretical possibility' (2001 91). We can see this thinking in statements such as this from his 1812 *Elements of Chemical Philosophy*: 'Most of the substances belonging to our globe are constantly undergoing alterations in sensible qualities, and one variety of matter becomes as it were transmuted into another' (4:1). The phrase 'as it were' might acknowledge the inadequacy of the term 'transmutation' to describe precisely what is happening, but Davy repeatedly throughout his career alluded to the idea that the purpose of chemistry was to study the 'continual transmutations and changes of external objects' (quoted in Davy J. 1:170).

Davy was certainly the primary means by which a good number of nineteenth-century poets and novelists learnt about chemistry, including Samuel Taylor Coleridge, William Godwin, Eleanor Anne Porden, Mary and Percy Shelley and William Wordsworth (see Levere 2002; Ruston 2005; Ruston 2013; Johns Putra; Sharrock). Throughout his life Davy refused to believe John Dalton's theory of atomic weight, which had been set out in his 1808 *A New System of*

Chemical Philosophy. This may account for the persistence of Davy's alchemical world view in literary texts. As Mark Morrisson has written, Dalton's work 'laid out the dominant view that would hold until the end of the nineteenth century, that atoms were the smallest particles and neither divisible nor alterable. Each was a distinct fundamental particle. Alchemical views that elements could be transmuted, then, were held to be an intellectual mistake' (2011 19). Evidence of Davy's pre-eminence comes in Anna Barbauld's poem *Eighteen Hundred and Eleven*, where Davy is listed among those who have made England great alongside Pitt, Fox, Nelson, Garrick, Franklin and Priestley (168). Using a gendered image similar to those Davy himself used, Barbauld writes that 'Nature's coyest secrets were disclosed' (l. 202) to him. Davy's influence continued well into the nineteenth century. For example, James Secord's *Visions of Science* examines the legacy of Davy's *Consolations in Travel*, which sold 'exceptionally well' (28) throughout the century.

Davy's primary definition of chemistry was that it was the study of change, transformation and even transmutation. The recognition that matter can change state while not changing its chemical properties was the 'decisive innovation' of the previous century, according to Jan Golinski. It was acknowledged that 'substances could be made into gases by the addition of heat without changing their chemical nature' (Golinski 2003 376). For Percy Bysshe Shelley, whose interest in chemistry is well documented, 'Poetical abstractions' are not 'beautiful and new' because they have never existed before 'in the mind of man or in Nature, but because the whole produced by their combination has some intelligible and beautiful analogy with those sources of emotion and thought and with the contemporary condition of them' (134; Ruston 2005). When Shelley declares in the preface to *Prometheus Unbound* (1820) that poetry creates 'by combination and representation' (134), he alludes to this new understanding that there is a finite quantity of matter in the world, which may be constantly transforming but is not being newly created or destroyed. Dalton had confirmed this:

> No new creation or destruction of matter is within the reach of chemical agency. We might as well attempt to introduce a new planet into the solar system, or to annihilate one already in existence, as to create or destroy a particle of hydrogen.
>
> *(1:212)*

This is again like the metaphor used by Shelley to describe the 'power' of 'modern literature of England': 'The mass of capabilities remains at every period materially the same; the circumstances which awaken it to action perpetually change' (134). This way of thinking has far-reaching consequences; not least for its analogy with revolution but also for other fields of artistic endeavour. For Shelley (and for Davy), the creative power of both chemistry and poetry is in this ability to bring already existing elements into new connections and combinations with each other in order to achieve something new. Mary Shelley writes in a similar fashion when she reflects upon the creative process of writing *Frankenstein* in her 1831 introduction:

> Invention, it must be humbly admitted, does not consist in creating out of void, but out of chaos; the materials must, in the first place, be afforded: it can give form to dark, shapeless substances, but cannot bring into being the substance itself.
>
> *(1993 195; see also Ruston 2013 128)*

In other words, nothing is created from nothing (*ex nihilo*); the 'substance' is always present but is instead 'mould[ed] and fashion[ed]' (Shelley 1993 195) by the author. In this respect there are distinct parallels between the work of literature and of chemistry.

Davy proofread Wordsworth's preface to *Lyrical Ballads* for publication in 1800 (see Sharrock). In his 1802 *Discourse*, written in part as a response to Wordsworth, Davy writes that chemistry obliges people 'to use a language representing simple facts' (2:323). Similarly, Wordsworth had determined to write more accurately or precisely using what he called a 'philosophical language' (1:130) rather than a poetic language. The influence of Davy upon Wordsworth has been widely argued (see, for example, Ross; Jenkins; Hindle; Ruston 2013). Wordsworth mentions the 'Chemist' specifically twice in the 1802 preface. In one comparison, he uses the language of chemistry to characterize poetry: 'What then does the Poet? He considers man and the objects that surround him as acting and re-acting upon each other, so as to produce an infinite complexity of pain and pleasure' (Wordsworth 1:165). If we look at one of the most famous passages of the preface again from this new perspective, Wordsworth can be seen as using the metaphor of a chemical experiment:

> I have said that Poetry is the spontaneous overflow of powerful feelings: it takes its origin from emotion recollected in tranquillity: the emotion is contemplated till by a species of reaction the tranquillity gradually disappears, and an emotion, kindred to that which was before the subject of contemplation, is gradually produced, and does itself actually exist in the mind.
>
> *(1:176)*

The central trope is of liquid, the 'overflow of powerful feelings'. Wordsworth here describes a kind of distillation whereby the original, pure emotion is isolated from the tranquillity with which it was remembered. There are a number of references to 'states' of mind in the preface, which might also invoke ideas of the changing states of matter. Often the mind is described as being in a state of 'excitement' because it has been 'agitated'; this word has a specific sense, which suggests a cloudy liquid that needs to be restored to equilibrium and transparency.

Coleridge attended Davy's lectures in order to increase his 'stock of metaphors' and he made extensive notes (Paris 92; Ruston 2013 135–7). Levere has written extensively on Coleridge's engagement with Davy; chemical metaphors abound in his writings, not least in his famous definition of the secondary imagination, which 'dissolves, diffuses, dissipates, in order to re-create' (1993 1:304). For Coleridge, the imagination is 'the *modifying* power' (1956–71 2:1034) or a 'shaping and modifying power' (1993 1:293). The influence of Davy is clear here, not least in this language of 'power' that is creative and can transform. Davy's sense of an underlying unity in nature – which can be seen in Davy's idea that there is an ultimate atom constantly transmuting into different forms – appealed to both Coleridge and Wordsworth.[3] For these reasons his science, as much as the poetry that he also wrote, can be described as 'Romantic' (see Lawrence).

Using for the most part Davy's writings as evidence, Stuart Sperry writes that for John Keats chemistry was key. Sperry draws the link between Davy's sense of the constantly changing matter in the world and Keats's view of poetry: for Keats, 'poetry can be said to operate through the laws by which the mind, and more especially the imagination, assimilates and transmutes the impressions it derives from nature' (41). We know that Keats studied chemistry at Guy's Hospital. The effects of this study can be seen, for example, in the phosphorous flashes of Lamia's transformation (Roe 75; 'Lamia', Part 1, l. 152). In this episode, Sperry notes that the transformation Lamia undergoes corresponds 'to some of Keats's favourite analogies for poetic creation' (302). It is imagined as 'a violent chemical reaction' (Sperry 302) as a 'sort of chemical analysis and separation of elements takes place'. However, Sperry finds the episode 'a brilliantly comic if somewhat bitter parody of Keats's whole early sense of the nature of poetic creation' (303).

In particular, he reads it as a 'caricature of the "Pleasure Thermometer"' (Sperry 303), which Keats thought might measure the 'gradations of Happiness' (2002 57) that we could move through until we emerge 'Full alchymized', as he describes it in *Endymion* (1 780). Keats imagines progressing from love of physical things until we become refined and purified from all selflessness. From such evidence, Jennifer Wunder argues that Keats was influenced by alchemy as well as chemistry, as for him 'the processes of the imagination are derived from alchemical hermetic studies that were infused with elements of Neo-Platonism' (9).

Victor Frankenstein: from alchemist to chemist

In Mary Shelley's 1818 narrative of Victor Frankenstein's education, ancient alchemy is replaced by modern chemistry but the aspirations of the former continue and are realized in the latter.[4] The young Victor Frankenstein is a follower of alchemical lore. He chances upon a volume of Cornelius Agrippa's works and is soon hooked, but when he communicates his interest to his father, Agrippa's work is dismissed as 'sad trash' (23). With hindsight, Frankenstein speculates on whether 'the train of [his] ideas' would ever 'have received the fatal impulse that led to [his] ruin' had his father taken the pains to explain 'that the principles of Agrippa had been entirely exploded' (23), and, importantly for my argument, 'that a modern system of science had been introduced, which possessed much greater powers than the ancient'. Frankenstein misguidedly continues in his alchemical studies, procuring the whole of Agrippa's works and also reading Paracelsus and Albertus Magnus with the 'elixir of life' as his particular goal. It is not until he reaches the University of Ingolstadt that his studies are properly directed (23). In the 1831 edition of the novel, he acknowledges that there is something within him that is particularly drawn to this field of study, namely, his 'fervent longing to penetrate the secrets of nature' (26). When he does eventually find out about the work of modern philosophers their achievements attract him for the same reasons. When Frankenstein sees the tree 'utterly destroyed' (24) by lightning and, in the 1831 edition, later hears 'a man of great research in natural philosophy' explain 'a theory which he had formed on the subject of electricity and galvanism' (28), he is astonished. After this, the alchemists of whom he had become a 'disciple' are overthrown, though he does not at this point pursue this new interest in electricity and instead studies mathematics (29). While, therefore, modern science has displaced alchemy, it becomes clear later in the novel that Frankenstein sees in both equal and similar aspirations and ambitions.

Despite the now common popular misconception, Frankenstein is not a doctor; at university he studies natural philosophy 'in its general relations' with the detested Professor Krempe, who tells Frankenstein that he has wasted his time with the 'nonsense' that is alchemy, and on alternate days studies with Professor Waldman, who teaches him chemistry (29). It is Waldman and chemistry that truly capture and fire Frankenstein's imagination. Laura Crouch was the first to notice that the lecture Waldman gives closely resembles a passage of Humphry Davy's *Discourse, Introductory to a Course of Lectures*. Both Waldman's and Davy's lectures are passionate about what modern chemistry has achieved and can still achieve. For Waldman, modern chemists 'have acquired new and almost unlimited powers' (30). Frankenstein's determination is confirmed when Waldman speaks privately to him of how modern philosophers are 'indebted' (31) to the alchemists; the 'mastery' that alchemists promised is thus realized in Davy's modern chemistry.

Frankenstein is careful not to let Waldman see the extent of his 'enthusiasm' for his new 'intended labours' (1831 35), but Waldman uses a word resonant with occult association when he declares Frankenstein his 'disciple' (31). The term had been used to describe Frankenstein's fervent study of the alchemists only moments before in his narrative. Waldman shows something of Frankenstein's zeal, telling him that the reason he chose to study chemistry rather than any other science is because 'Chemistry is that branch of natural philosophy in which the greatest improvements have been and may be made' (31). These passages are key to the novel's explanation of why Frankenstein creates the creature. Crucially they point to modern chemistry as a continuation of, rather than a turning away from, alchemy, in terms of its goals and sense of ambition, as Frankenstein becomes equally excited by the recent and potential achievements of modern chemists as he had been by the alchemists.

Alchemy in *Zanoni* (1847), chemistry in *The Alkahest* (1834)

Interest in alchemists and alchemy per se continue in certain nineteenth-century texts, such as in Bulwer-Lytton's novel *Zanoni*, which is set during the French Revolution of the previous century.[5] In this section, I consider Bulwer-Lytton's novel, which retains the link between alchemy and the world of hermetic secret societies, and compare it to Balzac's *The Alkahest*. Balzac's novel, though written earlier, is set later, at the start of the nineteenth century, and here alchemy is rooted in modern chemistry. In contrast, Bulwer-Lytton repeatedly criticizes modern chemistry for not fully realizing what it might achieve. For him, the tendency towards that 'excellent thing . . . Matter of Fact!' (Bulwer-Lytton 271) in contemporary art and science is stopping the human species from reaching its full potential. He is at pains to argue that alchemy is 'not magic' (Bulwer-Lytton 227) and instead claims that there is an 'all-pervading and invisible fluid resembling electricity' (Bulwer-Lytton 229) that links everyone together, which is like mesmerism. The alchemists in his novel discover this by looking more carefully at that which modern chemists have supposedly neglected to take account of: minute and seemingly unimportant aspects of the natural world.

The key to the difference between Bulwer-Lytton's and Balzac's novels may be found in their respective genres. While Balzac's novel aims at realism, and this reveals his view that alchemical problems can be solved by modern chemistry, Bulwer-Lytton's novel eschews realism. Indeed, he rails against the contemporary world, where superstition has given way to 'the age of facts' (Bulwer-Lytton xiii). There is a sustained debate running throughout the novel concerning the difference between truth and fact, which broadly corresponds to the qualities Bulwer-Lytton attributes to poetry and to chemistry. Alchemy has the potential to bridge the two. For example, alluding to a character in Wordsworth's poem 'The Idiot Boy', the narrator informs the old gentleman who gives him the manuscript that when translated becomes the story of Zanoni: 'Our growing poets are all for simplicity and Betty Foy; and our critics hold it the highest praise of a work of imagination, to say that its characters are exact to common life' (Bulwer-Lytton xiv). The unnamed gentleman is incredulous at this notion.

Zanoni features real people from history and quotations from published texts as well as fantastic and supernatural elements, but the narrator of *Zanoni* is resolute that his tale does not fit the new realist genre. This can be seen in such instructions as 'mount on my hippogriff, reader' (Bulwer-Lytton 43). The gentleman who gives the narrator Zanoni's story describes its genre: 'It is a romance, and it is not a romance' (Bulwer-Lytton xviii). There is much in the story that fits the conventions of the Gothic, not least the framed narrative and its supposed origin in a manuscript. The reader is challenged to find 'truth' in Bulwer-Lytton's novel, though there is an acknowledgement that some will find it merely 'an extravagance' (xviii). If we take it that the narrator is voicing Bulwer-Lytton's views, then he thinks the Romantic impulse to portray in literature what Wordsworth called 'real life' (1:161) is equivalent to, and equally to be disparaged as, contemporary science's focus on the facts.

In the appendix to the novel, 'Zanoni Explained', we are told that Mejnour represents science 'that contemplates' while Zanoni represents art 'that enjoys' (Bulwer-Lytton 365). In this role

Zanoni exhorts Glyndon to see beyond the facts in front of him: 'see you not that The Grander Art, whether of poet or of painter, ever seeking for the TRUE, abhors the REAL' (Bulwer-Lytton 106). The two alchemists in the novel, Mejnour and Zanoni, represent respectively science and art: alchemy, therefore, unites the two. Alchemy sees beyond the real to perceive truth.

For Bulwer-Lytton, alchemy is also superior to modern chemistry because it creates: 'art is more godlike than science; science discovers, art creates' (106). Such sentiments accord with particular Romantic conceptions of art. As Percy Shelley wrote, following Tasso, in his *Defence of Poetry*, 'None deserves the name of Creator except God and the Poet' (506 n. 7). Here, Bulwer-Lytton perhaps demonstrates knowledge of modern chemistry, namely that no matter can be created or destroyed, but, in his view, this proves its limitations. If nothing material can be created or destroyed, only art can create. Perhaps alluding to Dalton's exposition of this theory, Zanoni explains further:

> The astronomer who catalogues the stars cannot add one atom to the universe; the poet can call a universe from the atom; the chemist may heal with his drugs the infirmities of the human form; the painter, or the sculptor, fixes into everlasting youth forms divine, which no disease can ravage, and no years impair.
>
> *(Bulwer-Lytton 106)*

Keats's 'Ode on a Grecian Urn' similarly argues that art can fix 'into everlasting youth forms divine', using the example of the urn's depiction of a youth 'for ever young' (l. 27). For Bulwer-Lytton, all that the chemist can do is restore health that has been lost. In contrast, poets, painters and, ultimately, alchemists are able to create new worlds and to enjoy eternal youth and wellbeing rather than simply catalogue and measure matter or change it into new forms.

Zanoni states repeatedly that the alchemist's lore is 'not Magic' (Bulwer-Lytton 217). Instead it is merely 'the art of medicine rightly understood'. Zanoni and his fellow alchemist Mejnour are the sole recipients of the 'noble secret' by which that which chemists call 'HEAT or CALORIC' will renovate the principle of life (Bulwer-Lytton 217–18). In other words, they possess the elixir of life, or the secret of immortality, found through a careful and painstaking study of medicine. Confirming sentiments such as Davy's quoted earlier in this chapter, Mejnour tells his pupil, Glyndon, 'Nature herself is a laboratory, in which metals, and all elements, are forever at change' (Bulwer-Lytton 228). It is because of the continual changes undergone by natural elements that the transmutation of metals is possible; as Mejnour asserts, it is 'Easy to make gold' and precious stones (Bulwer-Lytton 229).

Davy is named by Bulwer-Lytton as one of the 'soundest chemists' that the 'present century' has produced when he quotes from Isaac Disraeli's *Curiosities of Literature*, which purports to reveal Davy's views on alchemy: '"Sir Humphry Davy told me that he did not consider this undiscovered art an impossible thing, but which, should it ever be discovered, would certainly be useless"' (96; Disraeli 1:287). By the publication date of *Zanoni*, Davy's fortunes had waned somewhat in scientific circles but he was still a figure of considerable popular appeal. Here Bulwer-Lytton uses Davy primarily to approve of his reported open-mindedness. The narrator declares just before this: 'Real philosophy seeks rather to solve than to deny' (Bulwer-Lytton 96). Davy seems to be used to swipe at contemporary science: his view is that scientists should be trying to answer problems rather than deny the theories of alchemists. Levere has shown how a series of discoveries in the nineteenth century – isomerism, allotropy and analogies drawn between different groups of elements, such as the halogens – meant that 'the dreams of the alchemists were not entirely unfounded, and transmutation might be a reasonable goal for the chemical philosopher' (1977 198).

Referring to the achievements of modern science, specifically those of microscopic chemistry, Mejnour informs Glyndon that 'science brings new life to light' (Bulwer-Lytton 225). Mejnour refers to Democritus's 'Great System', which 'forbids the waste even of an atom' (Bulwer-Lytton 226). This constant transmutation of atoms into other forms becomes the type for life itself: 'Life is the one pervading principle, and even the thing that seems to die and putrefy but engenders new life, and changes to fresh forms of matter' (Bulwer-Lytton 225–6). Life is the single constant; even when living matter dies, the principle of life continues, transforming dead to living matter. Davy repeatedly made this point in his lectures (see, for example, 3:420, 3:414). It is this very notion that enables Victor Frankenstein to transgress the 'bounds' of life and death, which he views as merely 'ideal' (Shelley 1993 36) or as existing only in the imagination. Frankenstein finds inspiration for the creation of a living being when he witnesses 'the change of life to death, and death to life' (Shelley 1993 34). Mejnour makes the same point that what was once dead will transform eventually into living matter, as 'the very charnel house is the nursery of production and animation' (Bulwer-Lytton 226). As Davy had noted, organized living beings were dependent upon chemical processes from birth to death (2:313). Chemistry was the study of these processes but it could also, in some instances, cause and control them. The key message of *Zanoni*, however, is that modern chemistry has not yet sufficiently explored such ideas.

The main action of Balzac's *The Alkahest* takes place between 1809 and 1832. The novel concerns a man's monomaniacal attempt to find alchemy's 'Absolute' (or 'Alkahest'): the 'substance common to all created things' (103).[6] The novel shares a number of characteristics with *Frankenstein* and *Zanoni*, but far more than either of these, it is steeped in knowledge of contemporary French chemistry and chemical experiments. While there are different specific literary and scientific contexts in France and Britain during the period that this novel is set, it is the case that there was much exchanging of knowledge as well as competition between the two countries. Though the findings of the main character, Balthazar Claes, and others seem to confirm the idea of the 'Great Triad' (105) of the alchemists, the novel explicitly rejects the supernatural. Claes's discoveries are placed within the very real, contemporary context of work being done by named chemists. This befits a novel written in the realist mode, even while it concerns such a seemingly unrealistic subject matter.

Claes shares with Victor Frankenstein a desire for individual glory, which similarly results in the neglect of his family and leads to the death of his wife, Josephine. Claes dedicates his life to discovering how to create diamonds. The Catholic Josephine is appalled, accusing him of assuming 'the attributes of God' (109). Electricity is again crucial to understanding how creation takes place. *The Alkahest* endorses the idea that chemistry is primarily concerned with the modifications and transformations that all matter – both organic and inorganic – goes through, and that this can be analysed and mastered until, with Davy-like ambition, the chemist himself possesses the power to create.

The book is steeped in knowledge of contemporary chemistry. When he was younger, the central character, Claes, studied in Paris as Lavoisier's 'devoted disciple' (30). Josephine, in an attempt to win back his attention, teaches herself chemistry, reading the books of Fourcroy, Lavoisier, Chaptal, Nollet, Rouelle, Berthollet, Gay-Lussac, Spallanzani, Leeuwenhoek, Galvani and Volta (75). These lists and other details concerning Claes's experiments reveal the extent of Balzac's own chemical knowledge. In 1809, an unnamed Polish military officer stays with the Claes family and recognizes Balthazar's chemical training. He presents a useful summary of where chemistry is at this point and states that it is now known that all organic things can be reduced to four simple substances – nitrogen, hydrogen, oxygen and carbon – while inorganic things are made of fifty-three different kinds of elements, which have 'one originating principle' (101). He is convinced that nitrogen is not actually an element but that it can be further decomposed. If this was achieved it would leave only three 'essential principles' likened to the 'Ternary of the ancients and of the alchemists of the middle ages, whom we do wrong to scorn' (101). It is the case that some of the most exciting work in the nineteenth century was being done in Europe in organic chemistry, which, while it was 'less developed' than other branches of chemistry at this time, also therefore 'offered more opportunity' (Levere 2001 130). We can see Balzac's sense of its potential in this novel.

The Polish officer presents his theory to Claes by describing an experiment where cress seeds are grown in flour of brimstone and fed with distilled water. When the plant that is thus grown is analysed, a number of elements are identified that would not have been found in the component parts that were used to grow the cress (the seeds, the brimstone and the water). For him, this proves 'the existence of a primary element common to all the substances contained in the cress, and also to those by which we environed it' (103). Using the alchemical idea that there is unity in nature but applying it to the discoveries of modern chemistry, he concludes that the 'PRIME MATTER must be the common principle in the three gases and in carbon. The MEDIUM must be the principle common to negative and positive energy' (104). Claes's descent into monomania is motivated by his determination to prove the officer's theory.

In *The Alkahest*, carbon is thought to be key to finding the prime matter of all life that Claes is searching for. In Balzac's contemporary world there were important developments in the study of carbon, which was recognized as a constituent element of all organic substances. In England in 1815, William Prout argued instead that hydrogen was the 'primary' matter 'from which all the elements, and thus all bodies, were composed' (Levere 2001 138–9, 108). Jöns Jacob Berzelius criticized Prout's theory precisely on the grounds that it suggested 'transmutation would at least be a theoretical possibility' (Levere 2001 115). Berzelius instead looked to the structure of molecules to explain the phenomenon described by Balzac's Polish officer, the way that different things can be composed of the same elements. These examples make clear that discussions were still ongoing during the century about the possibilities of transmutation. As Levere writes, 'the idea of a prime matter was as tempting to many nineteenth-century chemists as it had been to medieval alchemists' (2001 116).

John Tresch notes that Claes's project to discover the prime matter or 'Absolute' and the means by which he attempts to achieve it 'had in fact been the topic of debate . . . in recent years, with Arago and Davy showing keen interest in the possibility' (62). Madeleine Fargeaud has shown that Balzac lifted, almost wholesale, passages from Berzelius's *Treatise on Chemistry*, one of which comments on 'the hopes of the alchemists in showing how certain flowers produce out of their own substance new compounds of metals' (Tresch 62). Nitrogen was also a current topic of debate in the chemistry of Balzac's day. Combinations of nitrogen had been particularly important to alchemists; nitric acid became known as *aqua fortis*, while nitric and hydrochloric acid was *aqua regia*. Nitrogen gas itself had been discovered in 1772 and was studied by Carl Wilhelm Scheele, Joseph Priestley and Henry Cavendish. Priestley gave it the name of 'dephlogisticated air' because it did not support combustion.[7] Lavoisier renamed it 'azote', from the Greek word for 'lifeless', but in 1790 Jean-Antoine Chaptal named it nitrogen. Claes was not the only one to be convinced that this element could be further decomposed. Davy – who had, by the date of the book's narrative, made his name isolating sodium and potassium – announced in a letter to Berzelius that he had performed an experiment that he thought proved 'the decomposition & composition of Nitrogen' (20 March 1809; Berzelius 2:11). In this letter, Davy, like Claes, speaks of the possibility of there being some 'elementary matter' to be found. He concludes from his experiment that 'oxgyene [sic], hydrogen and nitrogene may be different modifications of the *same substance*', thus suggesting there is unity in nature and that all of these substances may in fact be one.

In the same letter Davy writes 'The diamond I regard as an Oxide of the pure carbonaceous element'. Lavoisier had discovered that diamonds were made of carbon in 1772 but clearly Davy did not believe they were simply carbon in another form but instead a combination of carbon and oxygen. Davy had managed to isolate elements using electrochemistry, and, like him, Claes builds an 'immense galvanic battery' (211) with which to pursue his experiments. Claes comes closest to success with diamond as he tests the hypothesis that carbon holds the key to the transformation of matter. His success is thwarted by the lack of sunshine in July because he needs intense heat to do this work. Davy, in his efforts 'on the combustion of the diamond', was able to use the 'great lens in the Cabinet of Natural History' in Florence and found that by doing so he was 'able to complete, in a few minutes experiments which have been supposed to require the presence of bright sunshine for many hours' (5:479). Davy was able to use the best equipment that could be afforded, aided by the Grand Duke of Tuscany, but it is striking that Claes's experiments in *The Alkahest* are the same.

Claes is not represented as deluded within the narrative itself. After the peace brought about by Waterloo, he receives 'the circulation of discoveries of scientific ideas acquired during the war by the learned of various countries who had been unable to hold scientific communication' (195). He finds that 'learned men' now think as he does: 'that the difference existing between substances hitherto considered simple must be produced by varying properties of an unknown principle' (195). In other words, there must be some primary matter of life underlying all of nature. This also suggests that 'substances hitherto considered simple', such as carbon and nitrogen, will be found to be combinations of elements rather than true elements (195). Later in the novel, another character acknowledges that Claes 'is right scientifically. A score of men in Europe admire him for the very thing which others count as madness' (222). In a horrible ironic twist, boys' taunts of 'Down with sorcerers!' (302) provoke Claes's eventual death. He collapses and is unable to speak. He dies at the moment that he discovers the secret that he has lost all in pursuit of but which he is now unable to communicate. David Meakin makes the point that Balzac employs alchemical modes in his novel process, promising to reveal alchemy's secrets but not delivering on that promise (9). The novel asserts the existence of the 'Absolute' in its final pages and in doing so encourages the reader to continue the search for it.

The case of Jekyll and Hyde

Stevenson's *Jekyll and Hyde* can be likened to *Frankenstein* in its use of the scientific Gothic genre, where, in both cases, chemistry takes the role of the supernatural. In *Jekyll and Hyde* modern chemistry succeeds where alchemy had failed: specifically, again, it effects the transformation of matter from one form to another. Dr Jekyll is a Fellow of the Royal Society as well as a doctor of medicine, and even though his house, having been 'bought from the heirs of a celebrated surgeon', was equipped with an anatomical lecture theatre, laboratory, and dissecting tables, Jekyll's tastes are described as 'rather more chemical than anatomical' (44).[8] The theatre is now silent and, instead, 'tables [are] laden with chemical apparatus' (44). Indeed, Jekyll's discoveries are made specifically through the discipline of chemistry. He creates a potion, with ingredients sourced from a local wholesale chemist, and finds that he is able to separate man's seemingly singular, elementary nature into two. In his statement, he notes that his 'scientific studies' have been directed 'wholly towards the mystic and the transcendental' (107), perhaps alluding here to the alchemists or at least to a similar kind of ambition.

Jekyll uses chemical language throughout. When exploring his theory that 'man is not truly one, but truly two' (116), he dwells 'on the thought of the separation of these elements'. He is attempting to find out how they might be 'disassociated' (116), which could be thought of as a synonym for decomposition; he also uses the chemical term 'dissolution' (115) to describe the division of the self. Like *Frankenstein*, Glyndon in *Zanoni* and Claes in *The Alkahest*, Jekyll is motivated partly by the idea of expanding knowledge but also by the glory and power that this will bring. He tempts his friend Dr Lanyon to learn more about his experiments with these words: 'a new province of knowledge and new avenues to fame and power shall be laid open to you, here' (103). Lanyon, Jekyll complains, has too 'long been bound to the most narrow and material views' (104) and in doing so has denied the existence of 'transcendental medicine'. Lanyon, for his part, considers Jekyll's theories 'too fanciful', 'unscientific balderdash' (17), and becomes distressed by his 'scientific heresies' (31). As with *Frankenstein* and *The Alkahest*, there is no explicit evidence of the supernatural. Jekyll accounts for his eventual failure to be able to transform back from Hyde to an impurity in the original batch of salt he had acquired from the chemists, which he cannot replicate.

There are real place names used in *Jekyll and Hyde* and tantalizingly there really was a 'Messrs Maw' (74), wholesale chemists and surgical instrument makers, operating at 11 Aldersgate in London.[9] There is some effort to create a sense of realism in the description of Jekyll's experiments too, as in *The Alkahest*. For example, chemical apparatus is named, such as the 'glass saucers' (86), a 'phial' (96) and a 'graduated glass' (102). Jekyll records his trials in an 'ordinary version book' (97) and the components of the 'transforming draught' (124) are hinted at in the narrative. Lanyon describes a 'crystalline salt of a white colour' (96) and a tincture of a 'blood-red liquid, which was highly pungent to the sense of smell and seemed to me to contain phosphorous and some volatile ether', although there are also other ingredients that he cannot identify. When the powder and tincture are mixed together, they become more intensely red in colour and 'throw off small fumes of vapour'; at this point 'the ebullition ceased and the compound changed to a dark purple, which faded again more slowly to a watery green' (102–3). In fact, it is specifically the salt that gives Jekyll problems. After acquiring a new batch of salt, the 'ebullition', or bubbling, takes place but after the first change of colour the second does not follow. Jekyll concludes 'I am now persuaded that my first supply was impure, and that it was the unknown impurity which lent efficacy to the draught' (140). In *Jekyll and Hyde*, then, we see the ultimate expression of the chemist's ability to change and control matter. This change, from Jekyll to Hyde and back to Jekyll, is repeatedly called a 'transformation', but as time goes on the 'pangs of transformation grew daily less marked' and Jekyll finds his control diminished. He needs ever greater amounts of the mixture and, finally, cannot effect the change at all.[10]

Conclusion

As the texts discussed here illustrate, during the nineteenth century modern chemistry could be represented as being compatible with the aspirations of the ancient alchemists. While they may use different means to ancient alchemists, such as the electric battery in *The Alkahest*, chemists seemed still to be aiming at the same objectives: the control of nature and life, as the example of Davy shows clearly. Poets and novelists valued chemistry because it studied changes in matter and seemed to reveal hidden unities in nature. Understanding that life was created by the rearrangement of elements already in existence, which could be separated and made into new forms, can be seen to be at the heart of texts such as *Frankenstein*, *Zanoni*, *The Alkahest* and *Jekyll and Hyde*. In these texts, chemistry enabled its disciples to master these changes and even, ultimately, to create life.

In 1869 Dmitri Mendeleev's periodic table helped to confirm the idea that elements were unique and indivisible. By the end of the century, chemistry was more fully under the influence of physics: eventual agreement concerning atomic weights meant that elements were regarded as unique, distinct and 'fundamental particles' (Morrisson 2011 19). After this has been accepted, transmutation is regarded as impossible. While earlier in the century Balzac could still imagine that diamond might be decomposed to reveal the 'absolute', the substance that is common to all living things, Stevenson's story of transformation in 1886 owed more to literary ideas of the possibilities of chemistry than to realities of contemporary chemistry. By this time, the aspirations of the alchemists were identified instead with the new science of atomic physics.

The key to the differences in how literary authors use chemistry is in their genre. Where in *Frankenstein* and *Jekyll and Hyde* chemistry takes the place of the supernatural, Bulwer-Lytton's novel is a 'romance', which continues to refer to secret societies and spirits. While Mary Shelley, Balzac and Stevenson imagine that alchemy's aspirations have been realized in modern chemistry, Bulwer-Lytton criticizes chemistry for not being open enough to such ideas. Balzac writes in the realist genre of the 'Flemish School' but on a topic that most would think belonged to the world of scientific Gothic. This is because Balzac thinks that contemporary chemistry is close to achieving the goals of the alchemists: his story is set in a recognizably real world featuring known chemists and their experiments. For the Romantic poets, Davy's world view offered a potent metaphor for the poetic act of creation: one that redefined what creation itself meant.

Notes

1 Davy is deliberately echoing Francis Bacon here (see Mellor 89). I am grateful for Frank A. J. L. James's assistance with this chapter. He reminded me that Davy is also credited with the first use of the word 'Baconian' in the *OED*.

2 The allocation of priority of discovery in this case was contentious; see Miller.

3 Coleridge, in a letter of 10 [11] November 1823, famously described Davy as 'the Father and Founder of philosophic Alchemy' because of the way that he 'converted Poetry into Science' (1956–71 5:309). An unpublished letter from Gregory Watt to Davy begins with 'My Dear Alchymist' (RI MS HD/26/G/1). This manuscript is published here by courtesy of the Royal Institution of Great Britain.

4 Unless otherwise stated, I refer throughout to the 1818 edition of the novel, edited by Marilyn Butler.

5 David Meakin remarks that an 'occult revival . . . marked the second half of the nineteenth century' (39). There was a revival in interest in alchemy among the Hermetic Order of the Golden Dawn, which numbered among its members the eminent chemist Sir William Crookes (Morrisson 2011 22). Dracula is an alchemist and alchemy features in the novels and stories of a number of other nineteenth-century writers; see also Browning; Beddoes; Stoker; Haggard; Hawthorne. Christine Ferguson's chapter in this volume also discusses Bulwer-Lytton's novel.

6 *The Alkahest* appears to have been first translated into English as *Balthazar; or, Science and Love* by William Robson in 1859. In the preface to this edition, Robson compares this novel to William Godwin's *St Leon* (1799). For details of Balzac's reception in England, see Hunt. *La Recherche de l'absolu* is one of the novels 'which, almost always and everywhere, won complete approval from the contributors to British periodicals' (Hunt 287) and which confirmed that Balzac, in the words of one anonymous critic writing in 1843, was 'at once a painter of the Flemish School, and a high artist of human passion' (Hunt 289). In this chapter, I use the translation by Katherine Prescott Wormeley first published in 1896.

7 Phlogiston was believed to be an element contained within flammable materials that was released during combustion. Phlogiston largely lost its advocates after Lavoisier's work in the 1780s, though Davy had never been particularly persuaded of its existence (see Chang).

8 I refer throughout to the 1886 Longman edition of the novel.

9 There can be found advertising services and products in various London newspapers from this period. See the Nineteenth-Century British Newspapers Database.

10 Jekyll's contemporary, Sherlock Holmes, is described as a 'first-rate chemist' (6) in *A Study in Scarlet* (1887); when he first meets Watson he warns him 'I generally have chemicals about, and occasionally do experiments' (10).

Bibliography

Primary texts

Balzac, Honoré de. *Balthazar; or, Science and Love*. Transl. William Robson. London: Routledge, Warne, and Routledge, 1859.

Balzac, Honoré de. *Sephaphita; The Alkahest* [1896]. Transl. Katherine Prescott Wormeley. Reprint. Boston, MA: Little, Brown, 1916.

Barbauld, Anna Letitia. *Selected Poetry and Prose*. Eds. William McCarthy and Elizabeth Kraft. Toronto: Broadview, 2001.

Beddoes, Thomas Lovell. *Death's Jest Book*. Ed. Michael Bradshaw. Abingdon: Psychology Press, 2003.

Berzelius, Jöns Jacob. *Bref*. Ed. H. G. Söderbaum. 6 vols. Uppsala: Almqvist and Wiksell, 1912–32.

Brontë, Anne. *The Tenant of Wildfell Hall*. London: T. C. Newby, 1848.

Browning, Robert. *Paracelsus*. London: Effingham Wilson, 1835.

Bulwer-Lytton, Edward. *Zanoni* [1842]. London: G. Routledge and Sons, 1877.

Coleridge, Samuel Taylor. *The Collected Letters of Samuel Taylor Coleridge*. Ed. E. L. Griggs. 5 vols. Oxford: Clarendon, 1956–71.

Coleridge, Samuel Taylor. *Biographia Literaria*. Eds. J. Engell and W. Jackson Bate. 2 vols. Princeton, NJ: Princeton University Press, 1993.

Conan Doyle, Arthur. *A Study in Scarlet, The Sign of Four*. London: Harper, 1904.

Dalton, John. *A New System of Chemical Philosophy*. 2 vols. London: R. Bickerstaff, 1808.

Davy, Humphry. *The Collected Works of Sir Humphry Davy*. Ed. John Davy. 9 vols. London: Smith, Elder, 1839–40.

Davy, John. *Memoirs of the Life of Sir Humphry Davy*. 2 vols. London: Longman, Rees, Orme, Brown, Green, and Longman, 1836.

Dickens, Charles. *Nicholas Nickleby* [1838–9]. Reprint. London: Chapman and Hall, 1839.

Disraeli, Isaac. *Curiosities of Literature* [1791]. New edn. 3 vols. London: Routledge, Warnes, and Routledge, 1859.

Eliot, George. *Middlemarch: A Study of Provincial Life*. Edinburgh: William Blackwood, 1871–2.

Godwin, William. *St Leon*. Ed. Pamela Clemit. Oxford: Oxford University Press, 1994.

Haggard, Henry Rider. *She*. Ed. Daniel Karlin. Oxford: Oxford University Press, 2008.

Hardy, Emma. *Diaries*. Ed. Richard H. Taylor. Ashington: Mid-Northumberland Arts Group and Carcanet New Press, 1985.

Hawthorne, Nathaniel. 'The Birthmark'. *Mosses from an Old Manse*. London: Wiley and Putnam, 1846. 32–61.

Herschel, John. *A Preliminary Discourse on the Study of Natural Philosophy* [1831]. Rev. edn. London: Longman, Brown, Green, and Longmans, 1851.

Keats, John. *John Keats*. Ed. Elizabeth Cook. Oxford: Oxford University Press, 1990.

Keats, John. *Selected Letters*. Ed. Robert Gittings. Rev. Jon Mee. Oxford: Oxford University Press, 2002.

Mayhew, Henry. *The Wonders of Science; Or, Young Humphry Davy* [1854]. Reprint. London: David Bogue, 1855.

Paris, John Ayrton. *The Life of Sir Humphry Davy*. London: H. Colburn and R. Bentley, 1831.

Shelley, Mary. *Frankenstein, or, the Modern Prometheus*. London: H. Colburn and R. Bentley, 1831.

Shelley, Mary. *Frankenstein: The 1818 Text*. Ed. Marilyn Butler. Oxford: Oxford University Press, 1993.

Shelley, Percy Bysshe. *Shelley's Poetry and Prose*. Eds. Donald Reiman and Sharon B. Powers. New York, NY: Norton, 1977.

Stevenson, Robert Louis. *Strange Case of Dr Jekyll and Mr Hyde*. London: Longman, Green, 1886.

Stoker, Bram. *Dracula*. Ed. Maurice Hindle. London: Penguin, 1993.

Wordsworth, William. *Prose Works*. Eds. W. J. B. Owen and J. W. Smyser. 3 vols. Oxford: Clarendon, 1974.

Secondary texts

Brock, William. *The Fontana History of Chemistry*. London: HarperCollins, 1992.

Brown, Raymond Lamont. *Humphry Davy: Life Beyond the Lamp*. Stroud: Sutton, 2004.

Crouch, Laura. 'Davy's *A Discourse, Introductory to a Course of Lectures on Chemistry*: A Possible Scientific Source of *Frankenstein*'. *Keats–Shelley Journal* 27 (1978): 35–44.

Chang, Hasok. *Is Water H2O? Evidence, Realism and Pluralism*. Cambridge: Springer, 2012.

Fargeaud, Madeleine. *Balzac et la Recherche de l'Absolu*. Paris: Hachette, 1968.

Flaherty, Clare. 'A Recently Rediscovered Unpublished Manuscript: The Influence of Sir Humphry Davy on Anne Brontë'. *Brontë Studies* 38.1 (2013): 30–41.

Hunt, H. J. 'The "Human Comedy": First English Reactions'. *The French Mind: Studies in Honour of Gustave Rudler*. Oxford: Clarendon, 1952. 273–90.

Golinski, Jan. *Science as Public Culture: Chemistry and Enlightenment in Britain, 1760–1820*. Cambridge: Cambridge University Press, 1992.

Golinski, Jan. 'Chemistry'. *Eighteenth-Century Science*. Ed. Roy Porter. Vol. 4 of *The Cambridge History of Science*. 8 vols. Cambridge: Cambridge University Press, 2003, 375–96.

Hindle, Maurice. 'Humphry Davy and William Wordsworth: A Mutual Influence'. *Romanticism* 18.1 (2012): 16–29.

Jenkins, Alice. 'Humphry Davy: Poetry, Science, and the Love of Light'. *1798: The Year of the 'Lyrical Ballads'*. Ed. Richard Cronin. Basingstoke: Palgrave Macmillan, 1998. 133–50.

Johns Putra, Adeline. '"Blending Science with Literature": The Royal Institution, Eleanor Anne Porden and The Veils'. *Nineteenth-Century Contexts* 33.1 (2011): 35–52.

Knight, David. *Humphry Davy: Science and Power*. Cambridge: Cambridge University Press, 1998.

Lawrence, Christopher. 'The Power and the Glory: Humphry Davy and Romanticism'. *Romanticism and the Sciences*. Eds. Andrew Cunningham and Nicholas Jardine. Cambridge: Cambridge University Press, 1990. 213–27.

Levere, Trevor. 'The Rich Economy of Nature: Chemistry in the Nineteenth Century'. *Nature and the Victorian Imagination*. Eds. U. C. Knoepflmacher and G. B. Tennyson. Berkeley, CA: University of California Press, 1977, 189–200.

Levere, Trevor. *Transforming Matter: A History of Chemistry from Alchemy to the Buckyball*. Baltimore, MD: Johns Hopkins University Press, 2001.

Levere, Trevor. *Poetry Realized in Nature: Samuel Taylor Coleridge and Early Nineteenth-Century Science*. Cambridge: Cambridge University Press, 2002.

Meakin, David. *Hermetic Fictions: Alchemy and Irony in the Novel*. Keele: Keele University Press, 1995.

Mellor, Anne. *Mary Shelley: Her Life, Her Fictions, Her Monsters*. London: Routledge, 1988.

Miller, David Philip. *Discovering Water: James Watt, Cavendish and the Nineteenth-Century "Water Controversy"*. Aldershot: Ashgate, 2004.

Morrisson, Mark. 'Alchemy'. *The Routledge Companion to Literature and Science*. Eds. Bruce Clarke and Manuela Rossini. London: Routledge, 2011, 17–28.

Nye, Mary Jo. *Before Big Science: The Pursuit of Modern Chemistry and Physics, 1800–1940*. London: Prentice Hall, 1996.

Roe, Nicholas. *John Keats: A New Life*. New Haven, CT: Yale University Press, 2012.

Ross, Catherine. '"Twin Labourers and Heirs of the Same Hopes": The Professional Rivalry of Humphry Davy and William Wordsworth'. *Romantic Science: The Literary Forms of Natural History*. Ed. Noah Heringman. New York, NY: State University of New York Press, 2003, 23–52.

Ruston, Sharon. *Shelley and Vitality*. Houndmills: Palgrave Macmillan, 2005.

Ruston, Sharon. *Creating Romanticism: Case Studies in the Literature, Science and Medicine of the 1790s*. Houndmills: Palgrave Macmillan, 2013.

Secord, James. *Visions of Science: Books and Readers at the Dawn of the Victorian Age*. Oxford: Oxford University Press, 2014.

Sharrock, Roger. 'The Chemist and the Poet: Sir Humphry Davy and the Preface to *Lyrical Ballads*'. *Notes and Records of the Royal Society* 17 (1962): 57–76.

Sperry, Stuart. *Keats the Poet*. Princeton, NJ: Princeton University Press, 1973.

Tresch, John. 'Electromagnetic Alchemy in Balzac's *The Quest for the Absolute*'. *The Shape of Experiment*. Eds. Henning Schmidgen and Julia Kursell. Berlin: Max-Planck, 2007, 57–77.

Wunder, Jennifer. *Keats, Hermeticism, and the Secret Societies*. Aldershot: Ashgate, 2008.

Further reading

Budge, Gavin. 'Mesmerism and Medicine in Bulwer-Lytton's Novels of the Occult'. *Victorian Literary Mesmerism*. Eds. Martin Willis and Catherine Wynne. Amsterdam: Rodopi, 2006, 39–61.

Goethe, Johann Wolfgang von. *Elective Affinities*. Transl. David Constantine. Oxford: Oxford University Press, 1994.

Gray, Ronald. *Goethe the Alchemist*. Cambridge: Cambridge University Press, 1952.

Hagen, Margareth, and Margery Vibe Skagen, eds. *Literature and Chemistry: Elective Affinities*. Aarhus: Aarhus University Press, 2013.

Holmes, Richard. *The Age of Wonder: How the Romantic Generation Discovered the Beauty and Terror of Science*. London: HarperPress, 2008.

Knight, David. *Ideas in Chemistry: A History of the Science*. London: Athlone, 1992.

Labinger, Jay. 'Chemistry'. *The Routledge Companion to Literature and Science*. Eds. Bruce Clarke and Manuela Rossini. London: Routledge, 2011, 51–62.

Morrisson, Mark. *Occultism and the Emergence of Atomic Theory*. Oxford: Oxford University Press, 2007.

Schütt, Hans-Werner. 'Chemical Atomism and Chemical Classification'. *The Modern Physical and Mathematical Sciences*. Ed. Mary Jo Nye. Vol. 5 of *The Cambridge History of Science*. 8 vols. Cambridge: Cambridge University Press, 2002, 237–54.

Ziolkowski, Theodore. *The Alchemist in Literature*. Oxford: Oxford University Press, 2015.

Winter, Alison. *Mesmerized: Powers of Mind in Victorian Britain*. Chicago, IL: University of Chicago Press, 1978.

19

THERMODYNAMICS

Barri J. Gold
Muhlenberg College

The centrality of thermodynamics to scientific literacy has been clear for about as long as the concept of scientific literacy itself. Indeed, in his 1959 lecture identifying the growing division between the sciences and the humanities – 'The Two Cultures' – C. P. Snow recounts asking, when provoked by those bemoaning the illiteracy of scientists, whether any of them could describe the Second Law of Thermodynamics. 'The response,' he says, 'was cold: it was also negative' (Snow 16). And while scholars have done considerable work to cross the divide entrenched in twentieth-century education, thermodynamics remains an area of science to which literary scholars have paid relatively little attention.

Why we have been less inspired by thermodynamics than, say, evolutionary or medical sciences, I can only speculate. Certainly, in our own time, thermodynamics has neither the controversial status nor the personal immediacy of other sciences. More subtly, perhaps, the 'hard' sciences seem more driven by positivist assumptions, harder to address (assail?) through cultural criticism. It may be that we are more convinced, or even intimidated, by their claims to objectivity. Or maybe it's all the maths.

We have, however, long known thermodynamics to be one of the most influential scientific discourses of the nineteenth century, and that the Scottish and English were major players in its development. And in spite of rumours to the contrary, the basic principles of thermodynamics are readily accessible. Far simpler to summarize than, say, *Bleak House*, its laws can still be articulated as they were in 1865 by Rudolph Clausius:

1 The energy of the universe is constant.
2 The entropy of the universe tends towards a maximum.

(365)

While annotating these statements could keep a host of literary scholars, historians and scientists busy for a very long time, their simple statements remain relatively uncontested: we can neither create nor destroy energy, though we can change it among its various forms – heat, light, electricity, mechanical work, chemical potential and so on. We also remain convinced that the amount of *usable* orderly available energy is always decreasing – that entropy, a measure of disorderly energy, is increasing. These laws of thermodynamics have long preoccupied anyone interested in engine design, refrigeration, air conditioning, meteorology, space travel, hot tea or sick children. Victorian thermodynamics has been foundational in a range of sciences, including not only chaos and information theory, non-equilibrium thermodynamics and statistical mechanics, but also the ecological sciences and modern biology, including much contemporary thinking on evolution. No wonder then, that historians of science have been so interested in the advent and influence of thermodynamics. Among these, literary scholars are likely to find particularly useful the work of Crosbie Smith, Anson Rabinbach, P. M. Harman, Thomas Kuhn and Stephen Brush, as well as the essays in the collection *Energy and Entropy*, edited by Patrick Brantlinger.

A cast of characters

Nonetheless, historical work on thermodynamics has failed to produce an Ur-text or central figure, from whom literary scholars can immediately and authoritatively draw. We have no Darwin and no *Origin* to rally around. In this sense, the field lacks a centre (a feature it shares with ecocriticism, to which I will return). This is no fault of the historians; it is connected to the way thermo-dynamics emerged as a set of physical principles. As Thomas Kuhn remarks, 'the hypothesis of energy conservation . . . publicly announced by four widely scattered European scientists' (66) between 1842 and 1847, is an exemplary case of simultaneous discovery. And indeed, the hypothesis of energy conservation is only one part of the picture. The notion that energy is conserved depends on a concept of energy that remains controversial for decades. Conservation, moreover, must be reconciled with pervasive observations of energy loss. It is not until 1854 that William Thomson identifies the science that does so as 'thermo-dynamics'; not until 1865 that Rudolph Clausius coins the key term 'entropy' as the desultory counterpart to energy (Harman 45; Clarke 1996 73). In the interim, Thomson expresses at once his conviction that energy is conserved as well as his dismay that science cannot yet explain how:

> When 'thermal agency' is thus spent in conducting heat through a solid, what becomes of the mechanical effect which it might produce? Nothing can be lost in the operations of nature − no energy can be destroyed. What effect then is produced in place of the mechanical effect which is lost? A perfect theory of heat imperatively demands an answer to this question; yet no answer can be given in the present state of science.
>
> *(1882 1:118−19 n.)*

Such an extended incubation period makes it rather difficult for readers to know where to begin. Crosbie Smith's *The Science of Energy* is extremely helpful in this regard. Dividing the world of British energy science into two camps, he directs us to 'the Northern School' and 'the Metropolitans', who struggled for scientific authority and differed strongly in their attitudes towards scientific naturalism and the place of religion in science. The North British school, closely associated with Scottish Presbyterianism, has as its most famous practitioners William Thomson (later Lord Kelvin) and James Clerk Maxwell. No physicist worth her salt could fail to know these names, or that of their frequent correspondent, James Prescott Joule, a third generation brewer in Manchester whose experiments were critical to the development of thermodynamics. Friends and collaborators Balfour Stewart and P. G. Tait, whose work has fallen into relative scientific obscurity, will also be of particular interest to literary scholars, for their extensive use of metaphor and period-specific interests and assumptions.

On the other side, as it were, stands John Tyndall. Extremely well known for his popular lectures, Tyndall was a frequent target of North British critique and Maxwell's ready wit. A vocal advocate, along with Thomas Henry Huxley and Herbert Spencer, of Darwin's theory of evolution and a firm believer in the separation of science and religion, Tyndall has emerged as a figure of extreme positivism. His contributions to thermodynamics are also less obvious to contemporary eyes, because he held out for 'force', in preference to the term 'energy' that eventually dominates. Nonetheless, Tyndall's *Heat as a Mode of Motion* was in print for half a century, though it was sharply criticized as well as praised by members of the North British school, and his vast oeuvre leaves much still to investigate in the field of literature and thermodynamics.

Still, we would do well to distinguish between Tyndall's (and, also famously, Faraday's) 'force' and that currently used by physicists, in which 'force' is not conserved. According to Maxwell's 1876 poetic 'Report on Tait's Lecture on Force' (see Campbell and Garnett 646–8), it is not even 'a thing'; it is, at best, only a mathematical convenience; at worst, it is an unfortunate and misleading relic of eighteenth-century physics:

> The phrases of last century in this
>   Linger to play tricks—
> Vis Viva and Vis Mortua and Vis
>   Acceleratrix:—
>
> Those long-nabbed words that to our text books still
>   Cling by their titles,
> And from them creep, as entozoa will,
>   Into our vitals.
> But see! Tait writes in lucid symbols clear
>   One small equation;
> And Force becomes of Energy a mere
>   Space-variation.
>
> Force, then, is Force, but mark you! not a thing,
>   Only a Vector;
> Thy barbèd arrows now have lost their sting,
>   Impotent spectre!
> Thy reign, O Force! is over. Now no more
>   Heed we thine action;
> Repulsion leaves us where we were before,
>   So does attraction. . .

The play here, the gentle satire, Tait's heroics, the creepy-crawlies, the double entendre of repulsion and attraction, the irony (likely intentional) and ambiguity (undoubtedly so) of declaring, in 1876, the reign of force to be over, suggest the literary pleasures that await readers of Maxwell's serio-comic verse. As for Maxwell's 'Report': if this declaration of victory is a bit premature, it does correctly anticipate what will soon be a well-defined scientific object called 'energy', as well as its mathematical relations to 'force'. Maxwell's 'Report', along with Tait's 'Lecture On Force' and Tyndall's 1864 lecture 'On Force' and 1874 'Belfast Address', which Tait's lecture targets, make a nice cluster of works by which one might begin to explore the scientific culture surrounding the advent of thermodynamics.

Literature and thermodynamics

In order to study the broader literary life of thermodynamics, it is useful to divide the nineteenth century into periods, as thermodynamic discourse operated quite differently in its contact with the changing zeitgeist of Romantic, mid- and late Victorian moments. The historian Stephen Brush identifies three such periods, as each relates to the laws of thermodynamics:

> The first law of thermodynamics (conservation of energy), inspired in part by the philosophy of romanticism, provided an organizing principle for the science of the realist period. Likewise the second law of thermodynamics (dissipation of energy), which arose from the technical analysis of steam engines, provided a *dis*organizing principle which turned out to be highly appropriate for the neoromantic period.
>
> *(29)*

For convenience, we may refer to these simply as 'before' (roughly, the Romantic period) and 'after' (which I take to refer to the late Victorian, even the decadent, period), separated by a long and ill-defined 'during' period of incubation, transition, linguistic confusion and cultural crosstalk, during which the physical theory we now call 'thermodynamics' gradually emerged. Each of these, I suspect, will require a distinct methodology of literary study, as well as its own set of thermodynamic keywords, suited to the discourse of the moment.

The late Victorian, or 'after', period is, perhaps, the best accessible by the most common methods of studying literature and science, wherein we think of literature primarily as responding to, wrestling with and deploying scientific concepts, both literally and metaphorically. Only once the laws of thermodynamics are established and clearly articulated can they be widely popularized. Only once the term 'energy' takes on its recognizable scientific meanings and 'entropy' is coined as its counterpart, do these take a clear and traceable hold on the literary imagination. Literary scholars may then proceed to explore the influence of thermodynamics on literature without too much methodological anxiety.

Particularly influential in this area has been Bruce Clarke's *Energy Forms*. Fascinated by the inherent literariness of entropy itself, Clarke argues that entropy is an 'isomorph of allegory' (2001 26), adding the dimension of irreversible time to Newtonian physics even as allegory expands in time and space the instantaneity of metaphor. This likeness then provides the foundation for his expansive study of how 'thermodynamic entropy became cultural allegory' (Clarke 2001 27), among authors ranging widely in time and space, including Charles Howard Hinton, Edward Bulwer-Lytton, Camille Flammarion, D. H. Lawrence, Thomas Pynchon and Yevgeny Zamyatin.

More focused on the late Victorians, Allen MacDuffie explores some of the fascinating economic implications of thermodynamics. He identifies in Joseph Conrad, for example, a critique of the widespread misuses of thermodynamic principles to serve economic interests of expanding empire, while imperial discourse 'draws upon the rhetoric of thermodynamic energy-dissipation to describe unrealised profits' even though the physical pressures of entropy 'work[s] everywhere to undermine its claims to advancing progress, prosperity, and control' (MacDuffie 2009 95). Sarah Alexander further connects nineteenth-century physics to political economy (especially Marx and Engels) as she argues for a kind of literary realism linked to such nineteenth-century imponderables as the ether, principles of energy conservation and dissipation, entropy, four-dimensional space and the vortex theory of the atom.

There is, undoubtedly, more work to be done in this area, as thermodynamic discourse throughout the nineteenth century is rife with economic metaphor. The sun, with 'his amazing fund of vitality', figures as 'the great almoner, pouring forth incessantly his boundless wealth of heat' ([Anon.] 1878 323; Siemens 510). The new theory may then be regarded as 'a first attempt to open for the sun a creditor and debtor account' (Siemens 510). In less self-consciously fanciful depictions, the connection between energy theory and political economy work so seamlessly among the Victorians as to seem completely natural, as when kinetic energy is described as having 'been spent in raising [an object thrown upward] against the force of gravity' (Stewart and Lockyer 310). And we find the sun's resources to be limited, as 'the sun's energy is spent in producing the wood or coal, and the energy of the wood or coal is spent (far from economically, it is to be regretted) in warming our houses and driving our engines' (Stewart and Lockyer 322). Our second-law regrets are both physical and financial, as 'far from economically' refers ambiguously both to heat loss due to friction and to the monetary price of warming houses and driving engines.

The importance of metaphor to the discourse of thermodynamics is not, however, limited to economic metaphor. For a rich sample of such popularization – at once economic and religiously motivated – I recommend Stewart and Lockyer's 'The Sun as a Type of the Material Universe', published in two parts in *Macmillan's Magazine* in 1868. A remarkable condensation of the issues that have concerned scholars of thermodynamics in literature, this piece touches on questions of scientific analogies, economic metaphors of work and energy and the implications of energy science for both religion and the possibilities of free will. These last give the reader a taste of what is explored at length in Stewart and Tait's book *The Unseen Universe or Physical Speculations on a Future State*, which went into four editions and twelve reprints between its original 1875 publication and 1901.

The role of metaphor in thermodynamic discourse is not limited to its deployment (as in the examples above) in the popular dissemination of thermodynamic ideas. By the time we arrive at the 'after' period of popular dissemination, 'physicists had already borrowed the language and authority of social prophets' (Myers 308); indeed, the concept of entropy had built itself on the foundation of an ancient commonplace of decline, irreversibility, and disorder. And in an essay dear to the heart of literature and thermodynamics, Gillian Beer shows how thermodynamics' prediction of ultimate entropy dovetails with cultural mythology regarding the 'The Death of the Sun'. Beer tracks the anxiety-ridden readiness of Victorian culture to receive Thomson's predictions regarding the sun's approaching demise:

> It seems, therefore, on the whole most probable that the sun has not illuminated the earth for 100,000,000 years, and almost certain that he has not done so for 500,000,000 years. As for the future, we may say with equal certainty, the inhabitants of the earth, cannot continue to enjoy the light and heat essential to their life, for many million years longer, unless sources now unknown to us are prepared in the great storehouse of creation.
>
> *(1862 393)*

As the magnitude of the event overshadows the one million- to five million-year interim, the death of the sun in particular, and entropy more broadly, contributes considerably to widespread anxieties of *fin de siècle, fin du globe*.

Such metaphorical foundations and ready-made cultural anxieties suggest that there is much thermodynamic work to be done by scholars of both the 'before' and 'during' periods. Here, however, familiar science-influences-literature models cannot adequately account for the cultural exchange between literature and thermodynamics. The term 'thermodynamics' itself must be used, very carefully, as creative anachronism. We must also develop the language that will enable us to see thermodynamic ideas at work in culture before they are clearly named and in literature that cannot proclaim loudly its connections to thermodynamics: will we find 'energy' already structuring the discourse? Perhaps 'work'? Shall we see out references to the 'sun' or to 'heat' or to 'transformation' or 'decay' or 'unity'? And we will need to address our own, not unreasonable, concerns about reading into the text something that is not there, to establish that the connections we make are more than scholarly caprice, why they are important, what we hope to understand from them.

Nonetheless, the challenges attached to thermodynamically-informed studies of Romantic and mid-Victorian literature should be well worth the effort. Such studies must contribute to our thinking on broader relationships between literature and science. As I have suggested above, they must force us to rethink the rules of literature and science scholarship. Moreover, the historical literariness of both entropy and energy – the long life each enjoyed in the cultural imaginary before its consolidation as a scientific object – promises to make such studies highly productive.

As for 'energy' itself, the term had fallen out of favour with Isaac Newton and began the nineteenth century as the almost exclusive province of poets, as when William Blake declares in 'The Marriage of Heaven and Hell' that 'Energy is the only life' (l. 11) and 'Energy is eternal delight' (l. 12). Soon, scientists would speak of such things as 'the conservation of force'. In an 1847 address in the St Ann's Church reading room, Joule drew upon an outdated notion of *vis viva*, even though he knew perfectly well that the concept was more metaphorical than scientifically accurate:

> The force possessed by bodies in motion is termed by mechanical philosophers *vis viva* or *living force*. The term may be deemed by some inappropriate, inasmuch as there is no life properly speaking, in question; but it is *useful*, in order to distinguish the moving force from that which is stationary in its character, as the force of gravity.
>
> *(266–7)*

From our own 'after' perspective, there is not only no life, but also no force, properly speaking, as Maxwell's 'Report' declares. Or, at any rate, what is in question here is not forces, but kinetic and potential energy. Still, we can see in Joule how the notions that would eventually become central to thermodynamics are already taking shape. Indeed, Joule's lecture staunchly insists on the principle of conservation of energy, though it cannot yet name or support such a notion:

> You will at once perceive that the living force of which we have been speaking is one of the most important qualities with which matter can be endowed, and, as such, that it would be absurd to suppose that it can be destroyed or even lessoned, without producing the equivalent of attraction through a given distance of which we have been speaking . . . We might reason *a priori*, that such absolute destruction of living force cannot possibly take place, because it is manifestly absurd to suppose that the powers with which God has endowed matter can be destroyed any more than that they can be created by man's agency.
>
> *(268–9)*

Joule's conviction may be understood as some combination of religious belief and scientific intuition. And scientists, of course, are not the only ones who can know things before they know them. Indeed, in Tom Stoppard's *Arcadia*, Valentine Coverly claims that the second law of thermodynamics had been known 'by poets and lunatics from time immemorial' (Act 2 Scene 5), to which claim Hannah Jarvis later adds Byron's 'Darkness' by way of example (Act 2 Scene 6).

Several scholars have made important contributions to our understanding of the early literary exploration of what will eventually become the principles of thermodynamics. Stuart Peterfreund observes that energy re-emerges from its period of disrepute in science, 'for reasons primarily metaphysical, and especially religious, rather than physical' (24), and finds precursors of Victorian energy science in the 'one omnipresent Mind' (34) of Coleridge's 'Religious Musings', wherein Coleridge 'declares it to "Nature's essence, mind and energy!" adding that "'tis God / Diffused through all, that doth make one whole"' (40–1). Peter Allan Dale's *In Pursuit of a Scientific Culture* explores more broadly the closer-than-expected relationship between Romantic philosophy and Victorian science, identifying positivism as 'the true nineteenth-century successor to the Romantics' efforts at totalization' (6). According to Stephen Brush, the 'doctrine of the essential unity of all forces in nature leads directly to the law of conservation of energy' (20). And Ted Underwood's *The Work of the Sun* reveals the importance of emergent class concerns, as well as increasing associations of human labour and natural powers, in the development of modern notions of energy and work, particularly as these helped shape Romantic poetry.

Joule's cognizance of the usefulness of metaphor as scientific thinking also gestures towards a growing nineteenth-century debate over the meanings of science itself, especially of the use of mathematics to describe real-world systems. James Clerk Maxwell was deeply interested in such questions, and as early as his Cambridge undergraduate days wrote an essay entitled 'Are There Real Analogies in Nature?' (Campbell and Garnett 235). And though he generally leans towards 'yes', he nonetheless worries poetically about the limitations of the mathematics he did so well: 'Is our algebra the measure / Of that unexhausted treasure[?]' (Campbell and Garnett 617).

If Maxwell's poetry seems a bit off the beaten path, much that is thermodynamic finds its way into the literary favourites of the mid-Victorian period. George Levine, in *Darwin and the Novelists*, identifies his 'foray into thermodynamics [as] an attempt to keep [him] honest' (13). He reports having gone looking for Darwin in *Little Dorrit* and having repeatedly turned up Thomson instead. In Levine's reading, 'the strangeness and self-contradictions of the novel enact a conflict between two mythic structures, the progressive vision of Darwinism and the degenerative vision of thermodynamics' (156). But, even though much of Victorian tragic narrative is informed by what will eventually be called 'entropy', scholars interested in thermodynamics are also at work investigating the bright side of thermodynamic discourse. Tina Choi, for instance, identifies a more optimistic, first-law, conservationist sensibility in the attempts at closure so familiar in nineteenth-century novels: 'the novel's . . . resolution, like the rainfall, must be drawn from that bounded system itself, from the circulation of characters and other elements already present' (316).

Tennyson's *In Memoriam* and Dickens's *Bleak House*, both of which I investigate at length in *ThermoPoetics*, are extremely telling regarding the anxieties and hopes suffusing Victorian thermodynamic narrative. *Bleak House* is informed not only by the degenerative vision soon to be associated with entropy but also by the qualified optimism of thermodynamic engine design. More than a decade before the advent of the term 'entropy', Dickens describes a landscape dominated by entropic decay at every scale, beginning with a London enveloped in fog, with 'smoke lowering down from chimney-pots, making a soft black drizzle, with flakes of soot in it as big as full-grown snowflakes – gone into mourning, one might imagine, for the death of the sun' (13). I have argued for the importance of such usage in the ultimate consolidation and circulation of the scientific object we call 'entropy'. And yet, like James Watt, to whom he repeatedly alludes, Dickens finds ways to enable work to be accomplished, in spite of the surrounding drive to entropy. The childhood experiments in making 'steam engines out of saucepans' of Dickens's Rouncewell powerfully evoke the stories of James Watt, who at thirteen was supposed to have been inspired 'by the force of steam gushing from a boiling kettle to invent the steam engine' (Marsden 11). And, of course, Mr Rouncewell names his son Watt. James Watt's famous improvements to the steam engine strove to get more work for the amount of energy put in. By Dickens's time, 'work' itself is a key term in thermodynamics, as well as a term with complex and evolving social significance, functioning as 'a measure of a person's moral worth in just the same way as it was an indicator of an individual's economic value' (Morus 123). Mechanical work is itself a form of energy, quantified as the amount of force exerted over a distance, but linguistically, metaphorically and culturally connected to work of all sorts. Dickens signals his interest in the problematics of work throughout *Bleak House*, in the contrast between Esther's oft-reiterated desire 'to do some good to some one' (31) and the dubious philanthropy of Mrs Jellyby and Mrs Pardiggle. Perhaps the problematics of work are most broadly tested in the workings of Chancery, framed by the opening conversation between Lady Dedlock and Mr Tulkinghorn. When she observes that 'it would be useless to ask . . . whether anything has been done' (25), Tulkinghorn answers in a typically evasive manner, 'Nothing that YOU would call anything has been done to-day', Dickens makes clear that in an entropic system available energies must dissipate, but in the interim these may be used in less or more orderly, useful, productive ways, that 'work' increasingly implies energy well used, but that what this means is itself subject to perspective. In turn, Dickens suggests the evolution of the term 'work' that is currently taking place. And, though physics now defines 'work' quite precisely, Dickens suggests that we too must attend to the perspective dependence attached to any attempt to define such thermodynamic concepts as 'order' and 'usable energy'.

'Work' and its analogues form but one strand of Dickens's thermodynamic keywords. Also recurrent are 'energy' itself, along with the language and imagery of heat, order and disorder, dissipation and, of course, combustion. In Tennyson's *In Memoriam*, 'energy' appears only twice, and his phrase 'full-grown energies of heaven' (Canto 40Ca l. 20) sounds almost Blakean. But Tennyson did much to transform a Romantic legacy into a decidedly Victorian form, and *In Memoriam*, written between 1833 and 1849, serves as a literary instance of the simultaneous discovery we have come to expect of early thermodynamic discourse.

Cotemporaneous with the conceptual development of thermodynamics among scientists, *In Memoriam* may be said, like the science of energy itself, to transform evidence of loss into principles of conservation. Initially driven by the death of Tennyson's friend, Arthur Henry Hallam, *In Memoriam* moves from images of loss, entropy, even heat-death, to a climax of consolation that rests on principles of transformation and conservation. From a dream, like Byron's in 'Darkness', that 'there would be Spring no more, / That Nature's ancient power was lost' (Canto 69, l. 1–2), Tennyson identifies a principle of conservation that eventually governs the spiritual and physical world, even the mechanics of friendship, such that 'the all assuming [all-destroying] months and years / Can take no part away from this' (Canto 85, ll. 67–8). No longer 'somewhere in the waste' (Canto 22, l. 19) or even lost 'in light' (Canto 47, l. 16), the poet decides, in a move that anticipates *The Unseen Universe*, 'That friend of mine [now] lives in God' (Epilogue, 1.140). First-law optimism pervades *In Memoriam*'s cosmology, as Tennyson anticipates 'One God, one law, one element, / And one far-off divine event, / To which the whole creation moves' (Epilogue, ll. 142–4). Such an event suggests a moment in which currently interconvertible elements – heat, electricity, work, light and life – merge as one divine undifferentiated element. At the same time, the scientific aspiration attached to the discovery of energy transformation, the hope that we will discover a single law that governs all natural processes, merges with the renewal of faith, the reconciliation of science and religion: Tennyson's work thus helps us to unpack the convergence of belief, science, poetry, analogy and consolation behind the development of the laws of thermodynamics, the complex foundations of such pithy declarations as Joule's, in 1848, that it would be manifestly absurd to believe God's work is ever truly destroyed.

The future of thermodynamics

We have, perhaps, not yet come to the end of Tennyson's prescience, since his optimism goes even beyond convictions of energy conservation, proving not only conservative but also productive as 'star and system rolling past, / A soul shall draw from out the vast / And strike his being into bounds' (Epilogue, ll. 122–4). Such a counter-entropic development is hardly conceivable within science until twentieth-century chaos theory and implies much about Tennyson's thermodynamic optimism. Moreover, Tennyson suggests that chaos theory itself is part of a larger cultural wrestling with the implications of Victorian thermodynamics. But, as scholars, we are not yet clear on how to handle such poetic prescience.

Still, the early history of thermodynamic discourse suggests that it is not merely a distorted scholarly perspective that finds the seeds of contemporary science among the Victorians. We see elements of what is currently called non-equilibrium thermodynamics in the notion of a 'moving equilibrium' (448) put forth by Spencer in his *First Principles*. Similarly, we may find butterfly effects in Stewart and Lockyer's notion of great delicacy, such that 'a very small cause might be the parent of enormous effects, of a visible and mechanical nature' (327), or of infinite delicacy, to account for 'the very perfection . . . of animated beings [whose] motions cannot possibly be made the subject of calculation' (324). And even Thomson included a famous caveat to his predictions of the death of the sun, in the 'sources now unknown to us' that might be found in the 'great storehouse of creation' (1862 393). This hope was solidified as science in 1904 when Rutherford alluded to Thomson's 'prophetic utterance' (Brush 43) in a lecture to the Royal Society announcing his recent discovery of radium.

The most famous Victorian suggestion for evading apparently inevitable entropic decline, however, hails from Maxwell's prolific and poetic imagination. Later dubbed a demon by Thomson, Maxwell's creature first appears in a letter to Tait and is later described in a note to Maxwell's 1871 *Theory of Heat*:

> If we conceive a being whose faculties are so sharpened that he can follow every molecule in its course, such a being, whose attributes are still as essentially finite as our own, would be able to do what is impossible for us . . . Now let us imagine [that a vessel full of air] is divided into two portions, A and B, by a division in which there is a small hole, and that a being, who can see the individual molecules, opens and closes the hole, so as to allow only the swifter molecules to pass from A to B, and only the slower ones to pass from B to A. He will thus, without expenditure of work, raise the temperature of B and lower that of A, in contradiction to the second law of thermodynamics.
>
> *(Quoted in Hayles 42)*

Though still understood as purely fictional, Maxwell's demon has caught both the scientific and literary imagination. 'One of the most famous conundrums in the history of science' (Hayles 32), the demon powerfully suggests the differing implications of classical and statistical thermodynamics and facilitates the twentieth-century move from a version of chaos that figures as the opposite of order to one that understands chaos as indicative of deep, emergent, dissipative or otherwise alternative structure. It has also inspired such literary work as Thomas Pynchon's *The Crying of Lot 49*. But its emergence among such demonic contemporaries as Dr Jekyll, Dorian Gray and Dracula suggests something wider and yet to be fully explored at work in Victorian scientific culture.

Energy and evolution

We need not, however, look through the lens of twentieth-century physics and mathematics to find uncharted areas in literature and thermodynamics, as the relations within, between and among Victorian sciences certainly require more subtle attention than they have as yet received. At present, it seems almost standard procedure for literary scholars interested in thermodynamics to bemoan the paucity of attention we have paid to this Victorian science, especially when compared to evolution, in order to argue for the necessity of such scholarship or to account for the disparity of our scholarly vision. Moreover, within thermodynamically informed literary scholarship, we have tended to focus on 'entropy'. In some ways less complex than 'energy', the term has nonetheless seemed more interesting to the literary-minded, perhaps because of the fascination of apocalyptic predictions – a fascination undoubtedly at work in our own moment as among the Victorians. Such a focus leads us to place thermodynamics and evolution in opposite corners, as it were. So do Thomson's well-known objections to evolution, based on his calculations of the sun's lifetime, which, in the absence of radiation theory, simply did not allow enough time for natural selection to have taken place as Darwin hypothesized (1862 391–2).

Still, what I have called the 'affective relations' between evolutionary and thermodynamic discourses, as well as their scientific ones, are more complex than the model in which evolutionary optimism is pitted against the degenerative inevitability of the second law. Again, Tennyson's *In Memoriam* is a good place to begin to complicate this picture. As the two sciences converge within the final lines of the poem, physics, as I have argued, offers a solution to the problems of waste wrought by biology. Tennyson suggests that evolutionary narrative is able to transform waste into progress because it grows up alongside of and in conversation with the notion of transformation central to the development of thermodynamics, especially its first law (Gold 2002 462).

In this way, Tennyson suggests avenues for future research as well. There is, as yet, much to be done in exploring the complex interchange between thermodynamic and evolutionary narrative. In particular, I believe that this aspect of literary studies would benefit considerably from a better understanding of the scientific implications of one for the other. Nineteenth-century scholars are familiar with the multifarious uses of Darwin's 'struggle for survival' or Spencer's formulation 'survival of the fittest'. But to what extent did Darwin and his readers understand these as the struggle for the free energy of a system, as Frank Herbert does a century later in *Dune*? When, how and through what processes did we come to understand this fundamental connection? To what extent are the emergent energy concerns of the first half of the nineteenth century latent in emergent evolutionary discourse? To what extent are such concerns actively suppressed or sidestepped in the name of progress – evolutionary, political, economic, industrial? And, again, how do scholars deal with such latency as well as the active or unconscious suppression of available concepts?

Once we have worked out more thoroughly the commonalities between Victorian evolutionary and thermodynamic sciences, we may return to their dominant affective distinctions with more subtlety. A quick preliminary scan suggests that these affective distinctions may lie in how each science interprets waste. *In Memoriam* is manifestly concerned with the careless waste of resources in the reproduction of species, as the speaker famously fixes on the profligacy of Mother Nature and 'find[s] that of fifty seeds, / She often brings but one to bear' (Canto 55, ll. 11–12). The problem escalates to the loss of lives and the extinction of species, but the poem finds consolation sixty-three stanzas later in the notion that this loss enables man to 'Move upward, working out the beast, / And let the ape and tiger die' (Canto 118, ll. 27–8). Whatever evolutionary optimism we may find here depends critically on an implicit sense that such upward movement is worth the waste. It requires, moreover, a decidedly anthropocentric reading – within or without, the ape and tiger are implicitly expendable. Contemporary thermodynamics identifies 'advanced species' as those that use the most energy to support the very biological complexity we admire in them (this, before we even consider humanity's technologically driven uses). Classical thermodynamics, however, has long indicated that in any system of limited energy resources, such a model of evolution is not indefinitely sustainable. Nor are ethical or economic systems that seek to reproduce such models of progress. But if some uses of evolution and energy concepts drove the pursuit of industrial and imperial progress at the expense of some objectified and subjectable nature, others may yet provide us with ways to think about our ties to an ecosystem that encompasses humanity, nature, and the energy required by all living things.

Ecology

This brings me to a second area of future research. A focus on waste or conservation or resource management or sustainability or other such problematic but ecologically concerned terms brings us to the possibilities of thermodynamics for the emergent area of Victorian ecocriticism. For all that ecocriticism has focused primarily on nineteenth-century American literature and Romantic poetry, I believe that a Victorian ecocriticism, informed by thermodynamic discourse, has much to offer that is both distinctive and essential, especially as competition for the free energy of the system becomes a foremost global concern. For a somewhat longer, but still brief, introduction to how thermodynamic literary scholarship may intersect with ecocriticism, see Gold 2012.

Perhaps the most forward-thinking of scholars in this area, John Parham, observes that 'ecology conjoins two different nineteenth-century scientific frameworks, natural history – positing a "balance of nature" – and evolutionary theory, change driven by species competition' (2003 258). Parham traces the inception of the word 'ecology' to Ernst Haeckel's 1866 description: 'the investigation of the total relations of the animal both to its inorganic and organic environment; including above all, its friendly and inimical relations with those animals and plants with which it comes directly or indirectly into contact' (quoted in 2003 258). And as Victorian ecology comes into contact with Victorian energy science, we begin to see the roots of what, since the 1930s, has been called 'ecosystems theory', in which what Haeckel terms 'total relations' are explored 'in terms of purely material exchange of energy and of such chemical sub-stances as water, phosphorus, nitrogen, and other nutrients' (quoted in Parham 2003 258).

Concerns about what we call ecosystems are already at work among the Victorians and, indeed, among the Victorianists. Tina Choi, for example, finds a first-law, conservationist aesthetic in Henry Mayhew's desire for sustainability, in his wish to 'demonstrat[e] the self-sufficiency of the system' (310) manifest in *London Labour and the London Poor*. Choi finds this desire, central to Mayhew's sense of the nation's moral economy, to be thoroughly informed by Victorian thermodynamics. Indeed, she suggests that 'we might read his encyclopedic work as an attempt to negotiate the space between conservation and entropy, between complete recovery and the admission of loss'. For fictional considerations, we may look again to Dickens. In 'Dickens in the City', Parham identifies a thermodynamically informed 'social ecology' in Dickens's thick description of urban (among other) environments; scientific concepts inform at once Dickens's dismay at technological and other hazards to humans and the environment, and his hope that we might find a more sustainable future.

A more compact example of the Victorian literary antecedents of ecological systems theory, however, is readily available in the poetry of Gerard Manley Hopkins. If 'Binsey Poplars' is the most explicitly preservationist of Hopkins's poems, his 'God's Grandeur' may serve as our closing example of a Victorian literary ecosystem:

> The world is charged with the grandeur of God.
>   It will flame out, like shining from shook foil;
>   It gathers to a greatness, like the ooze of oil
> Crushed. Why do men then now not reck his rod?
> Generations have trod, have trod, have trod;
>   And all is seared with trade; bleared, smeared with toil;
>   And wears man's smudge and shares man's smell: the soil
> Is bare now, nor can foot feel, being shod.
>
> And for all this, nature is never spent;
>   There lives the dearest freshness deep down things;
> And though the last lights off the black West went
>   Oh, morning, at the brown brink eastward, springs —
> Because the Holy Ghost over the bent
> World broods with warm breast and with ah! bright wings.

Finding thermodynamics compatible with the Roman Catholicism to which he converted, Hopkins expresses a first-law conviction that 'nature is never spent' because it is 'charged' by God's Grandeur. Not unlike Joule's, this conviction persists for Hopkins in spite of the evidence of decay. Energies given by God may 'flame out', flaming out*ward*, dissipating or spreading, but also quite possibly burning *out*. Humans undoubtedly participate in the rapid increase of entropy, the searing, blearing and smearing that results from 'toil', from work. But somewhere 'deep down things' the 'dearest freshness' still resides and will return. Hopkins, moreover, displays an awareness of something like Haeckel's 'total relations'. He depicts an ecosystem, based on the energetic connectedness of all things, living and non-living, human and non-human, naturally occurring and made. And if we have lost our sense of this connectedness, our feet having been too long 'shod', morning brings hope for better relations and greater responsibility towards the energetic system of which we are but a part.

Bibliography

Primary texts

Alexander, Sarah. *Victorian Literature and the Physics of the Imponderable*. London: Pickering and Chatto, 2015.

[Anon.]. 'Age of the Sun and Earth'. *Cornhill Magazine* 38 (July–December 1878): 321–41.

Campbell, Lewis, and William Garnett. *The Life of James Clerk Maxwell*. London: Macmillan, 1882. Reprint, New York, NY: Johnson Reprint Corporation, 1969.

Clausius, Rudolph. *The Mechanical Theory of Heat With its Application to the Steam-Engine and to the Physical Properties of Bodies*. London: John Van Voorst, 1867.

Dickens, Charles. *Bleak House*. Ed. Nicola Bradbury. London: Penguin, 1996.

Hopkins, Gerard Manley. *The Major Works*. Ed. Catherine Phillips. Oxford: Oxford University Press, 2009.

Joule, James Prescott. *The Scientific Papers*. London: Taylor and Francis, 1884.

Siemens, William. 'A New Theory of the Sun: The Conservation of Solar Energy'. *The Nineteenth Century* (April 1882): 510–25.

Spencer, Herbert. *First Principles*. 6th edn. Reprint, Honolulu, HI: University Press of the Pacific, 2002.

Stewart, Balfour, and Norman Lockyer. 'The Sun as a Type of the Material Universe'. *Macmillan's Magazine* (July 1868): 246–57 (Part 1); (August 1868): 319–27 (Part 2).

Stewart, Balfour, and Peter Guthrie Tait. *The Unseen Universe or Physical Speculations on a Future State*. 9th edn. London: Macmillan, 1890. Reprint, Whitefish, MN: Kessinger, 2003.

Stoppard, Tom. *Arcadia*. London: Faber and Faber, 1993.

Tennyson, Alfred. *In Memoriam*. Erik Gray. New York, NY: Norton, 1973.

Thomson, William. 'On the Age of the Sun's Heat'. *Macmillan's Magazine* (March 1862): 388–93.

Thomson, William. *Mathematical and Physical Papers*. 6 vols. Cambridge: Cambridge University Press, 1882.

Secondary texts

Beer, Gillian. '"The Death of the Sun": Victorian Solar Physics and Solar Myth'. *Open Fields: Science in Cultural Encounter*. Oxford: Clarendon, 1996, 219–241.

Brantlinger, Patrick, ed. *Energy and Entropy: Science and Culture in Victorian Britain*. Bloomington, IN: Indiana University Press, 1989.

Brush, Stephen. *The Temperature of History: Phases of Science and Culture in the Nineteenth Century*. New York, NY: Burt Franklin, 1978.

Choi, Tina Young. 'Forms of Closure: The First Law of Thermodynamics and Victorian Narrative'. *ELH* 74.2 (2007): 301–22.

Clarke, Bruce. 'Allegories of Victorian Thermodynamics'. *Configurations* 4.1 (1996): 67–90.

Clarke, Bruce. *Energy Forms: Allegory and Science in the Era of Classical Thermodynamics*. Ann Arbor, MI: University of Michigan Press, 2001.

Dale, Peter Allan. *In Pursuit of a Scientific Culture: Science, Art and Society in the Victorian Age*. Madison, WI: University of Wisconsin Press, 1989.

Gold, Barri J. 'The Consolation of Physics: Tennyson's Thermodynamic Solution'. *PMLA* 117.3 (2002): 449–64.

Gold, Barri J. *ThermoPoetics: Energy in Victorian Literature and Science*. Cambridge, MA: MIT Press, 2010.

Gold, Barri J. 'Energy, Ecology and Victorian Literature'. *Literature Compass* 9.2 (2012): 213–24.

Harman, P. M. *Energy, Force, and Matter: The Conceptual Development of Nineteenth-Century Physics*. Cambridge: Cambridge University Press, 1982.

Hayles, N. Katherine. *Chaos Bound: Orderly Disorder in Contemporary Literature and Science*. Ithaca, NY: Cornell University Press, 1990.

Kuhn, Thomas S. 'Energy Conservation as an Example of Simultaneous Discovery'. *The Essential Tension: Selected Studies in Scientific Tradition and Change*. Chicago, IL: University of Chicago Press, 1977, 66–104.

Levine, George. *Darwin and the Novelists*. Chicago, IL: University of Chicago Press, 1988.

MacDuffie, Allen. 'Joseph Conrad's Geographies of Energy'. *ELH* 76.1 (2009): 75–98.

Marsden, Ben. *Watt's Perfect Engine: Steam and the Age of Invention*. New York, NY: Columbia University Press, 2002.

Morus, Iwan Rhys. *When Physics Became King*. Chicago, IL: University of Chicago Press, 2005.

Myers, Greg. 'Nineteenth-Century Popularizations of Thermodynamics and the Rhetoric of Social Prophecy'. *Energy and Entropy: Science and Culture in Victorian Britain*. Ed. Patrick Brantlinger. Bloomington, IN: Indiana University Press, 1989, 307–38.

Parham, John. 'Green Man Hopkins: Gerard Manley Hopkins and Victorian Ecological Criticism'. *Nineteenth-Century Contexts* 25.3 (2003): 257–76.

Parham, John. 'Dickens in the City: Science, Technology, Ecology in the Novels of Charles Dickens'. *19: Interdisciplinary Studies in the Long Nineteenth Century* 10 (2010). <www.19.bbk.ac.uk/articles/529>. Accessed 3 June 2011.

Peterfreund, Stuart. 'The Re-emergence of Energy in the Discourse of Literature and Science'. *Annals of Scholarship* 4.1 (1986): 22–53.

Rabinbach, Anson. *The Human Motor: Energy, Fatigue and the Origins of Modernity*. Berkeley, CA: University of California Press, 1990.

Smith, Crosbie. *The Science of Energy: A Cultural History of Energy Physics in Victorian Britain*. Chicago, IL: University of Chicago Press, 1998.

Snow, C. P. *The Two Cultures and the Scientific Revolution*. New York, NY: Cambridge University Press, 1959.

Underwood, Ted. *The Work of the Sun: Literature, Science, and Economy 1760–1860*. New York, NY: Palgrave Macmillan, 2005.

Further reading

[Anon.]. 'The Chemistry of a Candle'. *Household Words* (3 August 1850): 440–41.

[Anon.]. 'The Sun's Surroundings and the Coming Eclipse'. *Cornhill Magazine* 31 (January–June 1875): 297–315.

Brown, Daniel. *The Poetry of Victorian Scientists: Style, Science, and Nonsense*. Cambridge: Cambridge University Press, 2013.

Bruni, John. 'Thermodynamics'. *The Routledge Companion to Literature and Science*. Eds. Bruce Clarke and Manuela Rossini. New York, NY: Routledge, 2011, 226–38.

Bulwer-Lytton, Edward. *The Coming Race*. Peterborough: Broadview, 2008.

Carnot, Sadi. 'Reflections on the Motive Power of Fire'. Ed. and transl. R. H. Thurston. *Reflections on the Motive Power of Fire and Other Papers on the Second Law of Thermodynamics*. Ed. E. Mendoza. New York, NY: Dover, 1960. Reprint, Gloucester: Peter Smith, 1977. 1–59.

Clarke, Bruce. 'From Thermodynamics to Virtuality'. *From Energy to Information: Representation in Science and Technology, Art, and Literature*. Eds. Bruce Clarke and Linda Dalrymple Henderson. Stanford, CA: Stanford University Press, 2002, 17–33.

Darwin, Charles. *On the Origin of Species*. London: John Murray, 1859.

Flammarion, Camille. *Omega: The Last Days of the World* [1894]. Reprint, Lincoln, NE: University of Nebraska Press, 1999.

Gleick, James. *Chaos: Making a New Science*. New York, NY: Viking, 1987.

Herbert, Frank. *Dune: 40th Anniversary Edition*. New York, NY: Ace, 2005.

MacDuffie, Allen. 'Victorian Thermodynamics and the Novel: Problems and Perspectives'. *Literature Compass* 8.4 (2011): 206–13.

MacDuffie, Allen. *Victorian Literature, Energy, and the Ecological Imagination*. Cambridge: Cambridge University Press, 2014.

Maxwell, James Clerk. *The Scientific Papers of James Clerk Maxwell*. Vol. 2, ed. W. D. Niven. Cambridge: Cambridge University Press, 1890. Reprint, Mineola, FL: Dover, 2003.

Nixon, Jude. '"Death Blots Black Out": Thermodynamics and the Poetry of Gerard Manley Hopkins'. *Victorian Poetry* 40.2 (2002): 131–55.

Pynchon, Thomas. *The Crying of Lot 49*. Philadelphia, PA: J. B. Lippincott, 1965.

Rankine, William John Macquorn. *Miscellaneous Scientific Papers*. London: C. Griffin, 1881.

Schneider, Eric D., and Dorion Sagan. *Into the Cool: Energy Flow, Thermodynamics, and Life*. Chicago, IL: University of Chicago Press, 2005.

Serres, Michel. 'Turner Translates Carnot'. *Hermes: Literature, Science, and Philosophy*. Eds. Josue V. Harari and David F. Bell. Baltimore, MD: Johns Hopkins University Press, 1982.

Stewart, Balfour. *The Conservation of Energy*. New York, NY: D. Appleton, 1873.

Tait, Peter Guthrie. *Lectures on Some Recent Advances in Physical Science, With a Special Lecture on Force*. London: Macmillan, 1885.

Thomson, William. 'On the Dissipation of Energy'. *The Fortnightly Review* (1 March 1892): 313–21.

Tyndall, John. *Heat: A Mode of Motion*. New York, NY: D. Appleton, 1873.

Tyndall, John. 'Address Delivered Before the British Association Assembled at Belfast, With Additions, 1874'. *Victorian Web*. <www.victorianweb.org/science/science_texts/belfast.html>. Accessed 7 December 2015.

Wells, H. G. *'The Time Machine' and 'The War of the Worlds'*. New York, NY: Fawcett Premier, 1989.

Wilkinson, Ann. 'From Faraday to Judgment Day'. *ELH* 34.2 (1967): 225–47.

Wise, M. Norton. 'Time Discovered and Time Gendered in Victorian Science and Culture'. *From Energy to Information: Representation in Science and Technology, Art, and Literature*. Eds. Bruce Clarke and Linda Dalrymple Henderson. Stanford, CA: Stanford University Press, 2002, 39–58.

Young, Thomas. *A Course of Lectures on Natural Philosophy and the Mechanical Arts* [1807]. 2 vols. Reprint, London: Johnson Reprint Corporation, 1971. Vol. 1.

Zencey, Eric. 'Entropy as Root Metaphor'. *Beyond the Two Cultures: Essays on Science, Technology, and Literature*. Eds. Joseph W. Slade and Judith Yaross Lee. Ames, IA: Iowa State University Press, 1990, 185–200.

20

ELECTRICITY

Stella Pratt-Smith

'The Electric Light possesses no novelty', declared the *Illustrated London News*, one dark and wintry night in the depths of 1848 ('The Electric Light'). 'Year after year it has been exhibited at every course of philosophical lectures since the time of Sir Humphry Davy'. The anonymous journalist's words might seem an early illustration of *ennui* but, in fact, what they express is a critical revolution in the public imagination. As he goes on to report, displays of electricity were ever more prominent each passing day. New types of apparatus were set out in evermore conspicuous positions, arranged on 'the raised steps forming the entrance to the National Gallery and the Royal Academy', for example, and even perched on 'the summit of the Duke of York's Column'. The realization had dawned that, while electricity 'would be still a costly experimental toy', there was a great deal more to it than mere display or novelty. 'We have now to direct our attention more particularly to its practical application', the writer proposes. 'Really its practicability forms the whole subject for consideration'. It was a recognition made manifest by the 1850s, with a permanent display of electrical apparatus in Leicester Square, housed in a specially constructed building. Its substantial thirty-two-metre frontage and twenty-eight-metre rotunda indicated electricity's new significance, as well as the growing public interest in it (Beauchamp 22). It foretold how the phenomenon would be implicated across seemingly different spheres of experience, imagination and culture. Not only was it hard to miss; it was also here to stay.

There would be no return from the conviction that electricity could be used, or, indeed, the realization that it was a resource that *should* be used, rather than simply wondered at. It was a shift in understanding that fundamentally altered perceptions of human capabilities and even the place of our species in the world. Two centuries later, these qualities continue to make the topic of electricity a rich and interdisciplinary focus for research, one that engages with an unusually broad range of concerns, from its core exploration in the physical sciences to its role as a driving force of innovation and its potential as an imaginative resource, particularly for fiction authors. So, in reality, electricity had lost none of its novelty in 1848. It was just that the source of its novelty had shifted from the merely visual to the more complex and conceptual promise of its practical applications. As a *Punch* columnist suggests in the same year:

> The last new marvel is a Company for lighting our streets, our shops, our houses, and even our bed-candlesticks with electric fluid, so that we may sit, and read or write by

flashes of lightning, and go to sleep with a column of electric fluid doing duty for a rushlight in our room . . . The electric light now threatens to supersede all.

('Lights! Lights! I Say!')

The writer's excitement is palpable but so, too, are his hesitations. Alongside the exciting convenience that electricity might offer, there was the 'threat' that it also carried, not merely in the harnessing of such a volatile force but also in the change and uncertainty it might bring.

Popular 1840s commentary on what electricity was, how it worked and what it might become contributes to a much longer history of responses that might seem, initially, overwhelming to researchers. Viewing the history of electricity as a development of four phases helps us grasp, albeit very broadly, how engagement with it emerged and intensified over time. In the first stage, up to the eighteenth century, electricity was simply wondered at as a spectacular natural force. How it worked was only scrutinized more closely in the second stage of its history, which largely occurred after 1732 when experimental naturalist Stephen Gray began reporting on electrical conductivity.

In the closing decades of the eighteenth century, such experimenters as Luigi Galvani and Alessandro Volta harnessed the power of electricity sufficiently for public demonstration and early battery mechanisms. Nonetheless, electricity was still characterized as a singularly incomprehensible and 'intangible' phenomenon, and it was far from understood. In fact, it was very much misunderstood: as some sort of pervading fluid, related mysteriously to the still more mysterious ether, and thus quite possibly to other invisible and even occult forces. For this reason, the potential of Galvani's and Volta's work was increasingly the focus of discussion and experiment not just by Davy but also the period's wider circle of radical and pioneering experimentalists, who included Joseph Priestley, Thomas Beddoes (mentor to Coleridge and Southey), botanist Sir Joseph Banks (President of the Royal Society between 1778 and 1820) and the Scottish pioneering sexologist and showman James Graham (see Fulford, Lee and Kitson; Delbourgo).

The crucial transition to a third phase of understandings about electricity began in the early decades of the nineteenth century, with discoveries and demonstrations of electromagnetism by Hans Christian Oersted and Michael Faraday in 1820 and 1831 respectively, and the construction of more usable batteries by John Frederic Daniell and Sir William Grove in 1836 and 1839. Still, despite this apparent barrage of developments, it would take until the 1880s for the history of electricity to reach its final stage, when the phenomenon would be understood sufficiently to be used as a power supply. Sir William Armstrong's mansion in Rothbury, Northumberland, was the first to use electric light, on 17 January 1881, and, just a few years later, the first public supply of electricity was offered in Godalming, Surrey, in the form of street lighting (McNeil 369). Across these four phases, the direction and pace of scientific investigations was entirely transformed, as was the locus of public interest in electricity.

Electricity's therapeutic potential and medical uses in relation to the body, psychology and psychiatry provided another rich vein of speculation for late eighteenth- and nineteenth-century experimentalists. As indicated above, scholarship has emerged on such leading figures as James Graham; however, many contemporary and popular writings on the topic remain to be explored. In the early decades of the century, electric shocks were commonly seen as a form of entertainment; this history offers fascinating intersections between the electrical sciences and psychology, alongside electricity's adoption as a common treatment for sexual and emotional disorders and the psychiatric ECT treatments that continued well into the twentieth century. An additional related strand that would bear substantial investigation is the advertising of therapeutic electricity by private companies. The emergence of a simultaneously commercial and pseudo-scientific discourse, particularly in contemporary periodicals, offers copious illustrations, too, for scholars to consider.

The electrical investigations of the early nineteenth century are especially rewarding for interdisciplinary research, thanks in part to Faraday's 'Experimental Researches in Electricity' (1832–56), which were published in the *Philosophical Transactions* of the illustrious Royal Society of London. It was a period when scientific thinking was very much in flux and, as contemporary thinking about electricity altered and developed, Faraday continued to amend, contribute to and comment on the papers. As a result, they can be read as a form of scientific diary, a palimpsest of how scientific understanding progressed with regard to a particular phenomenon. They also epitomize Faraday's deeply altruistic and egalitarian approach to science, in which he explicitly hopes that the papers will encourage specialists and non-specialists to participate in new developments. By doing so, he proposed, the wider community of electrical enthusiasts might speed understandings of how electricity worked and its more extensive use.

It is no coincidence that, as Faraday reflected publicly on his groundbreaking electrical experiments, there was a steady stream of engagements with the subject in popular fiction and non-fiction. Widespread fascination with electricity even prompted a magazine devoted specifi-cally to its study in the form of the grandly named *Annals of Electricity, Magnetism and Chemistry and Guardian of Experimental Science*, an electrical enthusiasts' magazine published between 1836 and 1843 and edited by William Sturgeon (1783–1850). While it is rarely possible to trace precise or direct relationships between literary and scientific spheres, let alone their productions and developments, it would be equally hard to deny that an exchange of thoughts and ideas takes place continually, on a variety of levels. During the 1840s, as in any other period, the body of thinking about electricity was contributed to and disseminated among specialists and non-specialists alike, and distinctions between the two were frequently blurred. Undoubtedly, budding electrical practitioners and enthusiasts wanted to know what electricity was and how it worked. That is not to say, though, that they or other more specialist participants were not also interested in electricity's more elusive, mysterious or figurative qualities, ones that tended to dominate in the period's fiction writings.

So Faraday was busily speculating on the workings of electricity and documenting his thoughts and experiments but, at the same time and just as busily, popular writers were speculating more widely. One of the earliest popular fiction engagements with the idea of electricity is 'The Adventures of Mr. Ledbury and His Friend Jack Johnson' by Albert Richard Smith. Originally published in *Bentley's Miscellany* in 1842, Ledbury's comic adventures resemble those of Dickens's Pickwick, and electricity appears to function as little more than a source of fun, almost of farce. A travelling lecturer entertains people at the roadside with demonstrations of electrical sparks, and a British Association paper is awarded the tongue-in-cheek title of 'The Totality of Dependence in Phrenology and Fireworks Upon Metaphysical Electricity' (Smith 1856 57). Later in the story, an elderly lady claims that, rather than rheumatism, she has 'electricity of the nerves' (Smith 1856 230) to such an extent that she is simply 'one large living battery, Sir' (317). Considering the extent to which scientific investigations of electricity had already moved on by that point, the tale appears to fail rather than succeed in capturing recognitions of electricity's importance. It would be a mistake, though, for the literary historian to dismiss the story as nothing more than literary flotsam, an anachronistic or irrelevant engagement with ideas about electricity. Taking a wider view, the significance of the tale is precisely because it harks back to a time past, a time when electricity was viewed as nothing more threatening than amusement or spectacle. It constitutes a vital fragment in the century's ephemeral and nostalgic discourse about electricity, one in which fiction not only departs from but also continually strains against the scientific tide.

In stark contrast to Smith's naïve playfulness, Charles Dodd's 'A Day at an Electro-Plate Factory' (1844) is terribly earnest, as might be expected of its publication in the mouthpiece of the Society for the Diffusion of Useful Knowledge. It is one in an extended series of *Penny* articles that focused on the country's burgeoning industrial growth, all of which conform strenuously to the publication's agenda of readerly 'improvement'. Each article in the series runs to several pages with carefully Spartan illustrations of workers who seem, to modern eyes, as wooden as they are unrealistically spotless, calm and conscientious. The one on electroplating focuses intently on the minute stages of the electroplating process, but, rather than simply reporting these, Dodd attempts to convey how thrilling it is that even the dowdiest objects can now be transformed into glittering and exquisite treasures.

Also set in an electroplating factory and offering a similar blend of industrial production commentary and animated hyperbole is the anonymous series 'Road to Happiness in Six Steps', which was published in *Reynolds's Miscellany of Romance, General Literature, Science, and Art* in 1848, unsurprisingly in six parts. Contributing to the tradition of 'mental improvement' (Rauch 24) defined by such works as Samuel Smiles's *Self-Help* (1859), the author of the series was one Edwin F. Roberts, a little-known but, at the time, regular writer of historical romances for 'Penny Dreadfuls'. In the series, the protagonist rises from his respectable but impoverished origins because he proposes various manufacturing innovations that demonstrate his initiative, intelligence and honourable nature. These enable him to shine, much like the products he works with – a comparison the author accentuates – and he makes his deserved way to success in a manufacturing career. A *Bildungsroman* as sincere in intent as Dodd's 'Days Out' and considerably more didactic, the 'Road to Happiness' makes the electro-plate factory and, to some extent, electricity itself an extended metaphor for self-improvement, success and honour by means of hard work and honesty. The series, like the century itself, is infused with the idea of electricity being representative of individual and collective moral progression.

Clearly, the electroplating process held a special fascination for nineteenth-century readers; like electricity, it had seemingly magical powers in invisibly, silently and effortlessly transforming the ordinary and mundane. It gave value to the worthless and made the outmoded seem modern. More than that, though, it allowed for the apparently contradictory coexistence of nostalgic tradition with technological modernity, and this became one of electricity's most strikingly characteristic features. Electricity had always been considered to have secret powers; but now, in tandem with (rather than in contrast to) the awe-inspiring capacities of new scientific and industrial processes, it seemed nothing less than the stuff of fairy tale. Like Rumpelstiltskin, it could turn straw into gold. As part of an entrancing modern fairy tale, it had even greater potency, not least as a force that might potentially be available to all. The juxtaposition of such seemingly contradictory archaic and modern qualities gave electricity unique allure, something that fiction writers would take delight in for the rest of the century.

By the 1870s, though, a battle had ensued, one in which electricity had to fight with gas for its survival and popularity as a mainstream fuel. Gas companies were particularly concerned, as electricity seemed poised to supplant gas entirely as an illuminating medium and more widely, too, as a commercial and domestic source of power. This was despite the 'reassuring tone' of some practitioners, such as the one who protested in *The Engineer* magazine in 1877 that

> Electricity for domestic illumination would never, in our view, prove as handy as gas. An electric light would always require to keep it in order a degree of skilled attention which few individuals would possess.
>
> *(Quoted in Beauchamp 136)*

Electricity was as inherently dangerous as gas, and that was a significant obstacle to its adoption in mainstream technologies, one that had to be neutralized effectively before its power could be understood or controlled, let alone utilized or accepted. The mechanisms by which electricity gained its subsequent dominance, by way of scientific developments, technological applications, and imaginative allure, are essentially what research on the subject has sought to elucidate.

Electricity is invisible and untouchable; its effects can be made apparent but, as a phenomenon that can be neither seen nor touched, it can only ever be obliquely observed. To a greater or lesser extent, therefore, investigations of electricity have always relied on acts of creativity and imagination just as much as on empirical observation. The representation of electricity began with predominantly visual displays before moving towards the desire for more precise modelling through verbal description and, finally, the symbolism and enhanced precision of mathematical depiction. While understandings of electricity would eventually evolve into the electrical 'sciences', initially they were based on establishing a clear notion of a force, something that frequently resisted attempts at accurate conceptualization. For these reasons, the study of electricity's history is fundamentally interdisciplinary. It engages not just with literature and science but with a much broader range of disciplines too, from the history of science and mathematics, phenomenology and philosophy to sociocultural and literary studies on the body, gender and morality. Many of these approaches share similar reasoning strategies, in ways that break down binary oppositions between literature and science and, instead, demonstrate similar reliance on analogy, imagery and metaphor. Gillian Beer's *Darwin's Plots* (1983) and her subsequent *Open Fields* (1996) established a critical lead in exploring these correlations, with a comparison of work by such nineteenth-century novelists and poets as George Eliot and Gerard Manley Hopkins as well as such scientific figures as Hermann von Helmholtz and John Tyndall. What Beer demonstrated was that relationships between literary and scientific endeavours were characterized by 'interchange' and by 'transformation rather than translation', whereby 'scientific discourses overlap, but unstably' (Beer 1996 173). The timing of Beer's insights was particularly significant, for it preceded immediately the development of the large-scale digitization and computerized search facilities that enabled such projects as *Science in the Nineteenth-Century Periodical* (2007), endeavours that would revolutionize explorations of paper archives beyond the canon. The two research innovations heralded a unique expansion in literature and science as a conjoined research area, one that could truly exploit the veritable 'electronic harvest' (Secord 2005 463) of digitized archives and enhanced research tools.

The history of electricity repeatedly employs the combination of the visual, verbal and symbolic elements outlined above. Long before it became the celebrated and popular subject of nineteenth-century scientific exploration, though, electricity was at the heart of the eighteenth-century vitalism debate. Electricity and vitality were powerfully related in the public imagination for, as Nicholas Roe explains in his study of romanticism and science, 'in the vitality debate, the "electrical fluid" served as a helpfully ambiguous go-between, a material numen infusing matter with a less tangible life force' (8). Electricity's qualities as an invisible and energizing, yet potentially life-giving, force gave it singular significance, particularly within the Romantic proposition that 'the principle of life can move and affect matter without being, in some part, matter itself' (Ruston 2005 31). This is also why the idea of electricity – or, more accurately, galvanism, given its timing – is so central to the creation of life portrayed by Mary Shelley's *Frankenstein* (1818). What electricity provided, within her novel and beyond, was not just a physical or material phenomenon; it was a metaphor for recurrent tensions between the abstract nature of unseen electrical forces and physical, bodily manifestations of sensory information. Even on a metaphorical level, 'linked to a monster, electricity came to signify the unnatural birth of a dangerous and uncontrollable phenomenon' (Botting 1991 190). The intense connections established by Mary Shelley meant that, ever afterwards, fiction authors would struggle to avoid the theme of electricity's distorting and corrupting influence on both body and mind.

The emergence of the Gothic genre in literature during the same period as early nineteenth-century sciences meant that, as Martin Willis suggests, magic and occultism 'continued to infect' them, as a way of 'interpreting the complex interplay of the mystical and the scientific' (66). Mary Shelley's *Frankenstein* enacts not just the climate of late eighteenth- and nineteenth-century electrical investigations or the period's juxtaposition of scientific and cultural debates but also the oddly co-dependent contest between 'Romantic' and 'materialist' viewpoints. In the battle for legitimacy, Romantic science presented electricity as a universal and supernatural force at the core of cosmic principles and human spirituality, based on the German concept of *Urphänomen* as an absolute or primal force (Willis 71). In Willis's reading, materialist science endeavoured to prove that electricity was 'just another natural phenomenon' (68), which would 'place materialist method and philosophies at the centre of scientific authority'. However, at the same time, the persistent and confusing 'heterogeneity' (Willis 71) of electrical theory continued to create unbounded space for uncertainty, contradiction, and speculation.

In the first half of the nineteenth century, those worryingly uncertain spaces were especially beguiling and they made concepts of space critical to debates about what electricity was: was it a fluid that flowed or something entirely different? This was all the more unsettling in an era overwhelmingly concerned with slippages between what was visible and invisible, corporal, material or spiritual. The term 'space' referred to concepts well beyond mere physical dimensions; enquiries about space underpinned investigations as to the precise composition of the material world, the tangible and otherwise. As Alice Jenkins proposes in *Space and the 'March of Mind'* (2007), in the popularization of the physical sciences between 1815 and 1850 there was a recognition of 'the immaterial, conceptual space that contains and informs those other [physical] kinds of space' (4) and out of that change in perception emerged 'a new spatial imagination' (1). As a result, the investigation of spaces became the focus for simultaneously specialist and popular research across Britain, most significantly perhaps in the Mechanics' Institutes lectures of the 1820s and 1830s.

The increasingly diverse nature of participation in science brought the study of electricity out of the gentleman's laboratory and into the wider world of industrial manufacturing. New scientific understandings of electricity represented a 'whole new scientific vision' (Smith 1998 2) of progress that would gradually, yet entirely, transform society. Accompanying the increased diversity were equally new and popular forums for authorship, offered by expanding periodical and book presses, through which interpretations and the visionary scope of electricity continued to be incorporated, extended and altered.

Nonetheless, despite these more abstract, even esoteric, developments, explorations and portrayals of bodily experiences with electricity remained fundamental to the whole unsteady gamut of electrical theories, hypotheses and laws in the nineteenth century. As Carolyn Marvin suggests in her seminal study of nineteenth-century electrical communications, the body has always provided 'the most familiar of all communicative modes' (109), as 'a convenient touchstone by which to gauge, explore, and interpret the unfamiliar'. The role that references to electricity played in literature was a great deal more than decorative. As Jason Rudy has pointed out since, in *Electric Meters* (2009), how the body allows us to draw on different forms of knowing and understanding was the core focus of nineteenth-century 'Spasmodic' poetry. For Alfred Tennyson, Elizabeth Barrett Browning, Gerard Manley Hopkins and Algernon Swinburne, the 'interplay of physiology and electricity' (Rudy 8) was 'not simply a trope used to describe or to elaborate a poetic function'. In poetry especially, electricity could be a way of understanding how reading affected the body – to investigate the 'direct transfer of energy' (Rudy 2) in emotional and physical terms beyond theories and physical laws. This thinking contributes to persuasive arguments that 'we need to view rhetoric as an integral part of science' (Cantor 1989 161) and that 'rhetorical tropes and aesthetic forms did not merely decorate the science presented, but helped to construct it' (O'Connor 227).

The instrumentally useful contribution of electricity to scientific development became increasingly distinct as the nineteenth century progressed, especially in comparison to the preceding period. As Barri J. Gold suggests in *ThermoPoetics: Energy in Victorian Literature and Science* (2010), the movement towards a single concept of 'energy', which unified previously disparate ideas about heat, light, electricity, magnetism, gravitational attraction and mechanical 'work', was a particularly radical advance. The recognition was deeply embedded within the period's historical context, as part of a lively and dynamic 'conversation' (Gold 28) between the energy sciences, poetry and prose fiction. 'Literature has often, perhaps always, influenced science,' Gold contends, 'especially in the delicate, early stages of a scientific development, before a phenomenon has been named or a hypothesis adequately articulated' (15). This influence and electricity's uniquely elusive and 'protean quality' (Gold 16) are aptly demonstrated in Gold's highly persuasive reading of Dickens's *Bleak House* (1853), in which the novel emerges as a literary explication of both machine technologies and the scientific principle of entropy. The evident correlations between contemporary fiction and closely contemporaneous scientific and technological developments validate the proposition that the electrical sciences did not exist in a vacuum. Instead, they were very much integrated within nineteenth-century culture and society, and the contexts within which they emerged did not act only as background. They were key components in investigations of phenomena.

By the 1840s, radical changes were already taking place in the practice and purposes of science, on national and individual scales. Scientific progress was increasingly a matter of national pride, with technological advancement recognized as a reflection of growing British prosperity and crucial to the country's ability to compete as an international power. Public perceptions of electricity were intimately connected to aspirations about future progress; at the same time, electricity's transformative capacities were as closely linked to personal economic gain as they were to national manufacturing interests. Underlying both was the elemental cleanliness of electricity, something akin to a purifying fire that had the capacity to transform a 'dirty' industrial world of superstition, corruption and nostalgia, and bring about a more ordered, even utopian, existence, one based on modernity, comfort and convenience. Yet these seemingly contradictory impulses were neither separate nor opposed. While new forms of knowledge certainly constituted a revolution in cultures of scientific thought, the technological advances they enabled were not distinct from the sensationalism and mystery with which electricity had always been so commonly associated. Startling and innovative electrical applications simply added to the impression of an 'uncanny' modern world, which was still very much 'full of invisible, occult forces' (Bown, Burdett and Thurschwell 1). As William Wilson, originator of the term 'science fiction', suggested in 1851, 'the truths of Science' could be 'interwoven with a pleasing story which may itself be poetic and *true*' (139). Breakthroughs in understandings of electromagnetism were especially rich in 'interwoven' scientific and literary meanings. The range of possible engagements with concepts of electricity offered by such works as *The Poetry of Science* (1848) by Robert Hunt, *The Soul in Nature* (1852) by Hans Christian Oersted and *The Coming Race* (1871) by Edward Bulwer-Lytton supports Wilson's contention that, in the discovery of electromagnetism, there was 'more magic about its reality, than the wildest creations of child-fiction and legend have in their ideality' (143). Electricity's possibilities were not restricted to scientific and technological developments; they extended much further and more intriguingly. What they offered were captivating visions of the future that, rather than excluding the past, drew instead on its tenacious imaginative power.

As a research topic, one of electricity's most unique features is its potential for linking scientific and literary histories with broader technological and cultural interests. Its relevance is particularly marked in relation to awareness of present-day energy issues, on both individual and collective levels. Setting aside such metaphors as 'fields' of influence or being 'shocked', which portray electricity as an external force, we might consider the increasingly common references to being 'energized', 'depleted' or 'recharging our batteries', which disclose much greater instabilities in the conceptual distinctions between humans and machines. Metaphorical connections like these date back to popular nineteenth-century propositions, including that of leading electro-biologist Arthur Smee (fellow of the Royal Society), who suggested in 1850 that 'man is made up of a great number of voltaic elements, so arranged as to form one whole' (211) and that 'man acts by electricity' (216). Such comparisons elicited some backhanded compliments: as one contemporary reviewer remarks, 'Mr. Smee is a more practical man, he is not a mere dreamer . . . in fact, he is another Frankenstein' ([Anon.] 913). Established and well-worn though these images may seem, they reflect a continued, underlying resistance to the conceptual opposition between the human and non-human. The conflation of the two persists in the body's representation as an entity critically fuelled and regulated by electricity, and even dependent on it.

The popular perception of the body as electrical has considerably wider cultural implications, ones that affect not just how we think of ourselves but also how we think of machines. The powerfully evocative relationship between industrial metaphors and technology is aptly christened 'the industrial imaginary' by Tamara Ketabgian in *The Lives of Machines* (2011). Ketabgian describes how it is 'rooted in complex models of affect, community, intelligence, energy and life itself' (2). She traces its evolution through the work of Blake and Byron, Wordsworth and Goethe, to Charles Babbage, Samuel Butler, Harriet Martineau and Karl Marx. While her primary focus is on the work of George Eliot, Charles Dickens and Elizabeth Gaskell, she helpfully locates these writers' works in terms of other leading intellectual figures, including Arnold and Ruskin, Andrew Ure, Faraday, W. R. Grove and Helmholtz. She challenges the convention of machines as soulless and unfeeling, and successfully achieves her goal of providing 'a more inclusive history of technoculture' (Ketabgian 3). In so doing, she also reveals just how frequently and how long body parts and machines have been portrayed as endowed with agency and as part of machinery, rather than just extended or contained by it.

The significance of the industrial imaginary in literature continues to be evident well after the century of the Industrial Revolution, particularly in the birth and fruition of science fiction, dystopian narratives and fantasy literature. However, the industrial imaginary's remarkable afterlife lies also in its wider connotations for the twenty-first century electrical human. Collectively as well as individually, the use of electricity is ubiquitous in all but the least developed areas of the world, and its technological applications have contributed positively in a multitude of ways. We live in an era still thrilled by the productive potential of electricity but nineteenth-century warnings also ring increasingly true. Witness, for example, Cardinal Nicholas Wiseman's forecasting in 1853 the 'real danger of seeing the next generation brought up in the ideas of many of the present, that man is a machine, the soul is electricity, the affections magnetism, that life is a railroad, the world a share-market, and death a terminus' (quoted in Yeo 1993 184). In Wiseman's dystopic vision, electricity is classed along with other manifestations of modernity, as a 'real danger' that can affect man's apprehension of his own nature.

The persistent 'novelty' of electricity identified in the early nineteenth century has resulted in more than a century of its widespread and wholehearted adoption; but this has been accompanied by an almost universal reliance that is equally disturbing. While most would stop short of describing our present dependence on technologies as an addiction, at no time in history has electricity played a more crucial role in interactions between human society and the natural world. The extent to which we adapt electricity, its sources and its applications will not merely continue to define our relationship with the phenomenon; it has the potential also to redefine our future as a species.

Electricity is heavily implicated, as ever, in an array of environmental issues that are more intricately connected and international in orientation than they might at first appear. After all, much of modern life, from mobile technologies to water purification to cyber-security, functions only because of electricity, and only for as long as it continues to be available. In reality, the increasingly urgent search, via solar, water, wind and nuclear technologies, is for electricity. Perceptions of electricity are inherently and fundamentally interdisciplinary; they become ever more so, as anxiety increases about how to power the modern world. Research on electricity through the lens of literature and science builds on that and extends it beyond theoretical studies and disciplinary boundaries. In doing so, it illustrates the unified nature of what might seem otherwise distinct spheres and, just as importantly, it points out directions for humanities scholarship that are more topical and relevant than they have ever been.

Bibliography

Primary texts

[Anon.]. Review of Arthur Smee's 'Electrical Theory of Life'. *Literary Gazette* 1768 (7 December 1850): 913–14.

Dodd, Charles. 'A Day at an Electro-Plate Factory'. *The Penny Magazine of the Society for the Diffusion of Useful Knowledge* 13.806 (26 October 1844): 419–25.

'Lights! Lights! I Say!'. *Punch*, July–December 1848.

Roberts, Edwin F. 'The Road to Happiness in Six Steps'. *Reynolds's Miscellany of Romance, General Literature, Science, and Art* 1.11 (23 September 1848) to 1.16 (28 October 1848).

Smee, Arthur. *Instinct and Reason: Deduced from Electro-Biology*. London: Reeve and Benham, 1850.

Smith, Albert Richard. 'The Adventures of Mr. Ledbury and His Friend Jack Johnson' [originally published in *Bentley's Miscellany*, 1842]. London: Routledge, 1856.

'The Electric Light'. *Illustrated London News*, 9 December 1848.

Wilson, William. *A Little Earnest Book on a Great Old Subject*. London: Darton, 1851.

Secondary texts

Beauchamp, K. G. *Exhibiting Electricity*. London: Institution of Electrical Engineers, 1997.

Beer, Gillian. *Darwin's Plots: Evolutionary Narrative in Darwin, George Eliot and Nineteenth-Century Fiction*. London: Routledge and Kegan Paul, 1983.

Beer, Gillian. *Open Fields: Science in Cultural Encounter*. Oxford: Clarendon, 1996.

Botting, Fred. *Making Monstrous: Frankenstein, Criticism, Theory*. Manchester: Manchester University Press, 1991.

Bown, Nicola, Carolyn Burdett, and Pamela Thurschwell, eds. *The Victorian Supernatural*. Cambridge: Cambridge University Press, 2004.

Cantor, Geoffrey. 'The Rhetoric of Experiment'. *The Uses of Experiment: Studies in the Natural Sciences*. Eds. David Gooding, Trevor Pinch and Simon Schaffer. Cambridge: Cambridge University Press, 1989. 159–80.

Delbourgo, James. *A Most Amazing Scene of Wonders: Electricity and Enlightenment in Early America*. Cambridge, MA: Harvard University Press, 2006.

Fulford, Tim, Debbie Lee, and Peter J. Kitson, eds. *Literature, Science and Exploration in the Romantic Era: Bodies of Knowledge*. Cambridge: Cambridge University Press, 2004.

Gold, Barri J. *ThermoPoetics: Energy in Victorian Literature and Science*. Cambridge, MA: MIT Press, 2010.

Jenkins, Alice. *Space and the 'March of Mind': Literature and the Physical Sciences in Britain, 1815–1850.* Oxford: Oxford University Press, 2007.

Ketabgian, Tamara. *The Lives of Machines: The Industrial Imaginary in Victorian Literature and Culture.* Ann Arbor, MI: University of Michigan Press, 2011.

Marvin, Carolyn. *When Old Technologies Were New: Thinking About Electric Communication in the Late Nineteenth Century.* Oxford: Oxford University Press, 1988.

McNeil, Ian, ed. *An Encyclopaedia of the History of Technology.* London: Routledge, 1990.

O'Connor, Ralph. *The Earth on Show: Fossils and the Poetics of Popular Science, 1802–1856.* Chicago, IL: University of Chicago Press, 2007.

Rauch, Alan. *Useful Knowledge: The Victorians, Morality, and the March of Intellect.* Durham, NC: Duke University Press, 2001.

Roe, Nicholas. *Samuel Taylor Coleridge and the Sciences of Life.* Oxford: Oxford University Press, 2001.

Rudy, Jason R. *Electric Meters: Victorian Physiological Poetics.* Athens, OH: Ohio University Press, 2009.

Ruston, Sharon. *Shelley and Vitality.* Basingstoke: Palgrave Macmillan, 2005.

Secord, James. Review: 'The Electronic Harvest'. *The British Journal for the History of Science* 38.4 (2005): 463–7.

Smith, Crosbie. *The Science of Energy: A Cultural History of Energy Physics in Victorian Britain.* London: Athlone, 1998.

Willis, Martin. *Mesmerists, Monsters, and Machines: Science Fiction and the Cultures of Science in the Nineteenth Century.* Kent, OH: Kent State University Press, 2006.

Yeo, Richard. *Defining Science: William Whewell, Natural Knowledge and Public Debate in Early Victorian Britain.* Cambridge: Cambridge University Press, 1993.

Further reading

Bowler, Peter J., and Iwan Rhys Morus. *Making Modern Science: A Historical Survey.* Chicago, IL: University of Chicago Press, 2005.

Caleb, Amanda Mordavsky, ed. *(Re)Creating Science in Nineteenth-Century Britain.* Newcastle: Cambridge Scholars, 2007.

Cantor, Geoffrey, and M. J. S. Hodge. *Conceptions of Ether: Studies in the History of Ether Theories, 1740–1900.* Cambridge: Cambridge University Press, 1981.

Cantor, Geoffrey, Gowan Dawson, Graeme Gooday, Richard Noakes, Sally Shuttleworth and Jonathan R. Topham, eds. *Science in the Nineteenth-Century Periodical: Reading the Magazine of Nature.* Cambridge: Cambridge University Press, 2004.

Clifford, David, Elisabeth Wadge, Alex Warwick and Martin Willis, eds. *Repositioning Victorian Sciences: Shifting Centres in Nineteenth-Century Thinking.* London: Anthem, 2006.

Darrigol, Olivier. *Electrodynamics from Ampère to Einstein.* Oxford: Oxford University Press, 2002.

Faraday, Michael. *The Correspondence of Michael Faraday.* Ed. Frank A. J. L. James. 6 vols. London: Institution of Electrical Engineers, 1991–2012.

Flood, Raymond, Mark McCartney and Andrew Whitaker, eds. *James Clerk Maxwell: Perspectives on His Life and Work.* Oxford: Oxford University Press, 2014.

Gooday, Graeme. *The Morals of Measurement: Accuracy, Irony and Trust in Late Victorian Electrical Practice.* Cambridge: Cambridge University Press, 2004.

Gooday, Graeme. *Domesticating Electricity: Technology, Uncertainty and Gender, 1880–1914.* London: Pickering and Chatto, 2008.

Gooding, David. '"Magnetic Curves" and the Magnetic Field: Experimentation and Representation in the History of a Theory'. *The Uses of Experiment: Studies in the Natural Sciences.* Eds. David Gooding, Trevor Pinch and Simon Schaffer. Cambridge: Cambridge University Press, 1989, 183–223.

Harman, Peter M. *Energy, Force, and Matter: The Conceptual Development of Nineteenth-Century Physics.* Cambridge: Cambridge University Press, 1982.

Harman, Peter M., ed. *The Scientific Letters and Papers of James Clerk Maxwell.* Cambridge: Cambridge University Press, 2008.

Irwin, Alan, and Brian Wynne. *Misunderstanding Science?: The Public Reconstruction of Science and Technology.* Cambridge: Cambridge University Press, 1996.

Mahon, Basil. *Oliver Heaviside: Maverick Mastermind of Electricity.* London: Institute of Electrical Engineering, 2009.

Maxwell, James Clerk. *The Scientific Papers of James Clerk Maxwell*. Ed. W. D. Niven. New York, NY: Dover, 1965.

Morus, Iwan Rhys. *Frankenstein's Children: Electricity, Exhibition, and Experiment in Early-Nineteenth-Century London*. Princeton, NJ: Princeton University Press, 1998.

Morus, Iwan Rhys. 'The Two Cultures of Electricity: Between Entertainment and Edification in Victorian Science'. *Science and Education* 16.6 (2006): 593–602.

Otter, Chris. *The Victorian Eye: A Political History of Light and Vision in Britain, 1890–1910*. Chicago, IL: University of Chicago Press, 2008.

Secord, James. 'Extraordinary Experiment: Electricity and the Creation of Life in Victorian England'. *The Uses of Experiment: Studies in the Natural Sciences*. Eds. David Gooding, Trevor Pinch, and Simon Schaffer. Cambridge: Cambridge University Press, 1989, 337–83.

Smil, Vaclav. *Transforming the Twentieth Century: Technical Innovations and Their Consequences*. Oxford: Oxford University Press, 2006.

Sussman, Herbert. *Victorian Technology: Invention, Innovation, and the Rise of the Machine*. Santa Barbara, CA: ABC-CLIO, 2009.

21

TECHNOLOGY

Meegan Kennedy

Florida State University

Of all the changes – political, economic, military, colonial, spiritual, aesthetic, scientific and social – that marked the nineteenth century, the developments in technology must have been among the most noticeable to those living through these decades. A child born in 1901 entered a world filled with new technologies – railway travel, steam presses, electric light and the telegraph – making a material difference in daily life. Many of them developed from research in the sciences, but they gained importance through broader cultural trends: the expansion of scientific education and the education of the working classes; the growth of industry; the rise of public health, requiring vast investment in urban engineering; the expansion of empire, with projects such as the Great Trigonometrical Survey of India; and the rise of the exhibition as a site for publicizing new inventions (see Barrow).

We might organize a survey of nineteenth-century technology by motive power (such as animal, water, steam, electricity) and materials (coal, cotton, wool, glass, steel, India rubber); or we could categorize it by its product or purpose (industry, transportation, measurement, entertainment). Nineteenth-century scientists often worked on a particular material or physical problem. Michael Faraday, for example, experimented with making different types of optical glass and with the problem of storing electricity in battery form, which would be useful in many different contexts. Others, however, focused on one end-use regardless of materials, as microscopists worked with different forms of gas and electric lighting. More recent scholarship is organized by the end technology (railways, factories, presses, telegraphs) than by the motive power or the material resource at its source. This chapter organizes technologies by their broad purpose: power (primarily steam power in transport and industry), light (gas and electricity), communication, sanitation and the observing and recording technologies.

Power: the steam engine

Transport

Steam power, developed during the eighteenth century, became the dominant mode of power during the nineteenth century, driving railway locomotives and steamships, newspaper presses and factory looms. Steam engines could be powered by wood, oil or coal. Matthew Boulton and James Watt's engines were well established – Erasmus Darwin had celebrated their work in his 1791 poem *The Botanic Garden*[1] – and the advent of high-pressure designs in the early nineteenth century enabled self-propelling steam engines and steam-powered transportation. Mines in Cornwall, far from the coal of Wales, created a high-pressure, low-fuel 'Cornish engine' to pump water from mines and haul ore (see Marsden and Smith 34). Richard Trevithick demonstrated the first self-propelled steam locomotive in South Wales in 1804. The prolific George Stephenson developed *Blücher* to haul coal in 1814, *Locomotion* for the first public railway (the Stockton and Darlington line) in 1825, and the *Rocket* for the Rainhill speed trials, reaching a record twenty-nine miles per hour in 1829. The first steam passenger railway, the Liverpool and Manchester line, opened in 1830, with its first day blighted by the railways' first fatal accident: William Huskisson, an MP attending the event, fell into the path of the speeding *Rocket*. For good or ill, public travel was changed forever.

The 1844 Railway Act sparked a 'railway mania', producing 6000 miles of track in Britain by 1850. Isambard Kingdom Brunel, chief engineer of the Great Western Railway between London and Bristol from 1835, built bridges and tunnels to overcome engineering challenges such as rivers and mountains. With his father Marc's innovative tunnelling shield he completed the Thames Tunnel, a tourist attraction, in 1843. The railways transformed not only the British landscape but also nineteenth-century life, rapidly taking over the freight system; locomotives could handle speed or changes of elevation better than canal boats (for a useful comparison of transport technologies, see Evans). The middle and working classes enjoyed excursion trains, fresh country milk and the penny post, and the railways provided new employment. However, some neighbourhoods of London were gutted to make way for the lines and stations of the rail companies.

Steam engines made significant tracks on literature. Charles Dickens edited *The Daily News* (a radical paper supported by the railway tycoon Joseph Paxton) from 1845–46, when he was asked to place stories about railways. His *Dombey and Son* (1846–48) is one of the earliest and most famous novels relating to the railway. It details the devastation in 'Stagg's Gardens' when the railway construction comes through; it introduces the cindery, cheerful railwayman Mr Toodle; and it offers the graphic death of Carker, who falls under a train. *Mugby Junction*, stories collected in 1866 from Dickens and collaborators, with its famous ghost story 'The Signal-Man', centres around a fictional railway station and its branch lines. Dickens had been involved in the Staplehurst crash of 1865, where a train went off a bridge. He and his companions were not seriously injured at the time – his carriage remained precariously hanging from the structure – but he tended to the injured and dying in the riverbed below, later clambering back into the dangling carriage to retrieve his manuscript. He was so shaken that the serial part he was drafting for *Our Mutual Friend* ended short, and he never fully recovered from the shock.

Other novelists whose work mentions the railways include Elizabeth Gaskell (*Cranford* (1851–53) and *North and South* (1854) both include railway-linked fatalities, and *Cousin Phillis* (1864) involves a young man working on a new branch line); Mary Elizabeth Braddon (*Lady Audley's Secret* (1862) features a protagonist who researches a mysterious disappearance with the aid of a *Bradshaw* railway guide and much train travel); George Eliot (*Middlemarch* (1872) discusses the coming of the railway to an English village); Bram Stoker (*Dracula* (1897) demonstrates the vampire's use of railways to infiltrate England); and Rudyard Kipling (*Kim* (1900–01) notes the importance of trains in opening up British India). The railways play a role in many of Hardy's novels: a character in *Desperate Remedies* (1871) experiences an epiphany as he witnesses a train rushing through a cutting; *A Pair of Blue Eyes* (1873) features a recognition scene facilitated by the multiple-carriage arrangement of a train; while two of the main characters in *A Laodicean* are nearly killed by a train as they cross the rails. In Edith Nesbit's *The Railway Children* (first serialized 1905), the railway represents the mobility and adventure that it brought to the working and middle classes. Many British novelists, such as Anthony Trollope and Arthur Conan Doyle, feature railway travel as an incidental event; and non-British novels such as Leo Tolstoy's *Anna Karenina* (1878) and Émile Zola's *La Bête humaine* (1890) also focus on the railway.

Novels were not the only texts to consider steam power and the railways. Samuel Smiles's *The Life of George Stephenson* (1857) met with immediate success. Smiles followed this with *Self-Help* and a series on the 'Lives of the Engineers'. Similar works followed in the train of Smiles's success, lionizing 'inventors' as great men (see MacLeod 258). Poets also wrote about the railways. Most famously, Wordsworth fought the expansion of the railway in his beloved Lake District with his sonnet 'On the Projected Kendal and Windermere Railway' (1844). Tennyson was a passenger aboard the ill-fated Liverpool and Manchester run that killed Huskisson. However, he mistakenly thought that the train ran not on rails but in grooves. Thus, in 'Locksley Hall' (1835; published 1842) he wrote what became a famous metaphor for change, based on a misunderstanding of this technology: 'Let the great world spin forever down the ringing grooves of change' (l. 182). Alexander Anderson or 'Surfaceman', a Scottish railwayman, wrote poems including *Songs of the Rail* (1878).

The railways also affected the mode of publication of literature. Railway passengers provided a new reading public: the train moved more smoothly and allowed reading more easily than did a carriage. W. H. Smith opened bookstalls in railway stations, and publishers offered inexpensive railway editions, including the notorious 'yellowbacks' sold for two shillings. Yellowbacks were cheap books, with bright colours and a lively cover illustration, but they could contain either light reading or more serious fare, including topics related to science and technology. For example, H. S. Gibbs's *Autobiography of a Manchester Cotton Manufacturer* (1887) and the Rev. J. G. Wood's *Common Objects of the Microscope* (1890) were published as yellowbacks. Linda Hughes and Michael Lund argue that serialization developed as the railway did (6); indeed, periodical publication was called 'the *railway* style' by a critic in 1857 (quoted in Davis 213).

Owing to the great effect of the railways on Victorian life and literature, scholars have explored this technology extensively. Many begin with Wolfgang Schivelbusch's analysis of the changes that the railway brought to nineteenth-century culture (1986). Stephen Kern places these issues in a larger context (1983). Michael Freeman (1999) expands his analysis to popular culture and games with compelling visual examples, and Ian Carter surveys literary texts addressing the changes of the railway (2001). In other literary studies, Nicholas Daly (2004) examines 'the modernization of the senses' (34) in railways and the cinema, while Lynne Kirby focuses on how both the railway and the cinema destabilized the experience of spectatorship (1997). Jonathan Grossman shows how the railways followed coaching networks that also freed up, while regulating, British travel (2012). Grossman also considers how Dickens's serialized novels created a reading audience analogous to the collective travelling public; and Alison Byerly examines virtual travel and railway guides (2013). While Jill Matus explores railway trauma in the context of Victorian psychological theory (2009), Ben Marsden and Crosbie Smith (2005) put railways, steamers and telegraphs within an imperial context.

If steam replaced horses in powering vehicles on land, it replaced wind in powering vehicles on oceans and rivers, although early ships combined steam with sail power, partly to save on coal. Brunel metaphorically extended the Great Western Railway across the Atlantic to New York by connecting steam-powered land and sea transport. The replacement of the paddle wheel with the screw propeller made steamships more fuel-efficient and enabled long-distance cargo hauling; the first purpose-built transatlantic steamer, Brunel's *Great Western*, crossed the Atlantic in only two weeks in 1838. At the same time, naval ships were becoming 'ironclads' to resist attack; Brunel's immense *Great Britain* was the first propeller-driven ocean-going ironclad in 1843. Eventually, in 1906, the dreadnought combined steam power, heavy armouring and a battalion of heavy firepower. Steamships became instrumental in transatlantic passenger travel (most steamers could not store enough coal for longer voyages) and helped drive the Victorian economy; Joseph Conrad's *Heart of Darkness* (1899) sets his iconic critique of imperialism aboard a river steamer in the Belgian Congo.

Industry

Steam was also used to power industry. The steam-powered factory loom had been patented in 1785 by Edmund Cartwright; by 1850 there were nearly 250,000 power looms in Britain (Timmins 232 n. 31). The resulting economic difficulties for handloom weavers appear in Benjamin Disraeli's *Sybil; Or the Two Nations* (1845) and Eliot's *Silas Marner* (1861). Disraeli's *Sybil*, inspired by Thomas Carlyle, helped spark the genre known as 'condition of England' novels, which depict industrialization, the difficulties of life for the factory workers and, often, the concern that the labourers might take collective action. For example, Dickens's *Hard Times* (1854) and Gaskell's *Mary Barton* (1848) and *North and South* (1854) address the debates over factory and mine safety.

New industrial technologies included precision milling machines and the Bessemer process for purifying steel by oxidation, together with innovations in textiles. John Fisher patented a sewing machine in 1844 that sped up factory production of clothing and revolutionized homemade stitching. Aniline dyes – mass-produced artificial dyes derived from coal tar – changed textile and fashion history, introducing the bright, colour-fast fabrics of the modern era. William Henry Perkin discovered mauveine (a brilliant purple) in 1856 just as purple became a fashionable hue. Fuchsine (magenta) was derived in 1858, followed by aniline reds, blues and other colours named after their industrial birthplace: 'Manchester yellow', 'Manchester brown' (Kargon 198). But the speed of steam-driven machines and the chemicals associated with industrial processes spurred debates over occupational safety. In *North and South*, Gaskell chronicles the desire of some workers to keep the floating cotton 'fluff' in the factory air, because it filled their empty bellies; she suggests that this contributes to Bessy's fatal lung disease; she notes the controversial proposal to filter the factory chimneys' smoke; and she records desperate John Boucher's determined suicide in a shallow stream, his face discoloured from factory runoff.

Many scholars have examined the role of the nineteenth-century factory system on literature, but most do not look at specific technologies. Herbert Sussman offers a succinct introduction to the factory and other nineteenth-century contexts (1999), and Joseph Bizup compares pro- and anti-industrial views (2003). Anson Rabinbach (1992) explains how thermodynamics[2] and engine culture changed how people thought about the capabilities of the human body. Scholars who have examined steam technology and literature include Tamara Ketabgian and Barri J. Gold. Ketabgian (2011B) rejects the notion of a distinct human/mechanical divide in the Victorian era; rather, she argues, the steam engine was an important example of the 'living machine' in this period. She also provides a history of power machinery and its relation to literature (Ketabgian 2011A). Gold focuses on energy gain and loss, reading, for example, Dickens's *Bleak House* as an engine (2010). Other forms of power remained viable. Water power, which Wordsworth had critiqued in Book 8 of *The Excursion* (1814) for its effects on labour and the environment, also drives Eliot's *Mill on the Floss* (1860) (as Ketabgian's *Lives* reminds us). Steam was used, too, to power large electric dynamos (power generators).

Steam also powered printing presses; improvements appeared at a pace that must have seemed itself steam-powered. Friedrich Koenig's steam-powered printing press, with a speed of 400 sheets per hour, made news when the London *Times* purchased it in 1814. The American Richard Hoe developed the cylinder press (1832, up to 4,000 sheets per hour) and the rotary press (1843, up to 8,000 sheets per hour); and William Bullock, the perfecting continuous roll (web) press (1865, up to 12,000 sheets per hour). Steam presses were most useful initially in the newspaper business, with its rapid-fire editions and large circulation, but periodicals and inexpensive book series began adopting them as well. Aileen Fyfe (2012), focusing on the publisher William Chambers, analyses the effects of both steam-powered printing and steam-powered transportation (railways and steamers) on the Victorian book and periodical trade.

Light

In Elizabeth Gaskell's *Cranford* (1851–53), the ageing sisters thriftily conserve light, huddling by the window to delay lighting their candles as long as possible and using dim tallow candles rather than good wax candles unless they have guests. Vignettes like this illustrate how revolutionary it must have been for the Victorians to have good lighting. Oil and kerosene lanterns and lamps were the staple of households, workrooms and streets, but the nineteenth century saw three major new lighting technologies that materially affected Victorians' lives indoors and out: gaslight, oxyhydrogen light (or limelight) and electric arc light.

When the Gas Light and Coke Company of London set up a gas distribution network in London in 1812, gaslight's wider availability enabled its much wider use. Gaslight changed the face of the nineteenth-century city, as Lydgate, in Eliot's *Middlemarch*, suggests when he muses on how Bichat's medical theories 'acted necessarily on medical questions as the turning of gas-light would act on a dim, oil-lit street' (141). Gaslight was prized for its brilliant lighting of public spaces and for its warm, adjustable glow in private ones, but it fell out of favour in domestic settings due to concerns over heat, ventilation, inflammability and dingy residue on furnishings and artwork (Schivelbusch 1988 33–47, 50–52; see also Milan).

Another new form of lighting, less recognized today, was oxyhydrogen light, more familiarly known as Drummond light or limelight. Produced by directing ignited streams of (usually) oxygen and hydrogen on a ball of lime, this intense light was developed by Goldsworthy Gurney around 1820 and improved by Thomas Drummond in 1826. It was thought to be useful for lighthouses, theatres and magic lanterns. Its glare, heat and hazard eventually proved it unsuitable for domestic or scientific use. Lydgate, whom Eliot places in 1830, fantasizes about how his own discovery will illuminate medicine even further with a light 'as of oxyhydrogen' (1994 141) – but he would not have known of the failings of limelight quite yet.

Research in electricity by Humphry Davy and Michael Faraday sparked late-century innovations.[3] Electric arc lights required large currents and far outshone gaslight in illuminating the urban streets, although they were much too bright for homes. Thomas Edison's lab developed a successful carbon-filament electric incandescent bulb in 1879, providing a form of electric illumination suitable for interior use. His first electrical network, using high-resistance lamps wired in parallel, powered by direct current provided by large dynamos run on steam engines, took an additional three years to set up (Hunt 125–6).

Lighting technologies shape characters in novels as diverse as Dickens's *Dombey and Son*, where the wealthy Dombey redecorates his sombre drawing rooms with garish gaslight, and Richard Marsh's *The Beetle* (1897), where the alternation of dazzling light and darkness prevents the protagonist (and the reader) from perceiving the nature of the mysterious beetle. Zola's *Au Bonheur des Dames* (1883) profiled a fictional Parisian department store, with gaslight blazing through its plate glass windows. Yet considering how important lighting was to nineteenth-century activities, and how much it changed through the century, surprisingly few scholars have examined the technology of light and literature. Schivelbusch (1988), whose work on the railway has been so influential, also studies changes in lighting. He emphasizes the overlap between old and new technologies: each newly 'pure' or 'brilliant' light gets remapped as 'dirty' or 'dim' as newer ones push it aside. Brian Bowers (1998) also provides a fascinating and thorough history of artificial lighting, while Leslie Tomory (2012) focuses on the period when the gas industry built its networks. Chris Otter (2008) places gas and electric light in the context of modern liberalism and considers how these changed the urban experience; he resists a standard Foucauldian approach to propose a nuanced reading of the new experiences available in the lighted metropolis. However, Graeme Gooday (2008), studying public resistance towards domestic use of electricity in the late Victorian period, reminds us that new technologies are not necessarily embraced.

Communication

Electricity also powered new forms of communication, in particular the telegraph and telephone. The earliest technology termed 'telegraph' did not use electricity. Instead, the French optical telegraphs designed by Claude Chappe in 1793 were semaphores: towers with manually operated arms installed at set distances. In 1816 the Englishman Francis Ronalds demonstrated an electrostatic (pith ball) telegraph, but his invention met with little interest (Hunt 80).

Following inventors in France, Germany and America, in 1837 William Fothergill Cooke and Charles Wheatstone collaborated on an electric telegraph for use on the London and Birmingham Railway. The railway companies were important early adopters of telegraph technology; fast, reliable signalling could prevent accidents. In 1845 the Electric Telegraph Company bought the rights to Cooke and Wheatstone's telegraph, and banking, mercantile, shipping and other business interests began using the technology more widely. By 1850, British railways were lined by more than 2,000 miles of telegraph lines (Hunt 84); the system was nationalized in 1868 as part of the postal utility. The transmission of a message eventually involved printed telegraph forms and, in London and other cities, a pneumatic tube system in which the print messages were shot in capsules from a central to a branch office or even within the same building.

Electric telegrams rapidly became a coded technology. Cooke and Wheatstone's early telegraphs were designed with five wires moving five needles to point to letters on a diamond-shaped indicator panel painted with a grid of letters. Eventually, they converted to a code requiring only two needles. Between 1837 and 1840 in America, Samuel Morse and Alfred Vail were developing the Morse telegraph with its tapping-style 'Morse key' and accompanying 'Morse code' (Hunt 82–3). Experienced operators could identify the hand of a long-term acquaintance across the wires by hearing the quality of his or her taps (in the Morse telegraph) or watching his or her jerks (on the two-needle telegraph). Coding, inspired by a naval system, was also used for savings and secrecy; senders abbreviated place names and frequently used words and phrases by substituting code 'words' of nonsense letters.

Victorians considered the telegraph a world-changing technology; contemporary writers forecast a telegraph network connecting the far-flung corners of the empire. The first successful cross-channel telegram was sent in 1851 and, with the assistance of the steamer the *Great Eastern*, transatlantic cables were successfully laid in 1866. By 1880, nearly 90,000 miles of submarine cable had been installed, and double that by the turn of the century (Hunt 92). In 1898, Guglielmo Marconi transmitted wireless signals using 'Hertzian waves' across the English Channel and in 1901 from Cornwall to Newfoundland. The technology of radiotelegraphy eventually developed into the transmission of speech and radio broadcasting.

In contrast to the clear and lively historical, literary and scholarly record in telegraphy, telephony presents a foggier case. The issue of priority is vexed: both Elisha Gray and Alexander

Graham Bell (with Thomas Watson) applied to the patent office regarding telephone technology in 1876, following work by Innocenzo Manzetti, Antonio Meucci and Philipp Reis. Furthermore, much less literary or scholarly attention centres on telephony. Steven Connor (2001) investigates the mystique of the telephone in relation to ventriloquism; Picker (2003) offers a reading of 'telephonic discourse' in Eliot's *Daniel Deronda*, and Galvan (2010) considers the role of female telephone operators in comparison to telegraphy. Menke argues that the telephone, though underrepresented in Victorian literature, symbolizes an emergent media system in 'The Finest Story in the World' by Kipling, *The Story of a Modern Woman* (1894) by Ella Hepworth Dixon and *A Writer of Books* (1898) by George Paston (Emily Morse Symonds).

The literary texts that engage most overtly with electrical technology tend to be 'scientific romance', from Mary Shelley's *Frankenstein* (1818) to Marie Corelli's *Romance of Two Worlds* (1886). Corelli used her novel to promote her theory blending electrical, spiritualist and Christian concepts. George Griffith's 1898 short story 'A Corner in Lightning' considers the outcome when an enterprising capitalist attempts to monopolize the world's electrical forces. Telegraphs feature in a range of poems and novels. Anthony Trollope's 'The Telegraph Girl' (published 1877 with his essay 'The Young Women at the Telegraph Office') depicts telegraphy as a respectable employment for women. Henry James's 'In the Cage' (1898) focuses on a telegraph operator's curious readings of the messages she transmits from a fashionable neighbourhood of London. James interrogates privacy in modern life, the structures of communication and the delicate piecework of the artist.

Electricity was understood in the context of other difficult-to-discern phenomena – ether, light, magnetism and mesmeric 'vital fluid' – that the Victorians were attempting to disentangle. While James Clerk Maxwell's 1864 essay 'A Dynamical Theory of the Electromagnetic Field' analysed the relation between light and the electric and magnetic forces, Rudyard Kipling wrote a 1902 short story, 'Wireless', which suggests that wireless technology may allow operators to tap into spiritualist currents as well.

Thomas Hardy's novel *A Laodicean* (1881) features a New Woman protagonist who uses telegraph technology to link her medieval-era castle with the nearby town; the technologies of the telegraph, railway and photograph enable the twists of Hardy's plot in this example of what he called 'Novels of Ingenuity'. Telegrams appear at crucial moments in the plots of other novels, such as that of Margaret Oliphant's *Hester* (1844). While telegraphs eventually changed the reporting of the news, they were also featured in many periodical and newspaper articles. Dickens's article 'Wings of Wire' (1850) explained the two-needle arrangement and told 'telegraphic anecdotes and little romances' including a much-repeated one about a murderer whose execution is first stayed, then re-commanded by telegraph (244). Ella Cheever Thayer, an American telegraph operator, wrote an 1880 bestseller titled *Wired Love: A Romance of Dots and Dashes*. And the physicist Maxwell wrote 'Valentine by a Telegraph Clerk [Male] to a Telegraph Clerk [Female]', a clever comic verse that was widely republished.

Nineteenth-century communication technologies and the networks they created, given their relevance to the history and theory of texts, print culture and reading, have inspired a rich array of scholarly and critical works. Christopher Keep (2002) discusses telegraphy and other communications and transportation technologies (including the railway, steamship, typewriter and gramophone). Clare Pettit (2004) argues that literary and mechanical invention are inextricably entwined during the development of copyright and patent protections in the nineteenth century. She considers telegraphy and electricity as well as the Great Exhibition and other Victorian technologies. Laura Rotunno (2013) includes a chapter examining how telegraphy changed the way correspondence was read. Some scholars who work on telegraphy focus on the analogy made – by writers like Alexander Bain – between the telegraph and the human nervous system (see, for example, Rudy 82–3). Jay Clayton (2003) points out how inextricably physical was the performance of telegraphy (50–52). Alison Winter (1998) explores the notion of the nerve as electrical conduit, among other contemporary explanations for spiritualism. Pamela Thurschwell (2005) was among the first scholars to examine the links between spiritualism and technology at the turn of the century. Jill Galvan (2010) shows how both spiritualist and modern communications technologies relied on a gendered intermediary; Enns (2006) focuses on the American context. Other scholars focus on the system as a network, on its role as communications technology or on its opportunities for women during a time of changing gender roles. Laura Otis (2001) influentially argues that the network of the railway and especially the telegraph served as a metaphor for the nervous system. Richard Menke (2008) focuses on the transmission of writing (and then aural text) through networks, whether postal, telegraphic or wireless, and the relations between these networks and the structures of realist prose. Jason Rudy (2009) argues that the telegraph provided an important model for how poetry worked in readers – how it might 'electrically' transmit rhythm, meaning and affect in a culture sometimes hostile to these.

Sanitation

The steam engine, railway and telegraph were perhaps the most celebrated technologies of the nineteenth century, but the technologies of public health infrastructure shaped everyday life in equally important if less visible ways. These rose to the surface of public and literary consciousness during the reform years of the 1840s, when the statistical insights available through the 'numerical method' of Pierre Louis sparked growing alarm over the living – and dying – conditions of the Victorian urban populace. Certain influential texts used statistics to suggest improvements to public sanitation. Dr James Phillips Kay (later Sir James Kay-Shuttleworth) wrote *The Moral and Physical Condition of the Working Classes Employed in the Cotton Manufacture in Manchester* in response to the cholera epidemic of 1832, and Edwin Chadwick published his *Report on the Sanitary Condition of the Labouring Population of Great Britain* in 1842. The prevalent theory of disease transmission at the time was that it was miasmatic (due to 'bad air'), but the horrific living conditions of the poor, once recognized, convinced many to work for a cleaner urban environment. In 1848 Parliament passed the Public Health Act, supporting paved streets, clean water and sewerage ('drains'), and mandating adequate light, space and ventilation in housing. During an 1853–54 epidemic, Dr John Snow showed, by mapping cholera cases, the link between disease and a water source; when he removed the pump handle at the tainted well, he proved his point, but the miasmatic theory held firm. During the summer of 1858, 'The Great Stink' of the polluted Thames helped convince Parliament to fund Joseph Bazalgette's engineering masterpiece: a redesign of the London sewerage system, narrowing the Thames by building the Embankments, which created a scouring effect on this tidal river and created space for the sewer pipes below. In the 1866 cholera epidemic, neighbourhoods with clean water were largely spared, rebutting the miasmists and proving the contagionist theory of disease transmission.

The polluted waters of the pre-reform era flow under and through crucial scenes in Dickens's *Oliver Twist* (1837–39) and *Our Mutual Friend* (1864–65), and in other novels, most famously perhaps in Victor Hugo's *Les Miserables* (1862). The Victorian interest in sanitation also powers critical work by Mary Poovey (1995), Pamela Gilbert (2004) and others (see also Ellen Handy (1995) and the collection *Filth*, edited by William Cohen and Ryan Johnson (2004)). Jules Law (2010) examines related questions of purity and leakage (2010), while Eileen Cleere (2014) demonstrates how Victorian theories of sanitation undergird the aesthetic work of the period.

Observing

New technologies underwrote tremendous changes in how people perceived the world and recorded their perceptions. The telescope and microscope, developed centuries earlier, became considerably more accurate and reliable, shaping theories of literary realism such as Eliot's. The Röntgen ray allowed people to look inside the body. These new kinds of vision were recorded with unprecedented specificity, as time became standardized and machine-made watches made precise timekeeping easier.

Under George Biddell Airy as Astronomer Royal beginning in 1835, the Royal Observatory in Greenwich emphasized accurate instrumentation. The Airy Transit Circle of 1850 helped ensure more precision and fewer human errors as observers constructed the star charts that were crucial to British navigation. Airy also introduced daily meteorological readings using the thermometer, barometer and other instruments. The reflector (mirror-based) telescopes used by astronomers were large; William Parsons, the Third Earl of Rosse, completed one in 1845 nicknamed the 'Leviathan of Parsonstown'. Using its six-foot mirror, Rosse became the first man to view the spiral structure of a nebula.

In the field of microscopy, while eighteenth-century craftsmen had reduced the optical distortion called chromatic aberration (a blur of rainbow colours edging a magnified object) by forming lenses that combined two types of glass, Joseph Jackson Lister solved the problem of spherical aberration (a distortion of the shape of the object itself) in 1830 by constructing 'compound' microscopes – i.e. with several weak lenses rather than just using a single very strong lens. His advances made it possible for instrument makers to provide microscopes that were not only inexpensive but also powerful and precise enough to be useful for research. Microscopists denoted certain difficult-to-discern natural objects as 'test objects' and Friedrich Nobert, a German, started engraving closely lined test plates in 1845. An observer would use these to rank and standardize the settings through an individual instrument. Microscopy became increasingly important in science and medicine; it even offered popular entertainment, both in the home and projected in a mass culture setting. While the telescope and microscope changed the scale of vision and promised insights into the structure of matter, in 1895 William Röntgen discovered a new way to view another structure: the Röntgen ray – or X-ray – machine showed the scaffolding of the bones underlying the skin and muscle of the human frame.

As these technologies were expanding the range and kinds of human perception, other technologies allowed observers to make more careful recordings of what they had seen. Lorraine Daston and Peter Galison (2010) have argued that one of the hallmarks of the nineteenth-century concept of objectivity is not only to observe dispassionately but to record that observation accurately, precisely and completely. New tools, standards and processes accompanied such a task. Factory clocks had made British workers more conscious of the exact time of starting and stopping work, and the advent of the railways made the British more precise about timekeeping in general. Times in towns around England could vary by up to twenty minutes from London time; stations sometimes displayed two clocks or a clock with two minute hands to mark the differences in time. The Great Western Railway first instituted 'railway time' in England in November 1840 to standardize railway clocks for safety and efficiency. The telegraph could send a 'time signal' with Greenwich Mean Time along the length of the line; by 1855 most towns had adapted to the new system. However, GMT was not the legal standard time for all of Britain until 1880. The specificity of 'railway time' colours the station scenes of Hardy's *A Pair of Blue Eyes* and other novels.

Timekeeping mechanisms became more precise, also. The Royal Astronomer Airy worked with two important new turret clocks, at the Royal Exchange and Parliament (Big Ben), ensuring that they met surprisingly rigorous standards for such large public clocks – keeping accurate time to within one second (Bennett 280–81). Watchmakers adopted the lever escapement, a mechanism allowing more accurate timekeeping and a lighter pocket watch. Gradually British and many European watchmakers followed the Americans in mass-manufacturing inexpensive, reliable watches. British Army officers wore wristwatches during the Boer War to synchronize attacks, which helped make this style of watch popular. In 1802, the deist William Paley had used the analogy of a watch in his book *Natural Theology* in order to argue that the complex natural world was clearly God's creation, like a watch – then a luxury item – left running on a stone. A century later, cheap watches were mass-manufactured so that every man could carry time on his wrist.

In an attempt to mechanize the tedious work of creating and checking the numerical tables of calculations that scientists, engineers, insurers and others relied on, Charles Babbage designed a machine – a 'Difference Engine' – that would calculate polynomials automatically using a precision-milled metal apparatus. He secured funding from the government but the project stalled in 1833 before construction was finished. He also planned a programmable Analytical Engine (1834) and a second Difference Engine (1847–49). Babbage never saw his designs come to fruition, but successful working models of his engines have proved his methods correct. The Analytical Engine inspired Ada Lovelace, the poet Byron's daughter, to write about the possibilities of a programmable device beyond simple mathematics. As a polymath, Babbage produced a diverse body of work. His calculating engines had been forgotten by the early days of twentieth-century computing, but his work left its mark on Victorian literary history. Clayton (2003) argues that Babbage's inventiveness made him a model for the character of Daniel Doyce in *Little Dorrit* (96); the British Association for the Advancement of Science, which Babbage was instrumental in founding, inspired Dickens's satirical 'Mudfog Papers'. Douglas-Fairhurst (2002) points to how Babbage's work shaped Dickens's moral platitudes about influence and self-improvement (97–8). Lee (2012) traces the influence of Babbage's probability theory in Edgar Allan Poe's work (22).

Technologies of perception and measurement were frequent subjects in periodicals and newspapers. Questions about perception and the technologies of vision appear too in prominent novels of the period. Hardy's *Two on a Tower* (1882) uses astronomy as a vehicle to develop the preoccupation with an indifferent fate that we associate with his more typically naturalist novels. Eliot's *Middlemarch* uses a sustained trope of microscopy to consider the habits and limits of human perception. Many scholars have examined the Victorians' interest in visuality. Among those who consider visual technologies specifically, Isobel Armstrong's work (2008) stands out for her specific yet wide-ranging focus on glass manufacturing and glass technologies, which reach from plate glass to the microscope and optical toys. Jutta Schickore (2007) offers a history of late eighteenth and early nineteenth-century microscopists and their interest in error. Anna Henchman (2014) examines how the telescope shaped Victorians' theory of mind and use of literary perspective. Martin Willis (2011) explores the relation between telescopic or microscopic vision and the spiritualist visions and Gothic narratives that also animated *fin de siècle* culture.

Recording

The new recording devices of the nineteenth century can also claim an influence on modern culture. The sphygmograph and other medical recording machines allowed physicians to trace the changes of the body, while the photograph, the phonograph, the moving picture and the typewriter revolutionized how people could preserve and re-experience the events of everyday life. Although nineteenth-century physicians could not do much to cure most diseases, their ability to observe the body and diagnose disease improved significantly. Advances in cardiac medicine led the way, beginning with Rene Theophile Hyacinthe Laënnec's introduction of the stethoscope for mediate auscultation in 1816. After Julius Hèrisson invented the sphygmo-manometer in 1831, Karl von Vierordt invented a graphing version of the device in 1854 and Etienne-Jules Marey made a smaller, more practical sphygmograph in 1863; doctors could track and record the movements of the heart without opening up the body or inserting a tube into an artery as had been necessary before. Although British physicians resisted these technologies at first, some, such as John Elliotson with the stethoscope or John Burdon Sanderson with the sphygmograph, advocated for their use.

Photography was another technology that interested both scientists and the culture at large. There had been earlier attempts to use the sun to 'paint' images onto paper, but Nicéphore Niépce was the first who successfully developed heliography. He used a light-sensitive solution to fix an image on a pewter plate in 1826. By 1837, Louis Daguerre, who had worked with Niépce, used silver iodide to create the daguerreotype, which required much less exposure time. William Henry Fox Talbot worked out a method of 'photogenic drawing' using silver chloride, with a technique to fix the images by 1839. Talbot's method was published before the French government released details of Daguerre's technique, but daguerreotypes became popular because they rendered images with clear detail, even in miniature. In 1840, Talbot discovered the calotype, which used a photographic negative. This allowed him to print multiple copies quickly on paper instead of plate, but the images were not as clear. The sculptor Frederick Scott Archer developed the wet collodion process in 1851, which offered speedier development time and multiple detailed prints from a glass negative. About this time, André-Adolphe-Eugène Disdéri used the albumen paper developed in 1850 by Louis-Désiré Blanquart-Evrard to create small portrait photographs called *cartes de visite*, which became very popular. Richard Leach Maddox contributed to the development in 1878 of dry plates, which allowed cameras to become faster, smaller and lighter. When George Eastman introduced the Kodak camera in 1888 and celluloid film in 1889, photography became even more portable and affordable, opening up this avocation to many more.

The Victorians were not naïve about photography but considered it an art form to be manipulated. Julia Margaret Cameron produced photographs in the Pre-Raphaelite style to illustrate an 1874–75 edition of Tennyson's *Idylls of the King*; and the plot of Hardy's *A Laodicean* turns on a case of photographic fraud. Viewing photographs provided education and entertainment as well. In 1838, Wheatstone invented a mirror-based stereoscope, in which a viewer's eyes see separate images that the brain combines to perceive an apparently three-dimensional scene. Sir David Brewster proposed a lighter stereoscope in 1849 and Oliver Wendell Holmes an even more portable model in 1861. After display at the 1851 Great Exhibition, stereoscopes became a popular way for Victorians to experience travel and other photographs.

If the stereoscope produced the illusion of depth, other photographers were tinkering with the illusion of motion. In America, Eadweard Muybridge was performing motion studies using multiple cameras triggered one after another to produce sequential photographs of a moving animal, each image showing a successive motion of the limbs. In 1879, he created the optical illusion of motion by projecting the photographs in order rapidly, using a zoopraxiscope, which draws upon the same principles as various optical toys: Joseph Plateau's 1832 phenakistoscope, William George Horner's 1834 zoetrope, and Emile Reynaud's 1877 praxinoscope. In 1882, Marey (still as fascinated with the problem of recording motion as he had been in the 1860s with the sphygmograph) developed a 'chronophotographic gun' to take rapid sequential images on the same plate, twelve per second.

The 1890s brought rapid developments in moving picture technology. In 1891, Thomas Edison's associate W. K. L. Dickson invented a kinetograph (camera to create ribbons of sequential photographs on film) and kinetoscope (peepshow box to view the 'moving' images); and in 1894 Herman Casler patented the mutoscope, a peepshow using a flipbook design. However, projected images became more popular. In 1892 Reynaud offered the first public exhibition of projected moving pictures using a praxinoscope. Louis and Auguste Lumière invented the Cinématographe in 1894, which set the 16mm standard, and showed their first film publicly in 1895. In 1896 Dickson patented the American Biograph projection method and Edison started marketing the vitascope, another projector. England's first public projector was the theatrograph (animatograph), presented by Robert W. Paul in London in 1896.

Edison conceived of his cylinder phonograph (1877) and graphophone (1887) as dictating machines, but by 1897 the general public could purchase musical recordings. This divide between business and leisure use is suggested in Bram Stoker's 1897 novel *Dracula*, where Dr Seward takes notes on both his own phonograph and Lucy's. Edison's machines used tinfoil or wax cylinders; Emile Berliner's gramophone (1887) recorded on hard rubber discs and could be mass-produced. The Victrola, a playback-only gramophone, made phonography available to mass culture in 1906.

Writing by hand also changed during the nineteenth century. Cheap Birmingham nibs were available but they leaked and blotted; America led in both pen and typewriter technology. Duncan MacKinnon and Alonzo Cross patented stylographic pens – with wire tips – in 1875 and 1877; Lewis Waterman patented a fountain pen with a reliable draw of ink in 1884; and George Parker another in 1889. Surprisingly, the first mass-produced typewriter appeared before writers had reliable, smooth-writing pens. Remington typewriters, built to Christopher Sholes's design, were produced in 1874 in New York, featured the QWERTY keyboard and sat on a treadle stand like the sewing machines Remington also manufactured. Franz Wagner developed Underwood typewriters with a visible typing field and lowercase letters. Typing was at first considered too impersonal for business deals, but within two decades many offices hired 'type-writer girls' instead of male clerks. Grant Allen's *The Type-Writer Girl* (published in 1897 under a pseudonym) critiques the office environment for treating women as objects and machines. He is more pessimistic than Stoker in *Dracula*, where Mina's transcription work enables victory over the vampire. George Gissing also portrays a school for typewriter girls in *The Odd Women* (1893).

Many scholars study the culture of recording and communication devices. Friedrich Kittler's poststructuralist analyses are especially influential. Kittler (1985) compares the media environment in the text-centric Romantic period to 1900, when new media offered more ways to record and store information, and writing began to lose its hegemonic status. Influenced by Foucault and Lacan, Kittler argues that our inscription systems determine our experiences by establishing what data is recorded and stored, and how it is recorded and stored. Kittler (1986) further explores this argument about how specialized functions developed for written, audio and visual media (both still photography and cinematography). He maps these onto new technologies and onto Lacan's tripartite registers of the psyche.

Thomas Hankins and Robert Silverman (1995) study the sphygmograph, stereoscope, 'speaking machines' and other less well known technologies. Lisa Cartwright (1995) argues that the history of cinema is imbricated with the X-ray and other technologies for surveilling the body. Willis (2011) discusses ophthalmology's relation to spiritualism. Lisa Gitelman (1999) studies the relation between recording and transcription technologies in late nineteenth century American culture, arguing that this period is wrongly scanted in histories of digital culture. She examines various text-related technologies – among them the typewriter – in relation to Edison's phonograph, the changing nature of textual practices, questions of identity and authorship, and the way technologies move between public and private space. The first two chapters in her later study (2006) focus on the history of recorded sound. Gitelman argues that examining the introduction of new media within its historical archive helps us understand the construction of 'new media' digital networks in our own culture.

Literary Secretaries/Secretarial Culture (2005), edited by Leah Price and Pamela Thurschwell, includes an analysis of typewriter culture; Galvan (2010) also discusses this. *Media, Technology, and Literature in the Nineteenth Century* (2011), edited by Colette Colligan and Margaret Linley, examines the typewriter, telegraph, Braille and composition photography. Both Ivan Kreilkamp (2009) and John Picker (2003) consider the phonograph as a device and as a figure for trends in the larger culture. From the earliest days of recorded sound, literature was a subject for preservation. Matthew Rubery is pioneering scholarship on audio-recorded books; see *Audiobooks, Literature, and Sound Studies* (2011).

Jonathan Crary (1992) has influenced many scholars working on the relation between literature and optical technologies. In contrast to the 'myth of modernist rupture' (Crary 4), which depends on a stable observer, he argues that a more important rupture in modern visuality occurred around 1820, before photography, when visual research suggested perception could be simply an effect in a shifting field of relations. Thus the camera obscura ceases to be a model for reliable human vision about this time (Crary 27). With scientific and popular interest in phenomena such as the retinal afterimage, 'an individual body . . . [becomes] at once a spectator, a subject of empirical research and observation, and an element of machine production' (Crary 112). Many recent scholars focus their analysis on the relation between photography and visual or literary realism. Nancy Armstrong (2000) argues for the role of photography in structuring literary realism, whereas Helen Groth (2003) focuses on nostalgia instead. Jennifer Tucker (2006) examines scientific photography in its role as evidence. Daniel Novak (2008) argues that both Victorian photography and realism fragment and manipulate images, and that audiences must reassemble a coherent narrative in both cases.

Conclusion

Just as Brewster designed the kaleidoscope in 1817 while studying the polarization of light, nineteenth-century technologies provided entertainment alongside enlightenment. Science 'shows' ranged from Michael Faraday's informed presentations at the Royal Institution to the massive crowds at the Adelaide Gallery.[4] Tennyson's poem *The Princess* (1847, revised through 1853) depicts an aristocratic party with such entertainments – a cannon fired by 'knobs and wires and vials' (Prologue, l. 50), a set of telescopes, an electric shock round-robin, a 'clock-work steamer' (l. 56) and a 'petty railway' and 'fire-balloon' (l. 59), and 'twenty posts of telegraph' (l. 62) with 'mimic stations' (l. 64) – so that 'sport / with Science hand in hand went' (ll. 64–5). Iwan Rhys Morus (1998) shows how very much intertwined electricity and spectacle were.

Technological display dominated the 1851 'Great Exhibition of the Works of Industry of All Nations' in Hyde Park. Even the building for this exhibition represented a triumph of Victorian technology. An immense structure designed by gardener Joseph Paxton, the 'Crystal Palace' used a modular greenhouse-like approach with a cast-iron grid supporting standardized plate glass. (The transformations brought about by plate glass are examined by Isobel Armstrong and by Andrew Miller (1995)). Paxton's innovative design allowed the 'palace' to be built swiftly and economically; the grid was light but strong and the glass let in daylight to illuminate the exhibits. In its first six months, six million visitors toured the Crystal Palace; many more 'viewed' it through media coverage in the *Illustrated London News* and other periodicals (MacLeod 212–13). With further international expositions, comparisons with other nations' handiwork accelerated anxieties over Britain's scientific and technological achievements. The century was only half-finished, but already the nation was boasting of its past and looking to its future with not a little trepidation.

Notes

1   The poem also praised Josiah Wedgwood (who experimented with pottery glazing), James Brindley (canal innovator) and Richard Arkwright (whose cotton-industry patents include the water frame).
2   See also Barri J. Gold, 'Thermodynamics', in this volume.
3   See Stella Pratt-Smith, 'Electricity', in this volume.
4   See Iwan Rhys Morus, 'Staging Science', in this volume.

Bibliography

Primary texts

Allen, Grant. *The Type-Writer Girl*. Ed. Clarissa Suranyi. Peterborough: Broadview, 2003.

Anderson, Alexander. *Songs of the Rail*. London: Simpkin, Marshall, 1878.

Braddon, Mary Elizabeth. *Lady Audley's Secret*. Ed. Lyn Pykett. Oxford: Oxford University Press, 2012.

Chadwick, Edwin. *Report on the Sanitary Condition of the Labouring Population of Great Britain*. Ed. M. W. Flinn. Edinburgh: Edinburgh University Press, 1965.

Conrad, Joseph. *Heart of Darkness*. Ed. Owen Knowles. New York, NY: Penguin, 2007.

Corelli, Marie. *Romance of Two Worlds*. New edn. London: R. Bentley and Son, 1887.

Darwin, Erasmus. *The Botanic Garden*. London: J. Johnson, 1791.

Dickens, Charles. 'Wings of Wire'. *Household Words* 37 (7 December 1850): 241–45.

Dickens, Charles. *Our Mutual Friend*. Ed. Adrian Poole. London: Penguin, 1998.

Dickens, Charles. *Dombey and Son*. Ed. Andrew Sanders. London: Penguin, 2002.

Dickens, Charles. *Bleak House*. Ed. Nicola Bradbury. London: Penguin, 2003.

Dickens, Charles. *Little Dorrit*. Ed. Helen Small. London: Penguin, 2008.

Dickens, Charles. *Oliver Twist*. Ed. Philip Horne. London: Penguin, 2003.

Dickens, Charles. *Mugby Junction*. Foreword Robert Macfarlane. London: Hesperus, 2005.

Dickens, Charles. *Hard Times*. Ed. Paul Schlicke. Oxford: Oxford University Press, 2008.

Disraeli, Benjamin. *Sybil; Or the Two Nations*. Ed. Sheila Smith. Oxford: Oxford University Press, 2009.

Dixon, Ella Hepworth. *The Story of a Modern Woman*. Ed. Steve Farmer. Peterborough: Broadview, 2004.

Eliot, George. *Middlemarch*. Ed. Rosemary Ashton. New York, NY: Penguin, 1994.

Eliot, George. *Mill on the Floss*. Ed. A. S. Byatt. London: Penguin, 2003.

Eliot, George. *Silas Marner: The Weaver of Raveloe*. Ed. Terence Cave. Oxford: Oxford University Press, 2009.

Gaskell, Elizabeth. *North and South*. Ed. Patricia Ingham. London: Penguin, 1996.

Gaskell, Elizabeth. *Mary Barton*. Ed. Macdonald Daly. London: Penguin, 1997.

Gaskell, Elizabeth. *Cousin Phillis and Other Stories*. Ed. Heather Glen. Oxford: Oxford University Press, 2010.

Gaskell, Elizabeth. *Cranford*. Ed. Elizabeth Porges Watson. Introd. and notes Dinah Birch. Oxford: Oxford University Press, 2011.

Gibbs, H. S. *Autobiography of a Manchester Cotton Manufacturer: Or, Thirty Years' Experience of Manchester*. Manchester: Heywood, 1887.

Gissing, George. *The Odd Women*. Ed. Patricia Ingham. Oxford: Oxford University Press, 2008.

Griffith, George. 'A Corner in Lightning'. *Pearson's Magazine* 5 (March 1898): 264–71.

Hardy, Thomas. *A Laodicean* [1881]. Ed. John Schad. London: Penguin, 1998.

Hardy, Thomas. *A Pair of Blue Eyes*. Ed. Pamela Dalziel. London: Penguin, 1998.

Hardy, Thomas. *Desperate Remedies*. Ed. Mary Rimmer. London: Penguin, 1998.

Hardy, Thomas. *Two on a Tower* [1882]. Ed. Sally Shuttleworth. London: Penguin, 2000.

Hugo, Victor. *Les Miserables* [1862]. Transl. Lee Fahnestock and Norman McAfee. Afterword Chris Bohjalian. New York, NY: Signet, 2013.

James, Henry. 'In the Cage' [1898]. *Complete Stories, 1892–1898*. Eds. John Hollander and David Bromwich. New York, NY: Library of America, 1996, 835–923.

Kay, James Phillips. *The Moral and Physical Condition of the Working Classes Employed in the Cotton Manufacture in Manchester*. London: James Ridgway, 1832.

Kipling, Rudyard. 'The Finest Story in the World'. *Many Inventions*. London: Macmillan, 1893, 106–50.

Kipling, Rudyard. *Kim*. London: Macmillan, 1901.

Kipling, Rudyard. 'Wireless'. *Traffics and Discoveries*. London: Macmillan, 1904, 211–39.

Marsh, Richard. *The Beetle. A Mystery*. London: Skeffington, 1897.

Maxwell, James Clerk. 'A Dynamical Theory of the Electromagnetic Field'. *Philosophical Transactions of the Royal Society of London* 155 (1865), 459–512. <www.jstor.org/stable/108892>. Accessed 28 January 2016.

Maxwell, James Clerk. 'Valentine by a Telegraph Clerk [Male] to a Telegraph Clerk [Female]'. *The Life of James Clerk Maxwell: With a Selection from His Correspondence*. Ed. William Garnett. London: Macmillan, 1882, 630–1.

Nesbit, E[dith]. *The Railway Children*. London: Wells Gardner, Darton, 1906.

Oliphant, Margaret. *Hester*. London: Macmillan, 1883.

Paley, William. *Natural Theology or Evidences of the Existence and Attributes of the Deity*. London: Printed for J. Faulder, 1809.

Paston, George [Emily Morse Symonds]. *A Writer of Books* [1898]. Chicago, IL: Academy Chicago, 1999.

Shelley, Mary. *Frankenstein, or the Modern Prometheus: The 1818 Text*. Ed. Marilyn Butler. Oxford: Oxford University Press, 2009.

Smiles, Samuel. *The Life of George Stephenson, Railway Engineer*. London: John Murray, 1857.

Smiles, Samuel. *Self-Help*. London: John Murray, 1859.

Stoker, Bram. *Dracula*. London: Archibald Constable, 1897.

Tennyson, Alfred. 'Locksley Hall' [1835]. *Poems*. London: Edward Moxon, 1842.

Tennyson, Alfred. *The Princess, A Medley*. 3rd edn. London: Edward Moxon, 1850.

Tennyson, Alfred, and Julia Margaret Cameron. *Illustrations by Julia Margaret Cameron of Alfred Tennyson's 'Idylls of the King' and Other Poems. Miniature Edition*. London: Privately printed, 1875. <www.bl.uk/collection-items/tennysons-idylls-of-the-king-photographically-illustrated-by-julia-margaret-cameron>. Accessed 22 January 2016.

Thayer, Ella Cheever. *Wired Love: A Romance of Dots and Dashes*. New York, NY: W. J. Johnson, 1880.

Tolstoy, Leo. *Anna Karenina*. Transl. Richard Pevear and Larissa Volokhonsky. New York, NY: Penguin, 2004.

Trollope, Anthony. 'The Young Women at the Telegraph Office'. *Good Words* 18 (June 1877): 377–84.

Trollope, Anthony. 'The Telegraph Girl'. *Good Cheer* 19 (December 1877 supplement to *Good Words*): 1–19.

Wood, Rev. J. G. *Common Objects of the Microscope*. London: Routledge, 1890.

Wordsworth, William. 'On the Projected Kendal and Windermere Railway'. *Complete Poetical Works of William Wordsworth*. Ed. John Morley. London: Macmillan, 1893. 785.

Zola, Émile. *Au Bonheur des Dames*. Paris: G. Charpentier, 1883.

Zola, Émile. *La Bête humaine*. Transl. Roger Pearson. Oxford: Oxford University Press, 2009.

Secondary texts

Armstrong, Isobel. *Victorian Glassworlds: Glass Culture and the Imagination, 1830–1880*. New York, NY: Oxford University Press, 2008.

Armstrong, Nancy. *Fiction in the Age of Photography: The Legacy of British Realism*. Cambridge, MA: Harvard University Press, 2000.

Barrow, Ian. *Making History, Drawing Territory: British Mapping in India, c. 1756–1905*. New Delhi: Oxford University Press, 2004.

Bennett, J[ames] A. 'George Biddell Airy and Horology'. *Annals of Science* 37 (1980): 269–85.

Bizup, Joseph. *Manufacturing Culture: Vindications of Early Victorian Industry*. Charlottesville, VA: University of Virginia Press, 2003.

Bowers, Brian. *Lengthening the Day: A History of Lighting Technology*. New York, NY: Oxford University Press, 1998.

Byerly, Alison. *Are We There Yet? Virtual Travel and Victorian Realism.* Ann Arbor, MI: University of Michigan Press, 2013.

Carter, Ian. *Railways and Culture in Britain: The Epitome of Modernity.* Manchester: Manchester University Press, 2001.

Cartwright, Lisa. *Screening the Body: Tracing Medicine's Visual Culture.* Minneapolis, MN: University of Minnesota Press, 1995.

Clayton, Jay. *Charles Dickens in Cyberspace: The Afterlife of the Nineteenth Century in Postmodern Culture.* New York, NY: Oxford University Press, 2003.

Cleere, Eileen. *The Sanitary Arts: Aesthetic Culture and the Victorian Cleanliness Campaign.* Columbus, OH: Ohio State University Press, 2014.

Cohen, William A., and Ryan Johnson, eds. *Filth: Dirt, Disgust, and Modern Life.* Minneapolis, MN: University of Minnesota Press, 2004.

Colligan, Colette, and Margaret Linley, eds. *Media, Technology, and Literature in the Nineteenth Century.* Burlington, VT: Ashgate, 2011.

Connor, Steven. *Dumbstruck: A Cultural History of Ventriloquism.* New York, NY: Oxford University Press, 2001.

Crary, Jonathan. *Techniques of the Observer: On Vision and Modernity in the Nineteenth Century.* Cambridge, MA: MIT Press, 1992.

Daly, Nicholas. *Literature, Technology, and Modernity, 1860–2000.* Cambridge: Cambridge University Press, 2004.

Daston, Lorraine J., and Peter Galison. *Objectivity.* New York, NY: Zone, 2010.

Davis, Philip. *The Oxford English Literary History: Volume 8: 1830–1880: The Victorians.* Oxford: Oxford University Press, 2002.

Douglas-Fairhurst, Robert. *Victorian Afterlives: The Shaping of Influence in Nineteenth-Century Literature.* Oxford: Oxford University Press, 2002.

Enns, Anthony. 'Mesmerism and the Electric Age: From Poe to Edison'. *Victorian Literary Mesmerism.* Eds. Martin Willis and Catherine Wynn. New York, NY: Rodopi, 2006, 61–82.

Evans, Francis T. 'Roads, Railways, and Canals: Technical Choices in 19th-Century Britain'. *Technology and the West: A Historical Anthology from Technology and Culture.* Eds. Terry S. Reynolds and Stephen H. Cutcliffe. Chicago, IL: University of Chicago Press, 1997, 199–232.

Freeman, Michael. *Railways and the Victorian Imagination.* New Haven, CT: Yale University Press, 1999.

Fyfe, Aileen. *Steam-Powered Knowledge: William Chambers and the Business of Publishing, 1820–1860.* Chicago, IL: University of Chicago Press, 2012.

Galvan, Jill. *The Sympathetic Medium: Feminine Channeling, the Occult, and Communication Technologies, 1859–1919.* Ithaca, NY: Cornell University Press, 2010.

Gilbert, Pamela. *Mapping the Victorian Social Body.* Albany, NY: State University of New York Press, 2004.

Gitelman, Lisa. *Scripts, Grooves, and Writing Machines: Representing Technology in the Edison Era.* Stanford, CA: Stanford University Press, 1999.

Gitelman, Lisa. *Always Already New: Media, History, and the Data of Culture.* Cambridge, MA: MIT Press, 2006.

Gold, Barri J. *ThermoPoetics: Energy in Victorian Literature and Science.* Cambridge, MA: MIT Press, 2010.

Gooday, Graeme. *Domesticating Electricity: Expertise, Uncertainty, and Gender, 1880–1914.* London: Pickering and Chatto, 2008.

Grossman, Jonathan. *Charles Dickens's Networks: Public Transport and the Novel.* New York, NY: Oxford University Press, 2012.

Groth, Helen. *Victorian Photography and Literary Nostalgia.* New York, NY: Oxford University Press, 2003.

Handy, Ellen. 'Dust Piles and Damp Pavements: Excrement, Repression, and the Victorian City in Photography and Literature'. *Victorian Literature and the Visual Imagination.* Eds. Carol T. Christ and John O. Jordan. Berkeley, CA: University of California Press, 1995, 111–33.

Henchman, Anna. *The Starry Sky Within: Astronomy and the Reach of the Mind in Victorian Literature.* New York, NY: Oxford University Press, 2014.

Hankins, Thomas L., and Robert J. Silverman. *Instruments and the Imagination.* Princeton, NJ: Princeton University Press, 1995.

Hughes, Linda K., and Michael Lund. *The Victorian Serial.* Charlottesville, VA: University of Virginia Press, 1991.

Hunt, Bruce J. *Pursuing Power and Light: Technology and Physics from James Watt to Albert Einstein.* Baltimore, MD: Johns Hopkins University Press, 2010.

Kargon, Robert H. *Science in Victorian Manchester: Enterprise and Expertise*. Baltimore, MD: Johns Hopkins University Press, 1977.

Keep, Christopher. 'Technology and Information: Accelerating Developments'. *A Companion to the Victorian Novel*. Eds. Patrick Brantlinger and William B. Thesing. Malden: Blackwell, 2002, 137–54.

Kern, Stephen. *The Culture of Time and Space, 1880–1920*. Cambridge, MA: Harvard University Press, 1983.

Ketabgian, Tamara Siroone. 'Machines'. *Oxford Bibliographies in 'Victorian Studies'*. Ed. Juliet John. New York, NY: Oxford University Press, 2011 [2011A].

Ketabgian, Tamara Siroone. *The Lives of Machines: The Industrial Imaginary in Victorian Literature and Culture*. Ann Arbor, MI: University of Michigan Press, 2011 [2011B].

Kirby, Lynne. *Parallel Tracks: The Railroad and Silent Cinema*. Exeter: University of Exeter Press, 1997.

Kittler, Friedrich. *Aufschreibesysteme 1800/1900* [*Discourse Networks 1800/1900*] [1985]. Stanford, CA: Stanford University Press, 1990.

Kittler, Friedrich. *Grammophon Film Typewriter* [*Gramophone, Film, Typewriter*] [1986]. Stanford, CA: Stanford University Press, 1999.

Kreilkamp, Ivan. *Voice and the Victorian Storyteller*. Cambridge: Cambridge University Press, 2009.

Law, Jules. *The Social Life of Fluids: Blood, Milk, and Water in the Victorian Novel*. Ithaca, NY: Cornell University Press, 2010.

Lee, Maurice S. *Uncertain Chances: Science, Skepticism, and Belief in Nineteenth-Century American Literature*. New York, NY: Oxford University Press, 2012.

MacLeod, Christine. *Heroes of Invention: Technology, Liberalism, and British Identity, 1750–1914*. Cambridge: Cambridge University Press, 2007.

Marsden, Ben, and Crosbie Smith. *Engineering Empires: A Cultural History of Technology in Nineteenth-Century Britain*. New York, NY: Palgrave Macmillan, 2005.

Matus, Jill. *Shock, Memory and the Unconscious in Victorian Fiction*. Cambridge: Cambridge University Press, 2009.

Menke, Richard. *Telegraphic Realism: Victorian Fiction and Other Information Systems*. Stanford, CA: Stanford University Press, 2008.

Milan, Sarah. 'Refracting the Gaselier: Understanding Victorian Responses to Domestic Gas Lighting'. *Domestic Space: Reading the Nineteenth-Century Interior*. Eds. Inga Bryden and Janet Floyd. Manchester: Manchester University Press, 1999, 84–102.

Miller, Andrew. *Novels Behind Glass: Commodity Culture and Victorian Narrative*. Cambridge: Cambridge University Press, 1995.

Morus, Iwan Rhys. *Frankenstein's Children: Electricity, Exhibition, and Experiment in Early-Nineteenth-Century London*. Princeton, NJ: Princeton University Press, 1998.

Novak, Daniel. *Realism, Photography, and Nineteenth-Century Fiction*. Cambridge: Cambridge University Press, 2008.

Otis, Laura. *Networking: Communicating with Bodies and Machines in the Nineteenth Century*. Ann Arbor, MI: University of Michigan Press, 2001.

Otter, Chris. *The Victorian Eye: A Political History of Light and Vision in Britain, 1800–1910*. Chicago, IL: University of Chicago Press, 2008.

Pettit, Clare. *Patent Inventions: Intellectual Property and the Victorian Novel*. New York, NY: Oxford University Press, 2004.

Picker, John. *Victorian Soundscapes*. New York, NY: Oxford University Press, 2003.

Poovey, Mary. *Making a Social Body: British Cultural Formation, 1830–1864*. Chicago, IL: University of Chicago Press, 1995.

Price, Leah, and Pamela Thurschwell, eds. *Literary Secretaries/Secretarial Culture*. Burlington, VT: Ashgate, 2005.

Rabinbach, Anson. *The Human Motor: Energy, Fatigue, and the Origins of Modernity*. Berkeley, CA: University of California Press, 1992.

Rotunno, Laura. *Postal Plots in British Fiction, 1840–1898: Readdressing Correspondence in Victorian Culture*. New York, NY: Palgrave Macmillan, 2013.

Rubery, Matthew, ed. *Audiobooks, Literature, and Sound Studies*. New York, NY: Routledge, 2011.

Rudy, Jason R. *Electric Meters: Victorian Physiological Poetics*. Athens, OH: Ohio University Press, 2009.

Schickore, Jutta. *The Microscope and the Eye: A History of Reflections, 1740–1870*. Chicago, IL: University of Chicago Press, 2007.

Schivelbusch, Wolfgang. *The Railway Journey: The Industrialization of Time and Space in the Nineteenth Century*. Berkeley, CA: University of California Press, 1986. Reissued 2014.

Schivelbusch, Wolfgang. *Disenchanted Night: The Industrialization of Light in the Nineteenth Century*. Transl. Angela Davies. Berkeley, CA: University of California Press, 1988.

Sussman, Herbert L. *Victorian Technology: Invention, Innovation, and the Rise of the Machine*. Santa Barbara, CA: Praeger, 1999.

Thurschwell, Pamela. *Literature, Technology and Magical Thinking, 1880–1920*. Cambridge: Cambridge University Press, 2005.

Timmins, Geoffrey. *The Last Shift: The Decline of Handloom Weaving in Nineteenth-Century Lancashire*. Manchester: Manchester University Press, 1993.

Tomory, Leslie. *Progressive Enlightenment: The Origins of the Gaslight Industry, 1780–1820*. Cambridge, MA: MIT Press, 2012.

Tucker, Jennifer. *Nature Exposed: Photography as Eyewitness in Victorian Science*. Baltimore, MD: Johns Hopkins University Press, 2006.

Willis, Martin. *Vision, Science and Literature, 1870–1920: Ocular Horizons*. London: Pickering and Chatto, 2011.

Winter, Alison. *Mesmerized: Powers of Mind in Victorian Britain*. Chicago, IL: University of Chicago Press, 1998.

IV

Biological and human sciences

22

NATURAL HISTORY, EVOLUTION AND ECOLOGY

The biological sciences

John Holmes

University of Birmingham

In 1834, Dionysius Lardner, editor of the *Cabinet Cyclopedia*, brought out a companion volume to John Herschel's pioneering treatise in the philosophy of science, *A Preliminary Discourse on the Study of Natural Philosophy* (1831). In *A Preliminary Discourse on the Study of Natural History*, the roving naturalist William Swainson set out to define how natural history differs from natural philosophy. As a field of study, Swainson writes, 'it is the province of natural history to embrace all that concerns the three great divisions or kingdoms of nature,—the animal, the vegetable, and the mineral' (95). Astronomy, chemistry, physiology and geology, by contrast, fall into the domain of natural philosophy. Aside from the need to portion out the phenomena of the world so that the human mind can grasp them, Swainson struggles to explain why this distinction should be drawn along these precise lines. Geology, he suggests, is not part of natural history because, unlike mineralogy, 'it relates more to the situation, than to the analysis, of the component parts of our globe' (Swainson 95). It would be hard to say the same of chemistry, however, and Swainson inadvertently collapses the distinction between mineralogy and chemistry later in the same chapter (105). Similarly, while he assigns physiology to natural philosophy, 'comparative anatomy' (Swainson 99) remains a key development in the study of natural history. Methodologically, too, Swainson is hard pushed to distinguish the two overarching forms of science. Instead he is at pains to insist that the natural historian, just like the astronomer or chemist, is engaged not only in observation but in generalization from observation, and that, insofar as these generalizations are consistent with external nature and with themselves, they have the full force of demonstration (Swainson 105).

Swainson's attempt to define natural history as a science is revealing both in its conviction and its imprecision. His particular division between natural philosophy and natural history anticipates our current distinction between the physical and the life sciences, yet it does not map onto it. The *Oxford English Dictionary* distinguishes between natural philosophy, which it equates to 'natural science', and natural history, which it defines variously as 'a systematic account based on observation rather than experiment', 'the natural phenomena of a region as observed or described systematically' and 'the study of animals and other living organisms, esp. as presented in a popular rather than in a strictly scientific manner' (*OED Online*). Each of these definitions distinguishes natural history from science proper, but Swainson explicitly refutes any charge that natural history is mere description, claiming that it is as analytical as any other science, and that it too seeks to discover natural laws. Swainson's very insistence on this point reveals that there were challenges to natural history's status as a science in the nineteenth century, equivalent to the vaunting of the physical sciences as the 'hard' sciences today.

Natural history – encompassing zoology, botany and palaeontology, together with mineralogy – was at once the most ubiquitous and the most established of the sciences. But its very ubiquity told against it in an age in which science came to define itself through professionalization and the experimental method, as set out not by Swainson but by Herschel. So too did its ties to the establishment. Over the course of the century, Swainson's ill-defined science of natural history came to be supplanted by the precise, professional discipline of biology, with its sub-disciplines of morphology, physiology and ecology and the overarching theoretical frameworks provided by competing theories of evolution. The transition, signalled by the change in terminology, began early. The first use of the word 'biology' in English cited by the *OED* is by Thomas Beddoes, as early as 1799, concurrent with Gottfried Treviranus's adoption of *Biologie* in German and Jean-Baptiste Lamarck's of *biologie* in French (*OED Online*). But where Lamarck and his colleagues Georges Cuvier and Étienne Geoffroy Saint-Hilaire at the Jardin des Plantes established Paris as the foremost centre for the systematic study of the natural world in Europe, natural history in England remained largely the province of collectors and clergymen until well into the middle of the century.

The two English scientists who would contribute most to fulfilling Swainson's promise that natural history could be a systematic science were themselves products of this amateurish culture. Charles Darwin's researches laid the foundations for modern biology, but he was a country squire not a professional scientist employed by a university or a museum. As Hunterian professor at the Royal College of Surgeons and later superintendent of the natural history collections at the British Museum, Richard Owen was a consummate professional. But, as Nicolaas Rupke has shown, the patrons who ensured his success were Oxford and Cambridge dons (53–89). They epitomized the mental habits described by Swainson, not least another fundamental tenet of natural history – that it is the study of the 'works of God' (93), 'things which, in themselves, are perfect, because they emanate from the Fountain of Perfection' (96). As Tim Hayward has remarked,

> The idea of an essentially benign balance of nature may have been useful for reconciling the growth of natural science with the claims of religion, but it did not exactly encourage effective research into material causes since the final cause and the intermediate ends are already known.
>
> *(25)*

The transition from natural history to biology entailed the professionalization and the secularization of the life sciences. But it also involved a more profound shift in the understanding of the natural world itself. Early nineteenth-century natural history was predicated on a presumption of fixity in nature. As Swainson explained, although science participates in the mutability of all human endeavour, the products of nature itself are immutable (97–9). The aim of natural history was to uncover the unchanging patterns underlying the variety of life, such as the vertebrate archetype proposed by Owen. Its fundamental method was taxonomy, as modelled by Linnaeus, and its ambitions were encyclopedic, as demonstrated by the forty-four volumes of the *Histoire Naturelle* completed by the Comte de Buffon and his successors from 1749 to 1804. Where natural history affirmed the reality of types, from species up to kingdoms, Darwinian biology would be predicated instead on a recognition that nature is in perpetual flux, that species are lineages of diverse individuals and that even the different species themselves are all interrelated, both by common descent and in their changing interactions with one another. The implications for our own self-image of the replacement of the first of these two accounts of nature with the second were immense, as we ceased to be a kind apart and became instead a part of the natural world.

The two most compelling symbols of these different world views are the Natural History Museum in South Kensington and Darwin's image of the tree of life from *On the Origin of Species*. The tree of life (Figure 22.1) shows nature as a continual growth, capturing at once the diversity of its forms, their genetic relation to one another and their change over time.

The plan of the Natural History Museum (Figure 22.2), built by Alfred Waterhouse in close collaboration with Owen, shows instead a nature rigorously divided into separate kinds, like the drawers in a collector's cabinet, with extinct types segregated from living forms. In getting from one gallery to another, a visitor had to walk through the Central Hall.

This vast space, likened in the journal *Nature* to a cathedral ([Anon.] 549), was dedicated to the wonders of God's creation, with side chapels offering an index to all life. Where Darwin's metaphor left humanity as just another twig on one branch of the tree of life, Owen's building was crowned with a statue known as Adam, surmounting the gable end, at once the culmination of nature and beyond it.

The Natural History Museum opened in 1881, twenty-two years after the publication of *On the Origin of Species*. It was less a sign of the rude health of natural history on the old lines, however, than an anachronism. This chapter charts the transition within nineteenth-century literature and science from a natural history characterized by fixity and religious faith to a secular

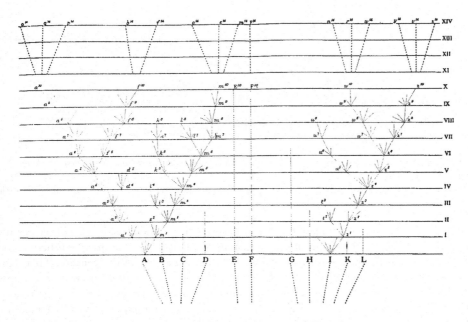

Figure 22.1 Darwin's image of the tree of life from *On the Origin of Species*

Reproduced with permission from John van Wyhe ed. 2002. *The Complete Work of Charles Darwin Online* (http://darwin-online.org.uk)

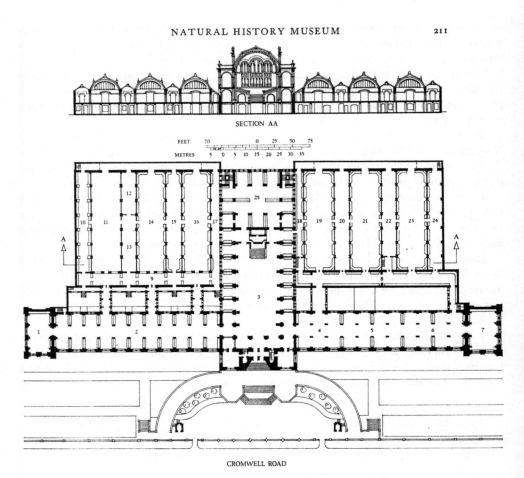

NATURAL HISTORY MUSEUM   211

SECTION AA

Fig. 33. Natural History Museum, plan and section in 1883.
1, 2, Bird Gallery; 3, Central Hall; 4, S.E. Gallery; 5, Geology and Palaeontology; 6, Fossil Mammalia; 7, Fossil Mammalia and Birds; 8, Fossil Reptilia; 9, Coral Gallery; 10, Bird Gallery for study; 11, Shell Gallery; 12, Students; 13, Star Fish Gallery; 14, Reptile Gallery; 15, Insect Gallery; 16, Fish Gallery; 17, Reserve Gallery; 18, Library and Geological Depart-ment; 19, Fossil Fishes, Cephalopoda and Pteropoda; 20, Geographical Collections; 21, Fossil Gasteropoda, Crustacea, Insecta, etc.; 22, Special Collections and Students' Use; 23, Fossil Corals, Sponges, Plants, etc.; 24, Stratigraphical Series and Ichnites of Birds and Reptiles; 25, British Natural History Museum (Zoology)

Figure 22.2 Natural History Museum, plan and section in 1883. *Survey of London: Volume 38, South Kensington Museums Area*
Originally published by London County Council, London 1975; reproduced with permission from British History Online

biology alert to flux and complex ecological relations. In the next section I will set out how the debate was framed in literary form by examining three books that inscribed the methods, implications and speculations of natural history at the beginning of the century. Then I will turn to one of the most profound shifts from one world view to another ever to have occurred, which was explored in the arts even as it was substantiated by the sciences: the recognition that living forms are not created but evolve. Finally, I will examine the part played by literature in the emergence of a no-less-crucial ecological understanding of the relationship between organisms and their environments. The thrust of my argument throughout is that the study of literature – including the literary forms of scientific writing – is integral to our understanding of natural history, both as a cluster of sciences in the nineteenth century and as a field of knowledge that bears on our place in the world, now as then.

The elements of nineteenth-century natural history

Three books published around 1800 establish the terms of reference for nineteenth-century natural history. Gilbert White set the template for the parson-naturalist in the letters and papers published as *The Natural History of Selborne* (1788). His premise was that, by carefully observing the natural history of even a small local area, one could contribute something substantial to the understanding of the natural world at large. His method was induction as defined by Francis Bacon, moving from recording observations to making tentative generalizations that could be tested in turn against further evidence. Where White's *Natural History* exemplified the method for studying nature, William Paley's *Natural Theology* (1802) provided a framework for interpreting it. Paley's book crystallizes a tradition of natural theology going back to the seventeenth century. Both books were widely read and immensely influential. Wordsworth bought a copy of White around 1790 (Wu 147), but it was principally from the 1830s, once cheap editions had become available, that *The Natural History of Selborne* became the model for the Victorian enthusiasm for natural history exemplified in the work of writers such as Philip Henry Gosse and Charles Kingsley (Merrill 22–5). The sculpted clarity of Paley's *Natural Theology* made it an ideal textbook for universities such as Oxford, Cambridge and King's College London, committed to promoting natural theology as the orthodoxy in science. This in turn helped to ensure the persistence of the tradition of natural theology through books such as the *Bridgewater Treatises* of the 1830s, several of them written by Oxbridge dons, and *The Animal Creation* (1865), a taxonomical guide by Thomas Rymer Jones, professor of natural history and comparative anatomy at King's. The third book that stands alongside these two was far less widely read and far more eccentric but in its own way no less representative. *The Temple of Nature; or, The Origin of Society*, by Charles Darwin's grandfather Erasmus Darwin, was published posthumously in 1803. Darwin's long didactic poem puts forward the first detailed account of the evolution of life in English, expanding on a briefer proposal along these same lines in his medical textbook *Zoonomia* (1794–96). Darwin's evolutionary ideas gained little traction, even among readers and writers who admired him, but what they represent is a current within natural history, even as early as the turn of the century, that pursued speculation, not the reaffirmation of conventional wisdom, and that saw nature as fluid not fixed.

As well as defining the field, these three books reveal a close relationship between literary form and ways of thinking about nature. White grappled with serious scientific questions – for example, whether swallows migrated or hibernated – but his commitment to Baconian procedure means that many of his observations seem disjointed. What gives his book coherence is its genre. As a series of letters, it recalls both the epistolary novels of the eighteenth century (although only White's side of the correspondence is given) and the essays of Bacon himself. Through his letters, White grounds the practice of natural history, and the acts of writing and reading about it, within patterns of social interaction and exchange. As the correspondence takes place between a curate, a gentleman and a judge, his book models the study of natural history as a cultural pursuit within polite society. At the same time, it suggests that it has moral value. The principle that the study of natural history is morally improving would become a perennial truth within nineteenth-century literature. It is implicit in Elizabeth Gaskell's novel *Wives and Daughters* (1864–66), for example, where an active interest in the lives of animals and plants is a moral barometer, and explicit in John Ruskin's call in his preface to the second edition of *Modern Painters* (1844) for the artist to study flowers with care and sympathy.

The poet, educator and political campaigner Anna Barbauld gives precise form to the moral gains to be made from natural history in her late poem 'The Caterpillar', published after her death in 1825. Where White was an Anglican clergyman, Barbauld was a Dissenter and a critic of the established social and political order. The speaker of her poem is a gardener who has recently killed as many caterpillars as possible, 'Nor felt the touch of pity' (l. 23), yet having found and watched a single survivor cannot now kill it:

> No, helpless thing, I cannot harm thee now;
> Depart in peace, thy little life is safe,
> For I have scanned thy form with curious eye,
> Noted the silver line that streaks thy back,
> The azure and the orange that divide
> Thy velvet sides . . .
>
> . . .
>
> Thou has curled round my finger; from its tip,
> Precipitous descent! with stretched out neck,
> Bending thy head in airy vacancy,
> This way and that, inquiring, thou hast seemed
> To ask protection: now, I cannot kill thee.

(ll. 1–6, 9–13)

The act of closely observing the caterpillar – its markings, textures, movements – creates for Barbauld a sympathy with it. That such acts of natural history bear not only on relationships between humans and animals but on our behaviour towards one another is a theme that is implicit in White's writing. In Barbauld's poem it becomes explicit through an increasingly intense analogy between the gardener and a soldier on the field of battle who cannot bring himself to kill an enemy survivor pleading for his life. Barbauld is no trite or easy moralist. Unlike White, she is not content to rehearse accepted moral codes, instead putting moral agency to the test. Her speaker concludes that the pity that leads both men to spare their enemies' lives is 'not Virtue, / Yet 'tis the weakness of a virtuous mind' (ll. 41–2), leaving us to resolve for ourselves how far we respect this distinction. Either way, the study of natural history is shown to have a moral correlative.

In spite of its Baconian empiricism, White's work is not free of theoretical assumptions. His natural history takes place within the intellectual framework of natural theology. By the logic of natural theology, discoveries in natural history provide evidence of God's purposes and character. At the same time, this evidence must square with what we already know of God's character from revelation and from his gift to humanity of rational enquiry. The classic formulation of the argument that nature bears the hallmarks of divine design is Paley's opening analogy in *Natural Theology* between a man finding a watch on a beach and knowing from its mechanism that it must have been designed and the natural historian studying the works of nature and making the same sure deduction. This analogy typifies the strengths of Paley's writing. Where White assembles vivid vignettes of observation, Paley musters a clear and sustained argument, as Darwin was to do in *On the Origin of Species*. There are inevitable weaknesses in his chains of reasoning, not least as he seeks to move from the deistic conclusions of the argument from design to an affirmation that Christian theism can be seen revealed in nature. But the precision and elegance of his prose, and his deft use of examples and analogies, such as the watch, that can be readily grasped, gave his book an authority vested, like that of White's *Natural History*, in its literary form.

Paley's argument is predicated on nature being in its essence unchanging: 'Though there may be the appearance of failure in some of the details of Nature's works,' he writes, 'in her great purposes there never are. Her species never fail' (249). Yet there was already a growing realization that this was not so. The discovery of extinct animals proved that the outward forms of nature had changed – first the mammoths, mastodons and megatheriums described by Cuvier and Blumenbach in the 1790s, then the pterodactyls, ichthyosaurs, plesiosaurs and dinosaurs that would be collectively characterized by Alfred Tennyson as 'Dragons of the prime' in *In Memoriam* (1850, Canto 66, l. 22). Even before these discoveries, Erasmus Darwin had begun to put the case for seeing natural history in terms not of creation but of evolution. In *The Temple of Nature*, Darwin explored this new vision of nature imaginatively. His poem, like White's and Paley's books, is a product of the eighteenth century. Both its form – heroic couplets – and its florid diction very soon seemed outmoded (on Darwin as a poet, see Harris; Page 17–37; Priestman). But his speculation that the history of life on earth was one of continual evolutionary change was a radical challenge to the assumptions of eighteenth-century natural theology. Noah Heringman (2004 224) and Martin Priestman (116) have argued that Darwin needed the imaginative reach and freedom of poetry in order to articulate and explore his audacious ideas to the full. At the same time, his choice to write in verse is at once a legitimization strategy and a disclaimer. On the one hand, he borrows the cultural authority of poetry to write his own answer to Alexander Pope's *Essay on Man* (1733–34); on the other hand, he can claim in the preface that his poem 'does not pretend to instruct by deep researches of reasoning' but 'simply to amuse' (Darwin, *Temple of Nature*, preface). The researches themselves are relegated to detailed footnotes, including over 140 pages of additional notes printed after the text of *The Temple of Nature* itself. These notes confirm that the poem is a Trojan Horse for Darwin's scientific speculation even as they elevate it into a classic text worthy of close scholarly commentary.

Darwin's scientific radicalism was directly linked to his political radicalism, from his championing of the French Revolution to his advocacy of freer and more equal sexual relations. His conviction that nature could teach lessons in political liberty and transformation was the radical mirror of White's establishment view that studying natural history could inculcate a polite moral order. Darwin's influence on younger writers was at once profound and contentious (King-Hele; Page 39–109; Priestman 217–56). Charlotte Smith was influenced as a poet by Darwin's example (George 11), as were Wordsworth and Coleridge (Amigoni 31–83), yet they all resisted and ultimately disparaged his influence, stylistically, intellectually and politically. For Lord Byron and Percy Shelley, Darwin's use of poetry as a free medium through which to explore the history and destiny of nature and society inspired and licensed their own speculations on these same themes in very different verse forms. The closet drama offered them a particularly vivid medium. In *Cain: A Mystery* (1821), Byron confronts extinction and its implications for our own place in God's supposed creation through a dialogue with the Devil. In *Prometheus Unbound* (1820), Shelley uses a radical reinterpretation of Greek myth and a plethora of voices dispersed throughout the natural universe to express inherent tendencies in nature and their political corollaries.

Strikingly, none of these writers adopted the evolutionary narrative that has come to be seen as Erasmus Darwin's most significant contribution to the history of ideas. Instead, it was his celebration of the dynamics of nature that gripped them. For Darwin, this principle of change and growth in nature was epitomized in sexual reproduction. For the Romantics, as Sharon Ruston (2011A), Denise Gigante and Robert Mitchell have shown, it was expressed rather in terms of a vital impulse driving life itself. Like Darwin's evolution, vitalism was another countervailing conception of nature to the static creationism of natural theology. Though both were highly speculative, each sought to ground itself in experimental practice and microscopic attention to detail that reached beyond the everyday observations of White or the schematic analyses of Paley. Neither would have much currency later in the century, although Samuel Butler attempted to revive Erasmus Darwin's speculations in a series of polemics against his grandson's theory of natural selection, beginning in 1878 with *Life and Habit*. But the freedom they embodied to hypothesize, and to attempt to test those hypotheses through experiment and scrupulous observation, would be fundamental to the transition from natural history to biology.

Evolution in Victorian literature and science

Butler aside, the legacy of Erasmus Darwin's writings has yet to be traced beyond the 1820s. But the political reaction against evolutionism – or transmutation, as it was known – set in very early in the aftermath of the French Revolution and was reinforced by the British scientific community. Charles Lyell devoted the bulk of the second volume of his *Principles of Geology*, published in 1832, to refuting the arguments of the French evolutionist Jean-Baptiste Lamarck, in part to enhance the respectability of his own contentious geological theories (Secord 2014 138–72). Evolutionary speculation was to gather steam again, however, with the publication of Robert Chambers's anonymous bestseller *Vestiges of the Natural History of Creation* in 1844. In *Victorian Sensation*, James Secord has documented exhaustively the extraordinary popular success of this evolutionary history of the natural world, most significantly among the respectable middle classes, for whom such ideas had been anathema up to this point. Although Chambers insisted that his account of the evolution of life was consonant with an elevated conception of the Christian god – as Erasmus Darwin had done, indeed – his ideas were attacked as materialism by the natural theologians and as mere speculation by the newly established class of professional scientists. *Vestiges* nonetheless introduced a far wider range of readers to evolution as a plausible narrative of the history of life on earth. This prompted Arthur Hugh Clough, for one, to speculate for himself on the implications of evolution for human beings in several lyrics from his early collection *Ambarvalia* (1849). Tennyson had been drawn to evolutionism since the late 1820s (Dean 2–5) but it was not until *Vestiges* had been published that he too worked through its implications in print. *The Princess* (1847) stages gendered debates on the potential for future evolution in the light of the past (Stott). *In Memoriam* (1850) grapples with the existential challenges posed by a knowledge of extinction (Fulweiler; Armstrong 252–83; Tomko; Rowlinson), as Byron had done in *Cain*, but with the difference that for Tennyson evolution offers a glimpse of an alternative future in which humans are not wiped out but transformed into progressively higher beings.

Aside from fixed laws of development, up until the 1850s the principle mechanism for evolutionary change was taken to be the inheritance of acquired characteristics. It was the assumption that characteristics acquired by one generation over the course of their lives could be inherited by the next that enabled Tennyson to urge himself and his readers to 'Move upward, working out the beast, / And let the ape and tiger die' in *In Memoriam* (Canto 118, ll. 27–8). The ape and tiger here are relics of our own earlier beasthood that we can work to outgrow once and for all. This was the mode of evolution that underpinned the social and psychological theories of Herbert Spencer, developed throughout the 1850s over several books and essays. But, for all the popularity of Chambers, Tennyson and Spencer, the watershed came with the publication of Charles Darwin's *On the Origin of Species* in 1859. Darwin's book initially provoked the same response from older scientists brought up in the tradition of natural theology as *Vestiges* had done. It was attacked in print by Adam Sedgwick, John Phillips and Owen, among others. Yet it very soon became apparent that Darwin's work had an empirical foundation that was far more robust than Chambers's. The relationships that Darwin established between different classes of animals could not be dismissed as mere speculation. Where Chambers and Erasmus Darwin had used narrative forms to entice readers into entertaining the possibility of evolution, Charles Darwin followed Paley, outdoing him in working through a minute and sustained argument.

Darwin's ideas had a defining impact on the study and interpretation of the natural world. He laid secure empirical foundations for the theory of evolution from common descent. At the same time, in proposing natural selection as the primary mechanism for evolution, he instigated intense debates that would last for decades, as scientists sought to establish the *verae causae* of evolutionary change. Alongside natural selection, contending mechanisms included the inheritance of acquired characteristics, radical mutation or saltation, and predetermined patterns of development or orthogenesis, divinely ordained or otherwise. Darwin himself contributed to these debates at length, but they were worked through too in the writings of countless other biologists and clerical and lay commentators. The stakes were high, as what was at issue was the very nature of nature itself, including human nature, and therefore the nature of society and societal change, not to mention of the creative processes that gave rise to human beings in the first place. Writers in many different genres, from scientific papers and political polemics to poems and naturalist novels, avowed comparably diverse views of the social implications of evolution, often lumped together as social Darwinism, although in reality many of them were predicated on non-Darwinian theories (see Hawkins). While many of these accounts professed to be descriptive, others were interventionist, seeking to shape the future of human evolution through what Darwin's cousin Francis Galton called 'eugenics'.

How the debates over evolution and its causes and ramifications played out in Victorian literature is the theme of a large and growing body of scholarship. The bulk of this work centres on Charles Darwin himself as a writer and on the relationship between his writings and ideas and Victorian fiction. Gillian Beer, George Levine, Peter Morton and James Krasner have collectively shown how Darwin's characteristic concepts, structures of argument and ways of seeing the world are reflected and refracted in the realist novels of George Eliot, Thomas Hardy and Joseph Conrad. (Full references for their work and that of other scholars mentioned below are given in the secondary texts section of the bibliography for this chapter.) In *Darwin and the Novelists*, Levine traces parallels too between *On the Origin of Species* and novels by Charles Dickens and Anthony Trollope written before Darwin published his theory. These critics' pioneering work on Darwin has been complemented by more recent scholarship, including David Amigoni's analysis of Darwin's contribution to debates on culture and authority reaching back to Wordsworth and Coleridge and forward to Butler and Edmund Gosse; Gowan Dawson's account of how Darwin and his allies became embroiled in debates over the materialism and even the obscenity of aestheticism; Robert Ryan's insightful study of the importance of Wordsworth's poetry to Darwin's attempts to define his own thinking and in the responses of both his advocates and his critics to it; and briefer but highly suggestive studies by Linda Bergmann of the rhetoric of *On the Origin of Species*, Tina Choi of contingency in *Vestiges*, *Origin*, and Eliot's *Adam Bede* (1859), and Mary Noble of Darwin's reading of novels as sources for his science.

Christine Ferguson, John Glendening and Virginia Richter have extended the analysis of Darwin's impact on fiction into discussions of popular novels by H. G. Wells, Bram Stoker, Rudyard Kipling, H. Rider Haggard and Robert Louis Stevenson, among others. Carolyn Burdett, Angelique Richardson and Sabine Ernst have traced how the New Woman novelists engaged with evolution and eugenics in their fiction and politics. William Greenslade has examined how debates over degeneration bore on late Victorian fiction, while Nicholas Ruddick has traced the emergence of a new genre of prehistoric fiction towards the end of the century. There is a substantial literature specifically dedicated to Hardy (Ebbatson 1–56; Otis 158–80; Richardson 2002; Richardson 1998), Wells (Parrinder; Hurley 55–113; Markley;

McLean), and George Meredith (Kelvin; Williams; Jones) as novelists who grasped and grappled with the implications of Darwinism, as well as individual studies of Conrad (O'Hanlon) and Stevenson (Reid).

Looking beyond fiction, Lionel Stevenson, Georg Roppen and I have written wide-ranging studies of how Victorian poets, including Tennyson, Browning, Meredith and Hardy, addressed the implications of evolutionary theory. There are accounts too of how particular poets engaged with evolution, including Gerard Manley Hopkins (Zaniello; Banfield), James Thomson (Forsyth), and W. B. Yeats (McDonald). The recovery of late Victorian women poets, obscured for much of the twentieth century, has brought to light their complex engagements with evolutionary science, ranging from the epic reach of Mathilde Blind (Groth 332–40; Fletcher; Rudy; Birch) to the deft comedy of Constance Naden (Moore; Murphy; Thain; Kaston Tange) and May Kendall (Holmes 2010B). Individual case studies are beginning to build up a picture too of how men who worked professionally in the natural sciences in the nineteenth century, including the physician Thomas Gordon Hake (Holmes 2004), the influential naturalist G. J. Romanes (Pleins) and the British Museum taxonomist Arthur O'Shaughnessy (Kistler 23–60), turned to poetry to work through their conceptions of and responses to their own science. Still more tightly focused analyses concentrate on individual poems, examining for example Tennyson's incorporation and circumvention of Darwinian evolution in *Idylls of the King* (1859–85) (Purton) and Meredith's recognition in *Modern Love* (1862) that Darwinism undercut the sexual double standard predicated on Eve's original sin (Holmes 2010A). Jane Goodall and Kirsten Shepherd-Barr have examined how evolution was played out on the Victorian stage, including popular theatre. Looking across to non-fictional prose, John Ruskin's fraught relationship with Darwin has been the subject of careful readings by Levine (2008 75–99), Jonathan Smith, M. M. Mahood (147–82), Sara Atwood and Clive Wilmer, while Piers Hale and Jessica Kuskey have placed William Morris's political writings in the context of evolutionary debate. Evolutionary thinking has been traced in Victorian exploration literature (Krasner; Schmitt), fairy tales (Talairach-Vielmas), and literary criticism (Holmes 2011), and there have been a number of attempts to define a new genre of prose histories of the earth as the evolutionary epic (Lightman 219–94; O'Connor; Amigoni and Elwick).

The sheer bulk of the scholarship on Victorian literature in relation to evolution testifies to the reach and intensity of the debates sparked off by Chambers and Darwin. The new evolutionary thinking radically altered Victorian presumptions about the natural world and our place within it. If the history of life on earth embodied God's purposes, then those purposes included extinction for some and transformation across countless generations for others. Furthermore, if Darwin's theory that evolution was driven by natural selection was correct, then it was not only God's ends that were ruthless but his means, too, as the emergence of new species was made possible by the premature deaths of those individuals born into older, unchanged, outcompeted forms.

The reality of extinction made the project of looking for a benevolent Providence in nature seem to many a fool's endeavour. As Michael Tomko has shown (113–14), even in *In Memoriam* Tennyson looks to wall the physical and spiritual worlds off from one another, to prevent the violence of 'Nature, red in tooth and claw' (Canto 56, l. 15) from being equated with the order of the universe as a whole. But where in *In Memoriam* evolution at least offers a counterweight to extinction, from the 1860s Tennyson was increasingly disinclined to put his faith in a creative process that, after Darwin, looked more brutal and less purposive (see Holmes 2009 62–74). Browning too warned against the risks of applying the logic of natural theology to a Darwinian universe. In 'Caliban Upon Setebos' (1864), he constructs an arbitrary God made in the dual image of nature and of Caliban himself, a brutish embodiment of the missing link (Karr; Holmes

2009 84–89). The problem, fundamentally, was that nature no longer lived up to the standards required of it by the rational, optimistic natural theology that the nineteenth century had inherited from the eighteenth. As the physiologist George Romanes, a committed but troubled Darwinian, put it in his poem 'Natural Theology',

> What man among you, had he made this earth,
> But all his brothers would condemn to die?
> The parentage of such a monstrous birth
> Would brand him with inhuman devilry.
>
> *(Quoted in England and Nixon 306)*

Not everyone agreed that it was necessary to abandon natural theology in the wake of Darwin (Lightman 39–94; England and Nixon), but even many of those who stuck with the project acknowledged that natural theology would itself have to evolve. For Kingsley, speaking on 'The Natural Theology of the Future' in January 1871, Darwin's nature was easier to square with the God of scripture, especially the Old Testament, whose actions appeared harsh and arbitrary and whose motives were ultimately unknowable, than with the enlightened God of eighteenth-century rationalism. But this was not a God many liberal Victorians were keen to reinstate.

If evolution called the nature of God into question, it also transformed the nature of human beings. Aside from simple distaste, there were other discomforts that arose from finding oneself closely related to apes (on humans and apes in Victorian literature and culture, see Beer 1996 115–45; Hodgson; Browne; Richter). For one thing, humanity's claim to be exceptional, and therefore to have a privileged perspective on and place in the universe, was undermined. As Thomson remarked in his bleak poem *The City of Dreadful Night* (1874), 'We bow down to the universal laws, / Which never had for man a special clause' (Canto 14, ll. 61–2). The distances between the different races of human beings contracted, too, although they could be and typically were reinscribed as differences in degrees of evolution rather than in kind. In *The Descent of Man and Selection in Relation to Sex* (1871) and *The Expression of the Emotions in Man and Animals* (1872), Darwin argued that even human morality was an evolved property derived from the instincts of social animals. His thesis undermined both the claim that the moral absolutes of Christianity had been ordained by God and the argument that our innate knowledge of these absolutes is evidence that we possess immaterial souls. Darwin remarked in *The Descent of Man* that, as

> few persons feel any anxiety from the impossibility of determining at what precise period in the development of the individual . . . man becomes an immortal being . . . there is no greater cause for anxiety because the period cannot possibly be determined in the gradually ascending organic scale.
>
> *(683)*

Such well-mannered avoidance smacks of casuistry, however. The biologist St George Mivart and the journalist Frances Power Cobbe were among those who had been happy to replace the myth of fallen man with the counter-myth of an evolutionary ascent. Once the full implications of Darwin's theory had become clear, however, they felt that the line needed to be drawn. In *On the Genesis of Species* (1871) and *Darwinism in Morals* (1872), respectively, they exposed the risks that Darwin had sought to conceal in a blanket of reassurance, even as they critiqued the arguments from which these risks arose.

By the early 1870s, it was apparent that the new evolutionary biology had implications not just for theology but for ethics, political theory, psychology, social science, even aesthetics. Literature in its various forms provided a uniquely rich medium for examining these ramifications, without authors being required to endorse one scientific theory or another. In *Middlemarch* (1871–72), Eliot depicts a society characterized by struggles as complex and competitive as those documented by Darwin in the natural world. She signals her allegiance to the new biology in a style described by Henry James in a review for the *Galaxy* as 'too often an echo of Messrs. Darwin and Huxley' (quoted in Carroll 359). She reveals too her understanding of what was at issue in the transformation of natural history into biology in the pairing of Camden Farebrother, a parson-naturalist of the 1820s, more drawn to specimen collection and books on natural history than to commentaries on the Bible, with Tertius Lydgate, a physician for whom studying organisms and tissues under the microscope is integral to his research into the nature of life itself. In her novels, Eliot hones the established techniques of realism – free indirect discourse, the omniscient, ostensibly dispassionate narrator and the careful recreation of the minutiae of everyday life – to create a form capable of demonstrating the need for a secular morality grounded in sympathy and duty.

Although it lacks a faith in the Christian God to underpin it, Eliot's morality is not radically at odds with Gaskell's liberal Christianity in *Wives and Daughters*. In *Tess of the D'Urbervilles* (1891), by contrast, Hardy mounts a far more radical challenge to Victorian mores from a Darwinian standpoint. He is scathing in his contempt for religious hypocrisy, the sexual double standard and a social order predicated on adherence to received opinions and doctrines. Tess's sense of shame at her sexual experience – for all that sex has been forced upon her – is central to Hardy's exposure of what he sees as the false foundations of Victorian moral values. The internalization by Tess and Angel Clare of 'an arbitrary law of society which had no foundation in Nature' (Hardy 353) leads not only to Tess's own self-castigation, but to the collapse of their marriage, Tess's initial acquiescence to life with her rapist Alec D'Urberville, her murder of him and ultimately her own execution.

Where Eliot and Hardy develop the established techniques of realist fiction to lay new foundations for morality in the age of evolution, Wells pioneers a new genre of scientific romance in which the writer and reader reach into the as-yet-non-existent future to explore possible outcomes of evolutionary processes. In her futuristic novel *The Last Man* (1826), Mary Shelley had looked forward to the late twenty-first century in setting a date for the extinction of humanity. By the time Wells wrote *The Time Machine* (1895), the expanse of time stretching in either direction had increased far beyond these early nineteenth-century limits. Expansive and at times rapturous narratives of evolution, including Chambers's *Vestiges*, Winwood Reade's *The Martyrdom of Man* (1872), and Edward Clodd's *The Story of Creation* (1888), had enabled Victorian readers to enter imaginatively into the sweep of evolutionary history. Through the conceit of time-travel, Wells carries this imaginative engagement on into the future as he anticipates the outcome of human evolution over 800,000 years and our extinction over thirty million. His subdivision of the human species into the hapless Eloi and the nightmarish Morlocks is at once a comment on late Victorian social relations and an extrapolation according to Darwinian evolutionary theory. In the first third of the twentieth century, Olaf Stapledon would extend this range still further, covering billions of years of evolution in his future history *Last and First Men* (1930); Bernard Shaw would attempt an alternative prognosis on Lamarckian principles in his unwieldy cycle of five plays *Back to Methuselah* (1921); while Arthur Conan Doyle would take the opposite approach in *The Lost World* (1912), collapsing evolutionary history by imagining the survival on a remote South American plateau of fauna from across geological time, including our own ape-man ancestors.

Satire further extended both the scope and the scrutiny of evolutionary thinking. In 1872, the sculptor and poet John Lucas Tupper explained in a letter to his close friend, the Pre-Raphaelite painter Holman Hunt, how some satirical sketches he had written exposed the fallacious logic of Darwin's theories of natural and sexual selection:

> The style is ostensibly Darwinian both as to form & thought ... Thus Darwin's principles (I believe, fairly used) are found able to account for such '*well known*' facts as these: Why moths, flies, &c never burn themselves in the candle. Why our commonest butterfly has the under surface of the wing blue and the upper green— why the wild pigeon or Rock is always grey or brown. Why there are no white moths— Why all wild animals destroy their first born—Why owls are always black—Why domestic oxen are hornless. Why the bat's wing is more richly ornamented than that of birds. Why Manx cats & guinea pigs have tails—Why men under civilized conditions have tails not unfrequently. Why Peacocks have melodious voices. and lastly Why there is a present tendency in the sexes to unsex themselves.
>
> *(Hunt and Tupper 164)*

By using counterfactuals and social trends to highlight how natural and sexual selection can be applied to explain any natural fact or fiction, Tupper exposes the circularity of Darwin's adaptationist arguments. As the same logic could apparently explain a radically different natural world from the one we inhabit, Tupper implies, it does not explain anything. Sadly, Tupper's satires are lost – they were rejected by Leslie Stephen, editor of the *Cornhill Magazine*, as too controversial (Hunt and Tupper 179) – but Butler used a similar device in his satirical utopia *Erewhon* (1872), where he extended Darwin's principle of evolution by natural selection to include machines. Where Tupper's aim had been to reduce Darwin's arguments to absurdity, Butler is less explicit in his intentions, with the result that his satire is all the more suggestive, inviting its reader to consider it as either a critique or a thought experiment without finally resolving it into one or the other.

Unlike prose, poetry rarely staged new evolutionary possibilities, and though many poets attempted to mythologize different scientific accounts of our evolutionary origins and prospects – often in imitation of poetry that Shelley had written under the star of Erasmus Darwin – these poems tend to be vague and even vacuous. But poetry was uniquely well-equipped to confront its readers with the existential implications of evolution. Victorian poetry at its best is not a celebration but rather a form of interrogation. Characteristic Victorian forms, including the dramatic monologue, the sonnet sequence and the parody, were used to examine how evolutionary thought challenged entrenched assumptions about sex and gender, about humanity's place in the universe and our relation to other animals, about God and the divine plan. But it was the lyric that took its readers most directly into the mind of another human being struggling to comprehend the human condition in a world of evolution.

In Memoriam set the tone and the bar for such poems, with its intense cry of pain in a world in which God and Nature appear to be at strife and its hesitant yet ultimately determined reaffirmation of faith, including faith in evolution itself. In the aftermath of *On the Origin of Species*, those poets who stared most unflinchingly at the nature Darwin had revealed would at once live up to and reject Tennyson's example. Where Hardy and Thomson seek to persist in the face of the existential horror of a world emptied partly or wholly of meaning, Meredith embraces Darwin's vision of nature as a positive good, confronting his own mortality in a world with no trace of a purpose or an afterlife by identifying not with his own bare life but rather with the tree of life itself. In his late Romantic 'Ode to the Spirit of Earth in Autumn', originally published alongside *Modern Love* in 1862, he exults in the collective life, urging 'Great Mother Nature' (l. 141) to 'Teach me to feel myself the tree, / And not the withered leaf' (ll. 154–5). In the more muted tones of 'In the Woods', published in the *Fortnightly Review* in 1870, he closes a series of nine lyrics by reminding his readers 'And we go, / And we drop like the fruits of the tree, / Even we, / Even so' (IX, ll. 12–15). In the dying fall of this poem we can hear the final acceptance of all that Darwin teaches about our place within nature.

Although Meredith offered a uniquely powerful positive vision of Darwinian nature in his poems, for many readers, including many scientists, it remained Tennyson's elegy that voiced most closely their own fears, doubts and hopes (see Holmes 2012). As William Henry Flower, then director of the Natural History Museum, put it at the end of his presidential address to the British Association for the Advancement of Science in 1889, there were 'the strongest grounds for the belief . . . that natural selection, or survival of the fittest, has, among other agencies, played a most important part in the present condition of the organic world' (29). But there were grounds too for seeing it as 'a universally acting and beneficent force continually tending towards the perfection of the individual, of the race, and of the whole living world' (Flower 29). Flower closed his talk with a quotation from *In Memoriam*:

> Oh yet we trust that somehow good
>   Will be the final goal of ill . . .
>
> That nothing walks with aimless feet;
>   That not one life shall be destroy'd,
>   Or cast as rubbish to the void,
> When God hath made the pile complete.

(Quoted in Flower 29)

For Flower, Tennyson's lines captured the creative purpose behind evolution. In effect, evolution was best explained not by Darwin on his own but by a synthesis of Darwin with Tennyson – not by science on its own but by science together with literature.

All told, the multiple voices, forms and perspectives of Victorian creative literature mean that it can reveal in far more depth, detail and variety the apprehension of the human condition in the age of evolution than the prose arguments of scientists and commentators can do on their own. Any attempt to chart the history of evolution as an idea that does not draw widely on this rich resource is proportionately diminished and insufficient. Moreover, as the existential and ethical ramifications of Darwinism have not radically changed since the later 1800s, the literature that confronts them – whether it be the rich prose of Darwin himself, the incisive polemics of his advocates such as T. H. Huxley, as Jeff Wallace has shown, or the poetry and fiction which engaged with their work – remains current, even vital, as we grapple with these same issues ourselves today.

The science of ecology in nineteenth-century literature

The transition from a belief that living forms were created *ex nihilo* to a recognition that they have evolved from one another by natural processes is one of the most fundamental shifts within western thought. From a twenty-first century perspective, however, a no-less-crucial consequence of Charles Darwin's work was the emergence of ecology as a science. The theory of evolution by natural selection is predicated on the insight that nature is characterized by perpetually changing interactions between countless individual organisms struggling to survive in diverse physical and climatic situations. The ecosystems that result from these interactions are inevitably fluctuating yet broadly self-sustaining. Darwin captured this vision of nature in the final paragraph of *On the Origin of Species* in his famous image of 'an entangled bank, clothed with many plants of many kinds, with birds singing on the bushes, with various insects flitting about, and with worms crawling through the damp earth' (397), all 'dependent on each other in so complex a manner'. Darwin's German advocate Ernst Haeckel coined the term 'ecology' (*Œcologie* in German) to identify the study of these relations in nature as a science in its own right:

> By ecology we mean the body of knowledge concerning the economy of nature – the investigation of the total relations of the animal both to its inorganic and to its organic environment; including above all, its friendly and inimical relations with those animals and plants with which it comes directly or indirectly into contact – in a word, ecology is the study of all those complex interrelations referred to by Darwin as the conditions of the struggle for existence.
>
> *(Quoted in McIntosh 7–8)*

From the outset, as Haeckel's definition shows, ecology was an expressly Darwinian science, predicated on a redefinition of what Linnaeus had first called the 'economy of nature' in 1749 (Egerton 80). Where Linnaeus saw the harmonious relation of natural beings to one another as ordained by God, Darwin saw instead the fraught though no less interdependent relations of the struggle for survival.

Historians of ecology tend to follow Haeckel in identifying Darwin as, in Donald Worster's words, 'the single most important figure in the history of ecology' (114). Frank Egerton argues that, while natural history increasingly worked with many components of what would become ecology, as 'Naturalists in the first half of the 1800s gradually transformed Linnaeus's static economy of nature into a dynamic one' (135), it was not until the work of Darwin and his circle that it had the range of data and the organizing theory needed to make it a scientific discipline. At the end of the eighteenth century, natural history had taken on a particular form through White's act of writing and publishing *The Natural History of Selborne*. Writing in another genre – the scientific travel memoir – helped to bring into focus the relations that Haeckel would define as ecological. Darwin's *Journal of Researches* (1839), better known as *The Voyage of the Beagle*, was central here, as were later works in the same genre by his close allies Joseph Hooker (*Himalayan Journals*, 1854), Henry Walter Bates (*The Naturalist on the River Amazons*, 1863), and Alfred Russel Wallace (*The Malay Archipelago*, 1869). Other newly emergent scientific ideas besides evolutionary biogeography contributed too, including developments in pathology and parasitology, as Egerton explains (178–98), and thermodynamic theories of the circulation of energy, as Allen MacDuffie explores in detail in *Victorian Literature, Energy, and the Ecological Imagination* (2014).[1]

Haeckel first defined ecology in the 1860s but it was not until the 1890s that the term caught on in the English-speaking world. Ecology itself was recognized as 'a branch of biology coequal with morphology and physiology and "by far the most attractive"' (McIntosh 29) by the pathologist J. S. Burdon Sanderson in his presidential address to the British Association in 1893. Given that ecology did not become established as a science under that name in Britain until the 1890s, it is an open question how helpful or appropriate it is to use the term in discussing nineteenth-century literature. There is a further complication in that the term 'ecology' itself has multiple meanings. As Robert McIntosh points out (16–17), it can refer to the scientific discipline and its working assumptions; to the object of its enquiry, that is, the ecology of a given organism; and, particularly in the adjectival form 'ecological', to a set of ethical values and political principles. These last are helpfully summarized by Hayward (31–2) as living in harmony with nature, overcoming anthropological prejudice and recognizing intrinsic value in beings other than humans. Then there are the close associations between ecology and ethology – the study of animal behaviour in the wild – as sciences and the blurred distinctions between ecological politics, environmentalism and conservation. Given this morass of different but not fully distinct meanings, it is perhaps no surprise that, as John Parham has remarked, 'Critical studies of literature and the environment have been somewhat lax in their terminology' (2010 13).

Taking 'ecology' in the sense in which Haeckel first defined it as a science, and as indicative of the complexity and dynamics of the natural world as revealed by Darwin and his collaborators, it is likely to be productive to read writers such as Meredith, Hardy and Wells, whose world view was overtly Darwinian, in ecological terms. As Nicholas Frankel explains, Meredith's extraordinary poem 'The Woods of Westermain' (1882) makes 'a discernible effort to dissolve the confines of self in the dynamics of environmental interaction' (631). Hardy's novels *The Return of the Native* (1878) and *The Woodlanders* (1887) are likewise open to study as accounts of human life within distinctive rural ecologies as well as economies. Both these writers engage in their poetry with another important ecological corollary of Darwin's theory of common descent: that non-human animals have their own experience of their lives that are necessarily both like and unlike those of human animals (Holmes 2009 154–84). Wells explores this too in his deeply disturbing scientific romance *The Island of Doctor Moreau* (1896), where Moreau experiments in turning animals into people by means of vivisection – an objective correlative for the possibilities of evolution itself. In a third Darwinian romance, *The War of the Worlds* (1898), he conducts another study in ecology by staging an ecological disaster precipitated by an invasive species. The Martians flee the environmental collapse of Mars, yet at the end of the book they are destroyed by their maladaption to Earth's bacterial ecology.

But while these late Victorian, Darwinian writers have an ecological view of natural relations, and themselves contribute to understanding them in these terms, what about writers who challenged Darwin's assumptions and their implications? It was not necessary to be a confirmed Darwinian to write ecologically in the later Victorian period. Richard Jefferies, for example, repudiated Darwin's materialism, yet in *The Life of the Fields* (1884) he traced with precision and sensitivity the ecological impact of modern land management on the non-human inhabitants of the rural landscape. But the further removed writers are from the secular framework of Darwinian science, the more the definition of ecology in the scientific sense needs to be stretched to accommodate them. Ruskin would seem to be an obvious candidate for an ecological writer, yet his intensely anthropocentric conception of the function of the natural world calls this into question. Parham and Brian Day have both advanced compelling readings of Gerard Manley Hopkins as an ecological poet, yet Day acknowledges that Hopkins's conception of nature as pervaded by Christ is remote from the science informing his work, while Parham recognizes that, although the syntax and sprung rhythm of Hopkins's poems are imbued with a sense of ecological interaction, their arguments tend rather to reaffirm natural theology.

The question of whether literature can be described as ecological in the scientific sense intensifies as we move back into the early decades of the century. The ethical and political imperative of environmentalism has led to a new interest in British romanticism and American transcendentalism as literary movements that foreshadow current concerns, and from which we can learn how to live without desecrating or destroying the rest of the living world. Some of the most concerted ecocritical studies of Romantic literature, including books by Jonathan Bate, Timothy Morton, Kevin Hutchings and Katey Castellano, are principally concerned with the politics of environmentalism and only peripherally interested in the science of ecology.

But other critics have sought to underwrite the claim that Romantic poetry can help us confront and survive the unfolding ecological catastrophe of our own moment by arguing that the Romantics share with us not only environmental ethics but also the conception of nature as an intricate web of interactions, including interdependencies, which we have inherited from Darwin and Haeckel. The precise nature and force of the claim made for reading the Romantic poets as ecological writers varies from critic to critic. For Karl Kroeber, James McKusick and Ashton Nichols, Romantic poetry combines a scientific habit of mind with an imaginative apprehension of the interrelations in nature, giving rise to a new kind of 'scientific lyricism' (Nichols xix). According to Kroeber, Romantic poetry was 'the first literature to anticipate contemporary biological conceptions' (1994 2), crucially the 'new biological, materialistic understanding of humanity's place in the natural cosmos'. In his view, this merits us defining romanticism as at least 'proto-ecological' (Kroeber 1994 5). McKusick goes further, arguing that the Romantics are 'the first full-fledged ecological writers in the Western literary tradition' (19), ahead of Darwin and his associates.

The more ambitious of these claims are not always sustained by readings of individual authors and texts. McKusick (35–76) shows how Wordsworth and Coleridge are thoroughly grounded in specific environments, and how each engages in environmental advocacy, but not how they anticipate the scientific conception of ecology defined by Haeckel. His readings of William Blake's *Jerusalem* (1804–20) and of Mary Shelley's *Frankenstein* (1818) and *The Last Man* as staging environmental apocalypses (McKusick 95–111) are persuasive in themselves, but again not predicated on ecology as a science. Kenneth Cervelli's interpretation of Dorothy Wordsworth as an ecological writer is convincing in that she is shown to be recording and constructing an ecology in her journals, and stimulating in proposing that she establishes links between domestic economy and the economy of nature, but the mental framework for her writing remains Linnaean natural history and not modern ecology. Most recently, Dewey Hall has argued that William Wordsworth's environmental politics derive from and are underpinned by the observational habits he picked up from White. The argument itself is plausible, but the only sustained reading of Wordsworth's poetry (as opposed to his prose) as revealing this tendency is of one passage from his very early and uncharacteristic poem *An Evening Walk* (1793) (Hall 12–16).

The aesthetic value of the Romantic poets' celebration of nature, and its ethical and imaginative power, do not depend upon their sharing our conception of how nature itself works. In scientific terms, Kroeber's description of the Romantic poets as 'proto-ecological' is both more convincing and more enabling than McKusick's claim that they are fully fledged ecologists. Ultimately, what matters is less the narrow issue of how far their poetry conforms to the science of ecology as it came to be defined than the wider issue of what insights their poetry holds for our own ecological understanding of the natural world. But to bring these insights to bear we need to recognize the ways in which the Romantics were writing from a different standpoint from our own.

Following Egerton's account of the gradual coalescence of ecology as a science, we can nevertheless ask what perspectives and ideas creative literature may have contributed to that process, directly or indirectly, before Haeckel defined the new science in the 1860s. On a conceptual level, perhaps the greatest contribution of romanticism to the new current of ideas that would give rise to – and still sustains – the science of ecology lies in the scepticism many Romantic poets expressed towards rigid hierarchies and taxonomies. In the first draft of *The Prelude*, completed in 1799, Wordsworth repudiated 'that false secondary power by which / In weakness we create distinctions, then / Believe our puny boundaries are things / Which we perceive, and not which we have made' (Part 2, ll. 251–4). These lines refer specifically to the human mind, however, and fifteen years later, in *The Excursion*, he would affirm in the voice of the Wanderer both taxonomy and the anthropocentrism of natural theology:

> 'Happy is he who lives to understand,
> Not human nature only, but explores
> All natures,—to the end that he may find
> The law that governs each; and where begins
> The union, the partition where, that makes
> Kind and degree, among all visible Beings;
> The constitutions, powers, and faculties,
> Which they inherit,—cannot step beyond,—
> And cannot fall beneath; that do assign
> To every class its station and its office,
> Through all the mighty commonwealth of things;
> Up from the creeping plant to sovereign man.'
>
> *(Book 4, ll. 332–43)*

Here Wordsworth sets the fixity of species within firm parameters. Beer has argued that Wordsworth's poetry had a shaping influence on Darwin's imaginative response to the natural world; at the same time, as she recognizes, Darwin rejected the divisions and hierarchy that Wordsworth imposed on nature in *The Excursion* (2010 8–9; see also Gaull). Indeed, Ryan suggests, very plausibly, that Darwin read and reread Wordsworth's poetry in the late 1830s 'partly because he realized that Wordsworth was the foremost spokesman for a vision of nature that he was about to repudiate' (15).

In this rejection, Darwin shows an affinity with his own grandfather but also with the younger Romantic poets Shelley and Keats, whom he had read enthusiastically as a young man, as Beer reveals (2010 7). The precise influence of Keats or Shelley on Darwin is probably irrecoverable at this distance. But it is at least plausible that their poetry helped him to imagine a world in which natural kinds did not have to be fixed or ranked. In *Queen Mab* (1813), Shelley rejected the hierarchy Wordsworth would affirm the following year, projecting a new-found paradise in which 'man has lost / His terrible prerogative, and stands / An equal amidst equals' (Canto 8, ll. 225–7) alongside birds and other animals. Theresa Kelley has shown how Shelley breaks down the division between plants and animals, including humans, in 'The Sensitive Plant' and 'The Triumph of Life' (210–15), while, for Onno Oerlemans, his anti-essentialism is revealed in his supposition in *Prometheus Unbound* – anticipated in *Queen Mab* – that individual identity might be found at all levels from the molecular to the global, anticipating both Richard Dawkins's selfish gene and James Lovelock's Gaia theory as well as Darwin's own 'dissolving of species fixity' (143). At the same time, Shelley's vision of perpetual change in 'Mont Blanc' is resolutely physical and non-teleological. 'What Shelley reminds us of here,' Oerlemans suggests, 'is what Charles Darwin also argued: that we see more clearly, we see more of the world, when we allow classification to be only a heuristic device' (147). The same insight is pivotal in Keats's 'Lamia' (1820), where the sage Apollonius's insistence that Lamia is a serpent effectively denies her her very existence, destroying his pupil Lycius's life too in the process, even as the poem as a whole gives the lie to any such taxonomical exclusivity. As Sharon Ruston has shown (2011B), the question of taxonomy bears too on the status of the Creature in Mary Shelley's *Frankenstein* – whether he is a monster, a new human species or a being whose existence undermines these very categories.

The rejection of fixed kinds by the Shelleys and Keats is complemented by the acute perceptions of nature in poems written in the 1830s and 1840s by two poets Darwin is much less likely to have read: John Clare and William Michael Rossetti. As Oerlemans (137–43), Mahood (112–46) and McKusick (77–94) have all shown, Clare's close and sympathetic observation of the natural world led him to recognize the importance of 'the interdependence of life forms' (Mahood 138). Oerlemans in particular draws a contrast between Coleridge and Clare in this regard. In his unfinished *Hints Towards the Formation of a More Comprehensive Theory of Life*, Coleridge sought to create a system of thought that would reify taxonomies and thereby human superiority. By contrast, Clare's more naïve empiricism 'emphasizes the individual over the species, and is a way of seeing particularity and complexity in the natural world' (Oerlemans 139). The same is true of Rossetti's early series of poems 'Fancies at Leisure', published in the magazine *The Germ* in 1850 alongside poetry and prose by his sister Christina and his brother Dante Gabriel (see Holmes 2015A 19–30).

The gradual shift from natural history to ecology is revealed in incremental differences between Clare's poetry and Rossetti's. Like Wordsworth in *An Evening Walk*, Clare has a tendency to see nature through White's eyes. In one of his letters, White describes 'a large white-bellied field-mouse with three or four young clinging to her teats by their mouths and feet' (126) as she fled from his gardeners. In his well-known poem on a mouse's nest, beginning 'I found a ball of grass', Clare too observes 'an old mouse . . . / With all her young ones hanging at her teats' (ll. 5–6). Like White, Clare describes the mouse as 'grotesque' (l. 7; White 126). Clare's poem 'Sand Martin' opens 'Thou hermit haunter of the lonely glen / & common wild & heath' (ll. 1–2). Again, this directly echoes White's description of the same bird as 'haunting wild heaths and commons' (143) in one of four letters from the *Natural History of Selborne* first published in the *Philosophical Transactions* of the Royal Society as monographs on the British hirundines or swallows. Francesca Cuojati (31–34) has shown that Clare disliked the educational exclusivity of the Linnaean system and its obliviousness to local particularity. All the same, like White's monographs, many of Clare's poems cast the individuals they describe as representative members of particular species (e.g. 'The Martin', 'The Hedgehog', 'The Badger').

Rossetti's own affinity for nature comes across in 'A Quiet Place':

> My friend, are not the grasses here as tall
> As you would wish to see? The runnell's fall
> Over the rise of pebbles, and its blink
> Of shining points which, upon this side, sink
> In dark, yet still are there; this ragged crane
> Spreading his wings at seeing us with vain
> Terror, forsooth; the trees, a pulpy stock
> Of toadstools huddled round them; and the flock—
> Black wings after black wings—of ancient rook
> By rook; has not the whole scene got a look
> As though we were the first whose breath should fan
> In two this spider's web, to give a span
> Of life more to three flies? See, there's a stone
> Seems made for us to sit on. Have men gone
> By here, and passed? or rested on that bank
> Or on this stone, yet seen no cause to thank
> For the grass growing here so green and rank?
>
> *(1850 76–7)*

Like Clare, Rossetti has an eye for the grotesque in nature, but where 'The Mouse's Nest' concludes with an unease at the ultimate incomprehensibility of other creatures' lives, Rossetti's poem reclaims and celebrates the 'rank' and revolting, even hinting that it may have a divine 'cause'. Both poets attend to specific details of the environment around them too, creating a powerful sense of place. Rossetti's direct address to a friend within the poem intensifies this illusion of immediacy, and with it the sense that the animals, plants, even rocks, that they encounter, are unique. The stone they sit on is 'this stone', the spider whose web they blow on is 'this spider', the bird they startle 'this ragged crane'. This recognition and celebration of the individuality of other living things, of their interaction with and effects upon one another, and of their shaping of the landscape around them, marks 'A Quiet Place' out as a truly eco-logical poem, notwithstanding its gesture towards a divine bounty in nature. None of this ecological thinking is foregrounded within the poem, however. Instead, it is implicit in Rossetti's use of language. And yet, unlike Clare, Rossetti was writing as a member of a movement – the Pre-Raphaelite Brotherhood – that deliberately set out to model its art on science (see Holmes 2015B). As Rossetti himself wrote in the *Spectator* in 1851, the Pre-Raphaelites were dedicated to 'investigation for themselves on all points which have hitherto been settled by example or unproved precept, and unflinching avowal of the result of such investigation' (956). His poem conducts such an investigation, in the process reaching an understanding of nature that is predicated on interactions between individual living things.

The Wordsworths and Clare, like White himself, achieve fresh insights into the details of nature within the parameters of the old natural history. By imitating in their poetry and painting the scientific rigour of attentive observation, the experimental method, and the scrupulous recording of data, the Pre-Raphaelites were able to make their own discovery of ecological relations in nature concurrent with but independent of Darwin and his circle. In choosing as their model not Swainson's natural history but Herschel's natural philosophy – specifically geology and chemistry, singled out by another of the Pre-Raphaelites, Frederick George Stephens, as 'the modern science' (171) – they made the same move that the scientists themselves made in transforming their own practice into the new science of biology. At the same time, their work reveals through the detail and clarity of their empirical observations – as the writing of the Shelleys and Keats does through its imaginative apprehension of new ways of thinking about nature – how literature and art did and still can contribute to our conception of ecology as a science.

The life sciences in the nineteenth century underwent a profound paradigm shift. From a conception of the natural world as ordained by God to be essentially unchanging, they moved to an understanding of nature as perpetually shifting under its own impetus. Living beings had evolved, not been created; their relations to one another were intricate and provisional, not fixed, with roots not in providence but in the struggle for life. This transformation in the understanding of nature was matched by a change in the culture and practices of biology as a science, as it became professionalized and secular. But it was not limited to science, nor could it be, as with the science a whole world view had to change. The literature of the nineteenth century captures this transition from a Linnaean to a Darwinian world view. Romantic and Victorian poetry and fiction not only reflect and reflect upon this change. They enact it, enabling the new vision of nature to take hold, extending its reach and probing its significance. At the same time, the very science itself took form through the skilled use of literary genre, from epistles and argument to travel writing and even poetry.

Note

1 See also Barri J. Gold, 'Thermodynamics', in this volume.

Bibliography

Primary texts

[Anon.]. 'The New Museum of Natural History'. *Nature* 23 (1881): 549–52.

Barbauld, Anna. *Selected Poetry and Prose*. Eds. William McCarthy and Elizabeth Kraft. Peterborough: Broadview, 2002.

Bates, Henry Walter. *The Naturalist on the River Amazons*. 2 vols. London: Murray, 1863.

Carroll, David, ed. *George Eliot: The Critical Heritage*. London: Routledge, 1995.

Chambers, Robert. *Vestiges of the Natural History of Creation and Other Evolutionary Writings*. Ed. James A. Secord. Chicago, IL: University of Chicago Press, 1994.

Clare, John. *Poems of the Middle Period 1822–1837*. Eds. Eric Robinson, David Powell, and P. M. S. Dawson. 5 vols. Oxford: Oxford University Press, 1996–2003.

Clodd, Edward. *The Story of Creation: A Plain Account of Evolution*. London: Longmans, 1888.

Cobbe, Frances Power. *Darwinism in Morals, and Other Essays*. London: Williams and Norgate, 1872.

Darwin, Charles. *Journal of Researches into the Geology and Natural History of the Various Countries Visited by H. M. S. Beagle*. London: Colburn, 1839.

Darwin, Charles. *On the Origin of Species by Means of Natural Selection*. Ed. Joseph Carroll. Peterborough: Broadview, 2003.

Darwin, Charles. *The Descent of Man and Selection in Relation to Sex*. Eds. James Moore and Adrian Desmond. London: Penguin, 2004.

Darwin, Charles. *The Expression of the Emotions in Man and Animals*. Ed. Paul Ekman. New York, NY: Oxford University Press, 2009.

Darwin, Erasmus. *The Temple of Nature; or, the Origin of Society: A Poem*. London: Johnson, 1803.

Doyle, Arthur Conan. *The Lost World*. Ed. Ian Duncan. Oxford: Oxford University Press, 2008.

Flower, W. H. *Essays on Museums and Other Subjects Connected with Natural History*. London: Macmillan, 1898.

Gaskell, Elizabeth. *Wives and Daughters*. Ed. Pam Morris. Harmondsworth: Penguin, 1996.

Hardy, Thomas. *Tess of the D'Urbervilles: A Pure Woman*. Eds. A. Alvarez and David Skilton. Harmondsworth: Penguin, 1978.

Hooker, J. D. *Himalayan Journals*. 2nd edn. 2 vols. London: Murray, 1855.

Hunt, William Holman, and John Lucas Tupper. *A Pre-Raphaelite Friendship: The Correspondence of William Holman Hunt and John Lucas Tupper*. Eds. James H. Coombs *et al*. Ann Arbor, MI: University of Michigan Research Press, 1986.

Kingsley, Charles. *Scientific Lectures and Essays*. London: Macmillan, 1880.

Lyell, Charles. *Principles of Geology*. Ed. James A. Secord. Harmondsworth: Penguin, 1997.

Meredith, George. 'In the Woods'. *Fortnightly Review* n.s. 8 (1870): 179–83.

Meredith, George. *The Poems of George Meredith*. Ed. Phyllis B. Bartlett. 2 vols. New Haven, CT: Yale University Press, 1978.

Mivart, St George. *On the Genesis of Species*. 2nd edn. London: Macmillan, 1871.

Paley, William. *Natural Theology or Evidence of the Existence and Attributes of the Deity, Collected From the Appearances of Nature*. Eds. Matthew D. Eddy and David Knight. Oxford: Oxford University Press, 2006.

Reade, Winwood. *The Martyrdom of Man*. London: Trübner, 1872.

Rossetti, William Michael. 'Fancies at Leisure'. *The Germ* (1850): 76–8.

Rossetti, William Michael. 'Pre-Raphaelitism'. *Spectator* 24 (1851): 955–7.

Ruskin, John. *Modern Painters*. 6 vols. London: Allen, 1896.

Shaw, George Bernard. *Back to Methuselah: A Metabiological Pentateuch*. Harmondsworth: Penguin, 1939.

Shelley, Mary Wollstonecraft. *Frankenstein, or, 'The Modern Prometheus': The 1818 Text*. Ed. Marilyn Butler. Oxford: Oxford University Press, 2008.

Shelley, Mary Wollstonecraft. *The Last Man*. Ed. Morton D. Paley. Oxford: Oxford University Press, 2008.

Shelley, Percy Bysshe. *The Major Works*. Eds. Zachary Leader and Michael O'Neill. Oxford: Oxford University Press, 2003.

Stapledon, Olaf. *Last and First Men*. London: Gollancz, 1999.

Stephens, Frederick George [as Laura Savage]. 'Modern Giants'. *The Germ* (1850): 169–73.

Swainson, William. *A Preliminary Discourse on the Study of Natural History*. London: Longman *et al.*, 1834.

Tennyson, Alfred. *The Poems of Tennyson*. Ed. Christopher Ricks. 2nd edn. 3 vols. Harlow: Longman, 1987.

Thomson, James. *The City of Dreadful Night*. Ed. Edwin Morgan. Edinburgh: Canongate, 1993.

Wallace, Alfred Russel. *The Malay Archipelago*. 2 vols. London: Macmillan, 1869.

Wells, H. G. *The Time Machine*. Eds. Patrick Parrinder, Marina Warner and Steven McLean. London: Penguin, 2005.

White, Gilbert. *The Natural History of Selborne*. Ed. Anne Secord. Oxford: Oxford University Press, 2013.

Wordsworth, William. *The Poetical Works of Wordsworth*. Eds. Thomas Hutchinson and Ernest de Selincourt. London: Oxford University Press, 1936.

Wordsworth, William. *The Prelude: The Four Texts (1798, 1799, 1805, 1850)*. Ed. Jonathan Wordsworth. Harmondsworth: Penguin, 1995.

Secondary texts

Amigoni, David. *Colonies, Cults and Evolution: Literature, Science and Culture in Nineteenth-Century Writing*. Cambridge: Cambridge University Press, 2007.

Amigoni, David, and James Elwick, eds. *The Evolutionary Epic*. Vol. 4 of *Victorian Science and Literature*. London: Pickering and Chatto, 2011.

Armstrong, Isobel. *Victorian Poetry: Poetry, Poetics, and Politics*. London: Routledge, 1993.

Atwood, Sara. 'The Soul of the Eye: Ruskin, Darwin, and the Nature of Vision'. *Nineteenth-Century Prose* 38 (2011): 127–46.

Banfield, Marie. 'Darwinism, Doxology, and Energy Physics: The New Sciences, the Poetry and the Poetics of Gerard Manley Hopkins'. *Victorian Poetry* 45 (2007): 175–94.

Bate, Jonathan. *Romantic Ecology: Wordsworth and the Environmental Tradition*. London: Routledge, 1991.

Bate, Jonathan. *The Song of the Earth*. London: Macmillan, 2000.

Beer, Gillian. *Open Fields: Science in Cultural Encounter*. Oxford: Oxford University Press, 1996.

Beer, Gillian. *Darwin's Plots: Evolutionary Narrative in Darwin, George Eliot and Nineteenth-Century Fiction*. 3rd edn. Cambridge: Cambridge University Press, 2009.

Beer, Gillian. 'Darwin and Romanticism'. *Wordsworth Circle* 41.1 (2010): 3–9.

Bergmann, Linda S. 'Reshaping the Roles of Man, God and Nature: Darwin's Rhetoric in *On the Origin of Species*'. *Beyond the Two Cultures: Essays on Science, Technology, and Literature*. Eds. Joseph W. Slade and Judith Yaross Lee. Ames, IA: Iowa State University Press, 1990, 79–98.

Birch, Katy. '"Carrying Her Coyness to a Dangerous Pitch": Mathilde Blind and Darwinian Sexual Selection'. *Women: A Cultural Review* 24.1 (2013): 71–89.

Browne, Janet. 'Constructing Darwinism in Literary Culture'. *Unmapped Countries: Biological Visions in Nineteenth Century Literature and Culture*. Ed. Anne-Julia Zwierlein. London: Anthem, 2005, 55–69.

Burdett, Carolyn. *Olive Schreiner and the Progress of Feminism: Evolution, Gender, Empire*. Basingstoke: Palgrave Macmillan, 2001.

Castellano, Katey. *The Ecology of British Romantic Conservatism, 1790–1837*. Basingstoke: Palgrave Macmillan, 2013.

Cervelli, Kenneth R. *Dorothy Wordsworth's Ecology*. New York, NY: Routledge, 2007.

Choi, Tina Young. 'Natural History's Hypothetical Moments: Narratives of Contingency in Victorian Culture'. *Victorian Studies* 51 (2009): 275–97.

Cuojati, Francesca. 'John Clare: The Poetics and Politics of Taxonomy'. *The Exhibit in the Text: The Museological Practices of Literature*. Eds. Caroline Patey and Laura Scuriatti. Bern: Peter Lang, 2009, 29–48.

Dawson, Gowan. *Darwin, Literature and Victorian Respectability*. Cambridge: Cambridge University Press, 2007.

Day, Brian J. 'Hopkins' Spiritual Ecology in "Binsey Poplars"'. *Victorian Poetry* 42 (2004): 181–93.

Dean, Dennis R. *Tennyson and Geology*. Lincoln: Tennyson Society, 1985.

Ebbatson, Roger. *The Evolutionary Self: Hardy, Forster, Lawrence*. Brighton: Harvester, 1982.

Egerton, Frank N. *Roots of Ecology: Antiquity to Haeckel*. Berkeley, CA: University of California Press, 2012.

England, Richard, and Jude V. Nixon, eds. *Science, Religion and Natural Theology*. Vol. 3 of *Victorian Science and Literature*. London: Pickering and Chatto, 2011.

Ernst, Sabine. '*The Woman Who Did* and "The Girl Who Didn't": The Romance of Sexual Selection in Grant Allen and Ménie Muriel Dowie'. *Grant Allen: Literature and Cultural Politics at the Fin de Siècle*. Eds. William Greenslade and Terence Rogers. Aldershot: Ashgate, 2005, 81–94.

Ferguson, Christine. *Language, Science and Popular Fiction in the Victorian Fin-de-Siècle: The Brutal Tongue*. Aldershot: Ashgate, 2006.

Fletcher, Robert P. '"Heir of All the Universe": Evolutionary Epistemology in Mathilde Blind's *Birds of Passage: Songs of the Orient and the Occident'. Victorian Poetry* 43 (2005): 435–53.

Forsyth, R. A. 'Evolutionism and the Pessimism of James Thomson (B. V.)'. *Essays in Criticism* 12 (1962): 148–66.

Frankel, Nicholas. 'The Ecology of Victorian Poetry'. *Victorian Poetry* 41 (2003): 629–35.

Fulweiler, Howard W. 'Tennyson's *In Memoriam* and the Scientific Imagination'. *Thought* 59 (1984): 296–318.

Gaull, Marilyn. 'From Wordsworth to Darwin: "On to the Fields of Praise"'. *Wordsworth Circle* 10 (1979): 33–48.

George, Sam. *Botany, Sexuality and Women's Writing 1760–1830: From Modest Shoot to Forward Plant.* Manchester: Manchester University Press, 2007.

Gigante, Denise. *Life: Organic Form and Romanticism.* New Haven, CT: Yale University Press, 2009.

Glendening, John. *The Evolutionary Imagination in Late Victorian Novels.* Aldershot: Ashgate, 2007.

Goodall, Jane R. *Performance and Evolution in the Age of Darwin: Out of the Natural Order.* London: Routledge, 2002.

Greenslade, William. *Degeneration, Culture and the Novel, 1880–1940.* Cambridge: Cambridge University Press, 1994.

Groth, Helen. 'Victorian Women Poets and Scientific Narratives'. *Women's Poetry, Late Romantic to Late Victorian: Gender and Genre, 1830–1900.* Eds. Isobel Armstrong and Virginia Blain. Basingstoke: Macmillan, 1999, 325–51.

Hale, Piers J. 'Labor and the Human Relationship with Nature: The Naturalization of Politics in the Work of Thomas Henry Huxley, Herbert George Wells, and William Morris'. *Journal of the History of Biology* 36 (2003): 249–84.

Hale, Piers J. 'Of Mice and Men: Evolution and the Socialist Utopia. William Morris, H. G. Wells, and George Bernard Shaw'. *Journal of the History of Biology* 43 (2010): 17–66.

Hale, Piers J. 'William Morris, Human Nature and the Biology of Utopia'. *William Morris in the Twenty-First Century.* Eds. Phillippa Bennett and Rosie Miles. Bern: Peter Lang, 2010, 107–27.

Hall, Dewey W. *Romantic Naturalists, Early Environmentalists: An Ecocritical Study, 1789–1912.* Farnham: Ashgate, 2014.

Harris, Stuart. *Erasmus Darwin's Enlightenment Epic.* Sheffield: Harris, 2002.

Hawkins, Mike. *Social Darwinism in European and American Thought, 1860–1945.* Cambridge: Cambridge University Press, 1997.

Hayward, Tim. *Ecological Thought: An Introduction.* Cambridge: Polity, 1994.

Heringman, Noah. *Romantic Rocks, Aesthetic Geology.* Ithaca, NY: Cornell University Press, 2004.

Hodgson, Amanda. 'Defining the Species: Apes, Savages and Humans in Scientific and Literary Writing of the 1860s'. *Journal of Victorian Culture* 4 (1999): 228–51.

Holmes, John. '*The New Day*: Dr Hake and the Poetry of Science'. *Journal of Victorian Culture* 9.1 (2004): 68–89.

Holmes, John. *Darwin's Bards: British and American Poetry in the Age of Evolution.* Edinburgh: Edinburgh University Press, 2009.

Holmes, John. 'Darwinism, Feminism, and the Sonnet Sequence: Meredith's *Modern Love'. Victorian Poetry* 48 (2010): 523–38 [2010A].

Holmes, John. '"The Lay of the Trilobite": Rereading May Kendall'. *19: Interdisciplinary Studies in the Long Nineteenth Century* 11 (2010). <www.19.bbk.ac.uk/article/viewFile/575/658>. Accessed 24 April 2015 [2010B].

Holmes, John. 'Victorian Evolutionary Criticism and the Pitfalls of Consilience'. *The Evolution of Literature: Legacies of Darwin in European Cultures.* Eds. Nicholas Saul and Simon J. James. Amsterdam: Rodopi, 2011. 101–12.

Holmes, John. '"The Poet of Science": How Scientists Read Their Tennyson'. *Victorian Studies* 54 (2012): 655–78.

Holmes, John. 'Poetry on Pre-Raphaelite Principles: Science, Nature and Knowledge in William Michael Rossetti's "Fancies at Leisure" and "Mrs. Holmes Grey"'. *Victorian Poetry* 53 (2015): 15–39 [2015A].

Holmes, John. 'Pre-Raphaelitism, Science and the Arts in *The Germ'. Victorian Literature and Culture* 43 (2015): 689–703 [2015B].

Hurley, Kelly. *The Gothic Body: Sexuality, Materialism, and Degeneration at the Fin de Siècle.* Cambridge: Cambridge University Press, 1996.

Hutchings, Kevin. *Romantic Ecologies and Colonial Cultures in the British Atlantic World, 1770–1850*. Montreal: McGill-Queens University Press, 2009.

Jones, Anna Maria. 'Eugenics by Way of Aesthetics: Sexual Selection, Cultural Consumption, and the Cultivated Reader in *The Egoist*'. *Literature Interpretation Theory* 16 (2005): 101–28.

Karr, Jeff. 'Caliban and Paley: Two Natural Theologians'. *Studies in Browning and His Circle* 13 (1985): 37–46.

Kaston Tange, Andrea. 'Constance Naden and the Erotics of Evolution: Mating the Woman of Letters with the Man of Science'. *Nineteenth-Century Literature* 61 (2006): 200–40.

Kelley, Theresa M. *Clandestine Marriage: Botany and Romantic Culture*. Baltimore, MD: Johns Hopkins University Press, 2012.

Kelvin, Norman. *A Troubled Eden: Nature and Society in the Works of George Meredith*. Edinburgh: Oliver and Boyd, 1961.

King-Hele, Desmond. *Erasmus Darwin and the Romantic Poets*. Basingstoke: Macmillan, 1986.

Kistler, Jordan. *Arthur O'Shaughnessy, a Pre-Raphaelite Poet in the British Museum*. London: Routledge, 2016.

Krasner, James. *The Entangled Eye: Visual Perception and the Representation of Nature in Post-Darwinian Narrative*. Oxford: Oxford University Press, 1992.

Kroeber, Karl. *Ecological Literary Criticism: Romantic Imagining and the Biology of Mind*. New York, NY: Columbia University Press, 1994.

Kuskey, Jessica. 'Bodily Beauty, Socialist Evolution, and William Morris's *News From Nowhere*'. *Nineteenth-Century Prose* 38 (2011): 147–82.

Levine, George. *Darwin and the Novelists: Patterns of Science in Victorian Fiction*. Cambridge, MA: Harvard University Press, 1988.

Levine, George. *Darwin Loves You: Natural Selection and the Re-Enchantment of the World*. Princeton, NJ: Princeton University Press, 2006.

Levine, George. *Realism, Ethics and Secularism: Essays on Victorian Literature and Science*. Cambridge: Cambridge University Press, 2008.

Levine, George. *Darwin the Writer*. Oxford: Oxford University Press, 2011.

Lightman, Bernard. *Victorian Popularizers of Science: Designing Nature for New Audiences*. Chicago, IL: University of Chicago Press, 2007.

MacDuffie, Allen. *Victorian Literature, Energy, and the Ecological Imagination*. Cambridge: Cambridge University Press, 2014.

Mahood, M. M. *The Poet as Botanist*. Cambridge: Cambridge University Press, 2008.

Markley, Robert. 'The Nightmare of Evolution: H. G. Wells, Percival Lowell and the Legacies of Frankenstein's Science'. *Frankenstein's Science: Experimentation and Discovery in Romantic Culture*. Eds. Christa Knellwolf and Jane Goodall. Aldershot: Ashgate, 2008, 183–99.

McDonald, Ronan. '"Accidental Variations": Darwinian Traces in Yeats's Poetry'. *Science in Modern Poetry: New Directions*. Ed. John Holmes. Liverpool: Liverpool University Press, 2012, 151–66.

McIntosh, Robert P. *The Background of Ecology: Concept and Theory*. Cambridge: Cambridge University Press, 1985.

McKusick, James C. *Green Writing: Romanticism and Ecology*. 2nd edn. New York, NY: Palgrave Macmillan, 2010.

McLean, Steven. *The Early Fiction of H. G. Wells: Fantasies of Science*. Basingstoke: Palgrave Macmillan, 2009.

Merrill, Lynn L. *The Romance of Victorian Natural History*. Oxford: Oxford University Press, 1989.

Mitchell, Robert. *Experimental Life: Vitalism in Romantic Science and Literature*. Baltimore, MD: Johns Hopkins University Press, 2013.

Moore, James R. 'The Erotics of Evolution: Constance Naden and Hylo-Idealism'. *One Culture: Essays in Science and Literature*. Eds. George Levine and Alan Rauch. Madison, WI: University of Wisconsin Press, 1987, 225–57.

Morton, Peter. *The Vital Science: Biology and the Literary Imagination, 1860–1900*. London: Allen and Unwin, 1984.

Morton, Timothy. *Shelley and the Revolution in Taste: The Body and the Natural World*. Cambridge: Cambridge University Press, 1994.

Morton, Timothy. *Ecology Without Nature: Rethinking Environmental Aesthetics*. Cambridge, MA: Harvard University Press, 2007.

Morton, Timothy. *The Ecological Thought*. Cambridge, MA: Harvard University Press, 2010.

Murphy, Patricia. 'Fated Marginalization: Women and Science in the Poetry of Constance Naden'. *Victorian Poetry* 40 (2002): 107–30.

Nichols, Ashton. *Beyond Romantic Ecocriticism: Toward Urbanatural Roosting*. New York, NY: Palgrave Macmillan, 2011.

Noble, Mary. 'Darwin Among the Novelists: Narrative Strategy and the Expression of the Emotions'. *Nineteenth-Century Prose* 38 (2011): 99–126.

O'Connor, Ralph. 'From the Epic of Earth History to the Evolutionary Epic in Nineteenth-Century Britain'. *Journal of Victorian Culture* 14 (2009): 207–23.

O'Hanlon, Redmond. *Joseph Conrad and Charles Darwin: The Influence of Scientific Thought on Conrad's Fiction*. Edinburgh: Salamander, 1984.

Oerlemans, Onno. *Romanticism and the Materiality of Nature*. Toronto: University of Toronto Press, 2002.

Otis, Laura. *Organic Memory: History and the Body in the Late Nineteenth and Early Twentieth Centuries*. Lincoln, NE: University of Nebraska Press, 1994.

Page, Michael R. *The Literary Imagination from Erasmus Darwin to H. G. Wells*. Farnham: Ashgate, 2012.

Parham, John. *Green Man Hopkins: Poetry and the Victorian Ecological Imagination*. Amsterdam: Rodopi, 2010.

Parrinder, Patrick. *Shadows of the Future: H. G. Wells, Science Fiction and Prophecy*. Liverpool: Liverpool University Press, 1995.

Pleins, J. David. *In Praise of Darwin: George Romanes and the Evolution of a Darwinian Believer*. New York, NY: Bloomsbury, 2014.

Priestman, Martin. *The Poetry of Erasmus Darwin: Enlightened Spaces, Romantic Times*. Farnham: Ashgate, 2013.

Purton, Valerie. 'Darwin, Tennyson and the Writing of "The Holy Grail"'. *Darwin, Tennyson and Their Readers: Explorations in Victorian Literature and Science*. Ed. Valerie Purton. London: Anthem, 2013, 49–63.

Reid, Julia. *Robert Louis Stevenson, Science, and the Fin de Siècle*. Basingstoke: Palgrave Macmillan, 2006.

Richardson, Angelique. '"Some Science Underlies All Art": The Dramatization of Sexual Selection and Racial Biology in Thomas Hardy's *A Pair of Blue Eyes* and *The Well-Beloved*'. *Journal of Victorian Culture* 3 (1998): 302–38.

Richardson, Angelique. 'Hardy and Biology'. *Thomas Hardy: Texts and Contexts*. Ed. Philip Mallett. Basingstoke: Palgrave Macmillan, 2002, 156–79.

Richardson, Angelique. *Love and Eugenics in the Late Nineteenth Century: Rational Reproduction and the New Woman*. Oxford: Oxford University Press, 2003.

Richter, Virginia. *Literature After Darwin: Human Beasts in Western Fiction, 1859–1939*. Basingstoke: Palgrave Macmillan, 2011.

Roppen, Georg. *Evolution and Poetic Belief: A Study in Some Victorian and Modern Writers*. Oslo: Oslo University Press, 1956.

Rowlinson, Matthew. 'History, Materiality and Type in Tennyson's *In Memoriam*'. *Darwin, Tennyson and Their Readers: Explorations in Victorian Literature and Science*. Ed. Valerie Purton. London: Anthem, 2013, 35–48.

Ruddick, Nicholas. *The Fire in the Stone: Prehistoric Fiction from Charles Darwin to Jean M. Auel*. Middletown, CT: Wesleyan University Press, 2009.

Rudy, Jason R. 'Rapturous Forms: Mathilde Blind's Darwinian Poetics'. *Victorian Literature and Culture* 34 (2006): 443–59.

Rupke, Nicolaas. *Richard Owen: Biology Without Darwin, A Revised Edition*. Chicago, IL: University of Chicago Press, 2009.

Ruston, Sharon. *Shelley and Vitality*. 2nd edn. Basingstoke: Palgrave Macmillan, 2011 [2011A].

Ruston, Sharon. '"What Was I?": *Frankenstein*, Natural History, and the Question of What it Means to be Human'. *La Questione Romantica* 3.1 (2011): 81–92 [2011B].

Ryan, Robert M. *Charles Darwin and the Church of Wordsworth*. Oxford: Oxford University Press, 2016.

Schmitt, Cannon. *Darwin and the Memory of the Human: Evolution, Savages, and South America*. Cambridge: Cambridge University Press, 2009.

Secord, James A. *Victorian Sensation: The Extraordinary Publication, Reception, and Secret Authorship of 'Vestiges of the Natural History of Creation'*. Chicago, IL: University of Chicago Press, 2000.

Secord, James A. *Visions of Science: Books and Readers at the Dawn of the Victorian Age*. Oxford: Oxford University Press, 2014.

Shepherd-Barr, Kirsten. *Theatre and Evolution from Ibsen to Beckett*. New York, NY: Columbia University Press, 2015.

Smith, Jonathan. *Charles Darwin and Victorian Visual Culture*. Cambridge: Cambridge University Press, 2006.

Stevenson, Lionel. *Darwin Among the Poets*. Chicago, IL: University of Chicago Press, 1932.

Stott, Rebecca. '"Tennyson's Drift": Evolution in *The Princess*'. *Darwin, Tennyson and Their Readers: Explorations in Victorian Literature and Science*. Ed. Valerie Purton. London: Anthem, 2013, 13–34.

Talairach-Vielmas, Laurence. *Fairy Tales, Natural History and Victorian Culture*. Basingstoke: Palgrave Macmillan, 2014.

Thain, Marion. '"Scientific Wooing": Constance Naden's Marriage of Science and Poetry'. *Victorian Poetry* 41 (2003): 151–69.

Tomko, Michael. 'Varieties of Geological Experience: Religion, Body, and Spirit in Tennyson's *In Memoriam* and Lyell's *Principles of Geology*'. *Victorian Poetry* 42 (2004): 113–33.

Wallace, Jeff. 'T. H. Huxley, Science and Cultural Agency'. *Darwin, Tennyson and Their Readers: Explorations in Victorian Literature and Science*. Ed. Valerie Purton. London: Anthem, 2013, 153–66.

Williams, Carolyn. 'Natural Selection and Narrative Form in *The Egoist*'. *Victorian Studies* 27 (1983): 53–79.

Wilmer, Clive. '"No Such Thing as a Flower [. . .] No Such Thing as a Man": John Ruskin's Response to Darwin'. *Darwin, Tennyson and Their Readers: Explorations in Victorian Literature and Science*. Ed. Valerie Purton. London: Anthem, 2013, 97–108.

Worster, Donald. *Nature's Economy: A History of Ecological Ideas*. 2nd edn. Cambridge: Cambridge University Press, 1994.

Wu, Duncan. *Wordsworth's Reading 1770–1799*. Cambridge: Cambridge University Press, 1993.

Zaniello, Tom. *Hopkins in the Age of Darwin*. Iowa City, IA: University of Iowa Press, 1988.

Further reading

Amigoni, David, and Jeff Wallace, eds. *Charles Darwin's 'The Origin of Species': New Interdisciplinary Essays*. Manchester: Manchester University Press, 1995.

Bellanca, Mary Ellen. *Daybooks of Discovery: Nature Diaries in Britain, 1770–1870*. Charlottesville, VA: University of Virginia Press, 2007.

Coriale, Danielle. 'Charlotte Brontë's *Shirley* and the Consolations of Natural History'. *Victorian Review* 36.2 (2010): 118–32.

Cosslett, Tess. *The 'Scientific Movement' and Victorian Literature*. Brighton: Harvester, 1982.

De Almeida, Hermione. 'Romanticism and the Triumph of Life Science: Prospects for Study'. *Studies in Romanticism* 43 (2004): 119–34.

Denisoff, Dennis, ed. 'Victorian Ecosystems: A Forum'. *Victorian Review* 36.2 (2010): 11–49.

Heringman, Noah, ed. *Romantic Science: The Literary Forms of Natural History*. Albany, NY: State University of New York Press, 2003.

Knoepflmacher, U. C., and G. B. Tennyson, eds. *Nature and the Victorian Imagination*. Los Angeles, CA: University of California Press, 1977.

Kroeber, Karl. 'Proto-Evolutionary Bards and Post-Ecological Critics'. *Keats–Shelley Journal* 48 (1999): 157–72.

Lightman, Bernard, and Bennett Zon, eds. *Evolution and Victorian Culture*. Cambridge: Cambridge University Press, 2014.

Norris, Margot. *Beasts of the Modern Imagination: Darwin, Nietzsche, Kafka, Ernst and Lawrence*. Baltimore, MD: Johns Hopkins University Press, 1985.

Parham, John. 'Was There a Victorian Ecology?'. *The Environmental Tradition*. Ed. John Parham. Aldershot: Ashgate, 2002, 156–71.

'Romanticism and Ecology'. *Wordsworth Circle* 27.3 (1997).

'Romanticism and the Sciences of Life'. *Studies in Romanticism* 43.1 (2004).

Ruse, Michael. *Darwinism as Religion: What Literature Tells Us About Evolution*. Oxford: Oxford University Press, 2017.

Ruston, Sharon. *Creating Romanticism: Case Studies in the Literature, Science and Medicine of the 1790s*. Basingstoke: Palgrave Macmillan, 2013.

'Science, Literature, and the Darwin Legacy'. *19: Interdisciplinary Studies in the Long Nineteenth Century* 11 (2010). <www.19.bbk.ac.uk/79/volume/0/issue/11>. Accessed 24 April 2015.

23

ARCHAEOLOGY AND ANTHROPOLOGY

Julia Reid

University of Leeds

Archaeological and anthropological enquiry traversed the borders of discipline, discourse and genre with notable freedom throughout the nineteenth century. Even by the end of the Victorian period, archaeology and anthropology were not yet fully professionalized, and literary writers, archaeologists and anthropologists engaged in productive dialogue, blurring the boundaries between scientific and literary writing. There is now a growing interdisciplinary scholarship recognizing these interconnections and exploring the importance of literary elements in these sciences and the place of archaeological and anthropological themes, concerns and motifs in novels and poetry. The present chapter argues that scientific and literary writers engaged in a common endeavour to explore connections between past and present. The first half of the chapter examines how literary and scientific writings on archaeology evoked the passage of time as a narrative of counterpoised loss and preservation. The second half demonstrates that anthropology brought to the surface archaeology's partially submerged concern with empire, nation and 'race'. It investigates the fraught preoccupation with civilization and savagery, with self and other, that ran through Victorian anthropology and literature.

Archaeology: 'From their dead Past thou liv'st alone'

The burgeoning of interest in archaeology across the nineteenth century expressed the era's peculiarly urgent attention to the past. Victorians' belief in their own modernity paradoxically entailed this absorption in the past and shaped the period's strongly historical imagination. The 'nineteenth century's unprecedented historicism,' Chris Brooks observes, 'was the corollary of its unprecedented consciousness of its own present' (3). Archaeology and the other historical sciences embody, in part, recuperative attempts to connect with a rapidly vanishing past: they constitute responses to the accelerating pace of social and cultural change. For scientific writers as well as novelists and poets, the work of excavation became a way of putting past and present into contact with each other and of tracing the endurance of the past into the present.

Historians emphasize that the nineteenth century saw archaeology's almost complete transformation into a professional discipline. William Stiebing evokes its passage from an 'adventuresome hobby' for amateur enthusiasts to an 'academic discipline' that was rigorously scientific, objective and professional (24). Virginia Zimmerman judges that Augustus Pitt Rivers's development of precise excavation techniques between 1880 and 1900 represented archaeology's 'final move away from antiquarianism and its associations with the Romantic view of the landscape' and its 'complete transformation into a science' (2008 101). These years, certainly, saw the emergence of modern archaeology as a discipline, as Pitt Rivers and others, including the Egyptologist W. M. Flinders Petrie, evolved systematic excavation methods, stratigraphic dating techniques and meticulous recording and preservation practices.

Yet throughout the period archaeological cultures were heterogeneous, as amateur and professional, elite and popular, and scientific and literary discourses coexisted and engaged in creative dialogue. Amateur archaeology flourished, with an impressive growth in the number of local societies. Archaeological audiences were multiplicitous. Flinders Petrie sensed the fractured nature of the readership and, in his popular works, sought to reach 'the large number of readers who feed in the intermediate regions between the arid highlands and mountain ascents of scientific memoirs, and the lush – not to say rank – marsh-meadows of the novel and literature of amusement' (1). 'High' and 'low' archaeological cultures were not as distinct as Petrie's hierarchical model suggested. Popular entertainment and educational discourses were fused, for example, in the public presentation of Egyptian artefacts in the early nineteenth century: as Sophie Thomas notes, the British Museum helped to fashion the Crystal Palace's Egyptian Court, a 'compelling blend of spectacle and public education' (19).

The literary and the scientific imagination were also entwined in nineteenth-century archaeological cultures, as Alexandra Warwick and Martin Willis have recently argued (1–2). Indeed, romanticism, romance and science were closely linked in the classic archaeological texts. The Assyrian archaeologist A. H. Layard's bestselling *Nineveh and Its Remains* (1849) is a work of travel writing or adventure fiction as much as an archaeological report. Infused with local colour, the text narrates a series of adventures, from robbery to the conflict between the Kurds and Chaldaeans (Layard 263, 126, 133, 140–41). Layard's narrative is imbued by the Romantic sublime, recording his awed 'intoxication of the senses' (63). He evokes the uncanniness of the scene as he approaches 'the time-worn ruins of Al Hather', which 'rose in solitary grandeur in the midst of a desert . . . as they stood fifteen centuries before' (Layard 75). The sense that barriers between past and present are dissolving emerges through the language of dreams and visions: recalling the buried sculptures he has seen, Layard is 'half inclined to believe that we have dreamed a dream, or have been listening to some tale of Eastern romance', and imagines that 'when the grass again grows over the ruins of the Assyrian palaces' others will 'suspect that I have been relating a vision' (333). The frequency of the 'dream' motif in archaeological writings, as Warwick contends, points to scientists' recognition that 'empirical observation' alone could not allow 'the comprehension of archaeological sites' (83).

Even with the rise of the professional excavation report in the 1880s, as David Gange observes, archaeologists continued to publish 'narrative and descriptive works' (47), and these were often better known at the time. Alongside his technical field reports on Egyptian sites, for instance, Petrie published popular works like *Ten Years' Digging in Egypt 1881–1891* (1892), which emphasized the archaeologist's emotional responses rather than professional objectivity. *Ten Years' Digging* appeals to popular interest in the heroic explorer and in the act of discovery. In one Gothic scene, Petrie's attempts to open a sarcophagus see him spending a 'gruesome day, sitting astride of the inner coffin, unable to turn my head under the lid without tasting the bitter brine in which I sat' (93). When he eventually retrieves the mummified body, Petrie's tone becomes reverent: 'Tenderly we towed him out to the bottom of the entrance pit . . . and then came the last, and longed-for scene, for which our months of toil had whetted our appetites, – the unwrapping of Horuta' (94).

The 'archaeological imagination', as Warwick and Willis term it, was not contained by the boundaries of genre, discipline, or discourse (1). Professional archaeologists, popularizers and literary writers participated in mutually influential dialogue. Literary writers engaged with the work of archaeological writers: as we shall see, Edward Bulwer-Lytton's *The Last Days of Pompeii* (1834) emphasizes his debt to the antiquarian William Gell's works on the Pompeian excavations, and Dante Gabriel Rossetti's 'The Burden of Nineveh' (1856, 1870) footnotes the work of Layard (Bulwer-Lytton 31; Rossetti 25). Equally important, and less fully explored to date, is the countervailing influence of literary texts on non-fictional treatments of archaeology. Percy Bysshe Shelley's sonnet 'Ozymandias' (1818) clearly articulated the interest in Egyptian archaeology aroused by Napoleon's Egyptian campaign. It is unsurprising that later literary tradition was influenced by Shelley's ironic rendering of the tyrant's inscription ('"My name is OZYMANDIAS, King of Kings. / Look on my works ye Mighty, and despair!"') and his suggestion that, in the 'decay / Of that Colossal Wreck', the sculptor's art alone survived (ll. 10–11, 12–13). More unexpectedly, 'Ozymandias' proved influential for later Egyptologists: E. P. Weigall's *Guide to the Antiquities of Upper Egypt* (1910) cited the poem to identify 'a fallen granite colossus' (Weigall 252; Janowitz 47).

At the heart of the archaeological imagination, uniting both its literary and its scientific articulations, was the desire to explore not only the past but also, crucially, the connection between past and present. The archaeological artefact, as Zimmerman demonstrates, embodied for Victorians a paradoxical reminder of both evanescence and preservation (2008 2–9). Archaeologists imagined the passage from past to present as marked by obliteration and loss; but they emphasized too the possibilities of survival and reconstruction. In *Ten Years' Digging*, Petrie evoked the destruction of the historical record – it was 'heartrending' to see 'the pile of papyrus rolls, so rotted that they fell to pieces with a touch' (33) – but also its resurrection through his labours. His excavations, he wrote, offer a 'glimpse of the prehistoric age in Egypt': 'we begin to see a great past rising before us, dumb, but full of meaning' (Petrie 145, 152).

Nowhere was this duality of annihilation and preservation more potent than in nineteenth-century responses to Pompeii and Herculaneum. Excavations at these sites, already a topic of fascination to British readers in the late 1700s, received a boost when Napoleon conquered the Kingdom of Naples in 1806 (Stiebing 152–3). As nineteenth-century Britons were well aware, the towns' catastrophic extinction ironically ensured their survival in the historical record (Easson 105; Zimmerman 2008 108–25). Among the many nineteenth-century literary authors to engage with the excavations were Felicia Hemans and Bulwer-Lytton. Hemans's poem 'The Image in Lava' (1828) meditates upon the impression of a woman and baby found in the sand at Herculaneum. The speaker contrasts the pair's sudden and agonizing destruction with the 'immortal' nature of the mother's 'love' for her child (ll. 41, 37). Hemans's insistence that 'I could pass all relics / Left by the pomps of old, / To gaze on this rude monument / Cast in affection's mould' (ll. 33–6) shares with much writing on Pompeii an emphasis on 'the individual and the quotidian' (Zimmerman 2008 108) and on 'Roman domesticity' (Easson 100). But for Hemans, this 'rude monument' also proffers a gendered moral: 'Empires from earth have pass'd, / And woman's heart hath left a trace / Those glories to outlast!' (ll. 35, 6–8).

Bulwer-Lytton's *The Last Days of Pompeii* (1834) similarly explores the twinned processes of destruction and survival. The novel's fictional account of Pompeii's 'last days' is punctuated by discussions of the archaeological 'trace' left by the characters, buildings and artefacts (Bulwer-Lytton 420). As Angus Easson observes, the novel offers a 'teasing mingling of fiction and reality' (106). Bulwer-Lytton repeatedly directs his readers to archaeological sites and museums where they will be able to see, for example, the impressions left in the sand by '[t]he skeletons which, re-animated for a while, the reader has seen play their brief parts upon the stage' (428, 420). The motif of reanimation was, Zimmerman shows, characteristic of nineteenth-century writing about Pompeii (2008 111). But Bulwer-Lytton shifts the emphasis onto the authority of the literary writer to 'people once more those deserted streets, to repair those graceful ruins, to reanimate the bones which were yet spared to his survey; to traverse the gulf of eighteen centuries, and to wake to a second existence – the City of the Dead!' (v).

Bulwer-Lytton himself and subsequent scholars have noted his indebtedness to Gell (Bulwer-Lytton 31; Easson 100–104; Zimmerman 2008 114). Equally striking is the author's battle for cultural authority with 'antiquaries' and archaeological writers, a struggle over who can best interpret the traces of the past (Bulwer-Lytton 27, 76 n.). Invoking Walter Scott, Bulwer-Lytton claims that the novelist's knowledge of the 'human passions and the human heart' makes him more 'at home with the past' than the 'learned' antiquary (viii–ix, x, 70, 73). In fact, Bulwer-Lytton's novel points to the generic instability of archaeological writing. The author describes his novel as a 'history' and hints that he could write a 'curious and interesting treatise' on the ancient world (Bulwer-Lytton 421, 423). He casts himself as an archaeological field worker, 'inspecting the strata' to adjudicate between 'theories' about Pompeii's destruction (Bulwer-Lytton 427). He asserts that his 'description of that awful event is very little assisted by invention, and will be found not the less accurate for its appearance in a Romance' (Bulwer-Lytton 427). Mediating between different genres, Bulwer-Lytton also playfully addresses a heterogeneous audience, directing 'the learned reader' and 'the reader who is *not* learned' respectively to original Latin sources and English translations (425).

In a subtler vein than Bulwer-Lytton's novel, Thomas Hardy's poetry and fiction explore the writer's ability to animate the past and to transcend time. Hardy's interest in archaeology is well known. He joined the Dorset Natural History and Antiquarian Field Club in 1881 (Radford 2013 212) and he wrote essays and letters to the newspapers on archaeological topics. Hardy's fictional and poetic landscapes are marked by history (Welshman 222–3). Traces of the past are everywhere, from the presence of Stonehenge at the end of *Tess of the D'Urbervilles* (1891) to the ancient British fort in his short story 'A Tryst at an Ancient Earthwork' (1885). Archaeological excavation brings the present into contact with the past: as the narrator in 'A Tryst' observes, 'by merely peeling off a wrapper of modern accumulations we have lowered ourselves into an ancient world' (Hardy 1977 325). Even in the absence of artefacts, the narrator believes that he can hear 'the lingering air-borne vibrations of conversations uttered at least fifteen hundred years ago' (Hardy 1977 321). Despite these 'lingering . . . vibrations', the story emphasizes obliteration and loss, a process that is compounded by unethical archaeology: a 'professed and well-known antiquary' violates the archaeological site, exhuming a human skeleton that disintegrates 'under his touch' (Hardy 1977 321, 323, 326). Similar anxieties about archaeological ethics, and about the effacement of the past, inform Hardy's poem 'The Clasped Skeletons' (1928), which opens with a poignant question about the excavation of an ancient British barrow near his house: 'O why did we uncover to view / So closely clasped a pair?' (ll. 1–2).

In the face of this obliteration, Hardy saw it as the poet's task to connect with the past and hence to transcend time. In the section of *Poems of the Past and Present* (1901) entitled 'Poems of Pilgrimage', two sonnets inspired by Hardy's visit to Italy dramatize an intimately entwined past and present. Hardy casts himself, Ian Ousby notes, as a 'poet-pilgrim' to the past (54). The first sonnet, 'In the Old Theatre, Fiesole', opens with the words 'I traced the Circus' (l. 1). The verb gestures towards both a physical act (walking round the amphitheatre) and an act of poetic reconstruction. The reanimation of the past in fact falls to a child who shows the speaker an 'ancient coin / That bore the image of a Constantine' (ll. 3–4). Evoking the idea of a resurrection, the poet notes that the child 'had raised for me' (l. 6) a shared past (see Zimmerman 2012 74). The past that is 'raised' by the child is a heritage shared across the ancient Roman Empire: Hardy need only 'delve' in his 'plot of English loam' to find 'Coins of like impress' (ll. 9–11). The next sonnet in the sequence, 'Rome: On the Palatine', intensifies this sense that past and present are intimately fused. Walking across the Palatine, the speaker reaches 'Caligula's dissolving pile' (l. 4), the present participle indicating effacement of the past yet also hinting at the melting of barriers between past and present. The visit to Caesar's house brings the past to life, as strains of music 'Raised the old routs Imperial lyres had led' (l. 12). The sonnet's final couplet evokes an ecstatic transcendence of time, describing how the music 'blended pulsing life with lives long done / Till Time seemed fiction, Past and Present one' (ll. 13–14).

Hardy's writings – like nineteenth-century writings about Pompeii – emphasize the relationship between the individual and history. More overtly political concerns about nation and empire also powerfully moulded archaeology, as scholars have recently explored. Archaeological practice and writing articulated a desire to understand Britain's place in the world as much as an interest in the excavated cultures. 'Biblical education', Gange observes, encouraged Britons to see 'a historical destiny passed down from the ancient Near East to themselves' (51), and the work of British archaeologists amplified this sense of connection with the ancient world. Speaking of his Assyrian excavations, for example, Layard recorded his 'wonder' that 'far distant, and comparatively new, nations should have preserved the only records of a people once ruling over nearly half the globe; and should now be able to teach the descendants of that people . . . where their monuments once stood' (316). Literary engagements with archaeology, too, address questions about nation, empire and Britishness. Dante Gabriel Rossetti's 'The Burden of Nineveh' and H. Rider Haggard's Egyptological Gothic fiction exemplify, in different ways, the use of archaeology to probe the relationship between imperial Britain and the ancient world.

Rossetti's poem 'The Burden of Nineveh' (1856, 1870) takes as its subject the 'winged lion or bull' discovered by Layard in the ancient Assyrian city of Nimroud, a site that the archaeologist identified with the Biblical Nineveh (Layard 50). Rossetti began writing the poem in 1850, a few weeks after the statue arrived in London, while the country was in the grip of 'Assyriamania' (Malley 2012 24, Stauffer 378). The poem acknowledges its debt to Layard (Rossetti 25). Layard had emphasized the artefact's power to 'conjure up strange fancies' (50) and to connect the present with the imagined past of remote antiquity. Rossetti too, passing 'A wingèd beast from Nineveh' in the 'swing-door' of the British Museum, is inspired to a waking dream about the statue's history (ll. 8–10). He addresses the bull as the sole link with the ancient past of the Assyrian priests: 'From their dead Past thou liv'st alone' (l. 48). However, Layard's account of appropriating and transporting Assyrian artefacts to the British Museum had expressed imperial self-confidence (89, 95, 106). By contrast, Rossetti's depiction of the relationship between the British Museum's Assyrian and Egyptian artefacts, which are 'All relics here together' (l. 107), emphasizes the misreadings of history. As Andrew Stauffer demonstrates, Rossetti's winged bull is 'a figure for imperial hubris and the confusions of history it engenders' (379). Assyria's aggressive imperialism, Stauffer observes, made it a 'dark mirror' of imperial Britain, with Nineveh's destruction prompting 'anxieties about England's future past' (370, 372). Rossetti's poem also looks uncertainly to the future, evoking a day when the bull god will once again set sail:

> In ships of unknown sail and prow,
> Some tribe of the Australian plough
> Bear him afar, – a relic now
>   Of London, not of Nineveh!
>
> *(ll. 177–80)*

Rossetti's description of London as 'this desert place' (l. 186) where visitors confuse the bull god for a British artefact resonates with T. B. Macaulay's famous portrayal of 'some traveller from New Zealand [who] shall, in the midst of a vast solitude, take his stand on a broken arch of London Bridge to sketch the ruins of St Paul's' (228). Like Macaulay's New Zealander, the future responses to the 'relic' imagined by Rossetti invoke an unsettling model of the cyclical decline and fall of empires.

By the end of the century, Rossetti's disquiet about imperial Britain's connections with the 'dead Past' assumed a new form. Imperial Gothic fiction used archaeology to explore the relationship between imperial Britain and the ancient world, amplifying and sensationalizing Rossetti's disquiet yet providing reassuring conclusions. Egyptological Gothic tales, by Haggard, Arthur Conan Doyle, Bram Stoker and others, were particularly popular during Britain's informal occupation of Egypt (1882–1914). These years also saw the flourishing of British Egyptology, which served to 'buttress . . . the British understanding of their own imperial power' (Deane 2008B 388). Haggard's short story 'Smith and the Pharaohs' (1912–13) dramatizes the threat posed by the resurgent past, as the mummies housed in the Cairo Museum are raised from the dead. In this tale, Haggard (who was himself a keen amateur Egyptologist) seems at first to evoke Egyptologists' uneasy conscience. Smith, an amateur Egyptologist, falls in love with the British Museum's bust of the ancient Egyptian queen Ma-Mee. He travels to Egypt and excavates her tomb, an act that, he reluctantly realizes, is a 'violat[ion]' (Haggard 1921 15). He finds himself locked in the Cairo Museum, where the reanimated mummies accuse him of grave-robbing. However, Ma-Mee reveals that Smith is the reincarnation of her lover, and he is pardoned because love prompts his actions. Ma-Mee ends by promising to Smith an everlasting 'union' (Haggard 1921 65). The representation of 'mummies as elusively seductive brides' in this and other tales, Bradley Deane argues, represents 'Pharaonic Egypt as a symbol of enduring power that could complement Britain's own' (2008B 406). Haggard's novel *She* (1886–87) can also be read as a 'mummy' fantasy that unsettles but eventually confirms Britain's imperial power. Ayesha's 'swathed mummy-like form' (Haggard 1991 142) is a problematic object of desire. As Nicholas Daly notes, Ayesha destabilizes the imperial/archaeological project: a 'collector in her own right', she is an 'exotic object' turned 'subject' (107–8). In due course, though, the explorers reassert their domination over her. The men's victory over Ayesha, like the antique potsherd's revelation of an Egyptian past 'embalmed in an English family name' (Haggard 1991 37), suggests that modern-day Britain is the true inheritor of an ancient Egyptian heritage. In *She*, as in his other Egyptological fiction, Haggard uses archaeology to meditate upon modern British imperialism and, ultimately, to vindicate white presence in modern-day Africa.

For Haggard and other literary writers as much as for the scientists, then, archaeology offered a means to explore urgent questions about the present as well as the past. Reanimating Egyptian mummies, Pompeiian citizens or Assyrian priests, literary and scientific writers emphasized not only the ravages of time but also their own ability to transcend mortal limits and to bring the past to life. Archaeological debate moved fluidly across the boundaries of fiction and non-fiction, as novelists, poets and travel writers played their part alongside archaeologists in sustaining the public appetite for archaeological narrative and spectacle.

Anthropology: 'a solid layer of savagery beneath the surface of society'

Anthropology brought to the fore the questions about nation, 'race' and empire that lurked beneath archaeological explorations of past and present. Anthropology and archaeology were in many ways entwined, both developing out of antiquarian cultures. The two disciplines remained conjoined in amateur societies, popular culture and exhibiting practice, which routinely presented archaeological artefacts and indigenous peoples as twin 'survivals' from the past. Mathilde Blind's poem 'The Beautiful Beesharean Boy' (1895) poignantly evokes the yoking together of archaeological and anthropological 'others', describing a boy from an Egyptian desert tribe:

> Shipped to the World's great Fair –
> The big Chicago Show!
> With mythic beasts and thin
> Beetles and bulls with wings,
> And imitation Sphinx,
> Ranged row on curious row!

(ll. 91–6)

Like archaeologists, anthropologists engaged in creative dialogue with literary writers. The late Victorian period saw the first steps towards professionalization, with anthropology's founding father, E. B. Tylor, appointed keeper of the University Museum at Oxford in 1883 (Stocking 264). Because the discipline still lacked full professional status, the boundaries between anthropology and literature were fluid. As a newly interdisciplinary scholarship has demonstrated, literary and anthropological discourses cross-fertilized one another (MacClancy 24–32). Scott Ashley points to the importance of creative writers in the 'prehistory of ethnography' (18), urging that the work of 'ethnographic *flâneurs* like [J. M.] Synge or . . . Robert Louis Stevenson' should be 'incorporated within the histories of anthropology, or the rich cultural context in which the discipline was founded risks being thinned'. Critics have also recently explored the literary qualities of anthropological texts, from Tylor's 'poetic Romanticism' (Logan 108) to the 'romantic' and 'poetic' elements infusing A. C. Haddon and C. R. Browne's ethnographic writings on the Aran Islands (Ashley 17, 11). Brad Evans locates the high-water mark of the relationship between literature and anthropology in a later period, the interwar years, 'when poets and anthropologists seemed to share the same project with regard to the elucidation of authentic cultures' (437). But this interdisciplinary dialogue was arguably at its richest during the Victorian period when, as we shall see, anthropologists and literary writers joined in exploring the connections between past, present and future, and between their own and other cultures.

A fraught preoccupation with the past ran through nineteenth-century anthropology and literature, shaping anthropological writings by Tylor, Andrew Lang and J. G. Frazer, and literary work by Walter Scott, George Eliot, Thomas Hardy, H. Rider Haggard, Robert Louis Stevenson and others. Tylor's foundational work, *Primitive Culture* (1871), which applied evolutionary methods to the field of culture, offered an undeniably hierarchical model of cultural progress. He proudly proclaimed that 'the science of culture is essentially a reformer's science' (Tylor 2:410). His influential doctrine of 'survivals' – customs and beliefs that have persisted 'by force of habit into a new stage of society' and provide 'proofs and examples of an older condition of culture' – evinces his debt to Enlightenment rationalism (Tylor 1:15). Through anthropology, he teaches, surviving superstition 'lies open to the attack of its deadliest enemy, a reasonable explanation' (Tylor 1:15). Building on the Scottish Enlightenment's stadial theory, Tylor calls his work 'a development-theory of culture' (2:100). For Tylor, the 'savage state in some measure represents an early condition of mankind', with present-day 'savages and barbarians' still 'produc[ing], in rude archaic forms, man's early mythic representations of nature' (1:28, 1:286). Tylor's scheme is temporally brutal: as Johannes Fabian argues, the nineteenth-century anthropologist engages in a 'denial of coevalness', casting the anthropological object as distant in time (35, 31). Anne McClintock too analyses Victorian culture's reading of the colonized 'other' as 'anachronistic', 'the living embodiment of the archaic "primitive"' (30). Gender ideologies also centrally shaped anthropology's hierarchical narrative, which equated femininity and primitivity and traced the gradual progress from matriarchy or matriliny to patriarchy (Reid 2015).

Despite the emphasis on unilinear, hierarchical development, however, Tylor's 'science of culture' harbours surprising tensions. Christopher Herbert emphasizes the duality of the Tylorian survival, which demonstrates both 'the transcendence of the primitive' and 'the opposite, its inescapable persistence' (432). Indeed, Tylor stresses similarities as well as distinctions between present and past, European self and racial other. The theory of the soul, he writes, 'unites, in an unbroken line of mental connexion, the savage fetish-worshipper and the civilized Christian', adding that 'there seems no human thought so primitive as to have lost its bearing on our own thought, nor so ancient as to have broken its connexion with our own life' (Tylor 1:453, 2:409). Tylor's project, Deane observes, is 'founded upon a complicated and unstable tension between past and present', in which nostalgia for a lost past coexists with an emphasis on 'taxonomies of cultural difference' (2008A 216–17).

The duality of the 'primitive survival' was at the heart of the nineteenth-century dialogue between anthropology and creative writing. An equivocal response to the past can be found, long before Tylor's day, in Walter Scott, who was also indebted to Scottish Enlightenment stadial theorists (Richards 125–30), and who passed on his ambivalence to a cluster of Scottish anthropologists, notably Lang and Frazer. In Scott, belief in progress is tempered by an elegiac and Romantic attraction to the savage past. Frank Osbaldistone in *Rob Roy* (1818) articulates this duality, describing himself as 'a supporter of the present government upon principle' (1995 37) but valuing the fierce 'loyalty and duty' of the rebellious Highlanders. A hierarchical model of cultural development – and an attendant principle of ' "salvage" ethnography' (112), in James Clifford's term – structures Scott's novels. Thus, *Waverley* (1814) aims to 'preserv[e] some idea of the ancient manners of which I have witnessed the almost total extinction' (Scott 1986 340). The act of preservation is only possible because the past no longer poses a threat. Yet Scott's negotiation of past and present was complex. As James Buzard discusses, Scott's 'performance of the role of autoethnographer on behalf of a "Scotland" he appears to have known himself to be fabricating' was 'highly self-conscious and ambivalent' (63). In *Waverley*, for example, the observation that the Highlanders' appearance 'conveyed to the south country Lowlanders as much surprise as . . . an invasion of African negroes, or Esquimaux Indians' (Scott 1986 214) might suggest the hierarchical comparative method, but it also offers a wry commentary on Lowland perceptions of Highlanders.

An equivocal relationship to the 'savage' past similarly underlies the work of those Scottish anthropologists who were influenced by Scott: Lang and Frazer. Both men were indebted as much to Scott as to Tylor, underlining the role of creative writers in the early history of anthropology (Crawford 1992 157). Indeed, Lang, who applied Tylor's anthropological method to comparative mythology, claimed that Scott 'first called attention in England to the scientific importance' of fairy tales (1873 619). Meanwhile, Frazer's epic work *The Golden Bough* is 'literature' as much as 'science', according to Robert Crawford (1990 28). Frazer's debt to Scott, I suggest, expresses their shared ambivalence towards progress. Critics commonly see Frazer as a 'rationalist' with an 'animus against religion', who saw 'human development' as 'linear' and 'progressive' (Connor 66–7). Certainly, the second edition of *The Golden Bough* (1900) advanced an apparently progressive account of the passage from magic through religion to science. Nostalgia for a lost world of belief, however, marks Frazer's depiction of the 'inevitable . . . breach[ing]' of religion's 'venerable walls' by the 'battery of the comparative method', an assault in which he participated: it was, he observed, a 'melancholy . . . task to strike at the foundations of belief' (1900 1:xxii). As Robert Fraser judges, Frazer's ostensible secularism is belied by an attraction to myth, ritual and religion (11–15). Frazer also amplifies and darkens Tylor's understanding of survivals. Like Tylor, he evokes the persistence of the past. But Frazer's account, by the second edition of *The Golden Bough*, is more threatening in tone: he warns that the 'solid layer of savagery beneath the surface of society' poses 'a standing menace to civilisation' (1900 1:74), and continues dramatically '[w]e seem to move on a thin crust which may at any moment be rent by the subterranean forces slumbering below'. Frazer's narrative, Herbert observes, offers a 'Gothic refraction' of Tylor's theory, intimating 'the uncanny latency of horrific primitive practices in modern-day Christianity' (432).

Throughout the nineteenth century, literary writers shared anthropologists' interest in narratives of development and survival, engaging with anthropological ideas to explore the relations between past, present and future, and between civilized and 'savage'. The remainder of the chapter first examines how Eliot and Hardy turned an anthropological gaze on English provincial and rural life, arguing that Eliot's fiction offers a critical scrutiny of anthropologists' pursuit of evolutionary origins, and that Hardy unsettles a unilinear model of temporal development. It then considers the late Victorian romance revival, scrutinizing the use and subversion of anthropological discourse by writers including Haggard and Stevenson.

Eliot's fascination with origins, progress, development and survivals aligns her with contemporary anthropologists. This preoccupation had diverse intellectual roots, stemming originally from her interest in the German critics Ludwig Feuerbach and David Strauss. She read Tylor and his fellow evolutionary anthropologists J. F. McLennan and John Lubbock, and she shared with these evolutionary anthropologists an important intellectual heritage (Eliot 1996A 19, 312–14; Eliot [n.d.] 31, 44–5). Eliot, Tylor and Frazer were all influenced by the French positivist Auguste Comte, whose 'Law of Three Stages' itself drew on Enlightenment stadial theory (Logan 69–71, 90; Richards 173). Eliot and Frazer also shared a love of Scott and were both interested in comparative philology, which 'practised a linguistic theory of survivals long before Tylor' (Richards 173). These complex lineages coalesce around a pursuit of evolutionary origins, a quest that impelled the anthropological project and that Eliot examined with an interested but critical eye.

Eliot's fiction offers a quasi-anthropological study of provincial life, as critics have often recognized. Her realism, according to P. M. Logan, served as 'a domestic form of Victorian ethnography' (68), viewing 'provincial life as if it were a less-developed form of her own advanced culture'. In *The Mill on the Floss* (1860), the 'Fetish' doll that Maggie Tulliver 'punished for all her misfortunes' (Eliot 1996C 28) exemplifies this anthropological approach. Fetishism – the anthropomorphic interpretation of the world long associated with the 'primitive' mind – was the first of Comte's 'Three Stages'; Tylor renamed it 'animism' (Logan 90). Eliot's depiction of Maggie's fetishism coincides with these contemporary constructions of primitive culture. *Silas Marner* (1861) offers a more sympathetic portrayal of animistic religion. Self-reflexively commenting on the realist novel, warning against inflated claims to objectivity, the narrator remarks of Silas's reluctance to abandon the old gods that 'The gods of the hearth exist for us still; and let all new faith be tolerant of that fetishism, lest it bruise its own roots' (Eliot 1996B 137). The sympathetic yet detached tone of Eliot's narrative voice has, indeed, led critics to read her novels as resonating more strongly with twentieth-century ethnography than with Victorian anthropology. Buzard and Clifford interpret Eliot's fiction as 'metropolitan autoethnography' (Buzard 12) and her narrators as engaging in 'participant-observation' *avant la lettre* (Clifford 114).

Turning a proto-ethnographic gaze inwards on English provincial communities, Eliot also reflects critically on the quest for origins that underlay nineteenth-century anthropology and comparative mythology. At the heart of her concern is the relationship between past and present: the 'vital connexion' between the 'world's ages' (Eliot 1997 198) explored in *Middlemarch*

(1871–2). The pursuit of origins embodied in Casaubon's 'Key to all Mythologies' is associated with the dead past of theology, with Casaubon 'the ghost of an ancient', trying to 'reconstruct a past world' (Eliot 1997 58, 16, 17). As Ian Duncan explains, Casaubon 'toils in the theological old regime of comparative mythology' (17), unaware of its transformation through German philology and biblical criticism – a transformation that was ultimately to pave the way for Tylor's anthropology. Casaubon's error is less his ignorance of German than his insistence on a backward-looking narrative of degeneration: 'all . . . mythical systems', he believes, are 'corruptions of a tradition originally revealed' (Eliot 1997 22). Casaubon's approach to the past is sterile, lacking the 'vital connexion' with the present and future felt by the artist Ladislaw. The word 'connexion' recurs in Dorothea's yearning for 'a binding theory which could bring her own life and doctrine into strict connexion with that amazing past, and give the remotest sources of knowledge some bearing on her actions' (Eliot 1997 79). In contrast with Casaubon's stultifying orientation towards past alone, Dorothea and Ladislaw embrace future evolutionary possibilities, mysteriously feeling the 'stirring of new organs' (Eliot 1997 461, 209). *Daniel Deronda* (1876) too meditates upon the evolution of religion, focusing on cultural and racial inheritance. It uses Tylorian language, describing the Cohens' charity towards Mordecai, for example, as 'a "survival" of prehistoric practice, not yet generally admitted to be superstitious' (Eliot 1967 449). For Duncan, the novel plays with evolutionary temporalities, contrasting the 'prehistory' embodied by Grandcourt and Gwendolen with the promise of Deronda as a rejuvenating 'future human type' (30). The opening of *Daniel Deronda* famously focuses readers' attention on the quest for origins, asserting that 'Men can do nothing without the make-believe of a beginning' (Eliot 1967 35). The search for a 'beginning', Eliot suggests here, is deeply problematic yet an inevitable part of human nature.

While Eliot focused on anthropology's quest for origins, Hardy was preoccupied by endings. He used archaeology, as we saw, to bring the present into contact with a near-obliterated past, and his interest in anthropology served a similar aim: 'to preserve . . . a fairly true record of a vanishing life' (1978A 477). This aim resonates with Clifford's '"salvage" ethnography', but Hardy was not quite an 'outsider' in relation to the obsolescent society he delineated (Clifford 112–13), and his works advanced a radical uncertainty about the salvage ethnographer's hierarchical model of cultural development. His anthropological lore drew as much on personal memory and local antiquarian writings as on Tylorian theory (Radford 2013 214–15). Respect rather than condescension, moreover, marked his 'record of a vanishing life'. His description of the reddleman in *The Return of the Native* (1878) as a 'nearly perished link between obsolete forms of life and those which generally prevail' (Hardy 1978B 59) was significantly echoed in his obituary of his mentor, the Dorset philologist and poet William Barnes: 'the most interesting link between present and past forms of rural life that England possessed' (Hardy 2001 66–7). *The Return of the Native*, indeed, unsettles the relations between past and present, querying evolutionary anthropology's progressivist narrative. The novel offers, in many ways, a Tylorian reading of peasant life, describing mumming, for instance, as a 'fossilized survival' (Hardy 1978B 178). But elsewhere it eschews Tylor's model of hierarchical development. In an image that tellingly pairs decay and preservation, Hardy describes the wind singing in the 'mummied heath-bells'; he imagines, too, a reaction that is mingled, at once primitive and civilized: 'an emotional listener's fetichistic mood', he suggests, might coexist with 'more advanced' thought (Hardy 1978B 105–6). The novel's plot also expresses doubt about the relations between past, present and future through its depiction of Clym Yeobright, the 'native' of the title, who is 'educated for an as yet non-existent future' (Beer 38). Despite his identification with progress, Clym longs to return to a past embodied in the heath: his solitary walks in the 'prehistoric' landscape see him 'seized' by the 'shadowy hand' of the 'past', which 'held him there to listen to its tale'

(Hardy 1978B 56, 449). Clym's attempt to return, as Gillian Beer and Andrew Radford argue, is marked by frustration and loss (Beer 38–53; Radford 2003 87–94). The novel's temporal uncertainties are never resolved. If Clym fails to return, the narrative itself arguably enacts a different kind of 'return', an imaginative renewal of Hardy's folk materials (Beer 53). This renewal could be understood as an act of ' "salvage" ethnography', yet the novel's accent on obliteration and obsolescence undermines its own attempt to reanimate a 'vanishing' past.

Where Eliot and Hardy directed their anthropological gaze inward on provincial and rural life, the late Victorian romance revival turned it outward to empire and adventure. Dramatizing colonial or historical encounters, Haggard, Rudyard Kipling, Conan Doyle, Stevenson and others explored Frazer's 'subterranean forces slumbering below' – the primitive survivals supposedly represented by racial others and European peasants, or to be found lurking in the depths of the civilized self. The imperial romance shared anthropology's central concern with the relations between civilization and savagery, and between past, present and future. Lang, who was both scientific popularizer and literary writer, played a catalysing role in the cross-fertilization of anthropological and literary discourses at the *fin de siècle*. He deployed an evolutionary vocabulary to champion the romance genre, hailing the love of romance and adventure as a 'survival of barbarism' (Lang 1887 689) and lauding its capacity to rejuvenate a jaded and effeminate modernity. He also lent anthropological support to romance novelists, acting, for example, as Haggard's informal ethnographic adviser (Reid 2011 2).

Romance novelists engaged in divergent and often complex ways with the relations of civilization and 'savagery'. The theory of survivals, as we have seen, was double-edged, suggesting both hierarchical progress and the endurance of the evolutionary past. Anthropology was, additionally, poised on the brink of a new cultural relativism at the *fin de siècle*, as its confident vision of hierarchical evolution was increasingly challenged by a nascent appreciation of cultural plurality (Reid 2006 141–2). Mary Kingsley, travel writer and ethnographer of West Africa, exemplifies the 'inconsistent pluralization of *culture*' (Buzard 6) that characterized this transitional period. Romance writers too were caught between Victorian anthropology's unilinear evolutionism and an incipient cultural relativism. Critics have emphasized the romance's complicity with imperialism's evolutionary hierarchies (Low 2–99, 264–5; McClintock 232–57). Certainly, the genre's deployment of anthropological discourses often served to legitimate imperial power and naturalize racial hierarchies. In Conan Doyle's *The Sign of Four* (1890), for instance, anthropological language casts the Andaman Islanders as a natural criminal type and works to divert attention from the potentially political motivations of colonial crime (68–9). However, other writers were more ambivalent in their scrutiny of the Tylorian survival, able to subvert as well as work within anthropological discourse.

Haggard's adventure fiction appears in some ways to challenge the ethnocentric assumptions which underlay Tylor's construction of the primitive survival. His novels evoke the persistence of 'primitive' forces within the supposedly civilized: the eponymous narrator of *Allan Quatermain* (1887) observes that 'in all essentials the savage and the child of civilisation are identical' and that '[c]ivilisation is only savagery silver-gilt' (Haggard 1995 10). Haggard's celebration of a shared 'primitive' masculinity apparently forges cross-cultural bonds. In *King Solomon's Mines* (1885), the Zulu Umbopa and Sir Henry Curtis exemplify an ideal manhood: they are described as 'two such splendid men', and Sir Henry strikingly chooses to 'dress . . . like a native warrior' (Haggard 1989 200, 199). Yet, as in many of Haggard's novels, binaries between savagery and civilization are only collapsed, and cross-cultural bonds are only valorized, in order to understand and regenerate British masculinity. Sir Henry's affinity with Umbopa, like his heritage of 'Danish blood' (Haggard 1989 11), serves to fashion a heroic British manliness rather than to propose racial equality. As Gail Ching-Liang Low judges, 'Haggard's romantic appropriation of Zulu military culture' is marked by 'narcissism', and the 'cross-cultural dressing works in one direction only' (9, 60). Kipling's novel *Kim* (1900–01) demonstrates a similarly ambivalent engagement with anthropology's evolutionary hierarchies, celebrating cross-cultural encounter but restricting the ability to cross between cultures and races to the colonizers. Kim, though he denies his own essential whiteness, is accorded an authority and mobility that are denied to Hurree Babu, the native ethnographer who quotes Herbert Spencer and 'collect[s] folk-lore for the Royal Society' but still 'dread[s] the magic' that he investigates (Kipling 180).

Stevenson engages more subversively with evolutionary anthropology's narrative of evolutionary progress. As my own work shows, throughout his oeuvre, from his Scottish fiction to his South Seas travel writing and imperial romances, Stevenson questioned the portrayal of racial others as 'primitive survivals' and dramatized the endurance of savagery within supposedly civilized societies (Reid 2006 111–73). *Kidnapped* (1886) unsettles a narrative of progress from a 'primitive' Highland culture to civilized modernity. The Lowland hero David Balfour initially sees Highlanders as barbarous survivals but moves towards a more sympathetic understanding, observing that '[i]f these are the wild Highlanders, I could wish my own folk wilder' (Stevenson 1994 101). From 1888 onwards, Stevenson's experiences in the South Seas led to an intensified mistrust of unilinear evolutionary narratives and an amplified perception of civilization's hidden savageries. The imperial romance 'The Beach of Falesá' (1892) undoes the ethnocentric assumptions embodied in the idea of superstition as 'primitive survival': superstitions in this tale are either imported by corrupt white traders as instruments of social control or valuable parts of a sophisticated and coherent Polynesian folk culture. Stevenson's South Seas travel writing, like his romance, celebrates cultural difference, pointing forward to the nascent relativism. *In the South Seas* (1896) prefigures twentieth-century ethnography's emphasis on the social value of superstition and ritual, observing, for example, that Polynesian taboo, far from being a 'meaningless and wanton prohibition', is 'more often the instrument of wise and needful restrictions' (Stevenson 1998 39, 40). Rejecting the characterization of superstition as irrational survival, Stevenson condemns missionaries who 'deride and infract even the most salutary *tabus*' and laments that 'so few people have read history and so many have dipped into little atheistic manuals' (1998 35, 65). The work condemns the harmful effects of colonialism on indigenous cultures and queries narratives of progress from savagery to civilization. Resembling an early ethnographic fieldworker, Stevenson advocates cultural immersion, yet he also acknowledges the barriers to sympathetic understanding. Trying to elicit folklore from a Pacific islander, he admits, 'I shall not hear the whole; for he is already on his guard with me' (Stevenson 1998 140).

Conclusion

'[I]magination, the power of inward vision, is as necessary to science as to poetry', wrote Frazer, contesting the idea that there was an epistemological break between science and literature (1927 301–2). Indeed, as we have seen, the borders between scientific and literary writing were blurred as novelists, poets, archaeologists and anthropologists engaged in a common endeavour to explore the relationship between past and present. Responding to contemporary perceptions of historical disjunction, these writers were centrally concerned to understand how far, and in what ways, the past persisted in the modern world. Their responses were complex, ambivalent and often at odds with each other. Archaeological writings – both literary and scientific – variously assume authority to reconstruct the past or evoke the futility of attempts to conquer time; at times they express imperial self-confidence but at other times (or even at the same time) they appear haunted by the archaeological past. Anthropological writings too are equivocal: absorbed yet unnerved by the primitive survival, that 'menace' which, Frazer feared, threatened to erupt through the 'thin crust' of civilization. Moving across science and literature, the debate about the survival raised fundamental questions about the relationship between past and present, savagery and civilization, self and other, as writers used but also at times subverted an ethnocentric narrative of progress and transcendence.

Bibliography

Primary texts

Blind, Mathilde. 'The Beautiful Beeshareen Boy'. *Birds of Passage: Songs of the Orient and Occident*. London: Chatto and Windus, 1895, 40–7. See *Victorian Women Writers Project*. <http://purl.dlib.indiana.edu/iudl/vwwp/VAB7026>. Accessed 25 September 2014.

Bulwer-Lytton, Edward. *The Last Days of Pompeii* [1834]. London: Routledge, 1877.

Doyle, Arthur Conan. *The Sign of Four* [1890]. London: Penguin, 2001.

Eliot, George. *Daniel Deronda* [1876]. Ed. Barbara Hardy. London: Penguin, 1967.

Eliot, George. *George Eliot's 'Daniel Deronda' Notebooks*. Ed. Jane Irwin. Cambridge: Cambridge University Press, 1996 [1996A].

Eliot, George. *Silas Marner: The Weaver of Raveloe* [1861]. Ed. Terence Cave. Oxford: Oxford University Press, 1996 [1996B].

Eliot, George. *The Mill on the Floss* [1860]. Ed. Gordon Haight. Intro. Dinah Birch. Oxford: Oxford University Press, 1996 [1996C].

Eliot, George. *Middlemarch* [1871–2]. Ed. David Carroll. Intro. Felicia Bonaparte. Oxford: Oxford University Press, 1997.

Eliot, George. *George Eliot's 'Middlemarch' Notebooks: A Transcription*. Eds. John Clark Pratt and Victor A. Neufeldt. Berkeley, CA: University of California Press, n.d.

Frazer, J. G. *The Golden Bough: A Study in Magic and Religion*. 2nd edn, revised and enlarged. 3 vols. London: Macmillan, 1900.

Frazer, J. G. 'Fison and Howitt' [1909]. Reprinted in *The Gorgon's Head and Other Literary Pieces*. London: Macmillan, 1927, 291–331.

Haggard, H. Rider. 'Smith and the Pharaohs' [1912–13]. Reprinted in *Smith and the Pharaohs and Other Tales*. New York, NY: Longmans, Green, 1921, 1–68.

Haggard, H. Rider. *King Solomon's Mines* [1885]. Ed. Dennis Butts. Oxford: Oxford University Press, 1989.

Haggard, H. Rider. *She* [1886–7]. Ed. Daniel Karlin. Oxford: Oxford University Press, 1991.

Haggard, H. Rider. *Allan Quatermain* [1887]. Ed. Dennis Butts. Oxford: Oxford University Press, 1995.

Hardy, Thomas. *The Complete Poems of Thomas Hardy*. Ed. James Gibson. London: Macmillan, 1976.

Hardy, Thomas. 'A Tryst at an Ancient Earthwork' [1885]. Reprinted in *Life's Little Ironies and A Changed Man*. Ed. F. B. Pinion. London: Macmillan, 1977, 317–26.

Hardy, Thomas. 'General Preface to the Wessex Edition of 1912'. *The Return of the Native*. Ed. George Woodcock. London: Penguin, 1978, 475–80 [1978A].

Hardy, Thomas. *The Return of the Native* [1878]. Ed. George Woodcock. London: Penguin, 1978 [1978B].

Hardy, Thomas. 'The Rev. William Barnes, B. D.' [1886]. Reprinted in *Thomas Hardy's Public Voice: The Essays, Speeches, and Miscellaneous Prose*. Ed. Michael Millgate. Oxford: Clarendon, 2001, 66–7.

Hemans, Felicia. 'The Image in Lava' [1828]. Reprinted in *Victorian Women Poets: An Anthology*. Eds. Angela Leighton and Margaret Reynolds. Oxford: Blackwell, 1995, 14–15.

Kipling, Rudyard. *Kim* [1900–1]. Ed. Alan Sandison. Oxford: Oxford University Press, 1987.

Lang, Andrew. 'Mythology and Fairy Tales'. *Fortnightly Review* 13.77 (1873): 618–33.

Lang, Andrew. 'Realism and Romance'. *Contemporary Review* 52 (1887): 683–93.

Layard, Austen Henry. *Nineveh and Its Remains: A Narrative of An Expedition to Assyria During the Years 1845, 1846, & 1847*. Abridged by the author. London: John Murray, 1867.

Macaulay, T. B. Review of *The Ecclesiastical and Political History of the Popes During the Sixteenth and Seventeenth Centuries*, by Leopold Ranke. *Edinburgh Review* 72.145 (1840): 227–58.

Petrie, W. M. Flinders. *Ten Years' Digging in Egypt 1881–1891*. New York, NY: Fleming H. Revell, 1892.

Rossetti, Dante Gabriel. 'The Burden of Nineveh' [1856; rev. 1870]. *The Pre-Raphaelites and Their Circle*. Ed. Cecil Y. Lang. 2nd edn. Chicago, IL: University of Chicago Press, 1968, 1975, 23–8.

Scott, Walter. *Waverley; or, 'Tis Sixty Years Since* [1814]. Ed. Claire Lamont. Oxford: Oxford University Press, 1986.

Scott, Walter. *Rob Roy* [1818]. Intro. Eric Anderson. London: David Campbell, 1995.

Shelley, Percy Bysshe. 'Ozymandias' [1818]. *The Complete Poems of Percy Bysshe Shelley*. Eds. Donald H. Reiman, Neil Fraistat, and Nora Crook. 3 vols to date. Baltimore, MD: Johns Hopkins University Press, 2003–12, 3:326.

Stevenson, Robert Louis. *Kidnapped* [1886]. Ed. Donald McFarlan. London: Penguin, 1994.

Stevenson, Robert Louis. *In the South Seas* [1896]. Comp. Sidney Colvin. Ed. Neil Rennie. London: Penguin, 1998.

Tylor, E. B. *Primitive Culture: Researches Into the Development of Mythology, Philosophy, Religion, Art, and Custom*. 2 vols. London: John Murray, 1871.

Weigall, Arthur E. P. *Guide to the Antiquities of Upper Egypt: From Abydos to the Sudan Frontier*. London: Methuen, 1910.

Secondary texts

Ashley, Scott. 'The Poetics of Race in 1890s Ireland: An Ethnography of the Aran Islands'. *Patterns of Prejudice* 35 (2001): 5–18.

Beer, Gillian. *Open Fields: Science in Cultural Encounter*. Oxford: Clarendon, 1996.

Brooks, Chris. 'Historicism and the Nineteenth Century'. *The Study of the Past in the Victorian Age*. Ed. Vanessa Brand. Oxford: Oxbow, 1998, 1–19.

Buzard, James. *Disorienting Fiction: The Autoethnographic Work of Nineteenth-Century British Novels*. Princeton, NJ: Princeton University Press, 2005.

Clifford, James. 'On Ethnographic Allegory'. *Writing Culture: The Poetics and Politics of Ethnography*. Eds. James Clifford and George E. Marcus. Berkeley, CA: University of California Press, 1986, 98–121.

Connor, Steven. 'The Birth of Humility: Frazer and Victorian Mythography'. *Sir James Frazer and the Literary Imagination: Essays in Affinity and Influence*. Ed. Robert Fraser. Basingstoke: Macmillan, 1990, 61–80.

Crawford, Robert. 'Frazer and Scottish Romanticism: Scott, Stevenson and *The Golden Bough*'. *Sir James Frazer and the Literary Imagination: Essays in Affinity and Influence*. Ed. Robert Fraser. Basingstoke: Macmillan, 1990, 18–37.

Crawford, Robert. *Devolving English Literature*. Oxford: Clarendon, 1992.

Daly, Nicholas. *Modernism, Romance and the Fin de Siècle: Popular Fiction and British Culture, 1880–1914*. Cambridge: Cambridge University Press, 1999.

Deane, Bradley. 'Imperial Barbarians: Primitive Masculinity in Lost World Fiction'. *Victorian Literature and Culture* 36 (2008): 205–25 [2008A].

Deane, Bradley. 'Mummy Fiction and the Occupation of Egypt: Imperial Striptease'. *English Literature in Transition, 1880–1920* 51.4 (2008): 381–410 [2008B].

Duncan, Ian. 'George Eliot's Science Fiction'. *Representations* 125.1 (2014): 15–39.

Easson, Angus. '"At Home" With the Romans: Domestic Archaeology in *The Last Days of Pompeii*'. *The Subverting Vision of Bulwer Lytton: Bicentenary Reflections*. Ed. Allan Conrad Christensen. Newark, DE: University of Delaware Press, 2004, 100–15.

Evans, Brad. 'Introduction: Rethinking the Disciplinary Confluence of Anthropology and Literary Studies'. *Criticism* 49.4 (2007): 429–45.

Fabian, Johannes. *Time and the Other: How Anthropology Makes its Object*. New York, NY: Columbia University Press, 1983.

Fraser, Robert. 'The Face Beneath the Text: Sir James Frazer in His Time'. *Sir James Frazer and the Literary Imagination: Essays in Affinity and Influence*. Ed. Robert Fraser. Basingstoke: Macmillan, 1990, 1–17.

Gange, David. *Dialogues with the Dead: Egyptology in British Culture and Religion, 1822–1922*. Oxford: Oxford University Press, 2013.

Herbert, Christopher. 'The Doctrine of Survivals, the Great Mutiny, and *Lady Audley's Secret*'. *Novel: A Forum on Fiction* 42.3 (2009): 431–6.

Janowitz, Anne. 'Shelley's Monument to Ozymandias'. *Philological Quarterly* 63.1 (1984): 477–90.

Logan, Peter Melville. *Victorian Fetishism: Intellectuals and Primitives*. Albany, NY: State University of New York Press, 2009.

Low, Gail Ching-Liang. *White Skins/Black Masks: Representation and Colonialism*. London: Routledge, 1996.

MacClancy, Jeremy. *Anthropology in the Public Arena: Historical and Contemporary Contexts*. Oxford: Wiley-Blackwell, 2013.

McClintock, Anne. *Imperial Leather: Race, Gender and Sexuality in the Colonial Context*. New York, NY: Routledge, 1995.

Malley, Shawn. 'Nineveh 1851: An Archaeography'. *Journal of Literature and Science* 5.1 (2012): 23–37. <www.literatureandscience.org/issues/JLS_5_1/JLS_vol_5_no_1_Malley.pdf>. Accessed 16 September 2014.

Ousby, Ian. 'Past and Present in Hardy's "Poems of Pilgrimage"'. *Victorian Poetry* 17.1/2 (1979): 51–64.

Radford, Andrew. *Thomas Hardy and the Survivals of Time*. Aldershot: Ashgate, 2003.

Radford, Andrew. 'Folklore and Anthropology'. *Thomas Hardy in Context*. Ed. Phillip Mallett. Cambridge: Cambridge University Press, 2013, 210–20.

Reid, Julia. *Robert Louis Stevenson, Science, and the Fin de Siècle*. Basingstoke: Palgrave Macmillan, 2006.

Reid, Julia. '"King Romance" in *Longman's Magazine*: Andrew Lang and Literary Populism'. *Victorian Periodicals Review* 44.4 (2011): 354–76.

Reid, Julia. '"She Who Must be Obeyed": Anthropology and Matriarchy in H. Rider Haggard's *She*'. *Journal of Victorian Culture* 20.3 (2015): 357–74.

Richards, David. *Masks of Difference: Cultural Representations in Literature, Anthropology and Art*. Cambridge: Cambridge University Press, 1994.

Stauffer, Andrew M. 'Dante Gabriel Rossetti and the Burdens of Nineveh'. *Victorian Literature and Culture* 33 (2005): 369–94.

Stiebing, William H., Jr. *Uncovering the Past: A History of Archaeology*. Oxford: Oxford University Press, 1993.

Stocking, George W., Jr. *Victorian Anthropology*. New York, NY: Free Press, 1987.

Thomas, Sophie. 'Displaying Egypt: Archaeology, Spectacle, and the Museum in the Early Nineteenth Century'. *Journal of Literature and Science* 5.1 (2012): 6–22. <www.literatureandscience.org/issues/JLS_5_1/JLS_vol_5_no_1_Thomas.pdf>. Accessed 15 September 2014.

Warwick, Alexandra. 'The Dreams of Archaeology'. *Journal of Literature and Science* 5.1 (2012): 83–97. <www.literatureandscience.org/issues/JLS_5_1/JLS_vol_5_no_1_Warwick.pdf>. Accessed 16 September 2014.

Warwick, Alexandra, and Martin Willis. 'Introduction: The Archaeological Imagination'. *Journal of Literature and Science* 5.1 (2012): 1–5. <www.literatureandscience.org/issues/JLS_5_1/JLS_vol_5_no_1_WarwickWillis.pdf>. Accessed 18 September 2014.

Welshman, Rebecca. 'Archaeology'. *Thomas Hardy in Context*. Ed. Phillip Mallett. Cambridge: Cambridge University Press, 2013, 221–30.

Zimmerman, Virginia. *Excavating Victorians*. Albany, NY: State University of New York Press, 2008.

Zimmerman, Virginia. '"Time Seemed Fiction": Archaeological Encounters in Victorian Poetry'. *Journal of Literature and Science* 5.1 (2012): 70–82. <www.literatureandscience.org/issues/JLS_5_1/JLS_vol_5_no_1_Zimmerman.pdf>. Accessed 16 September 2014.

Further reading

Holmes, John. 'Algernon Swinburne, Anthropologist'. *Journal of Literature and Science* 9.1 (2016): 16–39.

Malley, Shawn. '"Time Hath No Power Against Identity": Historical Continuity and Archaeological Adventure in H. Rider Haggard's *She*'. *English Literature in Transition 1880–1920* 40:3 (1997): 275–97.

Pearson, Richard. 'Archaeology and Gothic Desire: Vitality Beyond the Grave in H. Rider Haggard's Ancient Egypt'. *Victorian Gothic: Literary and Cultural Manifestations in the Nineteenth Century*. Eds. Ruth Robbins and Julian Wolfreys. Basingstoke: Palgrave, 2000, 218–44.

Radford, Andrew. 'Making the Past Wake: Anthropological Survivals in Hardy's Poetry'. *Science in Modern Poetry: New Directions*. Ed. John Holmes. Liverpool: Liverpool University Press, 2012, 167–80.

Sera-Shriar, Efram. *The Making of British Anthropology, 1831–1871*. London: Pickering and Chatto, 2013.

24

MEDICAL RESEARCH

Andrew Mangham
University of Reading

The links between medical research and literature have been discussed by a range of scholars in recent times, some arguing that the clinical precision of medicine influenced the novel's new-found preoccupation with detail and the trivialities of 'everyday life' (Rothfield; Kennedy) and others that medicine's languages of wonder and discovery inspired counter-realist genres such as the Gothic and the sensation novel (Caldwell; Smith; Talairach-Vielmas). Neither view appears to be incorrect as medicine grew exponentially in the nineteenth century to embrace a wide range of means for dissemination, including popular periodicals, and to develop terms and ideas that, in turn, lent themselves to an array of literary appropriations and thematic preoccupations. The introduction of nineteenth-century psychiatric theories, in particular, has inspired a number of reassessments of the period's fiction (Showalter; Taylor; Shuttleworth; Small). Concepts such as 'monomania', 'moral management' and 'unconscious cerebration' were seductive to authors such as Charlotte Brontë, whose intellect was weaned on Gothic tales from *Blackwood's Edinburgh Magazine*, and Wilkie Collins, whose uncanny stories of doubling, somnambulism and spontaneous insanity created a momentum that was both Gothic and modern.

But it is with methodologies rather than terminologies that this chapter is concerned. What George Levine says about realism in *The Realistic Imagination* (1981) can be applied more broadly to both medicine and literature in the nineteenth century:

> It was not a solidly self-satisfied vision based in a misguided objectivity and faith in representation, but a highly self-conscious attempt to explore or create a new reality. Its massive self-confidence implied a radical doubt, its strategies of truth telling, a profound self-consciousness.
>
> *(19–20)*

Medicine, like the novel, was a self-scrutinizing area of knowledge; it frequently assessed itself as to where it was going, what the implications of its research were and how best it might go about its practices. One of the problems with assigning to nineteenth-century medicine the kind of ideological manoeuvres discussed by Michel Foucault, among others, is that it suggests a strategic solidarity among its theorizers and practitioners, that there were no disagreements among the medical elite and that there was no discussion as to the best means of proceeding with its own researches or how the discipline might be conscious of its own need for social and professional control. Such a lack of self-reflexivity does not appear to have been the case, especially after scientific medicine, driven, in no small part, by the indefatigable labours of men like John and William Hunter and Astley Cooper, introduced important new questions about the links between the ethical commitments of medicine and the drive to improve knowledge using the empiricist principles that had dominated other sciences since the Scientific Revolution.

The intersections between literature and experimental medicine in the late eighteenth and nineteenth centuries is a tale of three cities. In Europe, the centres of medical excellence were Paris, London and Edinburgh; ideas circulated freely across all three cities, just as medical men would move from one capital to another in search of the latest theories and up-to-date practices. In his 1865 *Introduction à l'Étude de la Médecine Expérimentale*, the physiologist Claude Bernard noted:

> I have sometimes heard physicians express the opinion that medicine is not a science, because all our knowledge of practical medicine is empirical and born of chance, while scientific knowledge is decided with certainty from theories or principles ... New medical observations are generally made by chance; if a patient with a hitherto unknown affection is admitted to a hospital where a physician comes for consultation, surely the physician meets the patient by chance.
>
> *(190–1)*

In the preface to *The Morbid Anatomy of Some of the Most Important Parts of the Human Body* (1793), Matthew Baillie anticipated the view that medicine was imperfect because it relied upon chance encounters with the objects of its study:

> Although I have ventured to lay this work before the Public, yet I am very sensible of its imperfections. Some appearances are described which I have only had an opportunity of seeing once, and which, therefore, may be supposed to be described less fully and exactly than if I had been able to make repeated examinations ... There are others still, which I have only had an opportunity of examining in preparations ... All of these are sources of inaccuracy [which] may be said in some degree to be unavoidable.
>
> *(ix)*

But these sources of inaccuracy *were* avoidable, according both to Bernard and to Baillie's own uncle, the celebrity surgeon John Hunter. In reply to a research question from vaccination pioneer Edward Jenner, Hunter famously said 'Why think? Why not try the experiment?' (quoted in Guthrie 241). In an article for *Household Words* entitled 'The Hunterian Museum' (1850), Frederick Knight Hunt, journalist, surgeon and founder of the *Medical Times*, acknowledged how Hunter 'did not accept as truth, all that was told him; nor did he rest content with what his predecessors had done or said; but, intent upon the discovery of the facts, he went to work for himself' (279). And so did Bernard. Following Hunter and his own idol the Parisian physiologist François Magendie, he built an influential medical philosophy around the Baconian notion that 'thinking' could be weighed down by dogmatic errors. He warned that, in the study of medicine, 'we must keep our freedom of mind', adding:

> Men who have excessive faith in their theories or ideas are not only ill prepared for making discoveries; they also make very poor observations. Of necessity, they observe with a preconceived idea, and when they devise an experiment, they can see, in its

results, only a confirmation of their theory. In this way they distort observation and often neglect very important facts because they do not further their aim . . . Accordingly, we must disregard our own opinion quite as much as the opinion of others, when faced by the decisions of experience. If men discuss and experiment, as we have just said, to prove a preconceived idea in spite of everything, they no longer have freedom of mind, and they no longer search for truth. Theirs is a narrow science, mingled with personal vanity or the diverse passions of man.

(Bernard 38–9)

'Experimental reasoning,' he adds, 'is the only reasoning that naturalists and physicians can use in seeking the truth and approaching it as nearly as possible' (Bernard 31).

That medicine deals with people, as opposed to animals, vegetables or minerals, is an obvious fact that introduces a moral burden in medical research. The experimental method had been pioneered in branches of science that dealt with non-human subjects, and there emerged the thorny question of how one performs *medical* experiments without offending the basic human rights of the objects of study. Since the days of Hippocrates, the profession had made a vow to prioritize the health of patients over other considerations; it is, first and foremost, a palliative pursuit; experimentation can be only a secondary motive and – even then – there are clear objections to thrusting the 'knife here and there, to see what would come of it' (Foster 40). Logistically, for nineteenth-century practitioners, there was the problem of getting objects to study. Dead bodies were hard enough to come by at the start of the nineteenth century, living bodies (for experimentation) virtually impossible. Cases did not present themselves with any frequency. If a doctor wished to experiment on, say, a suppurated abscess, he would need to wait until a patient presented him- or herself with that condition and, naturally, not many would be willing to be experimented on at the possible cost of recovering quickly and more fully. The solution, according to infamous Nazi scientists, would be to look upon certain races as less than human and less deserving of basic human compassion. Almost eighty years earlier, doctors like Claude Bernard recognized that the only 'humane' solution was to be found in vivisection:

[The experimenter] no longer hears the cries of animals, he no longer sees the blood that flows, he sees only his idea and perceives only organisms concealing problems which he intends to solve. Similarly, no surgeon is stopped by the most moving cries and sobs, because he sees only his idea and the purpose of his operation. Similarly again, no anatomist feels himself in a horrible slaughter house; under the influence of a scientific idea, he delightedly follows a nervous filament through stinking livid flesh, which to any other man would be an object of disgust and horror.

(103)

Of course, to many commentators, vivisection itself was less than humane. It was such comments as these from Bernard that inspired the characterizations of the demon doctors Benjulia and Moreau from Wilkie Collins's *Heart and Science* (1883) and H. G. Wells's *The Island of Doctor Moreau* (1896) respectively. 'To this day,' says Moreau, 'I have never troubled about the ethics of the matter. The study of Nature makes a man at least as remorseless as Nature' (Wells 75). And 'knowledge is its own justification and its own reward', according to Benjulia:

The roaring mob follows us with its cry of Cruelty. We pity their ignorance. Knowledge sanctifies cruelty. The old anatomist stole dead bodies for Knowledge.

In that sacred cause, if I could steal a living man without being found out, I would tie him on my table, and grasp my grand discovery in days, instead of months.

(Collins 190)

From the 1870s, anti-vivisectionists would form a 'roaring mob' dead set on charging medical researchers with emotional insensitivity and cruelty to animals; Claude Bernard himself was the subject of a great deal of opposition and reports suggest that his wife left him because she could no longer put up with the horrors of his research.

Yet, on their home turf, vivisectors were challenged by their own colleagues who suggested that the development of experience, rather than experimentalism, was the true 'sacred cause' in medicine. In *The Scientific Revolution in Victorian Medicine*, A. J. Youngson notes that 'pressure to apply to medical problems the techniques and findings of physical and natural science' increased in the nineteenth century and 'this led to a struggle between the new attitudes of scientific observation, experiment, reasoning and innovation, and old attitudes of classical culture and conservatism' (16). While it is apparent that the medical profession faced some pressure to introduce the experimental techniques of natural and physical science into its own processes of learning, however, the main problem facing medical experimentalists was not 'old attitudes of classical culture and conservatism' but the theoretical question of whether the older system of experience through observation was better suited to the discipline.

When the revered physician Thomas Sydenham first met Hans Sloane, in 1684, he scoffed at the young Irishman's letter of introduction, which suggested that he was 'a ripe scholar, a good botanist, and a skilful anatomist' (quoted in Burch 30). 'This is all mighty fine,' said Sydenham,

> but it won't do – Anatomy – botany – Nonsense! Sir, I know an old woman in Covent Garden who understands botany better; and as for anatomy, my butcher can dissect a joint full and well; no, young man, all this is stuff; you must go to the bedside; it is there alone you can learn disease.

Learning by the bedside became known as the Sydenham method, named after the great 'English Hippocrates', whose method 'consisted in the careful observation and recording of the phenomena of disease . . . [He] diverted men's minds from speculation', according to one historian, and 'lead them back to the bedside, for there only could the art be studied' (Guthrie 202–4).

In 1819 the author of *Elements of Medical Logic*, Gilbert Blane, hoped to convert the Sydenham method into a more systematic methodology for the improvement of medical science. From where in medicine, he asked, 'is improvement to arise? The answer is, from accurate observation; in other words, from enlightened empiricism' (Blane 193). Rejecting the new trend for experimentalism, he praised 'those solid and applicable truths, the fruits of chaste observation and sober experience . . . the only parent of legitimate, substantial, and useful knowledge' (Blane 193), adding:

> by practice we learn to connect cause and effect, means and end, operations which in well tuned minds, are performed with promptitude and precision, by interpreting fairly the appearances of nature, and stripping them of those adventitious fallacies, which mislead ordinary minds . . . Whatever the attainment may be which we aim at, whether mental or manual, nothing but *practice will make perfect*, there being a certain expertness

in the exercises of the mind, as there is a slight of hand in mechanical operations attainable only by long and assiduous application.

(196–8)

Blane's language is highly revealing here. '*Practice will make perfect*', he emphasizes, because 'past experience of what may be termed simple *sequence*' (Blane 31) is the most secure means of attaining accurate medical knowledge.

This kind of learning was, for men like Claude Bernard, a passive and inadequate compilation of details: 'Pile up facts or observations as we may', Bernard says, and 'we shall be none the wiser. To learn, we must necessarily reason about what we have observed, compare the facts and judge them by other facts used as controls' (16):

> In my opinion, medicine does not end in hospitals, as is often believed, but merely begins there. In leaving the hospital, a physician . . . must go to his laboratory; and there, by experiments on animals, he will seek to account for what he has observed in his patients, whether about the action of drugs or about the origin of morbid lesions in organs or tissues. There, in a word, he will achieve true medical science.
>
> *(147)*

In *Vital Signs: Medical Realism in Nineteenth-Century Fiction*, Lawrence Rothfield clarifies 'the displacement of one genre (realism) by another (naturalism) by correlating it with the displacement of one form of scientific thought (that of clinical medicine) by another (that of experimental medicine)' (128). What I would like to consider here is how this correlation, though expressed most forcefully in a famous essay by Émile Zola (which I discuss below), predated French naturalism and served to inspire the narrative trajectories of British literature. Moving away from the question of whether a literary appropriation of a medical strategy is, necessarily, a realist manoeuvre (a topic skilfully considered by Rothfield), I wish to explore whether 'observation versus experiment' became an issue that impressed novelists of varying styles as much as it did medical practitioners of the nineteenth century. Rothfield notes that 'medicine enjoys by far the closest and most long-standing association with the issues of mimesis and knowledge so crucial to critical conceptions of realism' (12). The same point can be broadened to suggest that the epistemological problems that perplexed medicine also troubled the novel in general. 'How do we know what we know?' went the question in both medicine and literature. How can language be arranged in such a way that it best reflects what we do know? And how might these written forms be used to develop a better understanding of the social conditions of humans more generally?

Some authors were committed more to the experimental nature of the nineteenth-century novel than others. 'Realism', the style identified and celebrated by George Henry Lewes, shared medicine's wish to know things with certainty. His partner, George Eliot, perceived her fiction to be a scientific and philosophical laboratory. She wrote to Frederic Harrison in 1876:

> But my writing is simply a set of experiments in life – an endeavour to see what our thought and emotion may be capable of – what stores of motive, actual or hinted as possible, give promise of a better after which we may strive – what gains from past revelations and discipline we must strive to keep hold of as something more than shifting theory.
>
> *(Quoted in Carroll 24)*

In France, Balzac, Flaubert and Zola committed themselves to a similar project of literary analysis. In his well-known essay 'Le Roman expérimental' (1880), Zola leaned heavily on the ideas in Bernard's *Introduction* in order to suggest, similarly to Eliot, that the experimental method should be incorporated into the writing of fiction:

> We are neither chemists nor physicians nor physiologists; we are simply novelists who depend upon the sciences for support. We certainly do not pretend to have made discoveries in physiology which we do not practice; only, being obliged to make a study of man, we feel we cannot deny the efficacy of the new physiological truths . . . The novel has become a great inquiry on nature and on man. This is why we have been led to apply to our work the experimental method as soon as this method had become the most powerful tool of investigation.
>
> *(1893A 38)*

'Zola developed', according to Rothfield, 'a full-blown literary and intellectual persona of the writer as scientist' (128). Zola continued to outline how an author should be experimental with his subjects:

> And this is what constitutes the experimental novel: to possess a knowledge of the mechanism of the phenomena inherent to man, to show the machinery of his intellectual and sensory manifestations, under the influences of heredity and environment, such as physiology shall give them to us, and then finally to exhibit man living in social conditions produced by himself, which he modifies daily, and in the heart of which he himself experiences a continual transformation.
>
> *(1893A 20–1)*

The experimental novelist isolates a character and studies, through his plot, that individual's reactions to others, to events, environments and so on. Zola finds a fitting example in Balzac's *Cousine Bette* (1846):

> The general fact observed by Balzac is the ravages that the amorous temperament of man makes in his home, in his family, and in society. As soon as he has chosen his subject he starts from known facts; then he makes his experiment, and exposes *Hulot* to a series of trials, placing him amid certain surroundings in order to exhibit how the complicated machinery of his passions works. It is then evident that there is not only observation there, but that there is also experiment; as Balzac does not remain satisfied with photographing the facts collected by him, but interferes in a direct way to place his character in certain conditions.
>
> *(1893A 8–9)*

For Zola, then, the experimental method in literature involved 'provoked observation' (1893A 6); rather than a passive 'photographing' of phenomena, the author ought to expose his hero to a series of trials designed to test his character. Observing how the protagonist reacts becomes highly instructive – not only with reference to that character's individual psychology, but also to the environment(s) that he or she inhabits.

Zola acknowledges that the introduction of experimentalism into fiction requires an exploration that is primarily psychological. Dugald Stewart's *Elements of the Philosophy of the Human Mind* (1792), 'profoundly influen[tial to] nineteenth-century psychology' (Taylor and

Shuttleworth 141), was among the first texts to suggest that an experimental method could be applied to the study of human character. 'Bacon's philosophy,' Stewart wrote, 'was constantly present to my thoughts when I have dwelt in any of my publications on the importance of the Philosophy of the Human Mind' (quoted in Macintyre 251–2). Stewart was an admirer of the empirical method as it had been popularized by the philosophy of Francis Bacon. He dismissed the transcendentalism of German philosophy and favoured the more rigorous methodologies of the physical sciences. When challenged by Francis Jeffrey, one of the founders of the *Edinburgh Review*, on the links between Baconian experimentalism and the sort of observation that would be necessary to a study of the mind, Stewart said that 'the whole of a philosopher's life, if he spends it to any purpose, is one continued series of experiments on his own faculties and powers' (176).

In novels such as *Cousine Bette*, Zola adds, the author experiments on the faculties and powers of his hero. This is seemingly applicable to the nineteenth-century novel more widely, with its keen attention to its own structures, styles and revelations. Take, for instance, Anthony Trollope's 1858 novel *Doctor Thorne*. Here the provincial doctor is introduced in the second chapter among a speculation, on the part of the narrator, that the story might have begun better:

> As Dr. Thorne is our hero – or I should rather say my hero, a privilege of selecting for themselves in this respect being left to all my readers – and as Miss Mary Thorne is to be our heroine, a point on which no choice whatsoever is left to any one, it is necessary that they shall be introduced and explained and described in a proper, formal manner. I quite feel that an apology is due for beginning a novel with two long dull chapters full of description . . . but twist it as I will I cannot do otherwise.
>
> *(Trollope 17)*

The self-conscious tone of Trollope's narrative develops an explorative beginning to the novel; it produces the sense that author, reader, narrator and character are all on a journey of discovery and none of them has the kind of omniscient view that is associated, traditionally, with God: 'I can only plead for mercy,' says the narrator, 'if I am wrong' (Trollope 536). We are in a bewildering new world: 'the world, alas! Is retrograding; and according to the new-fangled doctrines of the day, a lady of blood is not disgraced by allaying herself to a man of wealth, and what may be called quasi-aristocratic position' (Trollope 456). These are actually the words of one of Trollope's snobbish aristocratic characters, Amelia de Courcy, and should not be read as a representation of the novel's political view, yet they do capture the overriding sense that the modern world is one where new discoveries are made every day and where a narrative like *Doctor Thorne* may be a key method of exhibiting and making sense of them.

Central to this objective of 'making sense' is Trollope's eponymous character. As is hinted by the beginning of the second chapter, Thorne appears to get put aside by the narrator on a number of occasions in spite of his being the text's potential hero. *Doctor Thorne* is actually a love story concerning the star-crossed youngsters Mary Thorne and Frank Gresham. The middle-aged doctor's role becomes important because his reactions to various situations that arise out of the class imbalance between the lovers offer a telling insight into the psychology of men like him, as well as into his society, and the characters that make it up. When it becomes apparent that her son has fallen in love with the (seemingly) penniless Mary Thorne, Lady Arabella Gresham ceases to ask for medical aid from the doctor and refuses her daughters the company of Mary. When Lady Arabella becomes seriously ill and is failed by the appropriately named Dr Fillgrave, however, she seeks the aid of Thorne – the man her pride had wronged but whose medical skills are undeniable:

'Well, doctor, you see that I have come back to you', she said, with a faint smile.

'Or, rather I have come back to you. And, believe me, Lady Arabella, I am very happy to do so. There need be no excuses. You were, doubtless, right to try what other skill could do; and I hope it has not been tried in vain'.

She had meant to have been so condescending; but now all that was put quite beyond her power. It was not easy to be condescending to the doctor: she had been trying all her life, and had never succeeded . . .

'So now we are friends again, are we not? You see how selfish I am'. And she put out her hand to him.

The doctor took her hand cordially, and assured her that he bore her no ill-will; that he fully understood her conduct – and that he had never accused her of selfishness. This was all very well and very gracious; but, nevertheless, Lady Arabella felt that the doctor kept the upper hand in those sweet forgivenesses. Whereas, she had intended to keep the upper hand, at least for a while, so that her humiliation might be more effective when it did come.

And then the doctor used his surgical lore, as he well knew how to use it. There was an assured confidence about him, an air which seemed to declare that he really knew what he was doing. These were very comfortable to his patients, but they were wanting in Dr Fillgrave. When he had completed his examinations and questions, and she had completed her little details and made her answer, she certainly was more at ease than she had been since the doctor had last left her.

(*Trollope* 478–9)

The scene is set up as one in which the reactions of both characters are crucial indicators of their personalities and each reaction says a great deal about the kinds of attitudes that both figures represent. Dr Thorne – patient, humble and a heroic man of industry – forgets his pride and offers Lady Arabella forgiveness and understanding. She – impatient, proud and frivolously obsessed with status and wealth – can offer none of the sympathies required in a world where 'a lady of blood [might] not be disgraced by allaying herself to a man of wealth'.

Such is the picture of modernity offered by another novel, similar in many ways to *Doctor Thorne* yet its polar opposite in many others: Mrs Henry Wood's sensational potboiler *East Lynne* (1861). Here the aristocratic but bankrupt Lady Isabel Vane marries the industrial entrepreneur Archibald Carlyle. The dramatic failure of their marriage implies that the old system of marrying in line with one's lineage is the better strategy, yet the interferences of Sir Francis Levison and Cornelia Carlyle are shown to be the true causes of the fatal breakdown between Isabel and Archibald. What is more, Lady Isabel's relationship with Sir Francis is hardly a harmonious one, in spite of the fact that they are both of noble birth.

In fact, *East Lynne* is an experimental novel and Lady Isabel its vivisected toad – laid bare, virtually, and open to the relentless scrutiny of narrator and reader. First Isabel is put into a marriage where the clues of her husband's seeming infidelities get agitated by her perceived domestic failures and the temptations of an aristocratic rake; then she makes the bold step of becoming governess to her own children and is forced to endure the spectacle of another woman in her role. Mrs Wood may share Trollope's unceremonious style of narration, yet she has little of his subtlety when it comes to outlining the findings of her experiment:

O reader, believe me! Lady – wife – mother! Should you ever be tempted to abandon your home, so will you awake. Whatever trials may be the lot of your married life, though they may magnify themselves to your crushed spirit as beyond the nature, the endurance of woman to bear, resolve to bear them; fall down upon your knees, and pray to be enabled to bear them – pray for patience – pray for strength to resist the demon that would tempt you to escape; bear unto death, rather than forfeit your fair name and your good conscience; for be assured that the alternative, if you do rush on to it, will be found worse than death.

(223)

Wood's heavy-handed moralizing suggests that the consequences of her heroine's choices were unlikely to be anything different, in spite of the text's insistence that 'married life' may offer 'trials'. A 'trial' may be a labour that one undertakes in order to succeed to rewards that have been promised, like the journey of Bunyan's pilgrim. Yet also it may be an experiment in which a number of outcomes may occur. The devoutly Anglican Wood was likely to have been thinking about the former meaning, yet the second definition applies to her text because the 'charming results' (223) she outlines in *East Lynne* require a trial, or an experiment, to have taken place. Wood unwittingly fulfils the observations laid out by Zola in his experimentalist manifesto and this is due in no small part to the fact that, by the mid-nineteenth century, the experimental legacy had become central to the strategies of fiction as much as to the aims and objectives of medicine.

Yet Mrs Wood's conservative pontification on the moral duties of ladies, wives and mothers highlights a problem with applying the experimental method to literature. In the *invented* world of the novel, where all variables, environments and reactions are determined by the author, is it possible to have the level of detachment necessary for a scientific exploration? In his *Introduction*, Claude Bernard had a keen sense that preconceived ideas could determine and thwart the results of any given investigation, as we have seen. Adapting Bacon's 'idols' idea, he pre-emptively supports the view of Levine that 'with the recognition of the inescapable presence of the interpreting self, it was incumbent on all scientists and thinkers the more rigorously to repress their own biases' (2002 3). As early as 1867, Zola was laying out a plan for literature as a scientific pursuit in *Thérèse Raquin*, but, unlike Bernard, he barely paid lip service to the question of how an author's preconceptions or biases might colour and shape his or her 'studies'. In the preface to the second edition of *Thérèse Raquin*, he writes:

> The reader will have started, I hope, to understand that my aim has been above all scientific. When I created my two protagonists, Thérèse and Laurent, I chose to set myself certain problems and to solve them . . . Those who read the novel carefully will see that each chapter is the study of a curious case of physiology . . . I have merely performed on two living bodies the analytical work that surgeons carry out on dead ones . . . [I] lost myself in a precise, minute reproduction of life, giving myself up entirely to an analysis of the working of the human animal . . . [I] turned [my] attention to human corruption, but in the same way as a doctor becomes absorbed in an operating theatre.

(Zola 2004 4–5)

In a letter Zola wrote to a friend in 1864, one year before the publication of Bernard's *Introduction*, he expressed some sense of 'the inescapable presence of the interpreting self' (quoted in Grant 28). There is a screen, he suggested, between the author and his subject, but 'the realistic screen is plain glass, very thin, very clear, which aspires to be so perfectly transparent that images may pass through it and remake themselves in all their reality'. 'I agree it's difficult,' Zola conceded, 'to describe a screen whose distinguishing quality is that is scarcely exists', and 'all screens must distort to a certain extent'. Yet, in *Thérèse Raquin*, the active process of experimentalism appears to do away with the problem of the distorting screen:

> In *Thérèse Raquin* I set out to study temperament, not character. That sums up the whole book. I chose protagonists who were supremely dominated by their nerves and their blood, deprived of free will and drawn into every action of their lives by the predetermined lot of their flesh. Thérèse and Laurent are human animals, nothing more.
>
> *(Zola 2004 4)*

Bernard would baulk at the idea of anything 'predetermined'. Zola's response to such an objection might claim that amorous characters such as Thérèse and Laurent do, almost always, have a predetermined fate, but this misses the larger question of how far an invented scenario might fail to create conditions equal to the 'real world'. Bernard had the idea of the scientist putting a question to nature and stepping back when she speaks her answers. In the world of the novel, however, there can be no stepping back because the author must determine the outcome as well as the set-up. Like Lady Isabel, Thérèse and Laurent have a narrative already mapped out for them.

Nevertheless, the novel, broadly speaking, always does more than what is intended by the author, as we know from poststructuralist criticism. If the nineteenth-century novel is experimental, it is so in a way that foreshadows the story, probably apocryphal, of Alexander Fleming's discovery of penicillin by accident. The laboratory is never completely a controlled environment and the crumbs that fall unintentionally through the cracks are a treasure in themselves. Aspects of Lady Isabel's story have been described, according to Lyn Pykett, as 'masochistic' (131), 'a prolonged, luxurious orgy of self-torture', and yet they 'also [afford] the reader the opportunity of spectating feelings of anxiety, separation, loss and claustrophobia which arise from middle-class women's experience of motherhood and domesticity'. Similarly, in *Thérèse Raquin* the masochistic orgy of the central characters' torments produces unexpected moments of aesthetic clarity. Laurent, a bad painter, develops new talents after he has killed his lover's husband:

> Admittedly, [these sketches] were naïve, but they had a strangeness about them and such power that they implied the most advanced aesthetic sense. You would have thought they were the product of experience. Never had Laurent's friend seen sketches exhibiting such high promise . . .
>
> 'I have only one criticism to make, which is that all your sketches look alike. Those five heads resemble one another. Even the women have a sort of violent look that makes them seem like men in disguise['] . . .
>
> 'He's right', [Laurent] murmured. 'They are all alike . . . They look like Camille'.
>
> *(Zola 2004 143–4)*

Clearly, Laurent's inability to shake his obsession with the man he murdered is part of his pre-ordained punishment, yet why are his pictures good? There emerges the suggestion that the scientific analysis undertaken by the author produces surplus elements – odd moments of beauty that go beyond the aims laid out by the preface's moral determinism.

Zola had a dislike for what he called 'phantasmagoria' (1893B 215), flights of fancy that characterized 'our age of lyricism, our romantic disease' (1893A 36). And yet, in describing the tortures of Thérèse and Laurent, the novel is extraordinarily phantasmagorical:

When the two murderers were under the same sheet and shut their eyes, they would imagine they could feel the damp corpse of their victim spread out in the middle of the bed, sending a chill through their flesh. It was like some grotesque barrier between them. They were seized by feverish delirium and the barrier would become an actual one for them; they would touch the body, they would see it lying like a greenish, rotten lump of meat and they would breathe in the repulsive odour of this heap of human decay. All their senses shared in the hallucination, making their sensations unbearably acute.

(Zola 2004 127)

The text works hard to make these hallucinations a matter of physiology rather than supernaturalism. It is 'things', ghastly 'things', rather than translucent chimeras or will-o'-the-wisps, that trouble the murderers. Making full use of the ideas of the sensationalist empiricists of his own country, Zola has the hallucinations assault Thérèse's and Laurent's senses as well as their feelings. Yet the novel's indebtedness to the kinds of phantasmagoria dismissed by Zola is clear. 'Even when the murderers thought they had completed the killing and could indulge the sweet pleasures of their love, the victim would return to chill their marriage bed' (Zola 2004 129). Never was a line more indebted to Gothic literature than this one. Indeed, the tortures of Thérèse and Laurent become carnivalesque; they appear grotesque, exaggerated and melodramatic. Like Laurent's sketches, however, they are powerfully artistic – not because they are experimental in the sense laid out by Bernard and Zola, but because they develop a form of naturalism that highlights how the complexities of portraying human nature in fiction produces more than the determinism of experimentalism can account for.

In medical literature of the nineteenth century, one of the recurrent objections against the introduction of the experimental method was that organic tissues cannot be manipulated and experimented with in the same way as inorganic materials. 'The fundamental laws of life,' Blane noted in his *Elements of Medical Logic*, 'are essentially distinct from those of inanimate matter' (154). He adds:

The simplicity of the laws of inanimate nature, admits of the most certain inferences, whereas the indefinite action and re-action of the numerous faculties peculiar to life . . . add greatly to the difficulty and uncertainty of experiment and observation . . . But this is not all; for constitutions being gifted with various degrees of each of these faculties, an endless variety is found to take place among individuals, giving rise to that uncertainty in the results of medicine, which has brought upon it the character of a conjectural art.

(Blane 202)

Such formed the basis of Blane's advocacy of observation over experimentalism: study without interference will allow the practitioner to glean the full complexities of the living, moving materials he is concerned with. In Collins's *Heart and Science*, the bedside observations of one medical man get juxtaposed with the vivisections of Dr Benjulia in order to highlight this very idea. While treating a man in Montreal, Ovid Vere comes across an unfinished treatise on brain disorder. It begins:

The information which is presented in these pages is wholly derived from the results of bedside practice; pursued under miserable obstacles and interruptions, and spread over a period of many years. Whatever faults and failings I may have been guilty of as a man, I am innocent in my professional capacity, of ever having perpetrated the useless and detestable cruelties which go by the name of Vivisection . . .

A celebrated physiologist, plainly avowing the ignorance of doctors in the matter of the brain and its diseases, and alluding to appearances presented by post-mortem examination, concludes his confession thus: 'We cannot even be sure whether many of the changes discovered are the cause or the result of the disease, or whether the two are the conjoint results of a common cause'.

So this man writes after experience in Vivisection . . .

In medical investigations, as in all other forms of human inquiry, the resulting view is not infrequently obtained by indirect and unexpected means. What I have to say here on the subject of brain disease, was first suggested by experience of two cases, which seemed in the last degree unlikely to help me.

(Collins 307–8)

Having been beaten to his grand discovery by Vere's Sydenham method, Benjulia releases his tortured animals and kills himself. 'Vivisection,' writes Collins, 'had been beaten on its own field of discovery' (324). The conquest is a result of the 'indirect and unexpected' events that may occur in the life of a scientist or a doctor. Not all of these may be accommodated within the plot laid out by an experimentalist, Collins suggests, and the best discoveries are those that are made through a combination of technical knowledge and professional experience.

Like nineteenth-century medicine, then, the novel was a place of experimentation. Concerned with what it means to experiment, both within its own forms and within the world of science, the novel made a crucial contribution to the question of how we begin to understand ourselves and our cultures scientifically. While it is true that the new drive towards experimentalism in medicine seemed to inspire an approach that put the central character in the novel through a number of predetermined trials, it is also the case that literature exposed the limits of this approach by featuring moments of exegesis that simply did not belong to the experimental mindset. The literary world, like the real world it seems, has facets that cannot be brought into line by experimentation, yet the determination to observe, to interpret and to express reality through the trajectories of fiction highlights a resolve to combine the strengths of medicine with those of the novel.

Bibliography

Primary texts

Baillie, Matthew. *The Morbid Anatomy of Some of the Most Important Parts of the Human Body* [1793]. Philadelphia, PA: Hickman and Hazzard, 1820.

Bernard, Claude. *Introduction à l'Étude de la Médecine Expérimentale* [1865]. Transl. Henry Copley Greene as *An Introduction to the Study of Experimental Medicine*. New York, NY: Dover, 1957.

Blane, Gilbert. *Elements of Medical Logic* [1819]. Hartford, CT: Huntington and Hopkins, 1822.

Collins, Wilkie. *Heart and Science* [1883]. Ed. Steve Farmer. Toronto: Broadview, 1996.

Foster, Michael. *Claude Bernard*. New York, NY: Longman, Green, 1899.

Hunt, Frederick Knight. 'The Hunterian Museum'. *Household Words* 2 (1850): 277–82.

Stewart, Dugald. 'Philosophical Essays' [1810]. *Selections from the Edinburgh Review*. Ed. Maurice Cross. Paris: Baudry's European Library, 1835, 172–214.

Trollope, Anthony. *Doctor Thorne* [1858]. Oxford: Oxford University Press, 1953.

Wells, H. G. *The Island of Doctor Moreau* [1896]. Ed. Patrick Parrinder. London: Penguin, 2005.

Wood, Mrs Henry (Ellen). *East Lynne* [1861]. London: Everyman, n.d.

Zola, Émile. 'Le Roman Expérimental'. *The Experimental Novel and Other Essays*. Transl. Belle M. Sherman. New York, NY: Cassell, 1893. 1–56 [1893A].

Zola, Émile. 'The Novel'. *The Experimental Novel and Other Essays*. Transl. Belle M. Sherman. New York, NY: Cassell, 1893, 209–90 [1893B].

Zola, Émile. *Thérèse Raquin* [1867]. Transl. Robin Buss. London: Penguin, 2004.

Secondary texts

Burch, Druin. *Taking the Medicine: A Short History of Medicine's Beautiful Idea, and Our Difficulty Swallowing It*. London: Vintage, 2010.

Caldwell, Janis McLarren. *Literature and Medicine in Nineteenth-Century Britain: From Mary Shelley to George Eliot*. Cambridge: Cambridge University Press, 2004.

Carroll, David. *George Eliot and the Conflict of Interpretations: A Reading of the Novels*. Cambridge: Cambridge University Press, 1992.

Grant, Damian. *Realism*. London: Methuen, 1970.

Guthrie, Douglas. *A History of Medicine*. London: Thomas Nelson, 1945.

Kennedy, Meegan. *Revising the Clinic: Vision and Representation in Victorian Medical Narrative and the Novel*. Columbus, OH: Ohio State University Press, 2010.

Levine, George. *The Realistic Imagination: English Fiction From 'Frankenstein' to 'Lady Chatterley'*. Chicago, IL: University of Chicago Press, 1981.

Levine, George. *Dying to Know: Scientific Epistemology and Narrative in Victorian England*. Chicago, IL: University of Chicago Press, 2002.

Macintyre, Gordon. *Dugald Stewart: The Pride and Ornament of Scotland*. Brighton: Sussex Academic Press, 2003.

Pykett, Lyn. *The 'Improper' Feminine: The Women's Sensation Novel and the New Woman Writing*. London: Routledge, 1992.

Rothfield, Lawrence. *Vital Signs: Medical Realism in Nineteenth-Century Fiction*. Princeton, NJ: Princeton University Press, 1992.

Showalter, Elaine. *The Female Malady: Women, Madness and English Culture, 1830–1980* [1985]. London: Virago, 2000.

Shuttleworth, Sally. *Charlotte Brontë and Victorian Psychology*. Cambridge: Cambridge University Press, 1996.

Small, Helen. *Love's Madness: Medicine, the Novel and Female Insanity, 1800–1865*. Oxford: Oxford University Press, 1996.

Smith, Andrew. *Victorian Demons: Medicine, Masculinity and the Gothic at the Fin de Siècle*. Manchester: Manchester University Press, 2004.

Talairach-Vielmas, Laurence. *Wilkie Collins, Medicine and the Gothic*. Cardiff: University of Wales Press, 2009.

Taylor, Jenny Bourne. *In the Secret Theatre of Home: Wilkie Collins, Sensation Narrative and Nineteenth-Century Psychology*. London: Routledge, 1988.

Taylor, Jenny Bourne, and Sally Shuttleworth, eds. *Embodied Selves: An Anthology of Psychological Texts*. Oxford: Oxford University Press, 1998.

Youngson, A. J. *The Scientific Revolution in Victorian Medicine*. London: Croom Helm, 1979.

Further reading

Foucault, Michel. *The Birth of the Clinic* [1963]. Transl. A. M. Sheridan. London: Routledge, 2007.

Goellnicht, Donald C. *The Poet-Physician: Keats and Medical Science*. Pittsburgh, PA: University of Pittsburgh Press, 1984.

Hurren, Elizabeth T. *Dying for Victorian Medicine: English Anatomy and its Trade in the Dead Poor, c. 1834–1929*. Basingstoke: Palgrave Macmillan, 2012.

Levine, George. *Realism, Ethics and Secularism: Essays on Victorian Literature and Science*. Cambridge: Cambridge University Press, 2008.

Porter, Roy. *The Greatest Benefit to Mankind: A Medical History of Humanity from Antiquity to the Present*. London: Harper Collins, 1999.

Sleigh, Charlotte. *Literature and Science*. Basingstoke: Palgrave Macmillan, 2011.

Sparks, Tabitha. *The Doctor in the Victorian Novel.* Farnham: Ashgate, 2009.

Stiles, Anne. *Popular Fiction and Brain Science in the Late Nineteenth Century.* Cambridge: Cambridge University Press, 2012.

Vrettos, Athena. *Somatic Fictions: Imagining Illness in Victorian Culture.* Stanford, CA: Stanford University Press, 1995.

Waddington, Keir. *An Introduction to the Social History of Medicine: Europe Since 1500.* Basingstoke: Palgrave Macmillan, 2011.

Wood, Jane. *Passion and Pathology in Victorian Fiction.* Oxford: Oxford University Press, 2001.

25

SCIENCES OF THE MIND

Suzy Anger

University of British Columbia

Oscar Wilde, recounting the thoughts of Lord Henry in *The Picture of Dorian Gray* (1891), writes: 'He began to wonder whether we could ever make psychology so absolute a science that each little spring of life would be revealed to us' (52). That there is a science of mind was fairly well agreed upon by the end of the Victorian period, even if that assumption continued to require some bolstering. In the face of long-standing objections that subjective mental states are not amenable to scientific methods of inquiry, the mind definitively became an object of empirical study in the nineteenth-century. By the 1890s, psychological laboratories were established on the Continent and in the United States, while Britain, somewhat tardy in the area, was taking the first steps towards launching its own experimental laboratories. Professional organizations in the sciences of the mind had been formed and academics were beginning to pursue the institutionalized study of psychology in the universities. The Psychological Society, founded in London in 1901, limited membership to those who taught or published in psychology, marking the arrival of a professional discipline of elites. Journals in medical and scientific psychology and neurology were launched in Britain from the mid-century onwards. As early as 1843, John Stuart Mill had confidently claimed that 'there is a distinct and separate Science of Mind' (556). Yet a surprisingly diverse group with wide-ranging commitments and approaches laid claim to that terrain over the course of the nineteenth-century: physiologists, anatomists, neurologists, medical psychologists but also mesmerists, phrenologists, spiritualists and investigators of ghosts and telepathy.

That varied range of positions on what constitutes the science of the mind had also entered pervasively and sometimes extravagantly into literary texts, both 'high' and 'low', shaping content and form. Israel Zangwill's short story 'The Memory Clearing House' (1892) imagines a new technology that commodifies mental experience, as memories are marketed and exchanged. H. G. Wells, developing the genre of scientific romance, portrays an elderly man, a 'profound student of mental science' (1896B 496), who transfers his consciousness into the brain of a youth. Thomas Hardy, using the language of medical psychology, depicts characters suffering from nervous disease and mental degeneration in *Jude the Obscure* (1895). New Woman writers, such as George Egerton, Mona Caird and Sarah Grand, discuss the hereditary nature of character and mental ability in their *fin de siècle* fiction. May Kendall, drawing on the language of physiology, begins her poem 'The Materialist' (1894) with the lines 'So, here is the neural matter / Life's frail citadel / This the brain that men would flatter' (59). Suffice it to say that by the 1890s the concepts and language of scientific psychology had become common cultural discourse.

Reciprocal influences

The exchanges between literary and psychological texts in the nineteenth century bear out Gillian Beer's two-way traffic model, which sees the influence between science and literature moving in both directions. Many British literary authors had extensive knowledge of the debates in the science of mind, and psychological ideas became both subject matter and informing conceptions in the fictional presentation of action, consciousness and character. The ongoing debates were frequently carried on in generalist journals, such as the *Fortnightly Review* and *Nineteenth Century*, making ideas from the new science the subject of popular discussion. Perhaps because psychological ideas and categories are necessary for everyday understanding, lines between the science of the mind and literary and popular cultures were difficult to draw. The new psychology helped shape literary authors' notions of the self, while scientists borrowed conceptions and categories from literary representations, particularly from the culturally important novel.

Given that both literature and psychology reflect and construct our imagination of ourselves, it is not surprising that the concerns of nineteenth-century literature and mental science coincided. One principal site of shared interest was the epistemology of mind. Debates on methodology arose, investigators asking whether the mind can be examined objectively, using the methods of the natural sciences, or whether testimony about introspection is the sole route to psychology's data, and, if so, whether subjective methods can be scientific. Literary texts often underscore the difficulties involved in knowing the mind of another, and the same worries show up in the science. James Sully, a key figure in the development of child psychology, makes a claim at the start of his well-respected textbook on psychology that might have come directly from a novel written by George Eliot, known for her attention to misinterpretations of the thoughts and behaviour of others. Sully writes: 'Whenever we are interpreting another mind, we are carrying out a process of interpreting signs which is always to some extent precarious . . . there is a special danger of reading into the mind observed our own peculiar modes of thought' (1892 1:21). Nineteenth-century fiction importantly tackled the problem of how to represent mental states, with narrators sometimes offering external-third-person descriptions of characters' actions and thoughts and sometimes ostensibly first-person introspective representations of mind.

Both the psychology of literary authors and the psychology of their characters were considered productive subjects for psychological investigation. Charles Dickens's imagination was discussed in works on mental science by William Carpenter, Henry Maudsley and Alexander Bain, among others (Carpenter 1875; Maudsley 1883; Bain 1880). George Henry Lewes, George Eliot's psychologist partner, wrote of Dickens: 'I say that in no other perfectly sane mind . . . have I observed vividness of imagination approaching so closely to hallucination' (144). Psychologists frequently turned to analysis of literary characters as well, using those characters as examples of particular mental conditions. Alienist Forbes Winslow discussed Dickens's 'A Madman's Manuscript' from *The Pickwick Papers* (1836) in a treatise on insanity. Maudsley wrote about Hamlet's psychology, as did his father-in-law John Conolly in his *Study of Hamlet* (1863). Their colleague John Charles Bucknill published on Tennyson's *Maud* and produced a monograph on *The Psychology of Shakespeare* (1859), which attends to the bard's depiction of 'abnormal conditions of mind' (vi) and 'his knowledge of the mental physiology of human life' (vii).

The idea that a literary artist is a kind of psychologist was also frequently invoked in criticism. The French literary critic Hippolyte Taine writes, in his *History of English Literature* (1871), 'What is a novelist? In my opinion he is a psychologist, who naturally and involuntarily sets psychology at work' (2:390). Hannah Lynch in an 1891 study describes George Meredith as 'a scientific psychologist' (5). James Sully pronounces in the journal *Mind* that 'In the eyes of the psychologist the works of George Eliot must always possess a high value by reason of their large scientific insight into character and life' (1881 378). Sully acknowledged in his memoirs (1918) that his conversations with Robert Louis Stevenson on dreams and artistic creation were an important influence on the formation of his views on dream psychology and the unconscious.

In recent decades, scholars have explored widely this reciprocal influence between the sciences of the mind and literary texts in the nineteenth century. Alan Richardson has demonstrated the influence of biologically oriented psychology and brain science on Romantic period writers, drawing connections between scientists who studied the brain and physiology, such as Erasmus Darwin and Charles Bell, and the poets Samuel Taylor Coleridge, William Wordsworth and John Keats. The historians of neuroscience Edwin Clarke and L. S. Jacyna have traced the influences in the opposite direction, contending that in the emerging neurosciences, 'changes in ideas of the function and structure of the nervous system during this period were stimulated by the romantic philosophy of nature that exerted a major influence upon biological thought in the first half of the nineteenth century' (1). Literary scholars have considered the Romantics' concern with the imagination and the psychology of sense perception (Sha; Jackson). Others have demonstrated particular authors' knowledge of developments in psychology, examining Coleridge's extensive concern with the mind (Ford), attending to the influence of Keats's medical and psychological knowledge on his poetry (Goellnicht; De Almeida), or examining Thomas De Quincey's theories of the unconscious (Iseli). Michelle Faubert has considered the relations between John Ferriar's psychological writing on apparitions and James Hogg's fiction, and examined the work of doctors who composed poetry about Romantic psychology.

In Victorian studies, the intense critical attention to the relations between the sciences of the mind and literature in recent years has resulted in studies on both particular topics in psychology and individual authors. Together, Sally Shuttleworth and Jenny Bourne Taylor have helped shaped the field for literary scholars through Taylor's examination of Wilkie Collins and physiological psychology, Shuttleworth's consideration of George Eliot and the new psychology, her exploration of *Charlotte Brontë and Victorian Psychology* and their jointly edited anthology of Victorian writing on psychology, *Embodied Selves*, with its fine introductory essays.

Other examinations of individual authors include studies on Eliot, in connection with Lewes's writing on physiological psychology (Postlethwaite; Davis); Dickens, fascinated by the debates on mind (Bernard; Vrettos 1999/2000; Stolte); Hardy, who wrote extensive notebook entries on the mind (Keen; Wickens; Anger); Stevenson, influenced by evolutionary psychology (Block; Reid); and Wilde, conversant with the debates (Hasseler; Seagroatt; Cohn). Gregory Tate has examined Victorian poets' knowledge of the new psychology. Other scholars have considered particular terrain in the science of the mind, such as the prehistory of trauma theory – or nervous shock – in Victorian fiction (Matus); the representation of dreamy states of mind (O'Neill); the depiction of déjà vu (Vrettos 2002); metaphors of mind (Kearns); the new technology's overstimulation of the nerves (Daly); and mid-century theories of the physiology of reading (Dames 2007). Shuttleworth demonstrates in *The Mind of the Child* that 'key literary works play[ed] a formative role in the development of the frameworks of nineteenth-century child psychiatry' (3). Rick Rylance's interdisciplinary study *Victorian Psychology and British Culture 1850–1880* brings to bear the methods of the literary critic on the work of the important psychologists Alexander Bain, Herbert Spencer and G. H. Lewes, exhibiting the shared culture of literature and art.

As this brief and partial survey of some of the criticism makes clear, the nineteenth-century science of mind ranged over many topics, its strands too diverse to examine comprehensively here. Having sketched out a number of the main lines of influence, I discuss in more detail some of the especially important connections between mental science and literary texts. I look first at the science before turning to an examination of the interactions between literature and psychology in these specific areas (although the topics are often overlapping and not neatly categorized). I also consider some of the work in the history of science and literary criticism that has explored the parallel developments of fiction and the new psychology.

False starts: phrenology and mesmerism

In the earlier decades of the nineteenth-century, the theories of phrenology and mesmerism were widely popular, although both practices were always objects of scientific debate and critique. Phrenology, initially advanced by Franz Joseph Gall in Vienna at the end of the eighteenth century, was based on the idea that the brain possesses innate mental faculties, or 'organs' of the mind. These included pride, affection, poetical talent, secretiveness and benevolence. Each of the faculties was said to be located in a specific cerebral region, and the assumption was that the sizes of the different organs were detectable from the shape of – or the bumps on – the skull. Character and abilities could therefore be discovered through careful physical measurement of the skull. The notion of reading personality in the physical body was familiar from the Swiss theologian Johann Kaspar Lavater's late eighteenth-century system of physiognomy, based on the idea that character can be discerned in the shape and features of the face (although an eighteenth-century theory, physiognomy continued to exert influence on Romantic- and Victorian-period writing (Sibylle; Graham; Tytler; Percival and Tytler; Hartley)). Gall claimed to have discovered the twenty-seven organs through experimental investigation, but his scientific methods were far from reliable. Still, the assertions that the mind was physical rather than metaphysical and that human psychology was empirically explainable were important in making the mind an object of scientific study. Gall's disciple Johann Gaspar Spurzheim made converts to the science when he lectured on a revised system of phrenology in Great Britain. In Edinburgh, the lawyer George Combe became convinced of phrenology's validity, transforming it into a social creed in his bestselling book *The Constitution of Man* (1828), which made phrenology widely popular across Britain. Practising phrenologists did readings of people's skulls and provided advice on education and the hiring of servants. Phrenology journals flourished in the 1820s and 1830s, as did phrenological societies, although the British Association for the Advancement of Science refused to create a phrenological division – evidence of phrenology's precarious status as a scientific theory even at the time. Roger Cooter investigates the social context of phrenology as a scientific theory, attending to issues of politics and class in *The Cultural Meaning of Popular Science* (1984), while Stephen Tomlinson examines phrenology's role in theories of education. Other scholarship has shown that phrenology swiftly became entangled with criminology (Rafter), scientific racism (Wagner; Davie) and gender science (Schiebinger; Russett).

Given the popularity of the theory, its presence in countless literary texts – some satirical and some admiring – is to be expected. Jason Hall has examined the impact of phrenology on the Romantic poets. Anne K. Mellor has considered phrenology's influence on William Blake. Thomas Love Peacock satirizes it in his novel *Headlong Hall* (1815), with Mr. Cranium the phrenologist announcing: 'Every man's actions are determined by his peculiar views, and those views are determined by the organisation of his skull' (60). We know that Eliot (then Mary Anne Evans) and Charlotte Brontë had their skulls read, Eliot by Combe himself. Brontë alludes to phrenology often in her novels, showing her characters' reliance on the science to read character; in *Shirley* (1849), for instance, the manufacturer Hiram Yorke is described as possessing 'too little of the organs of Benevolence and Ideality' (36). Dickens's convict Magwitch in *Great Expectations* explains that as a starving, homeless child he was repeatedly sent to prison for stealing. Critiquing phrenology's connection to deterministic, biological accounts of criminality, Dickens has Magwitch claim: 'they measured my head . . . they had better a measured my stomach' (346). Although by the time she wrote 'The Lifted Veil' Eliot had already become convinced of phrenology's lack of validity, she includes a scene in which her protagonist, Latimer, is examined as a child by a phrenologist: 'The deficiency is there, sir – there . . . here is the excess. That must be brought out, sir, and this must be laid to sleep' (26). Phrenological discourse negotiated questions surrounding physical determinism and the efficacy of education in altering character, as well as concerns about the knowability of the self.

In the 1760s, Anton Mesmer had posited the existence of a magnetic fluid that acted on and between animal bodies, maintaining that this fluid could be used for healing. Although his theory was debunked in France by the 1780s, mesmerism influenced Romantic period writers and was revived in Britain in the 1830s to great popularity (Fulford 2004A; Crabtree). Its medical uses were reasserted, along with claims that moved it into the spiritual realm. A person in a mesmeric state, for instance, was said to exhibit extreme powers of mind: clairvoyance, transposition of the senses or mental causation across distances. Journals of mesmerism appeared in Britain, such as *The Zoist: A Journal of Cerebral Physiology and Mesmerism and Their Application to Human Welfare* (1843–56), founded by the prominent physician John Elliotson. Mesmeric demonstrations became popular entertainment, with the mesmerizer ostensibly incapacitating the mesmerized person's volition and controlling her (most frequently women were the subjects, raising concerns about impropriety) like an automaton. Mesmerism's legitimacy was questioned by many in the scientific community, famously by Thomas Wakley, founding editor of the medical journal *The Lancet*. Debates over mesmerism's validity continued into the 1850s, but some of its claims were made scientifically acceptable with James Braid's introduction of 'hypnotism' in 1843. Braid contended that mesmeric states were explainable as physiologically induced trance states produced by intensive concentration, thereby doing away with the need for a hypothetical mesmeric fluid.

Alan Gauld in *The History of Hypnotism* and Alison Winter in *Mesmerized: Powers of Mind in Victorian Britain* examine the development and widespread cultural influence of the science. The literary representation of mesmerism extends throughout the century. Coleridge was intrigued by mesmerism (Levere; Fulford 2004A), as was Shelley (Dawson; Leask), while William Godwin sought to discredit it (Ruston). Thomas De Quincey wrote an essay on 'Animal Magnetism' (1834), in which he accepted that a person in a mesmeric trance becomes clairvoyant (Burwick 2005). Charles Dickens became interested in mesmerism after hearing some of Elliotson's lectures on the subject and practised mesmerism on one of his illustrators, John Leech, as well as on a neighbour, Madame de la Rue, curing her, he believed, of a mental 'phantom'. Dickens also frequently includes mesmeric references in his fiction (Kaplan). In the culture, the theory of mesmerism was tied to concerns about volition: could a mesmerist gain control over the actions of the mesmerized, turning a person into a machine without awareness or moral sense? Often mesmerism was entangled with issues of gender, with women succumbing to masculine power. In 'Mesmerism' (1855), Robert Browning depicts a young woman mesmerically controlled by a man, as does George du Maurier in *Trilby* (1894). At the end of the century, in response to Charcot's much discussed work on hypnotism, stories of preying mesmerists return in popular fiction, years after mesmerism had been effectively discredited by orthodox science. The dynamics were often reversed in those late-century stories, however. In Richard Marsh's eccentric novel *The Beetle* (1897), an Egyptian woman (who sometimes takes the form of a man and sometimes of a giant beetle) arrives in London to seek revenge on a British scientist, transforming an out-of-work man into a criminal automaton with her mesmeric powers. Arthur Conan Doyle depicts a West Indian female mesmerizer who gains power over a scientific man in *The Parasite* (1894) (Willis and Wynne).

Even after phrenology and mesmerism came to be definitively debunked, conceptions from those sciences remained essential to studies of the mind. Questions about automatism and self-control were central to the new physiological psychology. Despite having fallen into disrepute by the mid-century, phrenology made a reappearance of sorts in late-century cranial measurement of the skull, which asserted a correspondence between skull size and intelligence and was used to confirm racial hierarchy, gender inferiority and inherited criminal tendencies. Robert Young has demonstrated the important context that phrenology provided for neuroscience, particularly brain localization, in the last decades of the century.

The new psychology

The new physiological psychology was an effort to naturalize the mind, moving it from the realm of religion and metaphysics into biology and shifting to investigations from physiological, medical and neurological perspectives (see Hearnshaw; Danziger 1982; Danziger 1997; Smith 1973; Bunn, Lovie and Richards). By the mid-nineteenth century, physiological psychology was well established in Britain. It was described by Sully as working towards 'the determination of the exact physiological conditions of a certain group of mental phenomena' (1886 21). Key ideas inherited from eighteenth-century mental philosophy continued to be important to Romantic- and Victorian-period discussion, such as debates over whether the mind possessed innate faculties or was instead formed by the association of sensations into complex ideas. Unusual mental states such as sleepwalking, delirium, hallucination, trance and catalepsy also continued to attract attention throughout the nineteenth century, but there was an increasing interest in describing the 'normal' mind, in relation to which the deviant, the aberrant or the insane could be understood.

The naturalizing impulse also gave rise to speculation on the persistent philosophical problem of the relations between mind and brain. Contributors to the literature on scientific psychology committed to theories, some dualist and some monist, which they believed explained those relations. These included epiphenomenalism, the view that consciousness is merely a by-product of neural processes, and the 'mind-stuff' theory, advanced by W. K. Clifford, which posits that every atom possesses mental properties, or 'mind-stuff', which – in sufficient combination – produces consciousness. The new psychology also raised related questions about the efficacy of the mental, asking whether non-physical mental events could cause physical effects. Some scientists contended that the principle of the conservation of energy entailed that mental events could not play a role in the physical world. Such conceptual inquiries cut across the various branches of psychological inquiry discussed below.

The unconscious, the automatic and mental unity

Questions about volition were essential to discussions of the unconscious and involuntary, topics that particularly intrigued nineteenth-century psychologists (Daston; Smith 2004; Smith 2013). Thomas Laycock's work on the 'Reflex Function of the Brain' (1845) extended Marshall Hall's research on spinal cord reflexes and the brain and was important to discussions of automatic processes, giving rise to theories of thought, action and the will. The physiologist W. B. Carpenter introduced the term 'ideomotor actions' to designate those ideas unconsciously formed that influence behaviour. He coined the phrase 'unconscious cerebration' in the fourth edition of *Human Physiology* (1852) to describe reflex and other modifications of thought that are outside of awareness. His ideas appeared in many discussions of artistic creation. The possibility of unwilled action and thought led physiological psychologists such as Carpenter to assert the power of a self-determining will that could be trained to keep automatic processes under control.

The recognition that humans are often motivated by unconscious processes became central to the literary depiction of the mind in the Romantic and Victorian periods. In his preface to 'Kubla Khan', Coleridge famously presents the poem as the product of unconscious, involuntary mental activity, produced in a dream. John Polidori wrote a medical thesis on somnambulism in 1815, and the ideas he developed there also inform his 1819 novella *The Vampyre* (Stiles, Finger and Bulevich). Carpenter's notion of unconscious memory is used to explain the theft of a diamond in Collins's mystery novel *The Moonstone* (1868). Eliot discusses Stephen's motives in *The Mill on the Floss* (1860) in these terms: 'Perhaps he was not distinctly conscious that he was impelled to it by a secret longing – running counter to all his self-confessed resolves . . . Watch your own speech, and notice how it is guided by your less conscious purposes' (453). Samuel Butler writes in his posthumously published (in 1903) autobiographical novel *The Way of All Flesh* (1873–84):

> How little do we know our thoughts – our reflex actions indeed, yes; but our reflex reflections! Man, forsooth, prides himself on his consciousness! . . . I fancy that there is some truth in the view which is being put forward nowadays, that it is our less conscious thoughts and our less conscious actions which mainly mold our lives and the lives of those who spring from us.
>
> *(18)*

Butler suggests that the majority of our thinking occurs unconsciously, and alludes as well to his belief in unconscious, or 'organic', memory.[1] In raising questions about the extent and efficacy of willed action, such theories also raise questions about human agency and responsibility. Concerns about automatism and reflex action can in part be connected to an increasingly industrialized machine culture. In 'Signs of the Times' (1829), Carlyle had already condemned the mechanical view that he saw holding sway, championing instead a dynamic view of mind. Butler, in 'The Book of the Machines' in his satirical novel *Erewhon* (1872), radicalizes the idea of mind as physical when he depicts machines that develop consciousness and attain mastery over the human race.

The most extreme variety of automatism emerged in discussion of 'conscious automatism'. This theory posits that the body, a physical machine, carries on by reflex action alone, without any interference from thoughts and emotions, which are ineffectual accompaniments. There were many advocates of the theory, but Thomas Huxley's presentation of it in his essay 'On the Hypothesis That Animals Are Automata' (1874) received the greatest attention, prompting hundreds of responses in the ensuing decades. Huxley did not join the group that valiantly tried to hold on to the power of the will, but instead conceded that consciousness might not count for much of anything at all. Yet the theory fascinated many in the culture, figuring in literary texts and *Punch* magazine satires. In her final work *The Impressions of Theophrastus Such* (1879), Eliot imagines a future world in 'Shadows of the Coming Race' in which an 'immensely more powerful unconscious race' (200) will have replaced humans, who were hindered by believing in the efficacy of consciousness, and particularly the illusion that they possess free will.

Related to these questions about consciousness was a concern with the unity of the mental. The belief in unity was attacked by theorists of mind such as Maudsley, who writes in 'The Physical Conditions of Consciousness' (1887): 'As a matter of observation, consciousness is evidently capable of all sorts of disintegrations, mutilations, divisions, distractions so numerous and various in kind and degree as to prove that the conscious Ego has not real identity and unity' (505–6). Similar ideas emerged earlier in the century in research on the split brain, as mental scientists sought to understand the significance of the two cerebral hemispheres, work that is splendidly explored by Anne Harrington in *Medicine, Mind, and the Double Brain*. The notion of double or dual consciousness gained both popular and scientific attention, with discussion of several cases involving patients who ostensibly shifted between two separate mental lives, unaware of their second self (see also Azam; Hacking; Eigen). Sensational newspaper stories of crimes performed while sleepwalking or by an alternative personality (double consciousness) fascinated readers. Conceptions of multiple selves entered into literary works, the most well-known instances being the explorations of the divided self in James Hogg's *The Private Memoirs and Confessions of a Justified Sinner* (1824) and Robert Louis Stevenson's *Dr. Jekyll and Mr. Hyde* (1886). Hogg's Robert Wringham describes himself as having been under the 'delusion that I was two persons' (234), adding that '[t]he most perverse part of it was, that I rarely conceived *myself* to be any of the two persons'. Jekyll pronounces in his final confession: 'I hazard the guess that man will be ultimately known for a mere polity of multifarious, incongruous, and independent denizens' (108). These concerns were extended into late-century research on alternations of personality in hysteria (Micale 1995 and Micale 2009).

Memory

Questions of consciousness intersect with questions of memory, topics crucial to both the science of the mind and literature in the nineteenth century. Memory was increasingly acknowledged to depend upon physiological processes, and the experimental study of it emerged in the last decades of the century. Kurt Danziger's *Marking the Mind* provides a good starting point for the history of these developments. Many earlier theorists of memory maintained that everything was recorded in the brain, even if it is not easily accessible to consciousness. Carpenter for a time promoted such a theory of memory, expressed in a passage that is directly quoted in Collins's *The Moonstone*:

> [E]very sensory impression which has once been recognised by the perceptive consciousness, is registered (so to speak) in the brain, and may be reproduced at some subsequent time, although there may be no consciousness of its existence in the mind during the whole intermediate period.
>
> *(Carpenter 1853 781; Collins 1999 385–6)*

A stronger version of the view is expounded as late as 1891 in George du Maurier's novel *Peter Ibbetson*:

> Evidently our brain contains something akin both to a photographic plate and a phonographic cylinder, and many other things of the same kind not yet discovered; not a sight or a sound or a smell is lost; not a taste or a feeling or an emotion.
>
> *(227)*

Du Maurier's allusions to new technologies draws attention to the way in which conceptions of memory have relied on metaphor, tracking the prominent technologies of the historical moment (Draaisma).

Around the middle of the century, attention turned to problems of determining what actually gets recorded in memory in the first place, and how memories are revised over time. Are aspects of experience forgotten, never recorded, revised or inaccessible to consciousness? Are there pseudo-memories or hallucinations of memories? Frances Cobbe, a popular writer on psychological topics, strikingly describes memory in her 1886 essay 'The Fallacies of Memory' as 'a finger mark traced on shifting sand, ever exposed to obliteration when left unrenewed; and if renewed, then modified, and made, not the same, but a fresh and different mark' (157). Cobbe proposes a theory of memory according to which our narratives of the past are in constant revision. (See Taylor 2000 on Victorian theories of memory fallacy).

The Victorian autobiographical novel, such as Dickens's *David Copperfield* (1850), is explicitly concerned with the workings of retrospect, both on the level of narrative and in direct references to memory (Dames 2001). Such novels question their protagonists' ability to portray past experience accurately and draw attention to the constructing nature of memory. Dickens's novella *The Haunted Man* (1848) dramatizes an exploration of the functioning of memory. The story depicts a chemistry teacher who bargains with a spectre to erase all his painful memories, with the assurance that he will lose 'nothing but the intertwisted chain of feelings and associations, each in its turn dependent on, and nourished by, the banished recollections' (335). The memory obliteration – simultaneously a symbolic representation of the mechanism of repression (pre-Freud), an exploration of the workings of association and a depiction of memory pathology – leads to the chemist's loss of emotion and compassion, suggesting (as did other literary works of the time) that memory is a moral force. The science of mind became increasingly concerned with impairments of memory, such as amnesia and aphasia. As the century proceeded, such pathologies were primarily understood as consequences of physiological changes, disease or injury. Towards the end of the century, a series of scientific romances developing the themes of memory erasure and exchange appeared. In Walter Besant's story 'The Memory Cell' (1900), for example, a professor conceives of memory as a disease and imagines a technology for erasing memories and substituting false memories. Such stories question the links between memory and personal identity.

Abnormality, insanity and the asylum

Psychiatry, as we now call it, developed in the nineteenth century. Mental illness became a biological and medical problem, prompting study of abnormal mental states, such as hallucination and catalepsy, and investigations of morbid and criminal psychology. Cities and urbanization were understood as the cause of an epidemic of nervous disorders and insanity. Mental science was increasingly interested in categorizing pathologies, and constructed and theorized various disorders such as 'monomania' and 'kleptomania' (the latter being particularly associated with women). Such conditions quickly entered into literary representation. Heathcliff in Emily Brontë's *Wuthering Heights* (1847) is described as having 'monomania' (288), while Braddon's Lady Audley tries to convince others that the man who charges her with murder is himself a monomaniac. The anonymously published story 'A Delicate Situation' (1885), which appeared in *All the Year Round*, is told from the perspective of a kleptomaniac's husband. Good starting points for the now very large literature on the history of insanity include Scull 1979, Scull 2005, and Bynum, Porter and Shepherd.

In the middle decades of the nineteenth century, a slew of psychological texts written from a medical perspective appeared in Britain. In 1848 Forbes Winslow founded the *Journal of Psychological Medicine and Mental Pathology*, the first of the professional journals devoted to the study of mind. Professional organizations such as the psychological section of the British Medical Association were formed. Interest increased in medico-legal issues as well. The M'Naghten Rules were formulated following the 1843 murder trial of Daniel M'Naghten. They describe the conditions required for legal insanity, allowing for a verdict of not guilty by reason of insanity and provoking questions about responsibility. Medical experts played an increased role in criminal trials and forensic psychiatry emerged as a profession (see Smith 1981; Eigen).

Asylum reform and new methods for the treatment of the mentally ill came to be recognized as important policy issues. John Conolly, following Daniel Hack Tuke, advocated for what was termed moral management of 'lunatics', crusading for compassionate treatment and improvements in asylum care that included limits on the use of physical restraints. The nineteenth-century asylum housed people with a wide array of conditions, including those that would now be recognized as dementia, epilepsy and Down's syndrome. Later in the century, evolutionary psychology and heredity theories became dominant, resulting in a more pessimistic attitude about the inevitability of insanity and the futility of treatment. The genre of sensation fiction, described as early as 1863 in a notorious review by Henry Longueville Mansel as 'preaching to the nerves' (481) – that is, as acting on the physiology of the reader – made wide use of the tropes of insanity and asylum incarceration. Sensation fiction often blurs the distinction between madness and sanity, as in Collins's *The Woman in White* (1859), which involves the wrongful confinement of women, or Braddon's *Lady Audley's Secret* (1862), which concludes with the confinement of a wrongly admired would-be murderess. Charles Reade's *Hard Cash* (1863) condemns the corruption of 'mad doctors' who sign certificates for asylum confinement for a fee; the novel satirizes Conolly in the character Dr Wycherley, who pontificates on Hamlet's madness.

The gendered aspects of insanity, nervous disease and hysteria have been intelligently explored by Oppenheim (1991), Micale (1995; 2009) and Tomes, among others, in the history of science, and by Small, Martin, Wood, Logan, Pedlar and Mangham, to name just a few useful studies in literary criticism. (See also Laycock 1840). Literary portrayals of insanity are frequently sensationalized, and critics have examined the uses of madness in imaginative fiction to critique social conventions. Such representations figure prominently in nineteenth-century British literature, perhaps most famously in Charlotte Brontë's Bertha, locked away in an attic in *Jane Eyre* (1847). Romantic fiction presents psychological accounts of insanity, as in Hogg's *Confessions* (Faflak 2003). William Godwin, who included a chapter on 'The Mechanism of the Human Mind' in his *Enquiry Concerning Political Justice* (1793), explores insanity in *Caleb Williams* (1794) (Rabb) and depicts the mental deterioration of Charles Mandeville and his confinement in an asylum for the insane in *Mandeville* (1817). Portraits of the 'mad' appear in works such as William Wordsworth's poem 'The Mad Mother' (1798) and Collins's 'Mad Monkton' (1855), which depicts a man unable to escape 'the hereditary taint in his mental organization' (220). Frederick Burwick (1996) explores the connection between madness and creative inspiration in the Romantic period, while Ekbert Faas has discussed the portrayal of insane minds in the dramatic monologues of Robert Browning and Alfred Tennyson.

Evolutionary psychology, comparative psychology, degeneration and heredity theory

The evolutionary development of mind became a prominent area of scientific concern in the last few decades of the Victorian period. Darwin's claim at the end of *On the Origin of Species* (1859) that in the future '[p]sychology will be based on a new foundation, that of the necessary acquirement of each mental power and capacity by gradation' (488) was borne out in various branches of psychological study, as a vast amount of critical and historical work has by now demonstrated, for instance in the investigation of the 'lower races', heredity science, studies of criminality and genius and the comparative study of animal brains. An excellent starting point for the history of these developments is Robert Richards's *Darwin and the Emergence of Evolutionary Theories of Mind and Behavior*. The influence of evolutionary psychology on literary representation has been closely examined in studies by Block, Reid and Shuttleworth, among others.

Developmental accounts of mind appeared before Darwin's theory of evolution: for instance, Jean-Baptiste Lamarck's early nineteenth-century theories of the transformation of species. Herbert Spencer advanced his own Lamarckian-inspired system of the psychology of evolutionary associationism in his well-regarded (by his contemporaries) *The Principles of Psychology* (1855; revised 2nd edn 1870), maintaining that mind developed by a process of adaptation to its environment. Those theories continued to play an influential role after Darwin in the sciences of the mind, as in Hughlings Jackson's neurological work, and in imaginative literature, as in Butler's enthusiastic championing of a neo-Lamarckian view, according to which acquired characteristics are passed on to offspring and volition plays a role in the course of evolution.

Darwin asserted mental continuity between humans and non-human animals in *The Descent of Man* (1871), maintaining that the mental differences were a matter 'of degree and not of kind' (1:105), that is, denying that the human mind is metaphysically distinct from those of other animals. In *The Expression of the Emotions in Man and Animals* (1872), Darwin examined the evolution of emotions. George Romanes, a disciple of Darwin, extended his comparative work on the evolution of mind and language in books such as *Animal Intelligence* (1882) and *Mental Evolution in Man* (1888). Ideas from comparative psychology – animal psychology – were represented in contemporary literary works, with, for instance, Stevenson's Hyde exhibiting animal expressions of emotion and Wells's Moreau trying to speed up the endowment of language in vivisected animals. (On comparative psychology, see Boakes; Ritvo; Hearnshaw; Radick.)

Evolutionary ideas played a role in additional areas of psychological inquiry. Women's mental capacities were believed to have, from evolutionary pressures, developed to be different from and largely inferior to men's (Russett). In an article entitled 'Mental Differences Between Men and Women' (1887) published in the popular journal the *Nineteenth Century*, Romanes asserts: 'Seeing that the average brain-weight of women is about five ounces less than that of men, on purely anatomical grounds, we should be prepared to expect a marked inferiority of intellectual power in the former' (654–5). Hardy's doctor Fitzpiers echoes that belief in *The Woodlanders* (1887) while examining a portion of a brain under a microscope: 'a woman's typically weighs four ounces less than a man's' (46). Child psychology adopted theories of recapitulation of the species in explaining the mental development of children (see Shuttleworth 2010; Gurjeva). Many, advancing scientific racism, asserted that brains of non-white ('primitive') people were arrested at a low level of mental evolution (Richards G. 2007).

Darwin famously writes at the end of *The Descent* that man retains 'the indelible stamp of his lowly origins' (2:405). That idea was extended in the view that primitive aspects of mind remain in humans. Scientists like Hughlings Jackson maintained that earlier evolved parts of the human brain are the same as those of the lower animals, and that these shared parts function more automatically than those unique to human brains. The more recently evolved parts of the human brain relate to voluntary and conscious behaviour, but loss of control of the lower parts can occur, for instance, while dreaming or as a consequence of brain injury. That notion, together with degeneration theories – which proposed that the evolutionary process might result over generations in retrogression rather than progression and lead to adaptive pathologies such as nervous diseases and 'imbecility' – had a great influence on literary representation, as has been widely discussed in the criticism (see Pick; Greenslade). The emergence of criminology late in the century related criminal behaviour to atavism and inherited psychological qualities, notably in Havelock Ellis's *The Criminal* (1890), influenced by the work of the Italian criminologist Cesare Lombroso. Narratives depicting mental degeneration or hereditary criminality were common in the final decades of the century, as for instance in Wells's portrayal of the future race of the Eloi in *The Time Machine* (1895) or Bram Stoker's 1897 descriptions of Count Dracula.

Francis Galton's work on the inheritance of mental capacity, in works such as *Hereditary Genius* (1869) and *Inquiries Into Human Faculty and Its Development* (1883), was an inspiring force in literature. Developing statistical concepts such as regression, Galton advanced psychometrics, the science of measuring psychological characteristics. Galton also introduced the science of eugenics, contending that humans could work to direct the process of evolution, improving the 'stock' through selective breeding and other interventions. Galton even wrote his own eugenics romance *Kantsaywhere* (1910). Angelique Richardson has traced the shaping force of eugenics and conceptions of rational reproduction in New Woman fiction. Allusions to the inherited nature of character are pervasive in 1890s fiction. Dr Shadwell Rock in Sarah Grand's novel *The Heavenly Twins* (1893) has written a book on 'the heredity of vice' (662). In George Egerton's ambiguous short story 'Wedlock' (1894) a bricklayer discourses on the hereditary nature of alcoholism. Grant Allen was particularly obsessed with issues of inheritance, as the title of his award winning novel *What's Bred in the Bone* (1892) demonstrates. In his 'Child of the Phalanstery' (1884) he interrogates a future community that has made eugenics its religion. His story 'The Two Carnegies' (1885), taking its cue from Galton's studies of twins, describes identical twins with identical desires, diseases, tastes and motives, the plot problem being that they cannot both marry the same woman. Such mental determinism is also played out in fiction influenced by naturalism, such as George Gissing's.

Neurology and experimental psychology

Nineteenth-century sciences of the mind, as already noted, were deeply rooted in advances in knowledge of sensory-motor physiology and brain anatomy. Investigations of patients with brain trauma (such as the famous 1848 case of Phineas Gage, the railway worker whose personality altered following an accident that sent an iron rod straight through his brain) supported the contention that physiological and mental processes are correlated. In the second half of the century, neurology emerged as a specialized medical discipline. Hughlings Jackson studied neurological diseases, epilepsy, disorders of speech, aphasia and brain functions and made great advances in clinical neurology. *Brain: A Journal of Neurology* began publication in 1878 and the Neurological Section of the Royal Society of Medicine was established in 1886. The British neurologist David Ferrier made important contributions to the localization of brain function, demonstrating through electrical stimulation of the cerebral cortex of animals that certain parts of the cortex controlled specific muscles. (On the history of the neurosciences, see Clarke and Jacyna; Young; Finger; Jacyna; Harrington.)

The last quarter of the century also saw the emergence of the experimental study of psychology, with the first psychology laboratory, established by William Wundt in 1879 in Leipzig. Early laboratory studies quantified mental processes, measured mental events in time and examined sense perception and the perception of space, among other things. Hermann Ebbinghaus's experimental study of memory soon followed (see Danziger 1990; Boring). Although, as mentioned, Britain lagged behind in establishing its own psychological laboratories, the work abroad was extensively reviewed in the 1870s and 1880s in *Mind*, in the first issue of which the editor George Croom Robertson announced that the new journal sought to 'procure a decision of this question as to the scientific standing of psychology' (3).

Neurology and experimental psychological studies were frequently depicted in nineteenth-century literary works. As early as 1870 Dickens was credited with anticipating Hughlings Jackson's research on aphasia in his 1848 portrayal of *Dombey and Son*'s Mrs Skewton and since that time neurologists have found descriptions of a variety of neurological disorders in his novels ([Anon.] 1870). By 1955 the eminent British neurologist Sir Russell Brain praised in the *British Medical Journal* Dickens's skilfulness at detailing the changes in consciousness that followed upon head injuries and stroke. Conditions such as aphasia appear throughout late Victorian literature, as in Arthur Morrison's mystery story 'The Case of the Lost Foreigner' (1895), where the solution to the mystery depends on communication with a man who suffers from motor aphasia and agraphia. Collins alludes to Ferrier's work on localization in the introduction to his anti-vivisection novel *Heart and Science* (1883), and the novel implicitly connects Ferrier's methods with those of the gruesome vivisectionist Benjulia, whose research is pre-empted by that of a gentle physician who makes significant discoveries on brain disease through compassionate bedside observation of the ill (see Talairach-Vielmas; Otis 2007). In *Popular Fiction and Brain Science in the Late Nineteenth-Century*, Anne Stiles examines late-century writers of Gothic fiction and romance, such as Stoker and Marie Corelli, who incorporate neurological research into their works (see also Salisbury and Shall; Wood; Stiles and Finger).

Mental force, telepathy and ghost-seeing

Scientific psychology also made incursions into what seems to be non-scientific territory: the study of ghosts, telepathy and mental action at a distance. The Society for Psychical Research (SPR) was formed in 1879, its goal being the empirical investigation of phenomena such as hypnotism, mind reading, spiritualism and crisis apparitions (spectral sightings of people at the moment of their deaths). Respected scientists advanced explanations for telepathy as involving imponderable fluids. The chemist and physicist William Crookes suggested that Röntgen rays were the medium of psychic exchange. Later, the eminent physicist Oliver Lodge maintained that ether was the vehicle of telepathic communication with the dead. F. W. H. Myers, one of the founders of the SPR, posited the existence of a 'subliminal consciousness' that extended the powers of the mind. Mental science's attraction to the seemingly extraordinary is exhibited in the history of Britain's first psychological society (1875–79). Founder Edward Cox aimed to advance the science of psychology but rejected materialism and the attempt to reduce the psychological to the physiological; he examined 'psychic force', a concept first hypothesized by Crookes (Richards G. 2001). These unorthodox areas of mental science have been ably examined by historians Janet Oppenheim (1985), Alex Owen and Edward Brown; the literary connections have been traced in Roger Luckhurst's fine study *The Invention of Telepathy* (see also Thurschwell; Grimes). Related ideas were dramatized in stories such as Doyle's *The Parasite*, in which a materialist doctor announces (invoking Carlyle's 'Signs of the Times'):

> I had always looked upon spirit as a product of matter. The brain, I thought, secreted the mind, as the liver does the bile. But how can this be when I see mind working from a distance and playing upon matter as a musician might upon a violin?
>
> *(34)*

The story undercuts the authority of naturalist science in depicting a scientist forced to recognize mentality's independence from the physical. New technologies, invisible natural forces and consciousness are connected in Rudyard Kipling's story 'Wireless' (1902), where an ailing pharmacist, affected by electromagnetic radio waves, channels John Keats's poetry. Amy Levy in her short story 'The Recent Telepathic Occurrence at the British Museum' (1886) represents a case of crisis apparition, which might well have appeared in the SPR's collected volume of such experiences, *Phantasms of the Living* (1886).

Others such as Carpenter sought to show that spiritualism and telepathy could be accounted for psychologically and in accepted naturalist terms. The ostensibly extraordinary perceptual abilities of mind readers were explained as unconscious muscle reading. Reported communications with the dead by spiritualists were not necessarily to be understood as quackery or fraud but instead as mental disorder, hallucination or the effect of suggestion. Physiological explanations of spirit sightings were prominent in psychology throughout the century. Spectral illusion theory, which explained ghost sightings as hallucinations (psychological rather than actual), was advanced by medical practitioners in the first decade of the century, following the publication of C. F. Nicolai's memoir accounting for his sightings of apparitions in medical terms. Spectral illusion theory plays a role in many nineteenth-century literary texts, by Sheridan Le Fanu, Dickens, Kipling and others. In Charlotte Brontë's *Villette* (1853), Dr John invokes spectral illusion to explain Lucy's sighting of a ghost: 'I think it a case of spectral illusion: I fear, following on and resulting from long-continued mental conflict' (249). Shane McCorristine's interdisciplinary study *Spectres of the Self* traces the history of viewing ghost sightings as mental phenomena in British medical and literary texts.

Historians of psychology in recent decades have argued that the early construction of the discipline of psychology as an abandonment of metaphysics and a turn to experimental scientific psychology is a narrative that served the purposes of the new discipline, both justifying and shaping it. The once widely accepted account of the origins of scientific psychology in the second half of the nineteenth century has been revised and the complexities of that development examined (see Danziger 1990). In the epilogue to his *Briefer Course* (1892) abridgement of his influential *Principles of Psychology*, William James writes:

> When, then, we talk of 'psychology as a natural science', we must not assume that that means a sort of psychology that stands at last on solid ground. It means just the reverse; it means a psychology particularly fragile, and into which the waters of metaphysical criticism leak at every joint.
>
> *(467)*

The history of the interactions between the sciences of the mind and literary texts bears out the unruliness of the terrain. In Arthur Machen's *The Great God Pan* (1894), brain science becomes the naturalist foundation for access to the mystical. Machen's Dr Raymond performs a neurological experiment that breaks down the barriers between the material and spiritual worlds. '[A] slight lesion in the grey matter . . . a trifling rearrangement of certain cells, a microscopical alteration that would escape the attention of ninety-nine brain specialists out of a hundred' (Machen 4) allows the scientist to open the mind of his subject to the metaphysical. Machen's story is only one of many that makes naturalist science the launch pad for the spiritual.

Literary texts can perhaps go only so far with a determinist scientific psychology. The scientific project that seeks to find the causes of people's behaviour in events and conditions external to them collides with both the appeal and literary functions of the common-sense view that understands humans (characters) as autonomous agents. In the complex history of interactions between nineteenth-century literature and the new sciences of the mind, many writers come up against those incompatibilities. At the *fin de siècle*, psychologists such as James Ward began to assert that psychology should break away from natural science as a model, and focus instead on the reality experienced by the individual consciousness. Modernist fiction largely followed suit.

This is not to say that the interrelations between psychology and literature were severed after the nineteenth century. Even as dynamic psychology emerged as a shaping force on literature at the start of the twentieth century, psychology continued to acknowledge its debts to literary authors. As Sigmund Freud writes in a 1907 essay,

> Creative writers are valuable allies and their evidence is to be prized highly, for they are apt to know a whole host of things between heaven and earth of which our philosophy has not yet let us dream. In their knowledge of the mind they are far in advance of us everyday people, for they draw upon sources which we have not yet opened up for science.
>
> *(8)*

Note

1   See Otis 1994 on the theory of 'organic memory', first advanced by Ewald Hering in Germany, which posited that memories are passed on in the body, from parent to offspring; the theory provided a mechanism for the inheritance of habits, instincts and unconscious memories.

Bibliography

Primary texts

Allen, Grant. 'Child of the Phalanstery'. *Strange Stories*. London: Chatto and Windus, 1884, 301–20.
Allen, Grant. 'The Two Carnegies'. *The Cornhill Magazine* 4.21 (March 1885): 292–324.
Allen, Grant. *What's Bred in the Bone*. London: Tit-Bits Offices, 1891.
[Anon.]. 'A Delicate Situation'. *All the Year Round* 37.883 (31 October 1885): 198–203.
[Anon.]. 'Charles Dickens'. *British Medical Journal* 1 (18 June 1870): 636.
Azam, Eugène. 'Periodical Amnesia; or, Double Consciousness'. *Journal of Nervous and Mental Disease* 3 (1876): 584–612.
Bain, Alexander. 'Mr. Galton's Statistics of Mental Imagery'. *Mind: A Quarterly Review of Psychology and Philosophy* 5.20 (October 1880): 564–73.
Besant, Walter. 'The Memory Cell'. *Five Years Tryst and Other Stories*. London: Methuen, 1902, 175–94.
Braddon, Mary Elizabeth. *Lady Audley's Secret*. Ed. Lynn Pykett. Oxford: Oxford University Press, 2012.
Braid, James. *Neurypnology or The Rationale of Nervous Sleep Considered in Relation with Animal Magnetism*. London: John Churchill, 1843.
Brontë, Charlotte. *Villette*. Eds. Margaret Smith and Herbert Rosengarten. Oxford: Oxford University Press, 2000.
Brontë, Charlotte. *Jane Eyre*. Ed. Margaret Smith. Oxford: Oxford University Press, 2008.
Brontë, Charlotte. *Shirley*. Eds. Margaret Smith, Herbert Rosengarten and Janet Gezari. Oxford: Oxford University Press, 2008.
Brontë, Emily. *Wuthering Heights*. New York, NY: Harper, 1858.
Browning, Robert. 'Mesmerism'. *Selected Poems*. Ed. Daniel Karlin. New York, NY: Penguin, 2001, 85–9.
Bucknill, John Charles. '*Maud and Other Poems* by Alfred Lord Tennyson'. *Asylum Journal of Mental Sciences* 2 (October 1855): 95–104.
Bucknill, John Charles. *The Psychology of Shakespeare*. London: Longman, Brown, Green, Longmans, and Roberts, 1859.
Butler, Samuel. *Erewhon*. Ed. Peter Mudford. New York, NY: Penguin, 1974.
Butler, Samuel. *The Way of All Flesh*. Ware: Wordsworth, 1994.
Caird, Mona. *Daughters of Danaus*. London: Bliss, Sands, and Foster, 1894.
Carlyle, Thomas. 'Signs of the Times'. *Edinburgh Review* 49 (June 1829): 439–59.
Carpenter, William Benjamin. *Principles of Human Physiology*. 5th, American edn from the 4th, enlarged London edn. Philadelphia, PA: Blanchard and Lea, 1853.
Carpenter, William Benjamin. *Principles of Mental Physiology*. London: Henry S. King, 1875.
Clifford, W. K. 'On the Nature of Things in Themselves'. *Mind* 3 (1878): 57–67.
Cobbe, Frances Power. 'The Fallacies of Memory'. *Galaxy: A Magazine of Entertaining Reading* 1 (15 May 1866): 149–62.

Collins, Wilkie. 'Mad Monkton'. *The Queen of Hearts*. New York, NY: Harper, 1869, 147–225.

Collins, Wilkie. *The Woman in White*. London: Smith, Elder, 1871.

Collins, Wilkie. *Heart and Science*. Ed. Steve Farmer. Peterborough: Broadview, 1996.

Collins, Wilkie. *The Moonstone*. Ed. John Sutherland. New York, NY: Oxford University Press, 1999.

Combe, George. *System of Phrenology*. 3rd edn. Edinburgh: John Anderson, 1830.

Conolly, John. *A Study of Hamlet*. London: Edward Moxon, 1863.

Cox, Edward. *What Am I? A Popular Introduction to Mental Philosophy and Psychology: The Mind in Action*. Vol. 2. London: Longman, 1874.

Crookes, William. 'Presidential Address to the Society of Psychical Research'. *Proceedings of the Society for Psychical Research* 12 (March 1897): 338–55.

Darwin, Charles. *On the Origin of Species*. London: John Murray, 1859.

Darwin, Charles. *The Descent of Man and Selection in Relation to Sex*. 2 vols. London: John Murray, 1871.

Darwin, Charles. *The Expression of the Emotions in Man and Animals*. New York, NY: D. Appleton, 1873.

Dickens, Charles. *David Copperfield*. Ed. Nina Burgis. Oxford: Clarendon, 1981.

Dickens, Charles. *The Haunted Man and the Ghost's Bargain. A Christmas Carol and Other Christmas Books*. Ed. Robert Douglas-Fairhurst. New York, NY: Oxford University Press, 1987, 313–99.

Dickens, Charles. *Great Expectations*. Ed. Charlotte Mitchell. New York, NY: Penguin, 1996.

Dickens, Charles. *The Pickwick Papers*. Ed. James Kinsley. Oxford: Oxford University Press, 2008.

Doyle, Arthur Conan. *The Parasite*. New York, NY: Harper, 1895.

Du Maurier, George. *Trilby*. New York, NY: Harper, 1894.

Du Maurier, George. *Peter Ibbetson*. New York, NY: Harper, 1895.

Egerton, George. 'Wedlock'. *Discords*. London: John Lane, 1894, 117–44.

Eliot, George. 'The Lifted Veil'. *Blackwood's Magazine* 86 (July 1859): 14–48.

Eliot, George. 'Shadows of the Coming Race'. *The Impressions of Theophrastus Such*. New York, NY: Harper, 1879, 194–201.

Eliot, George. *The Mill on the Floss*. Ed. Gordon S. Haight. Oxford: Oxford University Press, 2008.

Ellis, Havelock. *The Criminal*. New York, NY: Scribner and Welford, 1890.

Freud, Sigmund. 'Delusions and Dreams in Jensen's *Gradiva*'. *The Standard Edition of the Complete Psychological Works of Sigmund Freud*. Transl. and ed. James Strachey. 24 vols. London: Hogarth, 1959, 9:7–93.

Galton, Francis. *Hereditary Genius: An Inquiry into Its Laws and Consequences*. London: Macmillan, 1869.

Galton, Francis. *Inquiries into Human Faculty and Its Development*. London: Macmillan, 1883.

Godwin, William. *Things as They Are, or, The Adventures of Caleb Williams*. 3 vols. London: B. Crosbie, 1794.

Godwin, William. 'Mechanism of the Human Mind'. *Enquiry Concerning Political Justice, and Its Influence on Morals and Happiness*. 3rd edn. 2 vols. London: G. G. and J. Robinson, 1798, 1:398–421.

Godwin, William. *Mandeville: A Tale of the Seventeenth-Century in England*. 3 vols. Edinburgh: Archibald Constable, 1817.

Grand, Sarah. *The Heavenly Twins*. Introd. Carol A. Senf. Ann Arbor, MI: University of Michigan Press, 1992.

Gurney, Edward, Frederic W. H. Myers, and Frank Podmore, eds. *Phantasms of the Living*. Vol. 1. London: Trubner, 1886.

Hardy, Thomas. *Jude the Obscure*. New York, NY: Harper, 1896.

Hardy, Thomas. *The Woodlanders*. Ed. Dale Kramer. Oxford: Oxford University Press, 2005.

Hogg, James. *The Private Memoirs and Confessions of a Justified Sinner*. London: Longman, 1824.

Huxley, Thomas. 'On the Hypothesis That Animals Are Automata'. *Fortnightly Review* 16.95 (1874): 555–80.

James, William. *Psychology: Briefer Course*. New York, NY: Henry Holt, 1920.

Kendall, May. 'The Materialist'. *Songs from Dreamland*. London: Longmans, Green, 1894.

Kipling, Rudyard. 'Wireless'. *Scribner's Magazine* 32.1 (August 1902): 129–43.

Laycock, Thomas. *A Treatise on the Nervous Diseases of Women*. London: Longman, Orme, Brown, Green, and Longmans, 1840.

Laycock, Thomas. 'On the Reflex Function of the Brain'. *The British and Foreign Medical Review* 19 (January 1845): 298–311.

Levy, Amy. 'The Recent Telepathic Occurrence at the British Museum'. *Woman's World* 1 (1886): 31–2.

Lewes, George Henry. 'Dickens in Relation to Criticism'. *Fortnightly Review* 17 (February 1872): 141–54.

Lodge, Oliver. *My Philosophy: Representing My View of the Many Functions of the Ether of Space*. London: Ernest Benn, 1933.

Lynch, Hannah. *George Meredith: A Study*. London: Methuen, 1891.

Machen, Arthur. *The Great God Pan: And the Inmost Light*. London: John Lane, 1895.

Mansel, Henry Longueville. 'Sensation Novels'. *Quarterly Review* 114 (April 1863): 481–514.

Marsh, Richard. *The Beetle: A Mystery*. Ed. Julian Wolfreys. Peterborough: Broadview, 2004.

Maudsley, Henry. 'Hamlet'. *Body and Mind: An Inquiry into Their Connection and Mutual Influence, Specially in Reference to Mental Disorders*. Rev. edn. London: Macmillan, 1873, 145–95.

Maudsley, Henry. *The Physiology of Mind*. New York, NY: D. Appleton, 1883.

Maudsley, Henry. 'The Physical Conditions of Consciousness'. *Mind* 12.48 (October 1887): 489–515.

Mill, John Stuart. *A System of Logic, Ratiocinative and Inductive*. London: Longmans, Green, 1941.

Morrison, Arthur. 'The Case of the Lost Foreigner'. *The Chronicles of Martin Hewitt*. New York, NY: D. Appleton, 1896, 228–67.

Myers, F. W. H. 'The Subliminal Consciousness. Chapter 1: General Characteristics and Subliminal Messages'. *Proceedings of the Society for Psychical Research* 7 (1892): 298–327.

Peacock, Thomas Love. *Headlong Hall*. London: T. Hookham, 1816.

Reade, Charles. *Hard Cash: A Matter-of-Fact Romance*. London: Chatto and Windus, 1927.

Robertson, George Croom. 'Prefatory Words'. *Mind* 1.1 (January 1876): 1–6.

Romanes, George J. *Animal Intelligence*. London: Kegan Hall, Trench, 1882.

Romanes, George J. 'Mental Differences Between Men and Women'. *The Nineteenth Century: A Monthly Review* (May 1887): 654–72.

Romanes, George J. *Mental Evolution in Man: Origin of Human Faculty*. London: Kegan Hall, Trench, 1888.

Spencer, Herbert. *The Principles of Psychology*. 2nd edn. London: William and Norgate, 1870.

Stevenson, Robert Louis. *The Strange Case of Dr. Jekyll and Mr. Hyde*. London: Longmans, Green, 1886.

Stoker, Bram. *Dracula*. New York, NY: Grosset and Dunlap, 1897.

Sully, James. 'George Eliot's Art'. *Mind* 6.23 (July 1881): 378–94.

Sully, James. 'Physiological Psychology in Germany'. *Mind* 1 (1886): 20–43.

Sully, James. *The Human Mind: A Textbook of Psychology*. 2 vols. New York, NY: D. Appleton, 1892.

Sully, James. *My Life and Friends: A Physiologist's Memories*. London: T. Fisher Unwin, 1918.

Taine, Hippolyte. *History of English Literature*. 5 vols. Edinburgh: Edmonston and Douglas, 1871.

Wells, H. G. *The Time Machine: An Invention*. New York, NY: Henry Holt, 1895.

Wells, H. G. *The Island of Doctor Moreau: A Possibility*. New York, NY: Storm and Kimball, 1896 [1896A].

Wells, H. G. 'The Story of the Late Mr. Elvesham'. *Idler* 9:4 (May 1896): 487–96 [1896B].

Wilde, Oscar. *The Picture of Dorian Gray*. Ed. Joseph Bristow. Oxford: Oxford University Press, 2008.

Winslow, L. Forbes. *Mad Humanity: Its Forms Apparent and Obscure*. London: C. A. Pearson, 1898.

Wordsworth, William. 'The Mad Mother'. *Lyrical Ballads and Other Poems*. Ware: Wordsworth, 2003, 63–5.

Zangwill, Israel. 'The Memory Clearing House'. *Idler: An Illustrated Monthly* (July 1892): 672–85.

Secondary texts

Anger, Suzy. 'Naturalizing the Mind in the Victorian Novel: Consciousness in Wilkie Collins's *Poor Miss Finch* and Thomas Hardy's *Woodlanders*'. *The Oxford Companion of the Victorian Novel*. Ed. Lisa Rodensky. Oxford: Oxford University Press, 2013, 483–506.

Bernard, Catherine A. 'Dickens and Victorian Dream Theory'. *Victorian Science and Victorian Values: Literary Perspectives*. Eds. James Paradis and Thomas Postlethwait. New York, NY: New York Academy of Sciences, 1981, 197–216.

Block, Jr., Ed. 'James Sully, Evolutionary Psychology, and Late Victorian Gothic Fiction'. *Victorian Studies* 25.4 (1982): 443–67.

Boakes, Robert. *From Darwinism to Behaviourism*. Cambridge: Cambridge University Press, 1984.

Boring, Edwin G. *A History of Experimental Psychology*. 2nd edn. New York, NY: Appleton, Century, Crofts, 1950.

Brain, Sir Russell. 'Dickensian Diagnoses'. *British Medical Journal* 2.4955 (24 December 1955): 1553–6.

Brown, Edward M. 'Neurology and Spiritualism in the 1870s'. *Bulletin of the History of Medicine* 57.4 (1983): 563–77.

Bunn, G. C., A. D. Lovie, and G. D. Richards, eds. *Psychology in Britain: Historical Essays and Personal Reflections*. Leicester: British Psychological Society, 2001.

Burwick, Frederick. *Poetic Madness and the Romantic Imagination*. University Park, PA: Pennsylvania State Press, 1996.

Burwick, Frederick. 'De Quincey and Animal Magnetism'. *The Wordsworth Circle* 36.1 (2005): 32–40.

Bynum, W. F., R. Porter, and M. Shepherd, eds. *The Anatomy of Madness. Essays in the History of Psychiatry.* 2 vols. London: Routledge, 1985–8.

Clarke, Edwin, and L. S. Jacyna. *Nineteenth-Century Origins of Neuroscientific Concepts.* Berkeley, CA: University of California Press, 1987.

Cohn, Elisha. '"One Single Ivory Cell": Oscar Wilde and the Brain'. *Journal of Victorian Culture* 17.2 (2012): 1–23.

Cooter, Roger. *The Cultural Meaning of Popular Science: Phrenology and the Organization of Consent in Nineteenth-Century Britain.* New York, NY: Cambridge University Press, 1984.

Crabtree, Adam. *From Mesmer to Freud: Magnetic Sleep and the Roots of Psychological Healing.* New Haven, CT: Yale University Press, 1993.

Daly, Nicholas. 'Railway Novels: Sensation Fiction and the Modernization of the Senses'. *ELH* 66.2 (1999): 461–87.

Dames, Nicholas. *Amnesiac Selves: Nostalgia, Forgetting, and British Fiction, 1810–1870.* New York, NY: Oxford University Press, 2001.

Dames, Nicholas. *The Physiology of Reading: Reading, Neural Science and the Form of Victorian Fiction.* Oxford: Oxford University Press, 2007.

Danziger, Kurt. 'Mid-Nineteenth-Century British Psycho-Physiology: A Neglected Chapter in the History of Psychology'. *The Problematic Science: Psychology in Nineteenth-Century Thought.* Eds. William R. Woodward and Mitchell G. Ash. New York, NY: Praeger, 1982, 119–47.

Danziger, Kurt. *Constructing the Subject.* Cambridge: Cambridge University Press, 1990.

Danziger, Kurt. *Naming the Mind: How Psychology Found Its Language.* London: Sage, 1997.

Danziger, Kurt. *Marking the Mind: A History of Memory.* Cambridge: Cambridge University Press, 2008.

Daston, Lorraine. 'British Responses to Psycho-Physiology, 1860–1900'. *Isis* 69 (1979): 192–208.

Dawson, P. M. S. '"A Sort of Natural Magic": Shelley and Animal Magnetism'. *Keats–Shelley Review* 1 (1986): 15–34.

Davie, Neil. 'Garder la race en tête: Phrénologie, race et discours colonial en Grande-Bretagne, *c.* 1810–1850'. *Racialisations dans l'aire anglophone.* Ed. Michel Prum. Paris: L'Harmattan, 2012. 17–49.

Davis, Michael. *George Eliot and Nineteenth-Century Psychology: Exploring the Unmapped Country.* Aldershot: Ashgate, 2006.

De Almeida, Hermione. *Romantic Medicine and John Keats.* New York, NY: Oxford University Press, 1991.

Draaisma, Douwe. *Metaphors of Memory: A History of Ideas About the Mind.* Transl. Paul Vincent. Cambridge: Cambridge University Press, 2001.

Eigen, Joel Peter. *Unconscious Crime: Mental Absence and Criminal Responsibility in Victorian London.* Baltimore, MD: Johns Hopkins University Press, 2003.

Faas, Ekbert. *Retreat into the Mind: Victorian Poetry and the Rise of Psychiatry.* Princeton, NJ: Princeton University Press, 1988.

Faflak, Joel. '"The Clearest Light of Reason": Making Sense of Hogg's Body of Evidence'. *Gothic Studies* 5.1 (2003): 94–110.

Faubert, Michelle. *Rhyming Reason: The Poetry of Romantic-Era Psychologists.* London: Pickering and Chatto, 2009.

Faubert, Michelle. 'John Ferriar's Psychology, James Hogg's Justified Sinner, and the Gay Science of Horror-Writing'. *Romanticism and Pleasure.* Eds. Michelle Faubert and Thomas Schmid. New York, NY: Palgrave, 2010. 83–108.

Finger, Stanley. *Origins of Neuroscience: A History of Explorations into Brain Function.* Oxford: Oxford University Press, 2003.

Ford, Jennifer. *Coleridge on Dreaming: Romanticism, Dreams, and the Medical Imagination.* Cambridge: Cambridge University Press, 1998.

Fulford, Tim. 'Conducting the Vital Fluid: The Politics and Poetics of Mesmerism in the 1790s'. *Studies in Romanticism* 43.1 (2004): 57–78 [2004A].

Gauld, Alan. *The History of Hypnotism.* Cambridge: Cambridge University Press, 1995.

Goellnicht, Donald. *The Poet-Physician: Keats and Medical Science.* Pittsburgh, PA: University of Pittsburgh Press, 1984.

Graham, John. 'Character Description and Meaning in the Romantic Novel'. *Studies in Romanticism* 5.4 (1966): 208–18.

Greenslade, William. *Degeneration, Culture, and the Novel.* Cambridge: Cambridge University Press, 1994.

Grimes, Hilary. *The Late Victorian Gothic: Mental Science, the Uncanny, and Scenes of Writing*. Aldershot: Ashgate, 2011.

Gurjeva, Lyubov G. 'James Sully and Scientific Psychology, 1870–1910'. *Psychology in Britain: Historical Essays and Personal Reflections*. Eds. G. C. Bunn, A. D. Lovie, and G. D. Richards. Leicester: British Psychological Society, 2001, 72–94.

Hacking, Ian. *Rewriting the Soul: Multiple Personality and the Sciences of Memory*. Princeton, NJ: Princeton University Press, 1995.

Hall, Jason Y. 'Gall's Phrenology: A Romantic Psychology'. *Studies in Romanticism* 16 (1977): 305–17.

Harrington, Anne. *Medicine, Mind, and the Double Brain: A Study in Nineteenth-Century Thought*. Princeton, NJ: Princeton University Press, 1987.

Hartley, Lucy. *Physiognomy and the Meaning of Expression in Nineteenth-Century Culture*. Cambridge: Cambridge University Press, 2001.

Hasseler, Terri A. 'The Physiological Determinism Debate in Oscar Wilde's *The Picture of Dorian Gray*'. *Victorian Newsletter* 84 (1993): 31–5.

Hearnshaw, L. S. *A Short History of British Psychology, 1840–1940*. London: Methuen, 1964.

Iseli, Marcus. *Thomas De Quincey and the Cognitive Unconscious*. Basingstoke: Palgrave Macmillan, 2015.

Jackson, Noel. *Science and Sensation in Romantic Poetry*. Cambridge: Cambridge University Press, 2008.

Jacyna, L. S. *Lost Words: Narratives of Language and the Brain, 1825–1926*. Princeton, NJ: Princeton University Press, 2000.

Kaplan, Fred. *Dickens and Mesmerism: The Hidden Springs of Fiction*. Princeton, NJ: Princeton University Press, 1975.

Kearns, Michael. *Metaphors of Mind in Fiction and Psychology*. Lexington, KY: University Press of Kentucky, 1987.

Keen, Suzanne. *Thomas Hardy's Brains: Psychology, Neurology, and Hardy's Imagination*. Columbus, OH: Ohio State University Press, 2014.

Leask, Nigel. 'Shelley's Magnetic Ladies: Romantic Mesmerism and the Politics of the Body'. *Beyond Romanticism: New Approaches to Texts and Contexts, 1780–1832*. Eds. Stephen Copley and John Whale. New York, NY: Routledge, 1992, 53–78.

Levere, Trevor H. 'S. T. Coleridge and the Human Sciences: Anthropology, Phrenology, and Mesmerism'. *Science, Pseudo-Science and Society*. Eds. Marsha P. Hanen, Margaret J. Osler, and Robert G. Weyant. Waterloo: Wilfred Laurier, 1980, 171–92.

Logan, Peter Melville. *Nerves and Narratives: A Cultural History of Hysteria in Nineteenth-Century British Prose*. Berkeley, CA: University of California Press, 1997.

Luckhurst, Roger. *The Invention of Telepathy, 1870–1901*. Cambridge: Cambridge University Press, 2001.

Mangham, Andrew. *Violent Women and Sensation Fiction: Crime, Medicine, and Victorian Popular Culture*. Basingstoke: Palgrave Macmillan, 2007.

Martin, Philip W. *Mad Women in Romantic Writing*. New York, NY: St Martin's, 1987.

Matus, Jill. *Memory, Shock, and the Unconscious in Victorian Fiction*. Cambridge: Cambridge University Press, 2009.

McCorristine, Shane. *Spectres of the Self: Thinking About Ghosts and Ghost-Seeing in England, 1750–1920*. Cambridge: Cambridge University Press, 2010.

Mellor, Anne K. 'Physiognomy, Phrenology, and Blake's Visionary Heads'. *Blake in His Time*. Eds. Robert N. Essick and Donald Ross Pearce. Bloomington, IN: Indiana University Press, 1978, 53–74.

Micale, Mark S. *Approaching Hysteria: Disease and Its Interpretations*. Princeton, NJ: Princeton University Press, 1995.

Micale, Mark S. *Hysterical Men: The Hidden History of Male Nervous Illness*. Cambridge, MA: Harvard University Press, 2009.

O'Neill, Anna. *Primitive Minds: Evolution and Spiritual Experience in the Victorian Novel*. Columbus, OH: Ohio State University Press, 2013.

Oppenheim, Janet. *The Other World: Spiritualism and Psychical Research in England 1850–1914*. Cambridge: Cambridge University Press, 1985.

Oppenheim, Janet. *Shattered Nerves: Doctors, Patients, and Depression in Victorian England*. Oxford: Oxford University Press, 1991.

Otis, Laura. *Organic Memory: History and the Body in the Late Nineteenth and Early Twentieth Centuries*. Lincoln, NE: University of Nebraska Press, 1994.

Otis, Laura. 'Howled Out of the Country: Wilkie Collins and H. G. Wells Retry David Ferrier'. *Neurology and Literature, 1860–1920*. Ed. Anne Stiles. New York, NY: Palgrave Macmillan, 2007, 27–51.

Owen, Alex. *The Place of Enchantment: British Occultism and the Culture of the Modern*. Chicago, IL: University of Chicago Press, 2004.

Pedlar, Valerie. '*The Most Dreadful Visitation*': *Male Madness in Victorian Fiction*. Liverpool: Liverpool University Press, 2006.

Percival, Melissa, and Graeme Tytler, eds. *Physiognomy in Profile: Lavater's Impact on European Culture*. Newark, DE: University of Delaware Press, 2005.

Pick, Daniel. *Faces of Degeneration: A European Disorder c.1848–c.1918*. Cambridge: Cambridge University Press, 1989.

Postlethwaite, Diana. *Making It Whole: A Victorian Circle and the Shape of Their World*. Columbus, OH: Ohio State University Press, 1984.

Rabb, M. A. 'Psychology and Politics in William Godwin's *Caleb Williams*: Double Bond or Double Bind?'. *Psychology and Literature in the Eighteenth Century*. Ed. Christopher Fox. New York, NY: AMS, 1987, 51–67.

Radick, Gregory. *The Simian Tongue: The Long Debate About Animal Language*. Chicago, IL: University of Chicago Press, 2007.

Rafter, Nicole Hahn. *The Criminal Brain: Understanding Biological Theories of Crime*. New York, NY: New York University Press, 2008.

Reid, Julia. *Robert Louis Stevenson, Science and the Fin de Siècle*. New York, NY: Palgrave, 2009.

Richards, Graham. 'Edward Cox: The Psychological Society of Great Britain (1875–1879) and the Meanings of an Institutional Failure'. *Psychology in Britain: Historical Essays and Personal Reflections*. Eds. G. C. Bunn, A. D. Lovie and G. D. Richards. Leicester: British Psychological Society, 2001, 33–53.

Richards, Graham. *Race, Racism, and Psychology: Towards a Reflexive History*. 2nd edn. New York, NY: Routledge, 2007.

Richards, Robert J. *Darwin and the Emergence of Evolutionary Theories of Mind and Behavior*. Chicago, IL: University of Chicago Press, 1987.

Richardson, Alan. *British Romanticism and the Science of the Mind*. Cambridge: Cambridge University Press, 2001.

Richardson, Angelique. *Love and Eugenics in the Late Nineteenth Century: Rational Reproduction and the New Woman*. Oxford: Oxford University Press, 2003.

Ritvo, Harriet. 'Animal Consciousness: Some Historical Perspective'. *American Zoologist* 40 (2000): 847–52.

Russett, Cynthia. *Sexual Science: The Victorian Construction of Women*. Cambridge: Harvard University Press, 1991.

Ruston, Sharon. *Creating Romanticism: Case Studies in the Literature, Science and Medicine of the 1790s*. New York, NY: Palgrave Macmillan, 2013.

Rylance, Rick. *Victorian Psychology and British Culture: 1850–1880*. New York, NY: Oxford University Press, 2000.

Salisbury, Laura, and Andrew Shall, eds. *Neurology and Modernity: A Cultural History of Nervous Systems: 1800–1950*. New York, NY: Palgrave, 2010.

Schiebinger, Londa. *The Mind Has No Sex? Women in the Origins of Modern Science*. Cambridge, MA: Harvard University Press, 1991.

Scull, Andrew. *Museums of Madness: The Social Organisation of Insanity in Nineteenth-Century England*. London: Allen Lane, 1979.

Scull, Andrew. *Most Solitary of Afflictions: Madness and Society in Britain 1700–1900*. New Haven, CT: Yale University Press, 2005.

Seagroatt, Heather. 'Hard Science, Soft Psychology, and Amorphous Art in *The Picture of Dorian Gray*'. *Studies in English Literature 1500–1900* 38.4 (1998): 741–59.

Sha, Richard. 'Toward a Physiology of the Romantic Imagination'. *Configurations* 17 (2009): 197–226.

Shuttleworth, Sally. *George Eliot and Nineteenth-Century Science: The Make-Believe of a Beginning*. Cambridge: Cambridge University Press, 1984.

Shuttleworth, Sally. *Charlotte Brontë and Victorian Psychology*. Cambridge: Cambridge University Press, 1996.

Shuttleworth, Sally. *The Mind of the Child: Child Development in Literature, Science, and Medicine 1830–1900*. Oxford: Oxford University Press, 2010.

Sibylle, Earl. *Blake, Lavater and Physiognomy*. Oxford: Legenda, 2010.

Small, Helen. *Love's Madness: Medicine, the Novel, and Female Insanity, 1800–1865*. Oxford: Clarendon, 1996.

Smith, Roger. 'The Background of Physiological Psychology in Natural Philosophy'. *History of Science* 11 (1973): 75–123.

Smith, Roger. *Trial by Medicine: Insanity and Responsibility in Victorian Trials*. Edinburgh: Edinburgh University Press, 1981.

Smith, Roger. 'The Physiology of the Will: Mind, Body, and Psychology in the Periodical Literature, 1855–1875'. *Science Serialized: Representations of the Sciences in Nineteenth-Century Periodicals*. Eds. Geoffrey Cantor and Sally Shuttleworth. Cambridge, MA: MIT Press, 2004, 81–110.

Smith, Roger. *Free Will and the Human Sciences in Britain, 1870–1910*. London: Pickering and Chatto, 2013.

Stiles, Anne. *Popular Fiction and Brain Science in the Late Nineteenth Century*. Cambridge: Cambridge University Press, 2012.

Stiles, Anne, and Stanley Finger, eds. *Literature, Neurology, and Neuroscience: Historical and Literary Connections*. Philadelphia, PA: Elsevier, 2013.

Stiles, Anne, Stanley Finger and John Bulevich. 'Somnambulism and Trance States in the Works of John William Polidori, Author of *The Vampyre*'. *European Romantic Review* 21.6 (2010): 789–807.

Stolte, Tyson. '"Putrefaction Generally": *Bleak House*, Victorian Psychology, and the Question of Bodily Matter'. *Novel: A Forum on Fiction* 44.3 (2011): 402–23.

Talairach-Vielmas, Lawrence. *Wilkie Collins, Medicine, and the Gothic*. Cardiff: University of Wales Press, 2009.

Tate, Gregory. *The Poet's Mind: The Psychology of Victorian Poetry, 1830–1870*. Oxford: Oxford University Press, 2012.

Taylor, Jenny Bourne. *In the Secret Theatre of the Home: Wilkie Collins, Sensation, Narrative, and Nineteenth-Century Psychology*. London: Routledge, 1988.

Taylor, Jenny Bourne. 'Fallacies of Memory in Nineteenth-Century Psychology: Henry Holland, William Carpenter, and Frances Power Cobbe'. *Victorian Review* 26.1 (2000): 98–118.

Taylor, Jenny Bourne, and Sally Shuttleworth, eds. *Embodied Selves: An Anthology of Psychological Texts, 1830–1890*. Oxford: Oxford University Press, 2003.

Thurschwell, Patricia. *Literature, Technology and Magical Thinking*. Cambridge: Cambridge University Press, 2001.

Tomes, Nancy. 'Feminist Histories of Psychiatry'. *Discovering the History of Psychiatry*. Eds. Mark S. Micale and Roy Porter. Oxford: Oxford University Press, 1994, 348–83.

Tomlinson, Stephen. *Head Masters: Phrenology, Secular Education, and Nineteenth-Century Social Thought*. Tuscaloosa, AL: University of Alabama Press, 2013.

Tytler, Graeme. *Physiognomy in the European Novel: Faces and Fortunes*. Princeton, NJ: Princeton University Press, 1982.

Vrettos, Athena. 'Defining Habits: Dickens and the Psychology of Repetition'. *Victorian Studies* 42.3 (1999/2000): 399–426.

Vrettos, Athena. 'Dying Twice: Victorian Theories of Déjà Vu'. *Disciplinarity at the Fin de Siècle*. Eds. Amanda Anderson and Joseph Valente. Princeton, NJ: Princeton University Press, 2002, 196–218.

Wagner, Kim A. 'Confessions of a Skull: Phrenology and Colonial Knowledge in Early Nineteenth-Century India'. *History Workshop Journal* 69 (Spring 2010): 27–51.

Wickens, G. Glen. *Thomas Hardy, Monism, and the Carnival Tradition: The One and the Many in 'The Dynasts'*. Toronto: Toronto University Press, 2002.

Willis, Martin, and Catherine Wynne, eds. *Victorian Literary Mesmerism*. Amsterdam: Rodopi, 2006.

Winter, Alison. *Mesmerized: Powers of Mind in Victorian Britain*. Chicago, IL: University of Chicago Press, 1998.

Wood, Jane. *Passion and Pathology in Victorian Fiction*. Oxford: Oxford University Press, 2006.

Young, Robert. *Mind, Brain and Adaptation in the Nineteenth Century: Cerebral Localization and Its Biological Context from Gall to Ferrier*. New York, NY: Oxford University Press, 1990.

Further reading

Bain, Alexander. *The Emotions and the Will*. London: John W. Parker, 1859.

Brown, Thomas. *Lectures on the Philosophy of the Human Mind*. 4 vols. Edinburgh: W. and C. Tait, 1820.

Budge, Gavin. *Romanticism, Medicine, and the Natural Supernatural: Transcendent Vision and Bodily Spectres, 1789–1852*. Basingstoke: Palgrave Macmillan, 2012.

Cohen, William. *Embodied: Victorian Literature and the Senses*. Minneapolis, MN: University of Minnesota Press, 2009.

Dixon, Thomas. *From Passions to Emotions: The Creation of a Secular Psychological Category*. Cambridge: Cambridge University Press, 2003.

Ellenberger, Henri. *The Discovery of the Unconscious: The History and Evolution of Dynamic Psychiatry*. New York, NY: Basic, 1970.

Elliotson, John. *Human Physiology*. London: Longman, Orme, Brown, Green, and Longmans, 1835.

Faflak, Joel. 'Was It for This? Romantic Psychiatry and the Addictive Pleasures of Moral Management'. *Romanticism and Pleasure*. Eds. Michelle Faubert and Thomas Schmid. New York, NY: Palgrave, 2010, 61–82.

Faflak, Joel, and Richard Sha, eds. *Romanticism and the Emotions*. Cambridge: Cambridge University Press, 2014.

Figlio, Karl M. 'Theories of Perception and the Physiology of Mind in the Late Eighteenth Century'. *History of Science* 12 (1975): 177–212.

Fulford, Tim. 'Radical Medicine and Romantic Politics'. *Wordsworth Circle* 35.1 (2004): 16–21 [2004B].

Graziano, Amy B., and Julene K. Johnson. 'Music in the Development of Nineteenth-Century Neurology'. *Music and the Nerves, 1700–1900*. Basingstoke: Palgrave Macmillan, 2014, 152–69.

Hoeldtke, R. 'The History of Associationism and British Medical Psychology'. *Medical History* 11.1 (1967): 46–65.

Hughlings Jackson, John. 'Remarks on Evolution and Dissolution of the Nervous System'. *Journal of Mental Science* 23 (1887): 25–8.

Jacyna, L. S. 'The Physiology of Mind, the Unity of Nature, and the Moral Order in Victorian Thought'. *British Journal for the History of Science* 14.47 (1981): 109–32.

Kitson, Peter. *Romantic Literature, Race, and Colonial Encounter*. Basingstoke: Palgrave Macmillan, 2007.

Lewes, Henry. *The Physical Basis of Mind, Being the Second Series of Problems of Life and Mind*. London: Trubner, 1877.

Mill, James. *Analysis of the Phenomena of the Human Mind*. 2 vols. London: Baldwin and Cradock, 1829.

Reed, Edward R. *From Soul to Mind: The Emergence of Psychology, From Erasmus Darwin to William James*. New Haven, CT: Yale University Press, 1997.

Ribot, Theodule-Armand. *English Psychology*. London: Henry S. King, 1873.

Richards, Graham. *Mental Machinery: The Origins and Consequences of Psychological Ideas, Part 1: 1600–1850*. London: Athlone, 1992.

Richardson, Angelique. *After Darwin: Animals, Emotions, and the Mind*. Amsterdam: Rodopi, 2013.

Smith, Roger. *Inhibition: History and Meaning in the Sciences of Mind and Brain*. Berkeley, CA: University of California Press, 1992.

Syson, Lydia. *Doctor of Love: James Graham and His Celestial Bed*. Richmond: Alma, 2008.

Vallins, David. *Coleridge and the Psychology of Romanticism*. New York, NY: St Martin's Press, 2000.

Vickers, Neil. *Coleridge and the Doctors: 1795–1806*. Oxford: Oxford University Press, 2004.

Warren, Howard Crosbie. *A History of the Association Psychology*. New York, NY: Scribner, 1921.

Wilson, Eric G. 'Matter and Spirit in the Age of Animal Magnetism'. *Philosophy and Literature* 30.2 (2006): 329–45.

Woodward, William R., and Mitchell G. Ash. *The Problematic Science: Psychology in Nineteenth-Century Thought*. New York, NY: Praeger, 1982.

26

SEXOLOGY

Anna Katharina Schaffner

University of Kent

From sin to pathology: the construction of the modern perversions

The nineteenth century saw the gradual replacement of predominantly Christian taxonomies of sexual sin with biological and psychological models, based on congenital, psychiatric and legal conceptions of the modern subject. Attempts to establish the sexually 'normal' peaked particularly in the last decades of the nineteenth and the early decades of the twentieth centuries, as evidenced by a rapid rise in studies dedicated to sexual pathologies. The scholarly activities of a cohort of predominantly German, French and English psychiatrists, doctors and psychologists led to the emergence of a *scientia sexualis* – a new scientific field of investigation that combined insights from medicine and forensic science, psychiatry and psychology, anthropology, biology and genetics. Sexology constituted a systematic attempt to identify, classify and contain the proliferation of the so-called sexual perversions. While questions relating to sexual behaviour were previously negotiated in the domains of theology, law and philosophy, and assessed with recourse to notions of sin, crime and moral failure, sexual deviance gradually became a concern of physicians and psychiatrists.

Pre-modern sexual deviance was essentially seen as a crime 'against nature'. The church delineated the parameters of what was natural, and the state and the community policed its boundaries (see, for example, Dabhoiwala; Peakman 2009; Porter and Hall). In the second half of the nineteenth century, however, normal and pathological emerged as the new epistemic yardsticks in the field. For Michel Foucault, the emergence of sexology marks a historic shift towards secular modernity: physicians and psychiatrists, he argues in *The History of Sexuality* (1976), reconfigured the religious ritual of confession into a secular, scientific search for the 'truth' about sex. The pervert ceased to be a sinner and instead became a patient. However, as this chapter seeks to demonstrate, although the majority of nineteenth-century sexologists presented their theories as strictly scientific findings, the boundaries between scientific and literary discourses on sexuality were much less firmly drawn than we might think.

The architects of pre-modern conceptions of the so-called perversions were ecclesiastical scholars. The most influential of these, the thirteenth-century Italian theologian Thomas Aquinas (1225–74), drew up the core Christian taxonomy of sexual sins in his *Summa Theologiae*. Aquinas defines any sexual act from which procreation cannot follow as 'unnatural vice'. He furthermore divides unnatural vice into different species of lechery. All sins of lechery are, first,

408

in conflict with right reason and, second, in conflict with the 'natural pattern of sexuality for the benefit of the species' (Aquinas 245). The species of lechery are self-abuse (that is, masturbation), bestiality, sodomy (sex with a person of the same sex) and deviations from the natural (genital) form of intercourse such as anal and oral sex. Aquinas then compares the different modalities of lechery and draws up a hierarchy. 'The gravity of a sin corresponds rather to an object being abused, than to its proper use being omitted', he reasons, and thus the lowest rank is held by the solitary sin of masturbation, while the greatest sin is that of bestiality as it crosses the species barrier (Aquinas 249).

It was only at the beginning of the eighteenth century that Aquinas's hierarchy of the worst sexual sins was challenged: masturbation, formerly classified as the least harmful of the unnatural vices, suddenly became the most vilified and feared act of sexual deviance. In 1712, an anonymous pamphlet entitled *Onania; or, the Heinous Sin of Self-Pollution, and All its Frightful Consequences, in Both Sexes, Considered, With Spiritual and Physical Advice to Those who Have Already Injur'd Themselves by This Abominable Practice*, authored by an English quack, was the first of a number of texts that triggered what was to become known as the masturbation pandemic in the eighteenth and early nineteenth centuries (see Laqueur). The anonymous author declares as the aim of his study the promotion of 'Virtue and Christian Purity' ([Anon.] [n.d.] 1) and the discouragement of 'Vice and Uncleanness' and, in particular, 'self-pollution'.

Apart from the fact that masturbation stands in the way of marriage, puts a stop to pro-creation and is generally 'displeasing to God', the author of *Onania* asserts that it also causes weakness and exhaustion, gonorrhoea, nocturnal effusions, seminal emissions, gleets, oozings, infertility and impotence ([Anon.] [n.d.] 8–9). Here, we can already observe a shift from a purely religious register to an increasingly medical one. Immoral action is declared to have material, organic consequences; the evocation of frightful medical scenarios, such as the loss of precious life energy, is deployed as a new pedagogic tool.

In 1760, a second influential treatise on masturbation appeared, authored by the Swiss physician Samuel-Auguste Tissot (1728–97). In the preface, Tissot renders explicit the change of strategy already present in the *Onania* pamphlet, a shift from moral appeals to the intimidating illustration of the physical consequences of immoral behaviour:

> My design was to write upon the disorders occasioned by masturbation, or self-pollution, and not upon the crime of masturbation: besides, is not the crime sufficiently proved, when it is demonstrated to be an act of suicide? Those who are acquainted with men, know very well that it is much easier to make them shun vice by the dread of a present ill, than by reasons founded upon principles, the truth of which has not been sufficiently inculcated into them.
>
> *(vii–viii)*

Here, Tissot openly acknowledges that the suggestion of physical ailments as a result of immoral practices is a much more effective tool for convincing human beings to shun evil than are appeals to religious and moral principles. This shift is significant not only because it illustrates a secular focus upon the here and now rather than the afterlife but also because it demonstrates how and why medical arguments were ever more regularly used as pedagogic, ideological and political tools.

Masturbation continued to be a key concern of both medical and moral writers in the early decades of the nineteenth century (see Laqueur). However, it was gradually replaced by a different set of so-called sexual perversions. These include homosexuality, sadism, masochism, fetishism, voyeurism and exhibitionism. They were first systematically defined by the German psychiatrist

Richard von Krafft-Ebing in his groundbreaking and bestselling study *Psychopathia Sexualis* (1886) and were later canonized by Freud in *Three Essays on the Theory of Sexuality* (1905). The preoccupation with the concept of 'perversion' is a specifically modern phenomenon, the product of various political, sociological, cultural and technological processes, which include the spread of secularization, capitalism, urbanization and industrialization, the emergence of the middle classes, the cult of progress and the rapid advance of secular sciences (see Griffin 45–6). To this list can be added a teleological belief in the Enlightenment promise of emancipation through reason, the materialization of a bourgeois notion of romantic love and what Anthony Giddens has described as the 'transformation of intimacy', the separation of the public and the private spheres, and the rise of consumer culture. Partly as a substitute for the rapidly increasing loss of faith in religious belief systems, the nineteenth century also saw the advent of positivism and scientism, which were characterized by the conviction that only empirical knowledge and the methods of science can yield 'proper' knowledge, that scientific theories should displace theological and metaphysical ones and that rational scientific principles should serve as general models for the organization of society. Nineteenth-century scientistic thought, including Saint-Simonian socialism, Comte's positivism, Marxism, Lamarckian evolutionary theory and most forms of social Darwinism, were generally optimistic in tone, in that they were based on the belief that it is possible to deploy scientific knowledge to create a better world (see Olson 294).

In the later decades of the nineteenth century, however, the myth of rational teleological progress began to be questioned. Partly owing to more refined statistical methods of analysis and an increase in media reports on these phenomena, and partly on account of a growth in crime rates, insanity, alcoholism, sexual diseases and prostitution, which were related to growing urban poverty, scientistic narratives turned darker, and the notion of degeneration and both physiological and societal decline began to dominate both medical and broader cultural debates (see Olson 294 and 301–2). Many commentators increasingly conceived of modernity not only as decadent but as a pathological condition, a 'perversion' of a better, healthier, more natural state. The conception of modernity as a state of decadence and decline proliferated both in the literary and the scientific spheres especially in the last quarter of the nineteenth century.

An important aetiological model for perversion, which was constitutive of the declinist zeitgeist, was the degeneration paradigm. The idea of degeneration as a negative, backward cultural development, which results in the gradual weakening of certain groups of individuals and increases from generation to generation, was introduced by the French physician Bénédict Augustin Morel (1809–73) in 1857 in a study entitled *Traité des dégénérescences physiques, intellectuelles et morales de l'espèce humaine et des causes qui produisent ces variétés maladives* (*Treatise on Physical, Intellectual and Moral Degeneracy in the Human Species and the Causes That Produce These Diseased Varieties*). Morel defines degeneration in the following terms:

> The clearest notion we can form of degeneracy is to regard it as a morbid deviation from an original type. This deviation, even if, at the outset, it was ever so slight, contained transmissible elements of such a nature that anyone bearing in him the germs becomes more and more incapable of fulfilling his functions in the world; and mental progress, already checked in his own person, finds itself menaced also in his descendants.
>
> *(Quoted in Nordau 16)*

Harry Oosterhuis argues that Morel 'translated the Christian doctrine of man's regression after original sin into a biological metaphor' (52–3) and that degeneration theory 'signaled a crisis in the social optimism that had characterized both liberalism and positivist science' (107). Indeed, it is possible to identify the advent and rapid proliferation of degeneration theory as the point at which the Enlightenment notion of unlimited progress turned sour, when modernity began to be construed as decadence.

Degeneration theory was popularized by Cesare Lombroso (1836–1909), the Italian criminal anthropologist most famous for the concept of the 'born criminal'. The French historian Hippolyte Taine (1828–93) transplanted the paradigm into the realm of French post-Revolution history, arguing that the degeneration of France's citizens led naturally to the revolutionary upheavals of 1848 and the Paris Commune in 1871. Émile Zola, who repeatedly emphasized the 'scientificity' of his literary methods, spread the fear of degeneration further in his popular late nineteenth-century naturalist novels such as *Nana* (1880), *Germinal* (1885) and *La Bête humaine* (1890). Another highly influential text based on degeneration theory is of course Max Nordau's *Degeneration* (1892–93), an international bestseller in which Nordau rejects contemporary art and music as decadent products of degenerate artists and, more significantly, as driving forces of further cultural degeneration. It was not until the publication of Freud's *Three Essays on the Theory of Sexuality* in 1905 that another powerful paradigm – that of arrested psycho-sexual development – finally challenged and replaced the hegemony of degeneration theory in sexological discourse and elsewhere.

However, not only was modernity increasingly conceived of as a 'perversion' in the sense of a change for the worse, a corruption of the natural function of social communities, but it also developed an obsession with what were then perceived as perverse sexualities. While earlier critics such as Jean-Jacques Rousseau and doctors such as Tissot perceived civilization as a polluting force contaminating the body and mind of the individual, sexologists who subscribed to the degeneration paradigm, such as Auguste Forel, instead saw the perverts as corruptors of the social body. The degenerates were construed as dangers to civilization, halting progress by their regression to pre-Christian rituals such as fetishism and even bestiality (see Rosario 88). As Vernon A. Rosario observes, what 'emerges from the antimasturbatory literature of the nineteenth century is the perception of "deviant" individuals as viruses of the social corps – polluting its national strength and purity' (40).

The modern perversions, which above all include homosexuality, sadism, masochism, fetishism, voyeurism and exhibitionism, are symptomatic of particular cultural anxieties and concerns. They were most famously defined by Freud in his *Three Essays* – undoubtedly the most influential and groundbreaking work on the topic. However, Freud drew substantially on the works of his predecessors in the field. The perversions that he firmly implanted into the modernist matrix had already been discussed by Krafft-Ebing, Alfred Binet, Havelock Ellis, Magnus Hirschfeld, Iwan Bloch and many others. They preoccupied the cultural imagination at the turn of the century, as evidenced by the dominant place they were given in sexological studies at the time and by the numerous representations of these phenomena in realist, naturalist, decadent and modernist literature. Think, for example, of Leopold von Sacher-Masoch's *Venus in Furs* (1870), Oscar Wilde's *The Picture of Dorian Gray* (1890) and *Teleny* (1893), Thomas Mann's *Death in Venice* (1912), Marcel Proust's *In Search of Lost Time* (1913–27) and D. H. Lawrence's *Women in Love* (1920).

Since the advent of modernity, and in particular in the later decades of the nineteenth century, a shifting regime of dominant perversions can be observed. Those variable core perversions feature more prominently than others in literary works, in debates in the media, in the arts and in medical, legal and psychiatric discourse. Moreover, to a certain extent they function as lightning rods for collective anxieties. In the wake of declining religious faith, Western societies attempted to address their spiritual crises by identifying culprits for their malaises. Like the Thebans in Sophocles's *Oedipus the King*, they turned to the ancient practice of scapegoating – this time, however, drawing upon the support of the positivist narratives of the scientists. Not only are transgressors always ideal scapegoat figures but, according to the sexologists, the perverts quite literally 'contaminated' the collective body: they spoiled the gene pool by passing on degenerate genetic materials and by sharing their fantasies they infested the imagination of others. The perverts were both polluted and polluters, wreaking havoc with their own and the social body, endangering the welfare of the state and ultimately the survival of the species by depriving sex of its procreative function.

Most sexologists seek primarily to explain the aetiologies of the perversions so that they can be understood, contained and possibly even cured. Frequently, moreover, medico-psychological studies go hand in hand with legalistic activism, for many sexologists attempted to change the law in favour of the so-called 'perverts'. Karl Heinrich Ulrichs's pamphlets, Ellis's *Sexual Inversion* (1897) and Krafft-Ebing's and, in particular, Magnus Hirschfeld's campaigns for the abandonment of §175 of the German penal code, which classified homosexuality as a crime, are cases in point. Others, such as Auguste Forel, were much less liberal in their aims and advocated radical eugenic measures to contain the spread of perversion.

In the course of the late nineteenth century, the cultural emphasis shifted from the problem of masturbation, which preoccupied the medical establishment in the eighteenth and at the beginning of the nineteenth century, to homosexuality (see, for example, Bauer; Brady; Cook; Nye; Porter and Hall; Rosario; White). The notion of sodomy, which could refer equally to anal sex, same-sex encounters, paedophilia and bestiality, was replaced by new definitions, including 'Urning' or 'Uranian', 'invert' and finally 'homosexual', coined in 1869 by the Hungarian Károly Mária Kertbeny (formerly Benkert) (1824–82), who wrote two pamphlets to appeal against Prussian sodomy laws. As is particularly evident in the sustained international interest in the Oscar Wilde trials in 1895, and in the Eulenburg affair,[1] homosexuality gradually took centre stage, becoming the most discussed of the perversions (see Hull 45–145). The homosexual was turned into the new emblem of moral corruption, a figure to be feared and contained, but also one that was celebrated as a challenger of the existing sexual order. The male homosexual appears in the works of numerous late nineteenth-century writers and modernist authors, including Oscar Wilde, John Addington Symonds, Jean Cocteau, E. M. Forster, André Gide, Christopher Isherwood, Henry James, D. H. Lawrence, Federico García Lorca, Thomas Mann and Marcel Proust. Lesbian love features prominently in works by Djuna Barnes, H. D., Radclyffe Hall, Katherine Mansfield, Gertrude Stein and Virginia Woolf, among others. The works of all of these writers attest to the cultural significance of the figure of the homosexual in the late decades of the nineteenth and the early decades of the twentieth centuries. Moreover, homosexuality was not only the most widely explored perversion in the literary works of the period, but it was also the one to which Krafft-Ebing dedicated most of his scholarly attention. 'Inversion' gradually became the key perversion in later, revised and expanded editions of *Psychopathia Sexualis*, as already indicated by the subtitle added to the second edition, *With Especial Reference to the Antipathic Sexual Instinct*. The vast majority of Krafft-Ebing's case studies in later editions of *Psychopathia Sexualis* dealt with varieties of 'contrary sexual feelings', and it was also the most frequent diagnosis of Krafft-Ebing's patients, followed by masochism, fetishism and sadism (see Oosterhuis 153).

An increasing preoccupation with sadism and masochism also becomes apparent at the end of the nineteenth century: both Krafft-Ebing and Freud consider sadism and masochism cardinal perversions. Krafft-Ebing describes them as 'fundamental forms of psycho-sexual perversion' (143) and Freud assigns them a 'special position among the perversions, since the contrast between activity and passivity which lies behind them is among the universal characteristics of sexual life' (159). The appeal of sadism as a metaphor of the battle for domination between the sexes, and as a trope capable of capturing tendencies of early twentieth-century capitalism, is obvious.

John K. Noyes offers a socio-political explanation for the emergence of masochism at the end of the nineteenth century 'both as a pathology and as a highly popular code of sexual imagery' (6). The invention of masochism, he argues, 'was a symptomatic move, an attempt to resolve some of the crises in liberal concepts of agency' (Noyes 8). It was indicative of an obsession with control and a general sense of crisis brought about by rapid technological developments. According to Noyes, the lure of masochism at that time can be explained in terms of an erotic reappropriation of control in an age dominated by imperialism, authoritarian regimes and the crumbling of established gender roles. The masochist turns that which induces fear and feelings of social disempowerment, the technologies of disciplinary punishment, into technologies of pleasure. Moreover, the concern with masochism, as one of the most obviously gendered perversions, also testified to a more general fear of feminization, which in turn is indicative of a crisis of masculinity.

In many French psychiatric studies in contrast, as Foucault points out, fetishism was considered the 'model perversion' (154). Binet, a student of Jean-Martin Charcot, coined the term in 1887, and considered all perversions as manifestations of a defective object choice. While literary representations of homosexuality date back as far as *The Epic of Gilgamesh*, representations of fetishism, voyeurism and exhibitionism are more recent phenomena, with fetishism first appearing in French eighteenth-century libertine literature such as the novels of the Marquis de Sade and Rétif de la Bretonne. The emergence of fetishism is closely related to the rise of consumer capitalism, the advent of department stores and an increased appetite for luxury goods, as is evident in Marx's notion of the commodity fetish. The interest in voyeurism and exhibitionism, too, seems to have risen as a result of a more rigorous division between public and private life. In contrast, prominent pre-modern perversions such as sodomy, bestiality and necrophilia were either redefined or gradually disappeared from the discussions. Bestiality (intercourse with animals) faded into the background, no doubt partly on account of growing urbanization (see Rydström). The overt concern with necrophilia – prominently present in nineteenth-century vampire literature such as Ludwig Tieck's 'Wake Not the Dead' (*c.* 1800), Théophile Gautier's 'The Dead in Love' (1836), Sheridan Le Fanu's *Carmilla* (1872) and Bram Stoker's *Dracula* (1897) – also lessened, although, as Lisa Downing has demonstrated, it lingered on in less immediately obvious manifestations such as Freud's conception of the death drive. Other perversions that Krafft-Ebing and other sexologists discussed in some detail also lost importance, including the love of statues, handkerchief fetishism and the compulsion to cut off women's plaits.

Not only was early sexology construed exclusively by male physicians and psychiatrists, but most modern perversions, with the exception of lesbianism and nymphomania, are predominantly applicable to men: it is significant that the vast majority of case studies of sadists, masochists, voyeurs, exhibitionists and fetishists in sexological texts feature male analysands. The stereotypical view of female sexuality was that it was passive, submissive, reactive and not very strongly developed. Women were frequently diagnosed as neurotic, especially as hysterical and frigid, but very rarely as perverse. Most of the sexologists are, for example, interested primarily in male masochists, female masochism being deemed a 'natural' condition. Krafft-Ebing's definition is paradigmatic of this tendency: he designates masochism as a pathological exaggeration of female psychological elements:

> Thus it is easy to regard masochism in general as a pathological growth of specific feminine mental elements – as an abnormal intensification of certain features of the psycho-sexual character of woman – and to seek its primary origin in the [female] sex.
>
> *(130)*

The male masochist is seen as pathological not just because he obtains libidinal gratification from pain or submission but because he transgresses gender roles by adopting a passive and thus 'feminine' position in the sexual act. Even the female sadist seems to exist principally to gratify the male masochist's desires, not as a perverse character in her own right. Almost all of the early sexologists produced highly gendered, heteronormative constructions of 'normal' and 'pathological' sexuality, which in most cases rest on essentialist assumptions about appropriate 'female' and 'male' behaviours and qualities (on the gendered nature of the perversions and specifically female perversions, see, for example, Apter; Kaplan; Mangham and Depedge; Welldon). While the gendered nature of the perversions remains mostly implicit in the accounts of pre-Freudian sexologists, it is made explicit by Freud, who famously considers neurosis as the 'negative' of perversion. Often, Freud writes in his *Three Essays*, perversion and psycho-neurosis are encountered in the same family, distributed among the sexes in such a way that the male members 'are positive perverts, while the females, true to the tendency of their sex to repression, are negative perverts, that is, hysterics' (236).

Havelock Ellis, Edward Carpenter and the literary imagination

When comparing French, German and English sexological traditions, it is striking that far fewer studies were written on the subject in Victorian Britain during the early decades of the discipline than were on the continent (see Hall; Mason; Porter and Hall for more information on the construction of Victorian sexualities). Before Havelock Ellis (1859–1939) published *Sexual Inversion* in 1897, not a single medico-psychiatric monograph had appeared on the subject of perversion. Ivan Crozier notes that, by and large, 'British psychiatry was not explicitly concerned with sexual perversion' (2008B 74). However, he also challenges the commonly held view that no attention at all had been paid to homosexuality in British medical discourse before Ellis's publication, while conceding that the existing discussion was conservative, apolitical, 'less theoretically sophisti-cated and less sexually explicit than Continental sexology' (Crozier 2008B 67). The 'pre-Ellisian' British sexological literature Crozier discusses consists exclusively of short essays and reviews, and tends to focus predominantly on homosexuality. There were no comprehensive or systematic studies on sexual perversion, and it is also noteworthy that the first more explicit piece on the subject of homosexuality, which appeared in an English medical journal in 1881, was written by a German (Krueg). There was clearly a reluctance to address the issue of perversion openly in British medico-psychiatric discourse, at least partly on account of concerns about propriety and professional status, as well as a generally less open intellectual climate when it came to the discussion of things of a sexual nature (see Crozier 2008B 100). An anonymous reviewer in the *Journal of Mental Science*, for example, who reviewed C. G. Chaddock's translation into English of Krafft-Ebing's *Psychopathia Sexualis* in 1893, deemed the subject of the perversions unsavoury and expressed concern about the general availability of the study in bookshops: 'Perhaps we are prudish, but we think that the production of this book by Ebing [*sic*] will not add to his reputation, nor will it do any possible good to the medical or psychological world' ([Anon.] 1893 51). Even the author of an entry on sexual perversion in the *Dictionary of Psychological Medicine* (1892) thought that most cases of perversion, 'in their disgusting details', were 'hardly worthy of the minute study that has been given them' (Norman), and argued that it was 'sufficient to look upon [aberrations] as varieties of masturbation' while directing people interested in more details to the works of Casper, Westphal, Krafft-Ebing, Tarnowsky, Lombroso, Charcot, Moll and others. The fact that Ellis's book on inversion was first published in Germany and was banned on the grounds of obscenity almost immediately after its publication in Britain is another case in point. Ellis decided to publish all subsequent writings on sex exclusively in the United States, and even there they were only allowed to be sold to medical professionals until the 1930s.

However, there was a dynamic non-medical discourse that took place at the fringes of the field, driven primarily by cultural reformers and political activists. I have shown elsewhere that even the purportedly 'properly' scientific studies of numerous nineteenth- and early twentieth-century German and French sexologists were substantially shaped by the literary imagination, in that they not only based many of their theories on case studies derived from literary works, which were treated as empirically solid 'evidence', but also adopted literary concepts and methodologies (see Schaffner). Yet the blurring of the boundaries between the literary and the scientific fields is perhaps even more pronounced in England. Heike Bauer rightly argues that early English sexology was essentially a literary affair (19). The most famous protagonists of English sexology, John Addington Symonds (1840–93), Edward Carpenter (1844–1929) and, of course, Ellis, approached the subject predominantly from non-scientific backgrounds. Symonds was a critic and a poet, Carpenter a poet and social activist, and before Ellis made his name as a sexologist he was a well-known literary critic and editor. Bauer points out that Ellis's first book-length publication was in fact an edition of the prose writings of Heinrich Heine, published in 1887, and he continued to publish criticism on authors as diverse as Thomas Hardy, Walt Whitman, Arthur Symons and Friedrich Nietzsche. He also edited the *Mermaid* series, which reissued sexually explicit Elizabethan and Jacobean plays, and wrote poetry himself. Ellis was not, as many critics assume, a fully qualified physician, but held only the lowest possible medical diploma, the Licentiate in Medicine, Surgery, and Midwifery of the Society of Apothecaries. He never practised as a doctor and remained a private scholar throughout his life. Crozier even suggests that Ellis obtained a professional degree solely to be able to speak with authority on matters of human sexuality – from the vantage point of a medical man (2008C 391).

Ellis's study *Sexual Inversion* was first published in Germany under the title *Das konträre Geschlechtsgefühl (Contrary Sexual Feeling)* in 1896, before it was published in Britain in 1897 as the first medical textbook on inversion. However, the second edition, published by Roland de Villiers in the same year, was banned in the Bedborough trial, in which George Bedborough, editor of *The Adult*, a 'Journal of Sex', was arrested for selling a scandalous, bawdy and wicked book, *Sexual Inversion*, to an undercover policeman, for which he was charged with obscene libel. *Sexual Inversion* constituted one of thirty-two individual studies in Ellis's series *Studies in the Psychology of Sex* (1897–1928), in which he addresses a wide range of topics, including sexual periodicity, auto-eroticism, the sexual impulse, sadism and masochism, the mechanisms of tumescence and detumescence, the psychic state in pregnancy, narcissism, eugenics, 'eonism' (cross-dressing) and 'undinism' (now more commonly known as urolagnia). Ellis originally planned to co-write *Sexual Inversion* with Symonds, who, before collaborating with Ellis, had privately published and distributed some pamphlets on homosexuality (see Brady for more details on Ellis's and Symonds's collaboration on this project). Symonds, however, died of influenza in 1893, and, owing to problems with his estate, his essay entitled 'A Problem in Greek Ethics', as well as other materials, ended up being relegated to the appendices in the first edition, and were cut down further in subsequent editions (see Bristow).

Ellis's study ends with a plea for legal reform, which was one of his key motivations for writing it in the first place (see Crozier 2008A 59). In 1885, the House of Commons passed the Criminal Law Amendment Act, which raised the age of consent from thirteen to sixteen, strengthened existing legislation against prostitution and, crucially, extended the law proscribing homosexual relations to include homosexual acts in public *and* in private (a provision of the Labouchere Amendment to the Criminal Law Amendment Act). Carnal knowledge *per anum* of a man, a woman or an animal was a felony punishable by penal servitude for life as a maximum and for ten years as a minimum, while the attempt at such carnal knowledge was punishable by ten years' penal servitude. The amendment, moreover, meant that 'gross indecencies' between males, even if committed privately, also counted as penal offences – a situation which Ellis wished to challenge. When Ellis wrote *Sexual Inversion*, Britain was still coming to terms with the aftermath of the Oscar Wilde trials in 1895. However, as he states in his autobiography, Ellis was also motivated by more personal concerns, for he found that some of his most esteemed friends (such as Edward Carpenter and particularly his wife Edith Lees Ellis) were homosexual.

Ellis adopts a positivist stance in his study, declaring his desire to obtain possession of 'the actual facts', from the investigation of which he wishes 'to ascertain what is normal and what is abnormal, from the point of view of physiology and of psychology' (Ellis and Symonds 91). Ellis's theory, like those in most other sexological works, is framed around a number of case studies. Unlike his Continental colleagues, Ellis was able to draw upon cases supplied by Symonds and his own friends and acquaintances – Symonds's own case history, for example, features in the study as Case XVIII. This might also, at least partially, explain that among his thirty-six cases Ellis found that 'twenty-four, or 66 per cent, possess artistic aptitude in varying degree' (Ellis and Symonds 195), while the average showing artistic tastes are around 30 per cent. The sociocultural background of these 'cases' was presumably rather different from those who featured in many Continental sexological studies, which were in some instances based on the analysis of subjects who had come into conflict with the law or had been institutionalized. In structure, the case studies resemble Krafft-Ebing's; that is, Ellis discusses family history, noticeable physical details and other circumstances that are out of the ordinary.

Ellis conceives of inversion as a predominantly congenital and relatively harmless abnormality. The normal sexual impulse, he argues, is inborn and organic. Particularly during puberty, however, it is susceptible to suggestion and association:

> [T]he soil is now ready, but the variety of seeds likely to thrive in it is limited . . .
> The same seed of suggestion is sown in various soils; in the many it dies out, in the few it flourishes. The cause can only be a difference in the soil.
>
> *(Ellis and Symonds 201)*

As evidenced by his use of the 'soil' metaphor, Ellis tends more towards the nature rather than the nurture side of the argument, but does not dismiss external influences entirely. In a related footnote, Ellis's language is even more figurative than in the previous passage, and once again he elaborates on a comparison with nature:

> The tentative and omnivorous habits of the newly-hatched chicken may be compared to the uncertainty of the sexual instinct at puberty; while the sexual pervert is like a chicken that should carry on into adult age an appetite for worsted and paper.
>
> *(Ellis and Symonds 201 n. 3)*

In a much-quoted passage he compares homosexuality to colour blindness:

> Just as the ordinary colour-blind person is congenitally insensitive to those red-green rays which are precisely the most impressive to the normal eye, and gives an extended value to the other colours – finding that blood is the same colour as grass, and a florid complexion blue as the sky – so the invert fails to see emotional values patent to normal

persons, transferring their values to emotional associations which for the rest of the world are utterly distinct.

(Ellis and Symonds 204)

Ellis not only frequently uses figurative language, in particular extended similes, but also col-zlaborated with a poet on what was primarily supposed to be a scientific work on homosexuality, granting ample space to a lengthy philosophical and cultural essay on the issue in the first edition of *Sexual Inversion*. Given his background as a literary critic and editor, it is not surprising that he, too, quotes extensively from literary sources: many of the early sexologists and, of course, Freud, make ample and frequent usage of literary works, which they use not only as illustrations, but even as case studies, often without acknowledging or theorizing their fictional status (see Schaffner). Unlike other sexologists such as Krafft-Ebing, however, Ellis only rarely uses literary representations as 'evidence' and as case studies. The sentence '[t]he fact that homosexuality is especially common among men of exceptional intellect was long since noted by Dante' is a rare exception (Ellis and Symonds 106). Ellis's discussions of Marlowe, Shakespeare and Whitman are essentially psychoanalytical-biographical in nature – he uses literary texts in order to diagnose their authors. Marlowe is diagnosed with sexual perversion on the basis of the presence of these topics in his literary works. Shakespeare, in contrast, is declared normal, in spite of his sonnets, which are excused as merely an 'episode' in the poet's life:

There is no other evidence in Shakespeare's work of homosexual instinct such as we may trace throughout Marlowe's, while there is abundant evidence of a constant preoccupation with women. Unlike Marlowe, Shakespeare seems to have been a man who was fundamentally in harmony with the moral laws of the society in which he lived.

(Ellis and Symonds 109)

Walt Whitman's case is more complicated: the celebration of 'manly love', particularly in the 'Calamus' and the 'Drum-tap' sections of *Leaves of Grass* (1855), prompted Symonds to write to the poet and to ask him 'frankly' about the 'precise significance' of said passages. Whitman responded indignantly in a letter dated 19 August 1890, which Ellis quotes in his Introduction:

That the 'Calamus' part has ever allowed the possibility of such constructions as mentioned is terrible. I am fain to hope that the pages themselves are not to be even mentioned for such gratuitous and quite at the time undreamed and unwished possibility of morbid inferences – which are disavowed by me and seem damnable.

(Ellis and Symonds 110)

Ellis, however, does not accept this response, arguing that Whitman was 'unaware' of sexual issues and 'lacking in analytical power'. On the basis of the frequent presence of the motif of manly love in Whitman's work, the absence of representations of sexual relationships with women and the fact that Whitman wrote ambivalent letters to young men and remained unmarried, Ellis concludes that the author's sexuality must correspond to that of his fictional voices and dominant themes. The passage below is an example of diagnostic biographical criticism *avant la lettre*, in which life and oeuvre are conflated:

It remains true, however, that 'manly love' occupies in his work a predominance which it would scarcely hold in the feelings of the 'average man' whom Whitman wishes to honour. A normally constituted person, having assumed the very frank attitude taken up by Whitman, would be impelled to devote far more space and far more ardour to the subject of sexual relationships with women and all that is involved in maternity than is accorded to them in *Leaves of Grass*. Some of Whitman's extant letters to young men, I understand, though they do not throw definite light on this question, are not of a character that easily permits of publication; and, although a man of remarkable physical vigour, he never felt inclined to marry. It remains somewhat difficult to classify him from the sexual point of view, but we can scarcely fail to recognise the presence of the homosexual instinct, however latent and unconscious.

(Ellis and Symonds 111)

Ellis's literary sources are predominantly French and English. For example, he quotes stanzas from Paul Verlaine's 'Ces Passions', taken from *Parallèlement* (1889) (Ellis and Symonds 111–12). In his discussion of female homosexuality, he refers to the following French works in a footnote: Diderot's *La Religieuse*, Balzac's *La Fille aux yeux d'or*, Gautier's *Mademoiselle de Maupin*, Zola's *Nana* and Belot's *Mademoiselle Giraud, ma femme*. Furthermore, he refers to Maupassant, Bourget, Daudet, Catulle Mendès, Lamartine and Swinburne (Ellis and Symonds 160).

Ellis does not believe in completely 'curing' homosexuals of their preferences but instead promotes acceptance, restraint and 'responsible' behaviour. In the final part of his study, he returns to literary sources once again, this time not treating them as a symptom that can yield insights into the sexual preferences of an author but instead advancing them as a possible cure, by suggesting Greek love and Whitman's ideal of male friendship as models for a moderate homosexual lifestyle. 'The "manly love" celebrated by Walt Whitman in *Leaves of Grass*, although it may be of a more doubtful value for general use, furnishes a wholesome and robust ideal to the invert who is insensitive to normal ideas' (Ellis and Symonds 214). Ellis ends with a lyrical plea for tolerance, and the reminder that all manifestations of the sexual drive are ultimately normal, for they are products of nature, a claim that is substantiated by a quotation from Shakespeare's *The Winter's Tale* (Act 4 Scene 4):

'O'er that art
Which you say adds to Nature, is an art
That Nature makes'
Pathology is but physiology working under new conditions. The stream of Nature still flows into the bent channel of sexual inversion, and still runs according to law.

(Ellis and Symonds 223)

In conjunction with punning on the 'bent' channel of inversion, Ellis mobilizes the most esteemed English poet to re-naturalize homosexuality. By strategically reaching into the literary field and appealing to shared cultural values, that is, ubiquitous respect for Shakespeare, Ellis seeks to authorize and legitimize his claims by switching modes, gesturing towards a form of older, non-empirical knowledge that belongs to the classical canon and situates him in an erudite, humanist tradition. Finally, in later works such as *Erotic Symbolism* (1927), Ellis draws further on literary concepts and hermeneutic techniques to develop his theories of sexuality, including the notions of symbolism, metonymy, and crystallization (see Schaffner 89–111).

Unlike Ellis, the Cambridge-educated poet Edward Carpenter belonged fully to the literary field and was interested primarily in social and educational reform. His key aim was not to theorize the perversions but rather to de-perversify homosexuality. In fact, in various ways he anticipated the revalorization of non-normative desires as a response to nineteenth-century scientific conceptions that took place in modernist literature and, later, in queer theory. Yet, like Ellis, Carpenter, too, explicitly engages with German sexology, for example in his study *Love's Coming of Age* (1896), which illustrates once again the significant sexological traffic between Germany and England.

Carpenter was an advocate of socialism, criticized inequality and the adverse effects of industrialization and propagated a return to a more natural form of living, most famously in his study *Civilization: Its Causes and Cure* (1889). He counted Symonds, Ellis and Ellis's wife, Edith Lees, among his friends, and was an important influence on E. M. Forster and D. H. Lawrence. As Emile Delavenay has demonstrated, Lawrence was familiar with Carpenter's works, and his ideas impacted both on Lawrence's fictions and on his theoretical texts (21–32, 238). In 1908, Carpenter, who was openly gay and a pioneering homosexual and women's rights activist, published *The Intermediate Sex*, one of the first English books on homosexuality that portrayed it neither as a medical nor as a moral problem but as an innate, positive and potentially socially transformative force. Although he engages with the works of Ulrichs, Moll and Krafft-Ebing, Carpenter adopts a decidedly anti-pathological stance, emphasizing physical and spiritual union rather than procreation as the aim of the sexual act.

Liberating sexuality, Carpenter argues, is the prerequisite for social reform, for same-sex love has the power to overcome class boundaries. In *Love's Coming of Age*, he maintains that the 'intermediate race', through its double nature – that is, its masculine and feminine characteristics – obtains 'a certain freemasonry of the secrets of the two sexes which may well favour their function as reconcilers and interpreters' (Carpenter 1911 140). 'Certainly it is remarkable', he asserts, 'that some of the world's greatest leaders and artists have been dowered either wholly or in part with the Uranian temperament'. Uranians, he specifies further in *The Intermediate Sex*, have the capacity to be the future engines of radical social change:

> Eros is a great leveller. Perhaps the true Democracy rests, more firmly than anywhere else, on a sentiment which easily passes the bounds of class and caste, and unites in the closest affection the most estranged ranks of society. It is noticeable how often Uranians of good position and breeding are drawn to rougher types, as of manual workers, and frequently very permanent alliances grow up in this way, which although not publicly acknowledged have a decided influence on social institutions, customs and political tendencies – and which would have a good deal more influence could they be given a little more scope and recognition.
>
> *(Carpenter 1908 114–15)*

To conclude, sexology not only arrived in Britain later than on the continent but was shaped even more substantially by theorists with literary interests and sensibilities than in Germany and France. Early British sexology was not established by psychiatrists, psychologists and physicians, but rather by a literary critic, who only held a low medical degree, and two poets. By deploying concepts and methodologies gleaned from the literary sphere for the analysis of sexual acts and desires, and by drawing on the biographies of authors and their works as case studies, Ellis's example in particular provides ample proof of the fertile cross-traffic between the fields of literature and science in the long nineteenth century and, on another level, the kinship between linguistic and sexual processes. Moreover, it explains why Freud, himself endowed with a sharp literary hermeneutic sensibility, referred more regularly to Ellis's than to any other sexological works in his *Three Essays*.

Note

1   The Eulenburg affair (1906–9), Wilhelminian Germany's biggest domestic scandal, was a politically motivated campaign led by the journalist Maximilian Harden, who publicly accused some of Emperor Wilhelm II's closest allies, most prominently Prince Phillip zu Eulenburg-Hertefeld and Count Kuno von Moltke, of being homosexuals.

Bibliography

Primary texts

[Anon.]. *Onania; or, the Heinous Sin of Self-Pollution, and All its Frightful Consequences, in Both Sexes, Considered, With Spiritual and Physical Advice to Those who Have Already Injur'd Themselves by This Abominable Practice. To Which is Subjoin'd, A Letter from a Lady to the Author, [Very Curious] Concerning the Use and Abuse of the Marriage-Bed, With the Author's Answer.* 4th edn. London: N. Crouch, n.d.

[Anon.]. Review of Krafft-Ebing, *Psychopathia Sexualis*, transl. C. G. Chaddock (Philadelphia, PA: F. A. Davis, 1893). *Journal of Mental Science* 39 (1893): 251–2.

Aquinas, St Thomas. *Summa Theologiae.* Vol. 43: *Temperance.* Transl. Thomas Gilby. London: Blackfriars in conjunction with Eyre & Spottiswoode; New York, NY: McGraw-Hill, 1968.

Carpenter, Edward. *The Intermediate Sex: A Study of Some Transitional Types of Men and Women.* London: George Allen and Unwin, 1908.

Carpenter, Edward. *Love's Coming of Age.* New York, NY: Modern Library, 1911.

Ellis, Havelock. *Erotic Symbolism. Studies in the Psychology of Sex: Complete in Two Volumes.* New York, NY: Random House, 1936, 1–114.

Ellis, Havelock. *Studies in the Psychology of Sex: Complete in Two Volumes.* New York, NY: Random House, 1936.

Ellis, Havelock, and John Addington Symonds. *Sexual Inversion: A Critical Edition.* Ed. Ivan Crozier. Basingstoke: Palgrave Macmillan, 2008.

Freud, Sigmund. *Three Essays on the Theory of Sexuality. The Standard Edition of the Complete Psychological Works of Sigmund Freud.* Ed. and transl. James Strachey. 24 vols. London: Vintage, 2001, 7:123–245.

Krafft-Ebing, Richard von. *Psychopathia Sexualis: With Especial Reference to the Antipathic Sexual Instinct. A Medico-Forensic Study.* Transl. from 12th German edn, Franklin S. Klaf. New York, NY: Arcade, 1998.

Krueg, Julius. 'Perverted Sexual Impulses'. *Brain* 4 (1881): 368–76.

Nordau, Max. *Degeneration.* Transl. George L. Mosse. Lincoln, NE: University of Nebraska Press, 1993.

Norman, Conolly. 'Sexual Perversion'. *A Dictionary of Psychological Medicine.* Ed. Hack Tuke. 2 vols. London: J and A Churchill, 1892, 2:1156–7.

Tissot, Samuel-Auguste. *Onanism: or, a Treatise Upon the Disorders Produced by Masturbation: or, the Dangerous Effects of Secret and Excessive Venery.* Transl. A. Hume. Based on the 3rd, revised edn. London: Wilkinson, 1767.

Secondary texts

Apter, Emily. 'Maternal Fetishism'. *Perversion: Psychoanalytic Perspectives – Perspectives on Psychoanalysis.* Eds. Dany Nobus and Lisa Downing. London: Karnac, 2006, 241–60.

Bauer, Heike. *English Literary Sexology: Translations of Inversion, 1860–1930.* Basingstoke: Palgrave Macmillan, 2009.

Brady, Sean. *John Addington Symonds (1840–1893) and Homosexuality: A Critical Edition of Sources.* Basingstoke: Palgrave Macmillan, 2012.

Bristow, Joseph. 'Symonds's History, Ellis's Heredity: *Sexual Inversion*'. *Sexology in Culture: Labelling Bodies and Desires.* Eds. Lucy Bland and Laura Doan. Cambridge: Polity, 1998, 79–99.

Cook, Matt. *London and the Culture of Homosexuality, 1885–1914.* Cambridge: Cambridge University Press, 2003.

Crozier, Ivan. 'Introduction: Havelock Ellis, John Addington Symonds and the Construction of Sexual Inversion'. Havelock Ellis and John Addington Symonds, *Sexual Inversion: A Critical Edition.* Ed. Ivan Crozier. Basingstoke: Palgrave Macmillan, 2008, 1–86 [2008A].

Crozier, Ivan. 'Nineteenth-Century British Psychiatric Writing About Homosexuality Before Havelock Ellis: The Missing Story'. *Journal of the History of Medicine and Allied Sciences* 63.1 (2008): 65–102 [2008B].

Crozier, Ivan. 'Pillow Talk: Credibility, Trust and the Sexological Case History'. *History of Science* 46/4.154 (2008): 375–404 [2008C].

Dabhoiwala, Faramerz. *The Origins of Sex: A History of the First Sexual Revolution*. London: Allen Lane, 2012.

Delavenay, Emile. *D. H. Lawrence and Edward Carpenter: A Study in Edwardian Transition*. London: Heinemann, 1971.

Downing, Lisa. *Desiring the Dead: Necrophilia and Nineteenth-Century French Literature*. Oxford: Legenda, 2003.

Foucault, Michel. *The Will to Knowledge: The History of Sexuality, Volume 1*. Transl. Robert Hurley. London: Penguin, 1998.

Giddens, Anthony. *The Transformation of Intimacy: Sexuality, Love and Eroticism in Modern Societies*. Cambridge: Polity, 1992.

Griffin, Roger. *Modernism and Fascism: The Sense of a Beginning Under Mussolini and Hitler*. Basingstoke: Palgrave Macmillan, 2007.

Hall, Lesley A. *Sex, Gender and Social Change in Britain Since 1880*. 2nd edn. Basingstoke: Palgrave Macmillan, 2013.

Hull, Isabel V. *The Entourage of Kaiser Wilhelm II, 1888–1919*. Cambridge: Cambridge University Press, 1982.

Kaplan, Louise J. *Female Perversions: The Temptations of Emma Bovary*. New York, NY: Doubleday, 1991.

Laqueur, Thomas W. *Solitary Sex: A Cultural History of Masturbation*. New York, NY: Zone, 2003.

Mangham, Andrew, and Greta Depedge, eds. *The Female Body in Medicine and Literature*. Liverpool: Liverpool University Press, 2012.

Mason, Michael. *The Making of Victorian Sexuality*. Oxford: Oxford University Press, 1994.

Noyes, John K. *The Mastery of Submission: Inventions of Masochism*. Ithaca, NY: Cornell University Press, 1997.

Nye, Robert A., ed. *Sexuality*. Oxford: Oxford University Press, 1999.

Olson, Richard G. *Science and Scientism in Nineteenth-Century Europe*. Urbana, IL: University of Illinois Press, 2008.

Oosterhuis, Harry. *Stepchildren of Nature: Krafft-Ebing, Psychiatry and the Making of Sexual Identity*. Chicago, IL: University of Chicago Press, 2000.

Peakman, Julie, ed. *Sexual Perversions, 1670–1890*. Basingstoke: Palgrave Macmillan, 2009.

Porter, Roy, and Lesley Hall. *The Facts of Life: The Creation of Sexual Knowledge in Britain, 1650–1950*. New Haven, CT: Yale University Press, 1995.

Rosario, Vernon A. *The Erotic Imagination: French Histories of Perversity*. Oxford: Oxford University Press, 1997.

Rydström, Jens. *Sinners and Citizens: Bestiality and Homosexuality in Sweden, 1880–1950*. Chicago, IL: University of Chicago Press, 2003.

Schaffner, Anna Katharina. *Modernism and Perversion: Sexual Deviance in Sexology and Literature*. Basingstoke: Palgrave Macmillan, 2011.

Welldon, Estela V. *Mother, Madonna, Whore: The Idealization and Denigration of Motherhood*. London: Free Association, 1988.

White, Chris, ed. *Nineteenth-Century Writings on Homosexuality: A Sourcebook*. London: Routledge, 1999.

Further reading

Amigoni, David, and Amber K. Regis, eds. *(Re)Reading John Addington Symonds*. English Studies 94.2 (2013).

Ball, Benjamin. *La Folie érotique*. Paris: J.-B. Baillière, 1888.

Binet, Alfred. 'Le Fétichisme dans l'amour'. *Études de psychologie expérimentale*. Paris: Octave Doin, 1888, 1–85.

Bland, Lucy, and Laura Doan, eds. *Sexology in Culture: Labelling Bodies and Desires*. Cambridge: Polity, 1998.

Bloch, Iwan. *Beiträge zur Aetiologie der Psychopathia sexualis*. 2 vols. Dresden: H. R. Dohrn, 1902–3.

Bloch, Iwan. *Das Sexualleben unserer Zeit in seinen Beziehungen zur modernen Kultur*. 10th–12th, rev. edn. Berlin: Louis Marcus, 1919.

Bloch, Iwan [as Eugen Dühren]. *Neue Forschungen über den Marquis de Sade und seine Zeit. Mit besonderer Berücksichtigung der Sexualphilosophie De Sade's auf Grund des neuentdeckten Original-Manuskriptes seines Hauptwerkes 'Die 120 Tage von Sodom'*. Berlin: Max Harrwitz, 1904.

Charcot, Jean-Martin, and Valentin Magnan. 'Inversion du sens genital'. *Archives de neurologie. Revue des maladies nerveuses et mentales* 3.7 (1882): 53–60 and 3.10 (1882): 296–322.

D'Arch Smith, Timothy. *Love in Earnest: Some Notes on the Lives and Writings of English 'Uranian' Poets From 1880–1930*. London: Routledge, 1970.

Dollimore, Jonathan. *Sexual Dissidence: Augustine to Wilde, Freud to Foucault*. Oxford: Clarendon, 1991.

Garnier, Paul-Emile. *Les Fétichistes, pervertis et invertis sexuels. Observations médico-légales*. Paris: J.-B. Baillière, 1896.

Hacking, Ian. 'Making up People'. *Forms of Desire: Sexual Orientation and the Social Constructivist Controversy*. Ed. Edward Stein. London: Routledge, 1992, 69–88.

Hekma, Gert. 'A History of Sexology: Social and Historical Aspects of Sexuality'. *From Sappho to De Sade: Moments in the History of Sexuality*. Ed. Jan Bremmer. London: Routledge, 1989, 173–93.

Hirschfeld, Magnus. *Sappho und Sokrates. Wie erklärt sich die Liebe der Männer und Frauen zu Personen des eigenen Geschlechts?* 2nd edn. Leipzig: Max Spohr, 1902.

Hirschfeld, Magnus. *Die Homosexualität des Mannes und des Weibes*. 2nd edn. Berlin: Louis Marcus, 1920.

Holmes, John. *Dante Gabriel Rossetti and the Late Victorian Sonnet Sequence: Sexuality, Belief and the Self*. Aldershot: Ashgate, 2005.

Kaan, Heinrich. *'Psychopathia sexualis'. Zur Reifizierung des Sexuellen im 19. Jahrhundert. Der Beginn einer Scientia Sexualis, dargestellt anhand dreier Texte von Hermann Joseph Löwenstein, Joseph Häussler und Heinrich Kaan*. Transl. Philipp Gutmann *et al*. Frankfurt: Peter Lang, 1998, 129–230.

Magnan, Valentin. 'Des anomalies, des aberrations, et des perversions sexuelles'. *Annales médico-psychologiques* 7.1 (1885): 447–74.

Michéa, Claude-François. 'Des déviations maladives de l'appétit vénérien'. *Union médicale. Journal des intérêts scientifiques et pratiques, moraux et professionnels du corps médicals* 3.85 (1849): 338–9.

Moll, Albert. *Handbuch der Sexualwissenschaften. Mit besonderer Berücksichtigung der Kulturgeschichtlichen Beziehungen*. 3rd, rev. edn. 2 vols. Leipzig: F. C. W. Vogel, 1926.

Moreau (de Tours), Paul. *Des aberrations du sens génésique*. 4th edn. Paris: Asselin and Houzeau, 1887.

Peakman, Julie. *The Pleasure's All Mine: A History of Perverse Sex*. London: Reaktion, 2013.

Pemble, John, ed. *John Addington Symonds: Culture and the Demon Desire*. Basingstoke: Macmillan, 2000.

Porter, Roy, and Mikulás Teich, eds. *Sexual Knowledge, Sexual Science: The History of Attitudes to Sexuality*. Cambridge: Cambridge University Press, 1994.

Roudinesco, Élisabeth. *Our Dark Side: A History of Perversion*. Transl. David Macey. Cambridge: Polity, 2009.

Sha, Richard C. *Perverse Romanticism: Aesthetics and Sexuality in Britain, 1750–1832*. Baltimore, MD: Johns Hopkins University Press, 2009.

Syson, Lydia. *Doctor of Love: James Graham and His Celestial Bed*. Richmond, VA: Alma, 2008.

Tarnowsky, Benjamin. *The Sexual Instinct and Its Morbid Manifestations: From the Double Standpoint of Jurisprudence and Psychiatry*. Transl. W. C. Costello and Alfred Allinson. Paris: Charles Carrington, 1898.

Weeks, Jeffrey. *Sex, Politics and Society: The Regulation of Sexuality Since 1800*. 2nd edn. London: Longman, 1989.

27

OCCULT SCIENCES

Christine Ferguson
University of Glasgow

The occult sciences were woven into the fabric of everyday life in nineteenth-century Britain. By no means the exclusive preserve of late Romantic all-male secret societies or, subsequently, of the urban bourgeoisie who formed the core membership of occult organizations such as the Theosophical Society or the Hermetic Order of the Golden Dawn (Campbell; Dixon; Wunder), they were available to the wider public in the fortunes told at local fairs, in the clairvoyant mirrors advertised in magazine classified columns, in the public lectures devoted to the speculative histories of alchemy and Rosicrucianism, in the dream-scrying techniques shared by word of mouth and, perhaps most of all, in the pages of popular novels, which, as the century progressed, became increasingly suffused with occult plots and tropes. The designation 'occult science' was liberally applied in this period to a dizzying gamut of old and new magical practices, including divination, geomancy, clairvoyance, palmistry, alchemy, tarot reading, ceremonial magic, astral projection, kabbalah, necromancy, angel invocation, demonology, astrology and many others (Hanegraaff 234). These eclectic forms of what Wouter Hanegraaff terms 'rejected knowledge' offered to reveal to their users a mysterious, hidden world that lay beyond normal sensory perception, one that no microscope could ever penetrate and in which the supernatural intermediaries and forces increasingly ousted by scientific naturalism were still very much alive and open to supplication. Yet it would be inaccurate to regard these speculative entities and their occult invokers as simply the antithetical and much-maligned others to the secular science of the era. Not content with their *de facto* banishment from the realm of scientific rationalism, many nineteenth-century occult practitioners, as this chapter will demonstrate, worked relentlessly to insist on the affinities, complicities and uncanny parallels between their own esoteric knowledge base and the emerging world view of secular scientific naturalism.

Their task was by no means an easy one. Although diffuse in number and often difficult to categorize, those sciences designated as occult in the nineteenth century shared one key characteristic: their frequent – but never total – rejection by members of the mainstream scientific establishment who desperately wanted to quarantine their own newly-professionalized field from the taint of superstition and irrationality.[1] We find a classic example of this boundary negotiation in Edward Burnett Tylor's landmark *Primitive Culture* (1871), a foundational work of cultural anthropology that dismisses occult science as 'contemptible superstition' (102) associated with 'the lowest known stages of civilization' (101) and characterized by 'a sincere but fallacious system of philosophy' (122). As a proponent of a new social-scientific discipline dedicated to studying the belief and cultural systems of so-called 'primitive' societies, Tylor was arguably more anxious than most of his fellow professionals to differentiate his own methodology and ontological first principles from those of the people he took as subjects. Yet, even Tylor was forced to admit that occult practice had by no means vanished from the modern West, in which scientific rationalism was now, allegedly, gaining dominance; it lived on in crude survivals such as clairvoyance, mesmerism and the seances of the so-called 'necromantic religion' (130) of modern spiritualism. Speaking of this American new religious import that had by the mid-century gained so many converts on British shores, he remarked:

> The world is again swarming with intelligent and powerful disembodied spiritual beings, whose direct action on thought and matter is again confidently asserted as in those [pre-Enlightenment and non-Western] times and countries where physical science had not as yet so far succeeded in extruding these spirits and their influences from the system of nature.
>
> *(Tylor 129)*

Although personally disdainful of occult belief, Tylor was intelligent enough never to dismiss its expressions as culturally insignificant (Stocking), recognizing instead that occultism still represented an important if atavistic means through which Britons of varying backgrounds sought to understand their place in the natural world.[2]

Naming and categorizing occult science

Once considered, as Nicholas Goodrick-Clarke (2008 4) and Martin Priestman (140) observe, an illegitimate topic of inquiry for serious scholars, nineteenth-century occultism has been subject to an explosion of critical interest in the last two decades as its historical significance for the professionalization of science, the emergence of modernist aesthetics and the development of feminist, socialist and anti-imperialist, but also at times deeply reactionary, politics has been increasingly acknowledged. (For representative examples of scholarly works in this vein, see Dixon; Ferguson; Galvan; Goodrick-Clark; Kontou 2008; Luckhurst; Luckhurst and Sausman; Morrison; Owen 2004A; Sword; Pasi.) This is not to suggest that the occult is now understood only as a butt of or catalyst for more exoteric and rational historical developments. On the contrary, the growing academic sub-discipline of Western esoteric history pioneered by Nicholas Goodrick-Clarke and others addresses occult science in its own right as an enterprise that has, since its inception, championed the cause of intellectual as well as spiritual enlightenment. The term 'occult', as Hanegraaff reminds us, has a long history, deriving in the late Middle Ages from the Latin *qualitates occultae* to refer to those properties of objects that are hidden to the naked or untrained eye (178–9). In this initial use, Hanegraaff explains, the occult had no particular connection to what we now think of as the mystical or the supernatural; rather, it identified natural phenomena that, although presently mysterious, might eventually be made explicable through study and investigation. '[F]ar from suggesting an "occultist" world view according to modern understandings of the term,' he writes, 'it was originally an instrument for disenchantment, used to withdraw the realm of the marvelous from theological control and make it available for scientific study' (Hanegraaff 180). It is perhaps a reflection of this disenchanting genealogy that hermetic inquiry, esoteric Freemasonry and the language of magic flourished rather than faded during the peak years of the European Enlightenment as thinkers sought to exercise their own faculties in investigating the material and spiritual world (Taylor; Wunder; Priestman). These precedents are important, demonstrating that the affinity between occult and scientific ways of knowing the world asserted by many if not all nineteenth-century practitioners was not simply a novel creation of their post-Enlightenment era but also a partial return to earlier conceptualizations of occult science. After all, as Hanegraaff notes, it was not until the sixteenth century that occultism gained its now-standard association with magic (180). This etymological history anticipates the tension between the desire to expose and to protectively conceal that would later emerge as a distinguishing feature of the discourse of nineteenth-century occult science, one further exacerbated by the growing social expectation that useful knowledge should always be made public.

Although many nineteenth-century occult practitioners referred to their study as 'science', they rarely shared the same understanding of that category and its epistemological criteria as the nation's leading scientific institutions. Routinely absent in the writing of self-professed occult scientists such as H. P. Blavatsky or Henry Steel Olcott, co-founders of the Theosophical Society, is the emphasis on method, falsifiability, reproducibility of results and open dissemination so crucial to the period's emerging paradigm of scientific investigation. Instead, occult science was for them more of a fixed body of knowledge acquirable through the adoption of a spiritually monist view of the universe. 'The term *occult*,' claimed Olcott, was simply the name 'given to the sciences relating to the mystical spirit of nature – the department of force or spirit' (1885 202). Empirical experiment could play a role in the pursuit of occult wisdom, but it was not necessarily required, and theosophists in particular became wary of phenomenal demonstration after the Society for Psychical Research denounced Blavatsky's purported psychic manifestations as fraudulent in 1885 (see Hodgson). Writing six years after this debacle, theosophist A. P. Sinnett had characters in his occult novel *Karma* claim that 'the psychic phenomena which has to do with outer facts . . . is [a] *cul de sac*' (66) and that 'it would degrade spiritual science to an extent quite revolting to its devotees, if it were pursued to a considerable extent for the sake of its lower victories on the material plain' (79). It is hardly surprising that an occultist like Sinnett, one whose movement's leading figure had just been accused of using conjuring tricks to produce fake spiritual manifestations, might seek to downplay the role of phenomenal display and material evidence in his belief system. But such anxieties about the ramifications of occult experiment were by no means confined to initiates. They surface also in surrounding popular Gothic fictions such as Robert Louis Stevenson's *Strange Case of Dr Jekyll and Mr Hyde* (1886) and Arthur Machen's *The Great God Pan* (1894), novellas that trace the horrific consequences of transferring arcane knowledge from the library to the laboratory. In the first, a seemingly respectable chemist degenerates into an inhuman troglodyte through the aid of an unstable, pseudo-alchemical formula; in the second, a pagan deity is unleashed on London in the shape of the homicidal *femme fatale* Helen Vaughan. Occult science, these narratives suggest, could be a distinctly hazardous enterprise, one best left to non-practising armchair eccentrics or indeed avoided entirely if the reputability and safety of the new scientific disciplines were to be maintained.

An ancient science for the modern age: the ethics of occult history

Although practical occult experimentation was often recognized as dangerous or unnecessary in nineteenth-century Britain, believers did not abandon the quest to fuse the arcane wisdom tradition with scientific naturalism. Instead, they pursued this alliance through other less controversial and less empirical tactics, recognizing that some form of imaginative consilience was necessary for the survival of the occult tradition in the modern age. In his 1891 study *The Occult Sciences* (1891), ceremonial magician A. E. Waite argued that, without scientific validation, 'the transcendental philosophy would be simply the revival of an archaic faith, and would be wholly unadapted to the necessities of to-day' (2). Understandably, then, he, like many of his co-believers, was frustrated by the ongoing refusal of contemporary investigators to recognize that the older occult sciences at least 'constitute[d] a sufficient ground for a new series of scientific inquiries' (Waite 1891 2). His sentiments provide an important corrective to those who would read nineteenth-century occultism simply as a nostalgic retreat from scientific modernity into a rarefied past. Were it to endure, occult science had to modernize, and if the secular scientific establishment was unwilling to assist in this effort via collaborative investigation, then occultists would forge the connection themselves by appropriating the language of, and claiming precedence for, emergent naturalist theories of species evolution and geological gradualism. (For differing accounts of Victorian occultism's appropriation of scientific naturalism, see Luckhurst; Thurschwell; Gomel; Galvan; Luckhurst and Sausman; Ferguson.) Blavatsky, for example, argued in *Isis Unveiled* that '[m]odern science insists upon the doctrine of evolution; so do human reason and the Secret Doctrine' (35). Darwin's thesis of gradual species evolution provided Theosophy with analogical support for its Eastern-influenced belief in the metempsychotic development of the human soul over consecutive reincarnations; similarly, in Lyell's extended estimate of the earth's geological age, the occult order found the time frame necessary to accommodate its theory of the slow development of the first five root races of man – there would, Blavatsky held, ultimately be seven in total – and the sinking of Atlantis. (Campbell and Ramaswamy provide fuller discussions of the role of the root races and evolutionary time in theosophical thought.) *Isis Unveiled* thus proclaims: 'The discoveries of modern science do not disagree with the oldest traditions. Within the last few years, geology . . . has found unanswerable proofs that human existence antedates the last glaciation of Europe – over 250,000 years!' (Blavatsky 4).

The modern scientific discoveries claimed by nineteenth-century occult scientists were creatively fused with the ageless wisdom of the *prisca theologia*, that is, the concept, in Goodrick-Clarke's words, of 'an ancient theology . . . deriving from such founder-figures and representatives as Moses, Zoroaster, Hermes Trismegistus, Plato, and Orpheus, who had supposedly bequeathed this unitary wisdom tradition to humankind in times immemorial' (Goodrick-Clarke 2008 7). Although long lost, this Ur-wisdom was nonetheless imagined as recoverable by appropriately trained and temperamentally suited seekers who knew how to crack its universal emblem code. This cipher is exotically rendered here in Waite's important translation of French ceremonial magician Eliphas Lévi's *Dogme et Rituel de la Haute Magie* (1856), published as *Transcendental Magic* in 1896:

> Behind the veil of all the hieratic and mystical allegories of ancient doctrines, behind the darkness and strange ordeals of all initiations, under the seal of all sacred writings, in the ruins of Nineveh or Thebes, on the crumbling stones of old temples and on the blackened visages of the Assyrian or Egyptian sphinx, in the monstrous or marvelous paintings which interpret to the faithful of India the inspired pages of the Vedas, in the cryptic emblems of our books on alchemy, in the ceremonies practised at receptions by all secret societies, there are found indications of a doctrine which is everywhere the same and everywhere carefully concealed.
>
> *(1)*

Such fanciful and loosely, if at all, evidenced perennialist syntheses of disparate cross-cultural beliefs obviously reflect what R. A. Gilbert has recognized as the pervasive amateurism of nineteenth-century occult scholarship, a body of writing replete with errors and plagiarisms and characterized by a frequent unwillingness to engage with primary source materials (Gilbert 1997 91; for more on plagiarism and suspicious source use in nineteenth-century occult writing, see Campbell on Blavatsky's plagiarism from William Howitt, Emma Hardinge Britten and

Hargrave Jennings, and Demarest (see Britten 2011) on Emma Hardinge Britten's covert borrowings from Jennings). Yet, even if historically incorrect, articulations of the *prisca theologia* such as these were nonetheless tremendously important to believers in the ethical position they enabled. Occultists used the *prisca theologia* to forge what they viewed as a much more ethical and empirical approach to the past than that offered by their sceptical scientific peers, who typically linked early human civilization with primitive backwardness and superstition. 'To deny Occultism, or Magic,' declared the first issue of the Glasgow-based *The Occult Magazine* in 1885, 'is not only to reject history, but to foolishly cast aside the testimony of witnesses thereof, extending through a period of more than four thousand years' (Mejnour 6). Blavatsky had made the same argument several years earlier, fulminating in *Isis Unveiled* that 'To believe that, for many thousands of years, one-half of mankind practised deception and fraud on the other half, is equivalent to saying that the human race was composed only of knaves and incurable idiots' (11). One could still champion the cause of progress, occultists claimed, without casting a suspicious eye on the motives or mental capacities of our ancient ancestors.

It was not only in its treatment of the past that occult science claimed to be more ethical than its secular rational counterpart. At a time when new biological, social, statistical sciences were combining to buttress the edifice of white supremacy under the auspices of eugenics and social Darwinism, some – if by no means all – occultists were deploying the ancient wisdom tradition to insist on a radical interracial equality. This tendency is particularly evident in those Eastern-influenced versions of occult science – Hargrave Jennings' phallism and Blavatsky's and Olcott's theosophy – that tapped into the interest in Asian religions spearheaded by William Jones and Richard Payne Knight in the late eighteenth century (Godwin J. 15–16). Hargrave Jennings, possibly the namesake inspiration for the demon-haunted Reverend Jennings in Sheridan Le Fanu's Swedenborgian Gothic tale 'Green Tea' (1869), was a self-taught occult scholar who wrote prolifically, albeit often highly inaccurately, on such topics as Freemasonry, Rosicrucianism and Buddhism. In his controversial and anonymously published *The Indian Religions* (1858), he notoriously defended the recent Indian rebellion – one that he scornfully refused to deem a 'mutiny' (Jennings 1858 166) – as a foreseeable and largely justifiable reprisal against long-standing religious oppression and commercial exploitation. British rule in India had constituted nothing more, Jennings contended, than

> a hundred years of active, unmitigated tyranny, unrelieved by any one trait of generosity, scarcely for once qualified, even accidentally, by a single act dictated by an unmixed, unselfish, and sincere desire to benefit the people whom Providence had delivered over into their charge.
>
> *(1858 158)*

The religions of the East, or rather Jennings's occult constructions of them, deserved respect rather than contempt; without a sea-change in attitude, the British empire would continue to reap the 'bitter fruits' (1858 159) of constant insurrection. An occult sensibility could prevent such conflicts by annihilating the arrogant sense of cultural, religious, and scientific superiority that triggered them.

The occult anti-imperialism championed by Jennings found a ready home in later theosophical thought, featuring regularly in the writings and platform speeches of the society's adherents as they vaunted the distinguished past and admirable contemporary spiritual practices of non-Western peoples. In a particularly vivid example of this orientation, Olcott, lecturing on occult science in British Ceylon in 1875, blasted European colonizers for putting too much metaphysical stock in their phenotype and their scientific expertise alike:

We modern Europeans have been so blinded by the fumes of our own conceit that we have not been able to look beyond our noses. We have been boasting of our glorious enlightenment, of our scientific discoveries, of our civilization, of our superiority to everybody with a dark skin, and to every nation east of the Volga and the Red Sea, or south of the Mediterranean, until we have come almost to believe that the world was built for our Anglo-Saxon race.

(1885 214)

The true practice of occultism, he concluded, would impel the West to 'approach the Eastern people in a less presumptuous spirit, and honestly confessing that we know nothing at all of the beginning or end of natural law, ask them to help us find out what their forefathers knew' (Olcott 1885 215). Unsurprisingly, sentiments such as these raised profound suspicions among colonial officials in India and Ceylon, ones that were only confirmed when, under the direction of Annie Besant, the Theosophical Society came to champion Indian independence in the early twentieth century (see Campbell and Dixon for more on theosophy's contribution to the Indian independence movement). These alignments provide an important counterweight to the more openly xenophobic treatments of non-Western and non-white cultures that dominated late Victorian popular Gothic novels such as H. Rider Haggard's *She* (1886–87), Bram Stoker's *Dracula* (1897) and Richard Marsh's *The Beetle* (1897), texts in which the East typically features as a source of threat and contamination rather than of valuable wisdom and ethical restoration. Occultism had the potential to heal and reverse these paranoid constructions, forming the basis of a potent anti-imperial science that would evidence the futility of racial domination and the necessity of cultural hybridity.

The dangerous circulation of (occult) scientific knowledge: Godwin's *St Leon* (1799) and Bulwer-Lytton's *Zanoni* (1842)

Hitherto we have been focusing on what might be considered the most progressive and populist aspects of nineteenth-century occult science – its openness, however eccentric and partial, to an alliance with contemporary scientific naturalism; its *ad populum* argument for the existence of a supernatural world on the basis of the sheer number of people over the space of the globe and through the course of history who have believed in one; and finally, its pioneering receptivity to non-First World spiritual traditions, one that remains a key feature of contemporary occult and New Age belief systems. We cannot conclude our discussion without considering the equally significant, if far less radical and democratic, representations of occult science best epitomized in the era's alchemical and Rosicrucian novels. Inspired by the German *Bundesroman* genre (Ziolkowski 69–70), these texts examine the actions of secret societies, or even of discrete individuals, who, for good or ill, possess and protect occult scientific knowledge from the general public. Their adept protagonists only very occasionally select new initiates, typically hapless or desperate men who are doomed rather than elevated by their sudden acquisition of occult wisdom. Perhaps the most famous Romantic example of this sub-genre is William Godwin's *St Leon* (1799), a tragic account of the near-fatal alienation suffered by a French nobleman whose entrustment with the secret of the philosopher's stone and the elixir of life severs him from his kin and community. Sworn to secrecy about the source of his wealth, and hounded and imprisoned by those who distrust its origins, St Leon lives to rue the day he exchanged the domestic comforts of family life for the solipsism of the occult. 'How unhappy the wretch,' he laments, 'who is without an equal; who looks through the world and cannot find a brother . . . who . . . lives the solitary, joyless tenant of a prison, the materials of which are emeralds and rubies!' (Godwin W. 210–11).

Occult science here encapsulates everything that Godwin's communitarian egalitarianism led him to reject, representing the horrific consequences of knowledge monopolization for both those who possess and those who are denied occult secrets.

Occultism fares little better in its later treatment by Percy Bysshe Shelley, whose bombastic *St Irvyne; or The Rosicrucian* (1810) features the pursuit of an outcast nobleman by a bandit-alchemist who tries aggressively to initiate him into the occult fold. By the end of the novel's ridiculous plot, both men are dead. For early nineteenth-century progressive radicals like Godwin and Shelley, occult science was a vicious dead end, a means of exploitation and tyranny antithetical to democracy and thus deserving of destruction.

Yet the very anti-egalitarianism that rendered occult science suspect for Godwin and Shelley would later be presented as its chief virtue in what is arguably the most significant, influential and also deeply reactionary Victorian fictional treatment of the topic: Edward Bulwer-Lytton's *Zanoni* (1842). A prolific popular writer and eventual peer of the realm, Bulwer-Lytton authored numerous occult-themed novels and stories over the course of his fifty-year long literary career, including 'The Haunted and the Haunters; or, The House and the Brain' (1859), *A Strange Story* (1862) and *The Coming Race* (1871); indeed, so frequently did he return to this shadowy fictional territory that he was often assumed, albeit incorrectly, to be an initiate of an esoteric order himself (Godwin J. 123). Within occult circles at least, *Zanoni* was the most influential of all of the author's supernaturally themed works, inspiring respectful homages in subsequent occult *Bildungsromans* such as Paschal Beverly Randolph's *The Wonderful Story of Ravalette* (1863) and Emma Hardinge Britten's *Ghost Land* (1876), and establishing Bulwer-Lytton, in J. Jeffrey Franklin's words, as the 'single person in the first half of the nineteenth century . . . representative of the period's enthusiasms, reservations, and deep-seated fears concerning occult spiritualities' (126). The novel is also a meditation on the dangers of modern democracy and the expansion of knowledge it promotes, arguing that occult scientific education must be confined to an elite and almost impossibly select few whose role must be to conceal rather than share its boons with the general populace. This obsession with secrecy, with the rigorous regulation and control of knowledge dissemination, was no less pronounced a characteristic of nineteenth-century occult thought than the previously-examined populist and modernizing tendencies with which it coexisted in uneasy tension. *Zanoni*'s paranoia about who has the right to possess and use (occult) scientific knowledge functions arguably as a parody of the boundary work simultaneously being performed by the professionalizing naturalist sciences. In the occult milieu of the novel, as in the scientific institutions of nineteenth-century Britain, scientific authority had to be protected from the incursions of improperly trained dilettantes and charlatans who would use it only for their own gain.

Zanoni was not unique in reflecting an anxiety about the application, authorization and dissemination of occult science; in fact, we find references to, and explanations for, the veiling of esoteric knowledge in most of the period's key occult texts. In many of these, however, concealment is defended as a historical necessity rather than as a defiantly exclusionary strategy. For Eliphas Lévi, for example, secrecy was the only way to protect occult practitioners from establishment persecution:

> The science was driven into hiding to escape the impassioned assaults of blind desire; it clothed itself with new hieroglyphics, falsified its intentions, denied its hopes. Then it was that the jargon of alchemy was created, an impenetrable illusion for the vulgar in their greed of gold, a living language only for the true disciple of Hermes.
>
> *(3–4)*

As Blavatsky contended in *Isis Unveiled*, this mystification, however unavoidable, had the unfortunate consequence of cementing secular prejudices against occult practice: 'The impenetrable veil of arcane secrecy was thrown over the science taught in the sanctuary. This is the cause of the modern depreciating the ancient philosophies' (5). Both Lévi and Blavatsky wrote in the conviction that times had changed, and that a loosening of the strictures on esoteric knowledge might now be both possible and desirable; even the very title of Blavatsky's first major occult treatise, *Isis Unveiled*, promises an act of portentous public revelation. *Zanoni* by contrast, although written within the far more popular genre of the occult romance, shares almost none of Blavatsky's optimism about the beneficial spread of arcane wisdom. Instead, it offers a dire warning about the consequences of expanding occult science's public sphere beyond even the extremely limited constituency of its two sole initiates, the Rosicrucian brother Zanoni and his teacher Mejnour.

Zanoni's obscurantist drive exists in awkward contradistinction with the massive range of occult practices it enumerates within its pages; the novel is after all, as Joscelyn Godwin writes, 'an encyclopedia of the occult sciences' (126), which dramatizes everything from mesmerism to alchemy and demonology. Indeed, it is very much the kind of novel that a nineteenth-century reader might have turned to as a primer in the secret arts. But *Zanoni* also insists that occult science only be pursued under controlled conditions by students of the rare character, integrity and experience now virtually unachievable in the violently democratized environment of its Western European setting. Presented as a translated cipher manuscript, it tracks the doomed love affair between the Rosicrucian adept Zanoni and the beautiful singer Viola Pisani from its origins in a Neapolitan opera house through to its bloody conclusion in Revolutionary France. Zanoni, having first abandoned his immortality to form a sexual relationship with Viola and father her child, then makes the further sacrifice of taking her place at the guillotine when she is entrapped by Robespierre's despotic regime. The still-imprisoned Viola dies soon afterwards, her cold face bearing a beatific smile suggestive of a post-life reunion with Zanoni in the upper spiritual spheres. The Revolutionary prison-house, this staging suggests, is ultimately a kinder place for pure-hearted spirits such as Viola than the turbulent Revolutionary world outside, where people mix without any regard to class or distinction. Indeed, the prison site had earlier won praise from *Zanoni*'s narrator for its subtle encouragement of a restored respect for rank among its inmates. There, he remarks, 'all ranks were cast, with an even-handed scorn. And yet there, the reverence that comes from great emotions restored Nature's first and imperishable, and most lovely, and most noble Law—THE INEQUALITY BETWEEN MAN AND MAN!' (Bulwer-Lytton 1877 392–3).

Bulwer-Lytton's virulently conservative novel does not condemn the Age of Reason for suppressing or ridiculing occult science; on the contrary, it lambasts its popularization of secret wisdom for an undeserving and tainted constituency. A forceful silencing of occult knowledge, it implies, would have been preferable to this reckless glamorization under the aegis of Enlightenment. Far from being mutually exclusive, *Zanoni* insists, democratic rationalism and certain strains of occultism are, in fact, volatile symbiotic partners. Their dangerous co-dependency is nowhere more evident than in the tragic conversion of young English artist Clarence Glyndon, a radical who meets Zanoni in Italy and begs to be admitted into the Rosicrucian fold. His desire for initiation, we are told, is a symptomatic outgrowth of the Revolutionary zeitgeist:

> It was then the period, when a feverish spirit of change was working its way to that
> hideous mockery of human aspirations, the Revolution of France. And from the chaos
> into which were already jarring the sanctities of the World's Venerable Belief, arose

many shapeless and unformed chimeras. Need I remind the reader, that while that was the day for polished skepticism and affected wisdom, it was the day also for the most egregious credulity and the most mystical superstitions,—the day in which magnetism and magic found converts among the disciples of Diderot,—when prophecies were current in every mouth,—when the salon of a philosophical deist was converted into an Heraclea, in which necromancy professed to conjure up the shadows of the dead ... Dazzled by the dawn of the Revolution, Glyndon was yet more attracted by its strange accompaniments, and natural it was with him, as with others, that the fancy which ran riot amidst the hopes of a social Utopia, should grasp with avidity all that promised, out of the beaten science, the bold discoveries of some marvelous Elysium.

(Bulwer-Lytton 1877 75–6)

This fascinating passage might strike us at first as a wholesale, and hence deeply paradoxical, condemnation of the occult science the novel has hitherto lauded through its eponymous protagonist; after all, here magic and mesmerism represent credulity and superstition, not viable or even superior forms of scientific knowledge. Such sentiments would indeed undo the narrative's occult ontology were it not for their framing stipulation. Fraudulent democratic mysticism may be widespread but it is not the only type of occult activity that exists in *Zanoni*; furthermore, it is easily distinguishable from more authentic forms through its site of practice. Any branch of occult science performed out in the open, in urban salons for bourgeois audiences, automatically marks itself as illegitimate.

It is no surprise then that the only valid forms of occult science in the novel, as practised by Zanoni and Mejnour, are witnessed by an extremely limited audience and are never successfully taught to would-be seekers. This lack of transmission seems at first highly regrettable in light of the potential utility of the Rosicrucian medicine they profess, one that can cure all ailments and extend human life for centuries, even millennia. 'This is not Magic,' explains Mejnour: 'All we do is but this—to find out the secrets of the human frame, to know why the parts ossify and the blood stagnates, and to apply continual preventatives to the effects of Time ... [I]t is the Art of Medicine rightly understood' (Bulwer-Lytton 1877 217). In *Zanoni*, however, the occult art of medicine is most beneficial when unapplied, and its restriction is presented as an act of compassion. Such at least is Mejnour's rationale when confronted by Glyndon with the question that perennially shadowed all practitioners of occult science in the fiercely utilitarian nineteenth century: '[I]f possessed of these great secrets,' he asks, 'why so churlish in withholding their diffusion[?] Does not the false and charlatanic science differ in this from the true and indisputable—that the last communicates to the world the process by which it attains its discoveries; the first boasts of marvelous results, and refuses to explain the causes[?]' (Bulwer-Lytton 1877 218). Mejnour replies:

Well said, O Logician of the Schools—but think again. Suppose we were to impart all our knowledge to all mankind, indiscriminately, alike to the vicious and the virtuous—should we be benefactors or scourges? Imagine the tyrant, the sensualist, the evil and corrupted being possessed of these tremendous powers; would he not be a demon let loose on earth? ... It is for these reasons that we are not only solemnly bound to administer our lore only to those who will not misuse and pervert it; but that we place our ordeal in tests that purify the passions, and elevate the desires.

(Bulwer-Lytton 1877 218)

Here the Genesis punishment for acquiring forbidden knowledge is invoked without any consoling promise of future remediation; once occult wisdom has been unleashed among the *demos*, there can be no return to Eden. Certainly, a select few might be chosen to act as safe repositories, but their appearance is obviously extremely rare, so much so as to provide the tacit rationale for Mejnour's and Zanoni's enforced immortality. Only by staying alive across the centuries can the Rosicrucian brother hope to find the few initiates worthy of receiving and preserving the hidden knowledge.

Zanoni thus dramatizes an occult milieu in which access to knowledge has become nihilistically, albeit necessarily, over-restricted. Even were the system more open, the narrative suggests, the acquisition of secret wisdom could still have little appeal to potential initiates, given the terrifying consequences of failure, ones manifested in the devastating aftermath of Glyndon's premature experimentation with the elixir of life. The English neophyte, who after his rejection by Zanoni has been studying with Mejnour in a remote Italian castle, becomes impatient to obtain results. Left alone for a few weeks as a covert test of character, he breaks into the occult master's study and attempts to concoct an immortality potion from partial instructions in a cipher manuscript. The experiment not only fails but also, like the alchemical investigations in Godwin's *St Leon*, reaps a horrible reward. Glyndon summons a demonic nemesis known as the Dweller on the Threshold, which haunts him for years to come, whispering in his ear poisonous words 'forbidden [for] the lips to repeat and the hand to record' (Bulwer-Lytton 1877 270) and ultimately causing the death of his sister Adela. Glyndon cannot save her from the spirit-produced epilepsy that eventually kills her; his brief training has left him with no practical or useful medical knowledge. Later, Zanoni helps to cure the artist of his deadly obsession, but only on the basis that Glyndon abandon any further attempts to penetrate behind the veil or to participate in the apparently equivalent pursuit of revolutionary action. 'Return, O wanderer!' Zanoni counsels. 'Return. Feel what beauty and holiness dwell in the Customary and the Old' (Bulwer-Lytton 1877 366).

The Zanoni–Glyndon plot's narrative arc ultimately reinforces the anti-populist and anti-democratic emphasis of the novel's surrounding frame, with both sequences combining to offer a staunch warning about the dangers of universal education and open scientific culture. *Zanoni*'s fascinating introductory paratext describes Bulwer-Lytton's alleged discovery of the main manuscript through a chance encounter in a virtually deserted Covent Garden occult bookshop, in which 'there were to be found no popular treatises, no entertaining romances, no histories, no travels, no "Library for the People", no "Amusement for the Million"' (xi). Instead, the shop's cantankerous proprietor, Mr D—, stocks a select variety of rare esoteric works that he actively defends from the few dilettantish customers who occasionally cross his threshold, refusing to sell, or even spontaneously buying back, his treasured stock when it attracts consumer interest. Mr D—'s self-defeatingly closed system is offered as an optimal model for the (non-)circulation of occult scientific knowledge whose necessity is dramatized in Glyndon's fate.

The fictionalized Bulwer-Lytton, one of a very few patrons tolerated in the shop, meets there a fellow seeker who approves of his interest in occult lore and offers him the Rosicrucian romance – namely, the interior manuscript of *Zanoni* – which he has written in 'an unintelligible cipher' (xviii). Fortunately, our narrator-editor finds a key, but new complications arise when, after translating one version of the text, Bulwer-Lytton discovers another one that has to be decoded anew and whose eccentric syntax and rhythm defy his best translative efforts. 'Truth compels me to confess,' he acknowledges, 'that, with all my pains, I am by no means sure that I have invariably given the true meaning of the cipher; nay, that here and there either a gap in the narrative, or the sudden assumption of a new cipher, to which no key was afforded, has obliged me to resort to interpolations of my own' (Bulwer-Lytton 1877 xix). As an exposé of the secrets of Rosicrucian science then, *Zanoni* presents itself as a partial or even fraudulent account in which key passages have, owing to their untranslatability, simply been replaced with the editor's inventions. Nonetheless, readers ultimately learn that such obfuscation achieves a great social good. To unleash too candid an account of esoteric wisdom into the public sphere would be to risk social insurrection. The value of occult science as imagined in *Zanoni* is calibrated directly in relation to its inapplicability.

Opposing the egalitarian and democratic ambitions of the more progressive nineteenth-century esoteric movements and of Godwin's *St Leon*, Bulwer-Lytton's romance clearly stands at the most conservative end of occult science's ideological spectrum. Yet its popular fictional genre complicates this positioning as it makes its appeal for an elite secretive occulture in a form geared towards mass audiences. *Zanoni* thus epitomizes the tensions between elitism and populism in nineteenth-century occult thought that would only intensify in the decades following the century's close, when the advent of World War I elicited new concerns about the legacy, and culpability, of (occult) scientific advancement for the unprecedented slaughter. In her remarkable 1915 spirit-soldier memoir *War Letters of the Living Dead Man*, American Theosophist Elsa Barker – or, rather, David P. Hatch, the recently departed California judge she allegedly channels – indicts the occult revival for contributing to the invasive and martial spirit of the age:

> Occult societies dot the world. In other days these societies were really secret, and no one had access to their knowledge until after tests were passed which proved fitness for further study and further secrets. But the doors of the unseen have been besieged by an army of intellectual enthusiasts who have not passed those tests . . . Democracy has even spread into occult orders, and sacred mysteries have been broadcast by those who put no curb upon their personal ambitions. The hosts of the unseen world have suffered invasion.
>
> *(173)*

Ostensibly a work of post-mortem war reportage, the text here yokes the recent spread and democratization of occult knowledge – processes to which it itself contributes – to Germany's martial incursions in Europe. In each case, a barbarous and brutal constituency has caused mayhem by over-spilling its boundaries. Barker's spirit correspondent asks in conclusion: 'Would you let a child loose in a gunpowder factory with a box of matches in his hand? That is what has been done in the last few years in the Western world' (174–5). Her sentiments here anticipate in occult form those of a later, secular twentieth-century scientist as he invoked the *Bhagavad Gita*, that most influential of Asian religious texts for Western esoteric thought, to reflect on his own awe-inspiring and devastating contribution to the technology of war. Describing his response to the 1945 Trinity nuclear detonation in New Mexico, physicist Robert Oppenheimer famously recounted: 'A few people laughed, a few people cried, most people were silent. I remembered the line from the Hindu scripture, the Bhagavad-Gita . . . "Now, I am become Death, the destroyer of worlds"' ('J. Robert Oppenheimer'). Although Barker, the fringe theosophical amanuensis, and Oppenheimer, the distinguished three-time Nobel nominee, may have shared nothing else, they evince in these moments a common concern that the quest for 'scientific' knowledge, whether exoteric or occult, might ultimately turn monstrous and annihilate the very population it was intended to serve.

Such cautions aside, occult science has never retreated into complete secrecy. The ancient wisdom tradition thrives today in the commercialized spiritualities of the New Age and in the offshoots and continuations of *fin de siècle* occult movements such as theosophy and the *Ordo Templi Orientis*, which continue to hold gatherings and attract members eager to learn the secrets of spiritual science. Students can study the history of esoteric thought at the University of Amsterdam's Center for the History of Hermetic Philosophy and Related Currents; they can also pursue practical training at London's College of Psychic Studies and the Faculty of Astrological Studies. The existence of the latter two in particular indicates that professionalization is by no means a unique priority of the non-occult sciences, even if occult practitioners still remain more ambivalent about, and less dedicated in their pursuit of, disciplinary institutionalization than their secular counterparts. Although often imagined as opposites, or even antagonists, in the nineteenth century, occult science and scientific naturalism are perhaps better understood as uneasy analogues, rehearsing a shared desire to expand understandings of the natural world while simultaneously controlling the potential applications of such knowledge and policing access to their own disciplinary authority. In fascinating ways, the nineteenth-century literature of occult science rehearses the complex boundary work simultaneously under way in the more central currents of professional British science, demonstrating the mutual commitment of occultists and scientific naturalists alike to both expanding and policing the constituencies of their discrete knowledge bases.

Notes

1   Any attempt to wholly oppose scientific naturalism and the occult in the nineteenth century must run aground of the number of high-profile scientific converts to phreno-mesmerism and spiritualism, including John Elliotson (1791–1868), Alfred Russel Wallace (1823–1913), C. F. Varley (1828–83), William Crookes (1832–1919) and Oliver Lodge (1851–1940).
2   Tylor's inclusion of spirit mediumship within the category of 'occult science' is somewhat controversial; occultism and spiritualism, though clearly interlinked, were increasingly recognized as differently inflected, and differently gendered, forms of numinous experience in the nineteenth century. In its association with the apparently passive and exoteric practice of mediumship, spiritualism was often regarded as a distinctively feminine pursuit (Owen 2004B 6–17), whereas occultism, understood as a scholastic ancient wisdom tradition that trained the practitioner's will, was encoded as masculine (Dixon 67). Tylor was by no means alone, however, in equating the two.

Bibliography

Primary texts

Barker, Elsa. *War Letters of the Living Dead Man: Written Down by Elsa Barker*. London: William Rider, 1915.
Blavatsky, H. P. *Isis Unveiled* [1877]. Ed. Michael Gomes. Wheaton: Quest, 1997.
Britten, Emma Hardinge, ed. *Ghost Land: or, Researches into the Mysteries of Occultism, Illustrated by a Series of Autobiographical Sketches*. Boston, MA: Published for the Editor, 1876.
Britten, Emma Hardinge. *Art Magic* [1876]. Ed. Marc Demarest. Forest Grove: Typhon, 2011.
Bulwer-Lytton, Edward. 'The Haunters and the Haunted; or, The House and the Brain'. *Blackwood's Edinburgh Magazine* (August 1859): 224–45.
Bulwer-Lytton, Edward. *A Strange Story*. London: Sampson, Low, 1862.
Bulwer-Lytton, Edward. *The Coming Race*. Edinburgh: W. Blackwood, 1871.
Bulwer-Lytton, Edward. *Zanoni* [1842]. London: G. Routledge, 1877.
Godwin, William. *St Leon: A Tale of the Sixteenth Century* [1799]. London: Henry Colburn and Richard Bentley, 1831.
Haggard, H. Rider. *She: A History of Adventure*. London: Longmans, Green, 1887.
Hodgson, Richard. *Report of the Committee Appointed to Investigate Phenomena Connected with the Theosophical Society*. London: Society for Psychical Research, 1885.
Howitt, William. *The History of the Supernatural, in All Ages and Nations, and in All Churches, Christian and Pagan: Demonstrating a Universal Faith*. London: Longman, Green, Longman, Roberts, and Green, 1863.
Jennings, Hargrave. *The Indian Religions, or the Results of the Mysterious Buddhism*. London: Guildford, 1858.

Lévi, Eliphas. *Transcendental Magic: Its Doctrine and Ritual*. Transl. and ed. A. E. Waite [1896]. London: William Rider, 1923.

Machen, Arthur. *The Great God Pan; and The Inmost Light*. London: J. Lane, 1894.

Marsh, Richard. *The Beetle: A Mystery*. London: Skeffington, 1897.

Mejnour. 'Rosicrucia'. *The Occult Magazine: A Monthly Journal of Psychical and Philosophical Research* 1.1 (January 1885): 4–7.

Olcott, Henry Steel. *Theosophy, Religion, and Occult Science*. London: George Redway, 1885.

Randolph, Paschal Beverly. *The Wonderful Story of Ravalette; Also, Tom Clark and His Wife*. New York, NY: Tousey, 1863.

Shelley, Percy Bysshe. *Zastrozzi and St Irvyne*. Ed. Stephen C. Behrendt. Peterborough: Broadview, 2002.

Sinnett, A. P. *Karma: A Novel*. London: Chapman and Hall, 1891.

Stevenson, Robert Louis. *The Strange Case of Dr Jekyll and Mr Hyde*. London: Longmans, Green, 1886.

Stoker, Bram. *Dracula*. London: Archibald Constable, 1897.

Tylor, E. B. *Primitive Culture: Researches into the Development of Mythology, Philosophy, Religion, Art, and Custom*. London: John Murray, 1871.

Waite, A. E. *The Occult Sciences: A Compendium of Transcendental Doctrine and Experiment*. London: Kegan Paul, Trench, Trubner, 1891.

Secondary texts

Campbell, Bruce F. *Ancient Wisdom Revived: A History of the Theosophical Movement*. London: University of California Press, 1980.

Dixon, Joy. *Divine Feminine: Theosophy and Feminism in England*. Baltimore, MD: Johns Hopkins University Press, 2001.

Ferguson, Christine. *Determined Spirits: Eugenics, Heredity, and Racial Regeneration in Anglo-American Spiritualist Writing, 1848–1930*. Edinburgh: Edinburgh University Press, 2012.

Franklin, J. Jeffrey. 'The Evolution of Occult Spirituality in Victorian England and the Representative Case of Edward Bulwer-Lytton'. *The Ashgate Companion to Nineteenth-Century Spiritualism and the Occult*. Eds. Tatiana Kontou and Sarah Wilburn. Farnham: Ashgate, 2012, 123–41.

Galvan, Jill. *The Sympathetic Medium: Feminine Channeling, the Occult, and Communication Technologies, 1856–1919*. Ithaca, NY: Cornell University Press, 2010.

Gilbert, R. A. *A. E. Waite: A Magician of Many Parts*. Wellingborough: Crucible, 1997.

Godwin, Joscelyn. *The Theosophical Enlightenment*. Albany, NY: State University of New York Press, 1994.

Gomel, Elana. '"Spirits in the Material World": Spiritualism and Identity in the *Fin de Siècle*'. *Victorian Literature and Culture* 35.1 (2007): 189–213.

Goodrick-Clarke, Nicholas. *The Occult Roots of Nazism: Secret Aryan Cults and Their Influence on Nazi Ideology*. Wellingborough: Aquarian, 1985.

Goodrick-Clarke, Nicholas. *The Western Esoteric Traditions: A Historical Introduction*. Oxford: Oxford University Press, 2008.

Hanegraaff, Wouter. *Esotericism and the Academy: Rejected Knowledge in Western Culture*. Cambridge: Cambridge University Press, 2012.

'J. Robert Oppenheimer'. *AtomicArchive.com*. A. J. Software and Multimedia. <www.atomicarchive. com/Movies/Movie8.shtml>. Accessed 15 September 2014.

Kontou, Tatiana. 'Introduction: Women and the Victorian Occult'. *Women's Writing* 15.3 (2008): 275–81.

Luckhurst, Roger. *The Invention of Telepathy, 1870–1901*. Cambridge: Cambridge University Press, 2001.

Luckhurst, Roger, and Justin Sausman, eds. *Marginal and Occult Sciences*. Vol. 8 of *Victorian Science and Literature*. Gen. eds. Gowan Dawson and Bernard Lightman. London: Pickering and Chatto, 2012.

Morrison, M. *Modern Alchemy: Occultism and the Emergence of Atomic Theory*. Oxford: Oxford University Press, 2007.

Owen, Alex. *The Darkened Room: Women, Power, and Spiritualism in Late Victorian England* [1989]. London: University of Chicago Press, 2004 [2004A].

Owen, Alex. *The Place of Enchantment: British Occultism and the Culture of the Modern*. Chicago, IL: University of Chicago Press, 2004 [2004B].

Pasi, Marco. *Aleister Crowley and the Temptation of Politics*. Durham: Acumen, 2014.

Priestman, Martin. *The Poetry of Erasmus Darwin: Enlightened Spaces, Romantic Times*. Aldershot: Ashgate, 2013.

Ramaswamy, Sumathi. *The Lost Land of Lemuria: Fabulous Geographies, Catastrophic Histories.* Berkeley, CA: University of California Press, 2004.

Stocking, George W., Jr. 'Animism in Theory and Practice: E. B. Tylor's Unpublished "Notes on Spiritualism"'. *Man* 6.1 (1971): 88–104.

Sword, Helen. *Ghostwriting Modernism.* Ithaca, NY: Cornell University Press, 2002.

Taylor, Anya. *Magic and English Romanticism.* Athens, GA: University of Georgia Press, 1979.

Thurschwell, Pamela. *Literature, Technology, and Magical Thinking, 1880–1920.* Cambridge: Cambridge University Press, 2005.

Wunder, Jennifer. *Keats, Hermeticism, and the Secret Societies.* Aldershot: Ashgate, 2008.

Ziolkowski, Theodore. *Lure of the Arcane: The Literature of Cult and Conspiracy.* Baltimore, MD: Johns Hopkins University Press, 2013.

Further reading

Atwood, Mary Ann. *A Suggestive Inquiry into the Hermetic Mystery, With a Dissertation on the More Celebrated of the Alchemical Philosophers.* London: T. Saunders, 1850.

Barrett, Francis. *The Magus, or Celestial Intelligencer; Being a Complete System of Occult Philosophy.* London: Lackington, Allen, 1801.

Barsham, Diana. *The Trial of Women: Feminism and the Occult Sciences in Victorian Literature and Culture.* Basingstoke: Macmillan, 1992.

Besant, Annie. *Karma.* London: Theosophical, 1895.

Besant, Annie. *The Ancient Wisdom: An Outline of Theosophical Teachings.* London: Theosophical, 1897.

Besant, Annie. *Esoteric Christianity: Or, The Lesser Mysteries.* London: Theosophical, 1901.

Blavatsky, H. P. *The Secret Doctrine: The Synthesis of Science, Religion, and Philosophy.* London: Theosophical, 1888.

Brandon, Ruth. *The Spiritualists: The Passion for the Occult in the Nineteenth and Twentieth Centuries.* London: Weidenfeld and Nicholson, 1983.

Butler, Alison. *Victorian Occultism and the Making of Modern Magic: Invoking Tradition.* Basingstoke: Palgrave Macmillan, 2011.

Caithness, Marie Sinclair. *Old Truths in a New Light, or, An Earnest Endeavour to Reconcile Material Science with Spiritual Science and With Scripture.* London: Chapman and Hall, 1876.

Chanel, Christian, John Patrick Deveney and Joscelyn Godwin, eds. *The Hermetic Brotherhood of Luxor. Initiatic and Historical Documents of an Order of Practical Occultism.* York Beach, ME: Weiser, 1995.

Collins, Mabel. *Light on the Path: A Treatise: Written for the Personal Use of Those who are Ignorant of the Eastern Wisdom and who Desire to Enter Within Its Influence.* London: Reeves and Turner, 1885.

Davies, Charles Maurice. *Mystic London: or, Phases of Occult Life in the British Metropolis.* London: Tinsley, 1875.

Davies, Charles Maurice. *The Great Secret and Its Unfoldment in Occultism: A Record of Forty Years' Experience in the Modern Mystery.* London: Redway, 1895.

Faivre, Antoine. *Western Esotericism: A Concise History.* Albany, NY: State University of New York Press, 2010.

Farr, Florence. *Egyptian Magic.* London: Theosophical, 1896.

Gilbert, R. A. *The Golden Dawn: Twilight of the Magicians.* Wellingborough: Aquarian, 1983.

Greer, Mary K. *Women of the Golden Dawn: Rebels and Priestesses.* Rochester, NY: Park Street, 1995.

Harrison, C. G. *The Transcendental Universe: Six Lectures on Occult Science, Theosophy, and the Catholic Faith.* London: James Elliot, 1894.

Howe, E. *Magicians of the Golden Dawn: A Documentary History of a Magical Order.* London: Routledge and Kegan Paul, 1972.

Jennings, Hargrave. *The Rosicrucians: Their Rites and Mysteries.* London: J. C. Hotten, 1870.

Jennings, Hargrave. *Phallicism, Celestial, and Terrestrial, Heathen and Christian.* London: Redway, 1884.

Kingsford, Anna Bonus. *Clothed With the Sun: Being the Book of Illuminations of Anna Kingsford.* London: George Redway, 1889.

Kontou, Tatiana. *Women and the Victorian Occult.* London: Routledge, 2011.

Kontou, Tatiana, and Sarah Willburn. *The Ashgate Research Companion to Nineteenth-Century Spiritualism and the Occult.* Farnham: Ashgate, 2012.

Mathers, Samuel Liddell MacGregor. *Fortune Telling Cards: The Tarot, Its Occult Signification, Use in Fortune-Telling, and Method of Play*. London: G. Redway, 1888.

Mathers, Samuel Liddell MacGregor. *The Key of Salomon the King. Clavicula Salomonis*. London: George Redway, 1889.

Olcott, Henry Steel. *People from the Other World*. Hartford, CT: American, 1875.

Old, Walter Gorn. *The New Manual of Astrology*. London: George Redway, 1898.

Oppenheim, Janet. *The Other World: Spiritualism and Psychical Research in England, 1850–1914*. Cambridge: Cambridge University Press, 1985.

Randolph, Paschal Beverly. *Seership! The Magnetic Mirror*. Boston, MA: Randolph, 1870.

Sinnett, A. P. *Some Fruits of Occult Teaching*. London: Theosophical Publishing Society, 1896.

Waite, A. E. *Azoth, or, The Star in the East*. London: Theosophical Publishing Society, 1893.

AFTERWORD

Bernard Lightman
York University

In 1869 the biologist Thomas Henry Huxley was asked by the editor of a new science journal to write the lead article for the first issue. Huxley decided to focus the piece on a poem by Goethe, a German poet and Romantic. This 'wonderful rhapsody' on nature, Huxley declared, had 'been a delight to me from my youth up'. To Huxley, 'no more fitting preface could be put before a Journal, which aims to mirror the progress of that fashioning by Nature of a picture of herself, in the mind of man, which we call the progress of Science'. After this endorsement of the poem's idealism, Huxley offered his translation of Goethe's poem and then remarked on the eternal character of art. For 'long after the theories of the philosophers whose achievements are recorded in these pages, are obsolete,' Huxley predicted, 'the vision of the poet will remain as a truthful and efficient symbol of the wonder and the mystery of nature' (Huxley, 10–11). Huxley was suggesting that Goethe's poetic rendering of nature was more enduring in comparison to the transitory theories of his scientific colleagues. This is the message that Huxley had for the readers of the very first issue of the journal *Nature*.

Huxley was not the only scientific naturalist who was taken with Goethe. For the physicist John Tyndall, Goethe summed up in his poetry the mystery underlying nature. 'To many of us who feel that there are more things in heaven and earth than are dreamt of in the present philosophy of science,' Tyndall proclaimed, 'but who have been also taught, by baffled efforts, how vain is the attempt to grapple with the Inscrutable, the ultimate frame of mind is that of Goethe' (Tyndall 1892 52). Like Huxley, Tyndall admired Goethe's poetic vision of nature. In his infamous 'Belfast Address' (1874), Tyndall declared that Goethe was not only a great poet but also an important contributor to natural history (Tyndall 1892 148). Tyndall expanded on this point in an essay on Goethe's theory of colours, comparing it to Newton's, while evaluating both in the light of modern science. Judged by contemporary standards, Tyndall admitted, Newton's theory was more accurate and it was the result of using strict scientific method. Nevertheless Tyndall pointed to errors made by Newton, who had adopted 'a wrong mechanical conception in his theory of light'. In the conclusion of the lecture Tyndall asserted that Goethe's poetic approach to nature was just as important as Newton's exclusively mechanical method. 'The feelings and aims with which Newton and Goethe respectively approached Nature were radically different,' Tyndall declared, 'but they had an equal warrant in the constitution of man' (Tyndall 1898 76–77). Here Tyndall elevated Goethe, the representative of German romanticism, to the same scientific status as Newton, the British hero of empiricism.

The receptiveness of leading scientific naturalists of the second half of the nineteenth century to a Romantic literary figure from an earlier age would have seemed somewhat puzzling to Victorianists active in the field before the 1980s. But, by surveying the recent scholarship, *The Routledge Research Companion to Nineteenth-Century British Literature and Science* has prepared its readers to appreciate the richer picture of the relationship between science and literature that has emerged in the last few decades. After all, the purpose of a research companion is to consolidate the field or to establish the default position. By providing a synthesis of the current state of play, a research companion allows scholars to push the field further in the future.

This synthesis is based on an important heuristic principle that has guided the newer research: during the nineteenth century there was a two-way traffic between science and literature. The traffic, then, did not consist simply in the influence of scientific ideas on literature – the one-way street model that provided the basis of the older scholarship – but involved ideas and structures moving in both directions. Pioneers such as Beer and Levine explored the traffic running between science and the realist novel. But busy traffic can also be seen in science fiction and even the sensation novel. As this volume has shown, the two-way traffic model is not limited to the novel. The dynamic is at work in virtually every genre of literature in the nineteenth century. Scientists and literary figures relied in their work and writing on very similar tools of narrative, visualization, description, structure and publishing format. Moreover, this complicated interpenetration of science and literature was distinctive of the nineteenth century for several reasons, many of them connected to the historical conditions shaping their mutual development.

Not only were the genres of literature in flux during this period, but what we now recognize as expert, metropolitan science was still in the process of being established. As a result, the lines between science and literature were blurred. The chapters in this research companion explore the many ways in which scientists and literary figures redefined those lines or even tried to erase them altogether. The blurring cuts across both literary genres and scientific disciplines. There were scientists, such as Maxwell, who were poets, and there were Romantic poets who drew on chemical philosophy. Significant exchanges between literary and psychological texts bear out the two-way traffic model. Science and literature also shared common methods, such as an emphasis on experiment and observation. There were even important connections between literary culture and institutionalized science. Figures from the world of literature were active participants in the scientific societies, institutions and associations that grew steadily in number throughout the century. This fostered intellectual exchange between literary and scientific figures. In sum – and here I use a more fluid metaphor than the one based on traffic – multiple literary streams mixed with numerous scientific tributaries within the great river of Victorian culture.

Such complicated, fluid relationships can only be studied by approaching them from multiple angles. Here, the editors have based their triangulation on three points: genres, disciplines and contexts. Broad definitions of genre and discipline provide us with an unusually comprehensive picture and facilitate the incorporation of new areas of scholarship. Chapters on fiction, poetry and prose would be *de rigueur* in any examination of science and literature. But the inclusion of literary criticism, biography and autobiography, theatre, science for the general reader and for children, and science in the periodical press draws on newer scholarly approaches and points towards novel resonances between science and literature. The exhaustive list of disciplines covered allows for the expansion of the scientific canon, encourages the consideration of nascent disciplines in formation (e.g. sexology and anthropology), and takes seriously the views of the historical actors on what constituted knowledge (e.g. occult sciences). The thorough analysis of multiple genres and disciplines is preceded by the first section, on 'contexts'. Although there are fewer chapters in this section, this by no means is intended to undermine the importance of analysing genres and disciplines within their cultural context. The discussion of genres and disciplines is contextualized throughout the volume.

The emphasis on context reveals the very close relationship that has developed between historians of nineteenth-century science and those who study the interactions between science and literature in the same period. This answers a question that readers may be tempted to ask of the editors: why is a historian of science writing an afterword to a literary research companion? The reality is that many literary scholars have become so historically informed that they are virtually indistinguishable from historians. The questions that they ask have become vitally important for historians. They are adept at integrating the work of historians of science into their literary analyses. In some of the chapters on genre the authors begin with a discussion of the history of science and then move on to the literature that has been informed by scientific ideas. But, by the same token, many historians have realized that the literary dimensions of scientific texts are a crucial element in any historical account of nineteenth-century science, and that there is much to be learnt from the scholars studying science and literature. As Jonathan Smith asserts, 'historians of science have turned more and more to the tools of literary analysis to understand how scientific prose does its work, and to argue increasingly for the essential contributions of the literary to the writing of science' (p. 151). Having a historian write the afterword for this research companion is symbolic of what has been a very productive partnership between two sets of scholars.

The study of Victorian science has truly become an interdisciplinary enterprise. This may be more evident in particular areas of research, for example in the case of the popularization of science. In his chapter on 'Science for the General Reader', Ralph O'Connor has declared that 'the literature of science for general readers is thus not the property of historians of science but is a rich and powerful resource for literary scholars too' (p. 167). The chapter on 'Staging Science' by Iwan Rhys Morus also points to the shared interest in the topic of performativity. Both historians and literary scholars, he argues persuasively, need to examine the parallels between scientific performances and theatre since both were products of hard work and careful preparation (p. 245). But the entire volume is an endorsement of what Pamela Gossin has referred to as the hybrid 'literary history of science' (p. 245). And it is only appropriate that historians and literary scholars have been forced to develop a nuanced interdisciplinary approach in order to understand the two-way traffic between science and culture that existed in the nineteenth century.

By pointing to future areas for research, the contributors to this volume indicate that gaps still remain in the scholarship based on the two-way traffic model. Perhaps the most striking is the one pointed to by Smith, who asserts that scholars in the field have neglected major nineteenth-century scientific figures (p. 145). Undoubtedly, the focus has been on Darwin and Huxley. Far less consideration has been given to Davy, Faraday, Lyell, Owen and Wallace. It could be added that the life sciences, especially evolution, have received the lion's share of attention. Moreover, little work has been done on technical works of science for obvious reasons. As exhaustive as the *Routledge Research Companion* is, there is still exciting new territory to explore. The contributors have provided researchers with solid ground from which to conduct their investigations. Scholars can now venture forth from this *terra firma* to seek out new instances of the busy two-way traffic between science and literature accompanied by a trusty and helpful companion.

Bibliography

Primary texts

Huxley, T. H. 'Nature: Aphorisms by Goethe'. *Nature* 1 (November 1869–April 1870): 10-11.
Tyndall, John. *Fragments of Science*. 8 edn. 2 vols. London: Longmans, Green, 1892, Volume 2.
Tyndall, John. *New Fragments*. New York, NY: D. Appleton, 1898.

INDEX